MECHATRONICS

ELECTRONIC CONTROL SYSTEMS
IN MECHANICAL AND ELECTRICAL
ENGINEERING

Sixth Edition

William Bolton

PEARSON

Harlow, England • London • New York • Boston • San Francisco • Toronto • Sydney
Auckland • Singapore • Hong Kong • Tokyo • Seoul • Taipei • New Delhi
Cape Town • São Paulo • Mexico City • Madrid • Amsterdam • Munich • Paris • Milan

Pearson Education Limited
Edinburgh Gate
Harlow CM20 2JE
United Kingdom
Tel: +44 (0)1279 623623
Web: www.pearson.com/uk

First published 1995 (print)
Second edition published 1999 (print)
Third edition published 2003 (print)
Fourth edition published 2008 (print)
Fifth edition published 2011 (print and electronic)
Sixth edition published 2015 (print and electronic)

ISBN: 978-1-292-07668-3 (print)
         978-1-292-08159-5 (PDF)
         978-1-292-08160-1 (eText)

British Library Cataloguing-in-Publication Data
A catalogue record for the print edition is available from the British Library

Library of Congress Cataloging-in-Publication Data
Bolton, W. (William), 1933–
   Mechatronics : a multidisciplinary approach / William Bolton. – Sixth edition.
      pages cm
   Includes bibliographical references and index.
   ISBN 978-1-292-07668-3
   1. Mechatronics. I. Title.
   TJ163.12.B65 2015
   621–dc23
                              2014041487

10 9 8 7 6 5 4
20 20 19 18

Cover illustration © Getty Images
Print edition typeset in 10/11pt, Ehrhardt MT Std by 71
Print edition printed in Malaysia (CTP-VVP)

NOTE THAT ANY PAGE CROSS REFERENCES REFER TO THE PRINT EDITION

Contents

Preface

The term **mechatronics** was 'invented' by a Japanese engineer in 1969, as a combination of 'mecha' from mechanisms and 'tronics' from electronics. The word now has a wider meaning, being used to describe a philosophy in engineering technology in which there is a co-ordinated, and concurrently developed, integration of mechanical engineering with electronics and intelligent computer control in the design and manufacture of products and processes. As a result, many products which used to have mechanical functions have had many replaced with ones involving microprocessors. This has resulted in much greater flexibility, easier redesign and reprogramming, and the ability to carry out automated data collection and reporting.

A consequence of this approach is the need for engineers and technicians to adopt an interdisciplinary and integrated approach to engineering. Thus engineers and technicians need skills and knowledge that are not confined to a single subject area. They need to be capable of operating and communicating across a range of engineering disciplines and linking with those having more specialised skills. This book is an attempt to provide a basic background to mechatronics and provide links through to more specialised skills.

The first edition was designed to cover the Business and Technology Education Council (BTEC) Mechatronics units for Higher National Certificate/Diploma courses for technicians and designed to fit alongside more specialist units such as those for design, manufacture and maintenance determined by the application area of the course. The book was widely used for such courses and has also found use in undergraduate courses in both Britain and in the United States. Following feedback from lecturers in both Britain and the United States, the second edition was considerably extended and with its extra depth it was not only still relevant for its original readership but also suitable for undergraduate courses. The third edition involved refinements of some explanations, more discussion of microcontrollers and programming, increased use of models for mechatronics systems, and the grouping together of key facts in the Appendices. The fourth edition was a complete reconsideration of all aspects of the text, both layout and content, with some regrouping of topics, movement of more material into Appendices to avoid disrupting the flow of the text, new material – in particular an introduction to artificial intelligence, more case studies and a refinement of some topics to improve clarity. Also, objectives and key point summaries were included with each chapter. The fifth edition kept the same structure but, after consultation with many users of the book, many aspects had extra detail and refinement added.

The sixth edition has involved a restructuring of the constituent parts of the book as some users felt that the chapter sequencing did not match the general teaching sequence. Thus the new edition has involved moving the system models part so that it comes after microprocessor systems. Other changes include the inclusion of material on Arduino and the addition of more topics in the Mechatronics Systems chapter.

The overall aim of the book is to give a comprehensive coverage of mechatronics which can be used with courses for both technicians and undergraduates in engineering and, hence, to help the reader:

- acquire a mix of skills in mechanical engineering, electronics and computing which is necessary if he/she is to be able to comprehend and design mechatronics systems;
- become capable of operating and communicating across the range of engineering disciplines necessary in mechatronics;
- be capable of designing mechatronic systems.

Each chapter of the book includes objectives, and a summary, is copiously illustrated and contains problems, answers to which are supplied at the end of the book. Chapter 24 comprises research and design assignments together with clues as to their possible answers.

The structure of the book is:

- Chapter 1 is a general introduction to mechatronics;
- Chapters 2–6 form a coherent block on sensors and signal conditioning;
- Chapters 7–9 cover actuators;
- Chapters 10–16 discuss microprocessor/microcontroller systems;
- Chapters 17–23 are concerned with system models;
- Chapter 24 provides an overall conclusion in considering the design of mechatronic systems.

An Instructor's Guide, test material and Powerpoint slides are available for lecturers to download at: www.pearsoned.co.uk/bolton

A large debt is owed to the publications of the manufacturers of the equipment referred to in the text. I would also like to thank those reviewers who painstakingly read through the fifth edition and made suggestions for improvements.

W. Bolton

Part I
Introduction

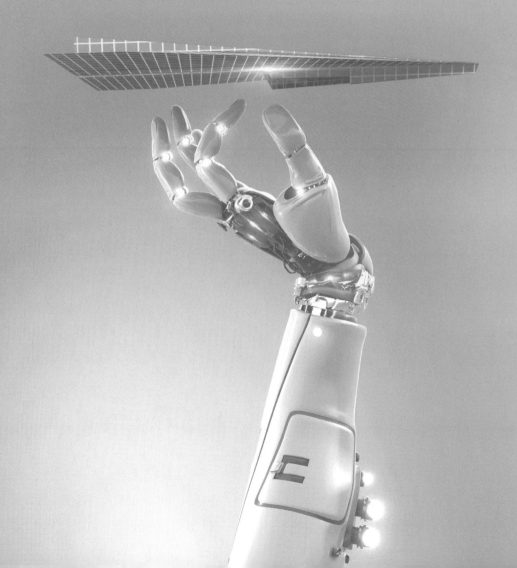

Chapter one Introducing mechatronics

Objectives

The objectives of this chapter are that, after studying it, the reader should be able to:
- Explain what is meant by mechatronics and appreciate its relevance in engineering design.
- Explain what is meant by a system and define the elements of measurement systems.
- Describe the various forms and elements of open-loop and closed-loop control systems.
- Recognise the need for models of systems in order to predict their behaviour.

1.1 What is mechatronics?

The term **mechatronics** was 'invented' by a Japanese engineer in 1969, as a combination of 'mecha' from mechanisms and 'tronics' from electronics. The word now has a wider meaning, being used to describe a philosophy in engineering technology in which there is a co-ordinated, and concurrently developed, integration of mechanical engineering with electronics and intelligent computer control in the design and manufacture of products and processes. As a result, mechatronic products have many mechanical functions replaced with electronic ones. This results in much greater flexibility, easy redesign and reprogramming, and the ability to carry out automated data collection and reporting.

A mechatronic system is not just a marriage of electrical and mechanical systems and is more than just a control system; it is a complete integration of all of them in which there is a concurrent approach to the design. In the design of cars, robots, machine tools, washing machines, cameras and very many other machines, such an integrated and interdisciplinary approach to engineering design is increasingly being adopted. The integration across the traditional boundaries of mechanical engineering, electrical engineering, electronics and control engineering has to occur at the earliest stages of the design process if cheaper, more reliable, more flexible systems are to be developed. Mechatronics has to involve a concurrent approach to these disciplines rather than a sequential approach of developing, say, a mechanical system, then designing the electrical part and the microprocessor part. Thus mechatronics is a design philosophy, an integrating approach to engineering.

Mechatronics brings together areas of technology involving sensors and measurement systems, drive and actuation systems, and microprocessor systems (Figure 1.1), together with the analysis of the behaviour of systems and control systems. That essentially is a summary of this book. This chapter is an introduction to the topic, developing some of the basic concepts in order to give a framework for the rest of the book in which the details will be developed.

Figure 1.1   The basic elements of a mechatronic system.

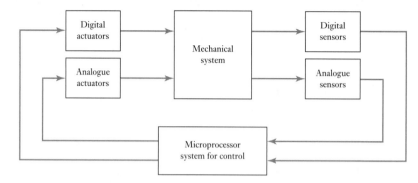

1.1.1 Examples of mechatronic systems

Consider the modern autofocus, auto-exposure camera. To use the camera all you need to do is point it at the subject and press the button to take the picture. The camera can automatically adjust the focus so that the subject is in focus and automatically adjust the aperture and shutter speed so that the correct exposure is given. You do not have to manually adjust focusing and aperture or shutter speed controls. Consider a truck smart suspension. Such a suspension adjusts to uneven loading to maintain a level platform, adjusts to cornering, moving across rough ground, etc., to maintain a smooth ride. Consider an automated production line. Such a line may involve a number of production processes which are all automatically carried out in the correct sequence and in the correct way with a reporting of the outcomes at each stage in the process. The automatic camera, the truck suspension and the automatic production line are examples of a marriage between electronics, control systems and mechanical engineering.

1.1.2 Embedded systems

The term **embedded system** is used where microprocessors are embedded into systems and it is this type of system we are generally concerned with in mechatronics. A microprocessor may be considered as being essentially a collection of logic gates and memory elements that are not wired up as individual components but whose logical functions are implemented by means of software. As an illustration of what is meant by a logic gate, we might want an output if input A AND input B are both giving on signals. This could be implemented by what is termed an AND logic gate. An OR logic gate would give an output when either input A OR input B is on. A microprocessor is thus concerned with looking at inputs to see if they are on or off, processing the results of such an interrogation according to how it is programmed, and then giving outputs which are either on or off. See Chapter 10 for a more detailed discussion of microprocessors.

For a microprocessor to be used in a control system, it needs additional chips to give memory for data storage and for input/output ports to enable it to process signals from and to the outside world. **Microcontrollers** are microprocessors with these extra facilities all integrated together on a single chip.

An embedded system is a microprocessor-based system that is designed to control a range of functions and is not designed to be programmed by the end user in the same way that a computer is. Thus, with an embedded system, the user cannot change what the system does by adding or replacing software.

As an illustration of the use of microcontrollers in a control system, a modern washing machine will have a microprocessor-based control system to control the washing cycle, pumps, motor and water temperature. A modern car will have microprocessors controlling such functions as anti-lock brakes and engine management. Other examples of embedded systems are autofocus, auto-exposure cameras, camcorders, cell phones, DVD players, electronic card readers, photocopiers, printers, scanners, televisions and temperature controllers.

1.2 The design process

The design process for any system can be considered as involving a number of stages.

1 *The need*
   The design process begins with a need from, perhaps, a customer or client. This may be identified by market research being used to establish the needs of potential customers.

2 *Analysis of the problem*
   The first stage in developing a design is to find out the true nature of the problem, i.e. analysing it. This is an important stage in that not defining the problem accurately can lead to wasted time on designs that will not fulfil the need.

3 *Preparation of a specification*
   Following the analysis, a specification of the requirements can be prepared. This will state the problem, any constraints placed on the solution, and the criteria which may be used to judge the quality of the design. In stating the problem, all the functions required of the design, together with any desirable features, should be specified. Thus there might be a statement of mass, dimensions, types and range of motion required, accuracy, input and output requirements of elements, interfaces, power requirements, operating environment, relevant standards and codes of practice, etc.

4 *Generation of possible solutions*
   This is often termed the **conceptual stage.** Outline solutions are prepared which are worked out in sufficient detail to indicate the means of obtaining each of the required functions, e.g. approximate sizes, shapes, materials and costs. It also means finding out what has been done before for similar problems; there is no sense in reinventing the wheel.

5 *Selections of a suitable solution*
   The various solutions are evaluated and the most suitable one selected. Evaluation will often involve the representation of a system by a model and then simulation to establish how it might react to inputs.

6 *Production of a detailed design*
   The detail of the selected design has now to be worked out. This might require the production of prototypes or mock-ups in order to determine the optimum details of a design.

7 *Production of working drawings*
   The selected design is then translated into working drawings, circuit diagrams, etc., so that the item can be made.

It should not be considered that each stage of the design process just flows on stage by stage. There will often be the need to return to an earlier stage and give it further consideration. Thus when at the stage of generating possible solutions there might be a need to go back and reconsider the analysis of the problem.

1.2.1 Traditional and mechatronics designs

Engineering design is a complex process involving interactions between many skills and disciplines. With traditional design, the approach was for the mechanical engineer to design the mechanical elements, then the control engineer to come along and design the control system. This gives what might be termed a sequential approach to the design. However, the basis of the mechatronics approach is considered to lie in the concurrent inclusion of the disciplines of mechanical engineering, electronics, computer technology and control engineering in the approach to design. The inherent concurrency of this approach depends very much on system modelling and then simulation of how the model reacts to inputs and hence how the actual system might react to inputs.

As an illustration of how a multidisciplinary approach can aid in the solution of a problem, consider the design of bathroom scales. Such scales might be considered only in terms of the compression of springs and a mechanism used to convert the motion into rotation of a shaft and hence movement of a pointer across a scale; a problem that has to be taken into account in the design is that the weight indicated should not depend on the person's position on the scales. However, other possibilities can be considered if we look beyond a purely mechanical design. For example, the springs might be replaced by load cells with strain gauges and the output from them used with a microprocessor to provide a digital readout of the weight on an light-emitting diode (LED) display. The resulting scales might be mechanically simpler, involving fewer components and moving parts. The complexity has, however, been transferred to the software.

As a further illustration, the traditional design of the temperature control for a domestic central heating system has been the bimetallic thermostat in a closed-loop control system. The bending of the bimetallic strip changes as the temperature changes and is used to operate an on/off switch for the heating system. However, a multidisciplinary solution to the problem might be to use a microprocessor-controlled system employing perhaps a thermo-diode as the sensor. Such a system has many advantages over the bimetallic thermostat system. The bimetallic thermostat is comparatively crude and the temperature is not accurately controlled; also devising a method for having different temperatures at different times of the day is complex and not easily achieved. The microprocessor-controlled system can, however, cope easily with giving precision and programmed control. The system is much more flexible. This improvement in flexibility is a common characteristic of mechatronics systems when compared with traditional systems.

1.3   Systems

In designing mechatronic systems, one of the steps involved is the creation of a model of the system so that predictions can be made regarding its behaviour when inputs occur. Such models involve drawing block diagrams to represent systems. A **system** can be thought of as a box or block diagram

Figure 1.2 Examples of systems: (a) spring, (b) motor, (c) thermometer.

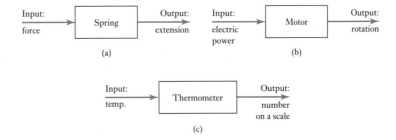

which has an input and an output and where we are concerned not with what goes on inside the box but with only the relationship between the output and the input. The term **modelling** is used when we represent the behaviour of a real system by mathematical equations, such equations representing the relationship between the inputs and outputs from the system. For example, a spring can be considered as a system to have an input of a force F and an output of an extension x (Figure 1.2(a)). The equation used to model the relationship between the input and output might be $F = kx$, where k is a constant. As another example, a motor may be thought of as a system which has as its input electric power and as output the rotation of a shaft (Figure 1.2(b)).

A **measurement system** can be thought of as a box which is used for making measurements. It has as its input the quantity being measured and its output the value of that quantity. For example, a temperature measurement system, i.e. a thermometer, has an input of temperature and an output of a number on a scale (Figure 1.2(c)).

1.3.1 Modelling systems

The response of any system to an input is not instantaneous. For example, for the spring system described by Figure 1.2(a), though the relationship between the input, force F, and output, extension x, was given as $F = kx$, this only describes the relationship when steady-state conditions occur. When the force is applied it is likely that oscillations will occur before the spring settles down to its steady-state extension value (Figure 1.3). The responses of systems are functions of time. Thus, in order to know how systems behave when there are inputs to them, we need to devise models for systems which relate the output to the input so that we can work out, for a given input, how the output will vary with time and what it will settle down to.

As another example, if you switch on a kettle it takes some time for the water in the kettle to reach boiling point (Figure 1.4). Likewise, when a microprocessor controller gives a signal to, say, move the lens for focusing

Figure 1.3 The response to an input for a spring.

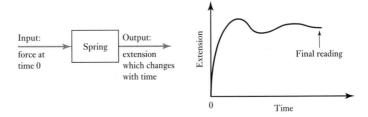

Figure 1.4   The response to an input for a kettle system.

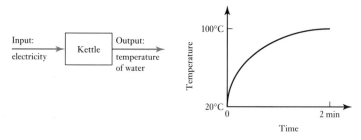

Figure 1.5   A CD player.

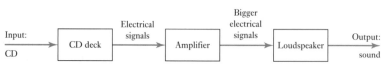

in an automatic camera then it takes time before the lens reaches its position for correct focusing.

Often the relationship between the input and output for a system is described by a differential equation. Such equations and systems are discussed in Chapter 17.

1.3.2   Connected systems

In other than the simplest system, it is generally useful to consider it as a series of interconnected blocks, each such block having a specific function. We then have the output from one block becoming the input to the next block in the system. In drawing a system in this way, it is necessary to recognise that lines drawn to connect boxes indicate a flow of information in the direction indicated by an arrow and not necessarily physical connections. An example of such a connected system is a CD player. We can think of there being three interconnected blocks: the CD deck which has an input of a CD and an output of electrical signals; an amplifier which has an input of these electrical signals, and an output of bigger electrical signals; and a speaker which has an input of the electrical signals and an output of sound (Figure 1.5). Another example of such a set of connected blocks is given in the next section on measurement systems.

1.4   Measurement systems

Of particular importance in any discussion of mechatronics are measurement systems. **Measurement systems** can, in general, be considered to be made up of three basic elements (as illustrated in Figure 1.6).

1   A **sensor** responds to the quantity being measured by giving as its output a signal which is related to the quantity. For example, a thermocouple is a temperature sensor. The input to the sensor is a temperature and the output is an e.m.f., which is related to the temperature value.

2   A **signal conditioner** takes the signal from the sensor and manipulates it into a condition which is suitable either for display or, in the case of a control system, for use to exercise control. Thus, for example, the output

Figure 1.6 A measurement system and its constituent elements.

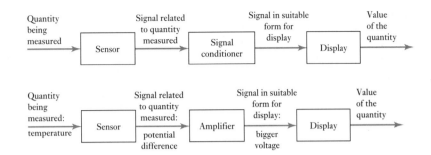

Figure 1.7 A digital thermometer system.

from a thermocouple is a rather small e.m.f. and might be fed through an amplifier to obtain a bigger signal. The amplifier is the signal conditioner.

3 A **display system** displays the output from the signal conditioner. This might, for example, be a pointer moving across a scale or a digital readout.

As an example, consider a digital thermometer (Figure 1.7). This has an input of temperature to a sensor, probably a semiconductor diode. The potential difference across the sensor is, at constant current, a measure of the temperature. This potential difference is then amplified by an operational amplifier to give a voltage which can directly drive a display. The sensor and operational amplifier may be incorporated on the same silicon chip.

Sensors are discussed in Chapter 2 and signal conditioners in Chapter 3. Measurement systems involving all elements are discussed in Chapter 6.

1.5   Control systems

A **control system** can be thought of as a system which can be used to:

1 control some variable to some particular value, e.g. a central heating system where the temperature is controlled to a particular value;
2 control the sequence of events, e.g. a washing machine where when the dials are set to, say, 'white' and the machine is then controlled to a particular washing cycle, i.e. sequence of events, appropriate to that type of clothing;
3 control whether an event occurs or not, e.g. a safety lock on a machine where it cannot be operated until a guard is in position.

1.5.1   Feedback

Consider an example of a control system with which we are all individually involved. Your body temperature, unless you are ill, remains almost constant regardless of whether you are in a cold or hot environment. To maintain this constancy your body has a temperature control system. If your temperature begins to increase above the normal you sweat, if it decreases you shiver. Both these are mechanisms which are used to restore the body temperature back to its normal value. The control system is maintaining constancy of temperature. The system has an input from sensors which tell it what the temperature is and then compare this data with what the temperature should be and provide the appropriate response in order to obtain the required

Figure 1.8 Feedback control: (a) human body temperature, (b) room temperature with central heating, (c) picking up a pencil.

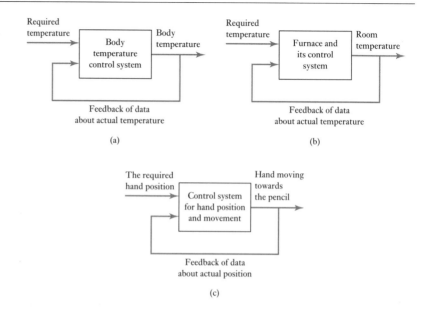

temperature. This is an example of **feedback control**: signals are fed back from the output, i.e. the actual temperature, in order to modify the reaction of the body to enable it to restore the temperature to the 'normal' value. Feedback control is exercised by the control system comparing the fed-back actual output of the system with what is required and adjusting its output accordingly. Figure 1.8(a) illustrates this feedback control system.

One way to control the temperature of a centrally heated house is for a human to stand near the furnace on/off switch with a thermometer and switch the furnace on or off according to the thermometer reading. That is a crude form of feedback control using a human as a control element. The term feedback is used because signals are fed back from the output in order to modify the input. The more usual feedback control system has a thermostat or controller which automatically switches the furnace on or off according to the difference between the set temperature and the actual temperature (Figure 1.8(b)). This control system is maintaining constancy of temperature.

If you go to pick up a pencil from a bench there is a need for you to use a control system to ensure that your hand actually ends up at the pencil. This is done by your observing the position of your hand relative to the pencil and making adjustments in its position as it moves towards the pencil. There is a feedback of information about your actual hand position so that you can modify your reactions to give the required hand position and movement (Figure 1.8(c)). This control system is controlling the positioning and movement of your hand.

Feedback control systems are widespread, not only in nature and the home but also in industry. There are many industrial processes and machines where control, whether by humans or automatically, is required. For example, there is process control where such things as temperature, liquid level, fluid flow, pressure, etc., are maintained constant. Thus in a chemical process there may be a need to maintain the level of a liquid in a tank to a particular level or to a particular temperature. There are also control systems which involve consistently and accurately positioning a moving part or maintaining a constant speed. This might be, for example, a motor designed to run at a constant speed or perhaps a machining operation in which the position, speed and operation of a tool are automatically controlled.

1.5.2 Open- and closed-loop systems

There are two basic forms of control system, one being called **open loop** and the other **closed loop**. The difference between these can be illustrated by a simple example. Consider an electric fire which has a selection switch which allows a 1 kW or a 2 kW heating element to be selected. If a person used the heating element to heat a room, he or she might just switch on the 1 kW element if the room is not required to be at too high a temperature. The room will heat up and reach a temperature which is only determined by the fact that the 1 kW element was switched on and not the 2 kW element. If there are changes in the conditions, perhaps someone opening a window, there is no way the heat output is adjusted to compensate. This is an example of open-loop control in that there is no information fed back to the element to adjust it and maintain a constant temperature. The heating system with the heating element could be made a closed-loop system if the person has a thermometer and switches the 1 kW and 2 kW elements on or off, according to the difference between the actual temperature and the required temperature, to maintain the temperature of the room constant. In this situation there is feedback, the input to the system being adjusted according to whether its output is the required temperature. This means that the input to the switch depends on the deviation of the actual temperature from the required temperature, the difference between them being determined by a comparison element – the person in this case. Figure 1.9 illustrates these two types of system.

An example of an everyday open-loop control system is the domestic toaster. Control is exercised by setting a timer which determines the length of time for which the bread is toasted. The brownness of the resulting toast is determined solely by this preset time. There is no feedback to control the degree of browning to a required brownness.

To illustrate further the differences between open- and closed-loop systems, consider a motor. With an open-loop system the speed of rotation of the shaft might be determined solely by the initial setting of a knob which

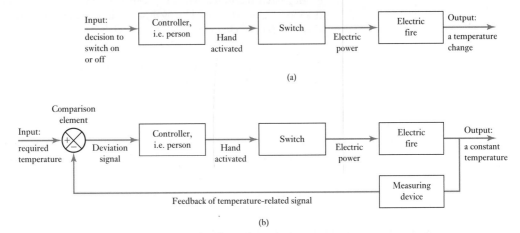

Figure 1.9 Heating a room: (a) an open-loop system, (b) a closed-loop system.

affects the voltage applied to the motor. Any changes in the supply voltage, the characteristics of the motor as a result of temperature changes, or the shaft load will change the shaft speed but not be compensated for. There is no feedback loop. With a closed-loop system, however, the initial setting of the control knob will be for a particular shaft speed and this will be maintained by feedback, regardless of any changes in supply voltage, motor characteristics or load. In an open-loop control system the output from the system has no effect on the input signal. In a closed-loop control system the output does have an effect on the input signal, modifying it to maintain an output signal at the required value.

Open-loop systems have the advantage of being relatively simple and consequently low cost with generally good reliability. However, they are often inaccurate since there is no correction for error. Closed-loop systems have the advantage of being relatively accurate in matching the actual to the required values. They are, however, more complex and so more costly with a greater chance of breakdown as a consequence of the greater number of components.

1.5.3   Basic elements of a closed-loop system

Figure 1.10 shows the general form of a basic closed-loop system. It consists of five elements.

1 *Comparison element*
   This compares the required or reference value of the variable condition being controlled with the measured value of what is being achieved and produces an error signal. It can be regarded as adding the reference signal, which is positive, to the measured value signal, which is negative in this case:

   error signal = reference value signal − measured value signal

   The symbol used, in general, for an element at which signals are summed is a segmented circle, inputs going into segments. The inputs are all added, hence the feedback input is marked as negative and the reference signal positive so that the sum gives the difference between the signals. A **feedback loop** is a means whereby a signal related to the actual condition being achieved is fed back to modify the input signal to a process. The feedback is said to be **negative feedback** when the signal which is fed back subtracts from the input value. It is negative feedback that is required to control a system. **Positive feedback** occurs when the signal fed back adds to the input signal.

2 *Control element*
   This decides what action to take when it receives an error signal. It may be, for example, a signal to operate a switch or open a valve. The control

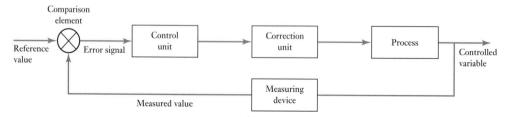

Figure 1.10   The elements of a closed-loop control system.

plan being used by the element may be just to supply a signal which switches on or off when there is an error, as in a room thermostat, or perhaps a signal which proportionally opens or closes a valve according to the size of the error. Control plans may be **hard-wired systems** in which the control plan is permanently fixed by the way the elements are connected together, or **programmable systems** where the control plan is stored within a memory unit and may be altered by reprogramming it. Controllers are discussed in Chapter 10.

3 *Correction element*
The correction element produces a change in the process to correct or change the controlled condition. Thus it might be a switch which switches on a heater and so increases the temperature of the process or a valve which opens and allows more liquid to enter the process. The term **actuator** is used for the element of a correction unit that provides the power to carry out the control action. Correction units are discussed in Chapters 7, 8 and 9.

4 *Process element*
The process is what is being controlled. It could be a room in a house with its temperature being controlled or a tank of water with its level being controlled.

5 *Measurement element*
The measurement element produces a signal related to the variable condition of the process that is being controlled. It might be, for example, a switch which is switched on when a particular position is reached or a thermocouple which gives an e.m.f. related to the temperature.

With the closed–loop system illustrated in Figure 1.10 for a person controlling the temperature of a room, the various elements are:

Controlled variable	–	the room temperature
Reference value	–	the required room temperature
Comparison element	–	the person comparing the measured value with the required value of temperature
Error signal	–	the difference between the measured and required temperatures
Control unit	–	the person
Correction unit	–	the switch on the fire
Process	–	the heating by the fire
Measuring device	–	a thermometer

An automatic control system for the control of the room temperature could involve a thermostatic element which is sensitive to temperature and switches on when the temperature falls below the set value and off when it reaches it (Figure 1.11). This temperature-sensitive switch is then used to switch on the heater. The thermostatic element has the combined functions of comparing the required temperature value with that occurring and then controlling the operation of a switch. It is often the case that elements in control systems are able to combine a number of functions.

Figure 1.12 shows an example of a simple control system used to maintain a constant water level in a tank. The reference value is the initial setting of the lever arm arrangement so that it just cuts off the water supply at the required

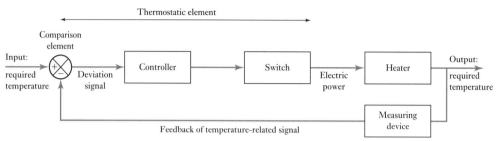

Figure 1.11   Heating a room: a closed-loop system.

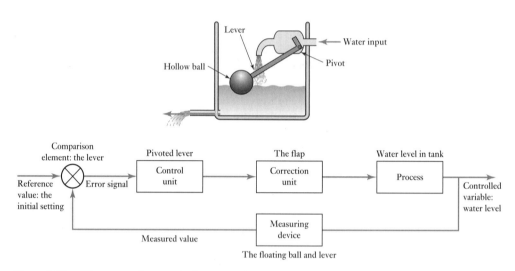

Figure 1.12   The automatic control of water level.

level. When water is drawn from the tank the float moves downwards with the water level. This causes the lever arrangement to rotate and so allows water to enter the tank. This flow continues until the ball has risen to such a height that it has moved the lever arrangement to cut off the water supply. It is a closed-loop control system with the elements being:

Controlled variable	–	water level in tank
Reference value	–	initial setting of the float and lever position
Comparison element	–	the lever
Error signal	–	the difference between the actual and initial settings of the lever positions
Control unit	–	the pivoted lever
Correction unit	–	the flap opening or closing the water supply
Process	–	the water level in the tank
Measuring device	–	the floating ball and lever

The above is an example of a closed-loop control system involving just mechanical elements. We could, however, have controlled the liquid level by means of an electronic control system. We thus might have had a level

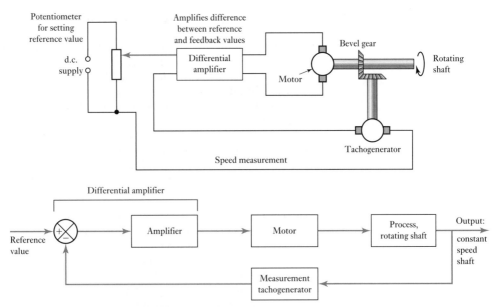

Figure 1.13 Shaft speed control.

sensor supplying an electrical signal which is used, after suitable signal conditioning, as an input to a computer where it is compared with a set value signal and the difference between them, the error signal, then used to give an appropriate response from the computer output. This is then, after suitable signal conditioning, used to control the movement of an actuator in a flow control valve and so determine the amount of water fed into the tank.

Figure 1.13 shows a simple automatic control system for the speed of rotation of a shaft. A potentiometer is used to set the reference value, i.e. what voltage is supplied to the differential amplifier as the reference value for the required speed of rotation. The differential amplifier is used both to compare and amplify the difference between the reference and feedback values, i.e. it amplifies the error signal. The amplified error signal is then fed to a motor which in turn adjusts the speed of the rotating shaft. The speed of the rotating shaft is measured using a tachogenerator, connected to the rotating shaft by means of a pair of bevel gears. The signal from the tachogenerator is then fed back to the differential amplifier:

Controlled variable	–	speed of rotation of shaft
Reference value	–	setting of slider on potentiometer
Comparison element	–	differential amplifier
Error signal	–	the difference between the output from the potentiometer and that from the tachogenerator system
Control unit	–	the differential amplifier
Correction unit	–	the motor
Process	–	the rotating shaft
Measuring device	–	the tachogenerator

1.5.4 Analogue and digital control systems

Analogue systems are ones where all the signals are continuous functions of time and it is the size of the signal which is a measure of the variable (Figure 1.14(a)). The examples so far discussed in this chapter are such systems. **Digital signals** can be considered to be a sequence of on/off signals, the value of the variable being represented by the sequence of on/off pulses (Figure 1.14(b)).

Where a digital signal is used to represent a continuous analogue signal, the analogue signal is sampled at particular instants of time and the sample values each then converted into effectively a digital number, i.e. a particular sequence of digital signals. For example, we might have for a three-digit signal the digital sequence of:

no pulse, no pulse, no pulse representing an analogue signal of 0 V;
no pulse, no pulse, a pulse representing 1 V;
no pulse, pulse, no pulse representing 2 V;
no pulse, pulse, pulse representing 3 V;
pulse, no pulse, no pulse representing 4 V;
pulse, no pulse, pulse representing 5 V;
pulse, pulse, no pulse representing 6 V;
pulse, pulse, pulse representing 7 V.

Because most of the situations being controlled are analogue in nature and it is these that are the inputs and outputs of control systems, e.g. an input of temperature and an output from a heater, a necessary feature of a digital control system is that the real-world analogue inputs have to be converted to digital forms and the digital outputs back to real-world analogue forms. This involves the uses of analogue-to-digital converters (ADC) for inputs and digital-to-analogue converters (DAC) for the outputs.

Figure 1.15(a) shows the basic elements of a digital closed-loop control system; compare it with the analogue closed-loop system in Figure 1.10. The reference value, or set point, might be an input from a keyboard. Analogue-to-digital (ADC) and digital-to-analogue (DAC) elements are included in the loop in order that the digital controller can be supplied with digital signals

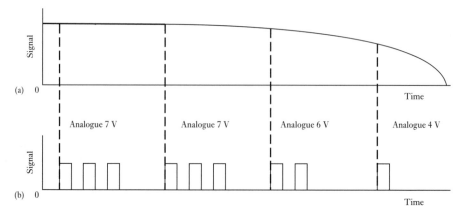

Figure 1.14 Signals: (a) analogue and (b) the digital version of the analogue signal showing the stream of sampled signals.

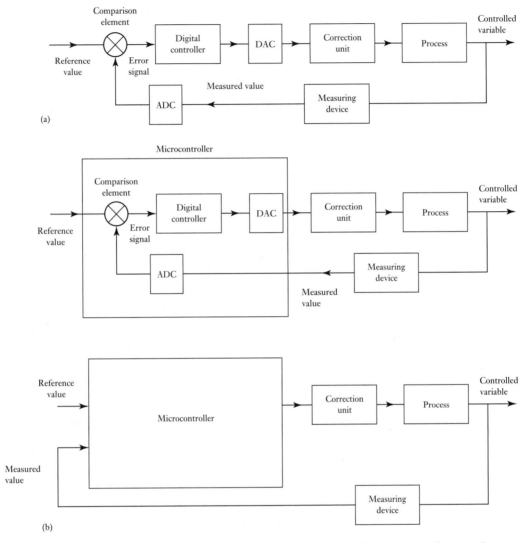

Figure 1.15 (a) The basic elements of a digital closed-loop control system, (b) a microcontroller control system.

from analogue measurement systems and its output of digital signals can be converted to analogue form to operate the correction units. It might seem to be adding a degree of complexity to the control system to have this ADC and DAC, but there are some very important advantages: digital operations can be controlled by a program, i.e. a set of stored instructions; information storage is easier, accuracy can be greater; digital circuits are less affected by noise and also are generally easier to design.

The digital controller could be a digital computer which is running a program, i.e. a piece of software, to implement the required actions. The term control algorithm is used to describe the sequence of steps needed to solve the control problem. The control algorithm that might be used for digital control could be described by the following steps:

>Read the reference value, i.e. the desired value.
>Read the actual plant output from the ADC.
>Calculate the error signal.
>Calculate the required controller output.
>Send the controller output to the DAC.
>Wait for the next sampling interval.

However, many applications do not need the expense of a computer and just a microchip will suffice. Thus, in mechatronics applications a microcontroller is often used for digital control. A microcontroller is a microprocessor with added integrated elements such as memory and ADC and DAC converters; these can be connected directly to the plant being controlled so the arrangement could be as shown in Figure 1.15(b). The control algorithm then might be:

>Read the reference value, i.e. the desired value.
>Read the actual plant output to its ADC input port.
>Calculate the error signal.
>Calculate the required controller output.
>Send the controller output to its DAC output port.
>Wait for the next sampling interval.

An example of a digital control system might be an automatic control system for the control of the room temperature involving a temperature sensor giving an analogue signal which, after suitable signal conditioning to make it a digital signal, is inputted to the digital controller where it is compared with the set value and an error signal generated. This is then acted on by the digital controller to give at its output a digital signal which, after suitable signal conditioning to give an analogue equivalent, might be used to control a heater and hence the room temperature. Such a system can readily be programmed to give different temperatures at different times of the day.

As a further illustration of a digital control system, Figure 1.16 shows one form of a digital control system for the speed a motor might take. Compare this with the analogue system in Figure 1.13.

The software used with a digital controller needs to be able to:

>Read data from its input ports.
>Carry out internal data transfer and mathematical operations.
>Send data to its output ports.

In addition it will have:

>Facilities to determine at what times the control program will be implemented.

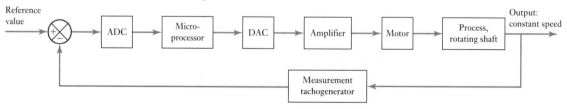

Figure 1.16   Shaft speed control.

Thus we might have the program just waiting for the ADC sampling time to occur and then spring into action when there is an input of a sample. The term **polling** is used for such a situation, the program repeatedly checking the input ports for such sampling events. So we might have:

Check the input ports for input signals.
No signals so do nothing.
Check the input ports for input signals.
No signals so do nothing.
Check the input ports for input signals.
Signal so read data from its input ports.
Carry out internal data transfer and mathematical operations.
Send data to its output ports.
Check the input ports for input signals.
No signals so do nothing.
And so on.

An alternative to polling is to use **interrupt control**. The program does not keep checking its input ports but receives a signal when an input is due. This signal may come from an external clock which gives a signal every time the ADC takes a sample.

No signal from external clock.
Do nothing.
Signal from external clock that an input is due.
Read data from its input ports.
Carry out internal data transfer and mathematical operations.
Send data to its output ports.
Wait for next signal from external clock.

1.5.5 Sequential controllers

There are many situations where control is exercised by items being switched on or off at particular preset times or values in order to control processes and give a step sequence of operations. For example, after step 1 is complete then step 2 starts. When step 2 is complete then step 3 starts, etc.

The term **sequential control** is used when control is such that actions are strictly ordered in a time- or event-driven sequence. Such control could be obtained by an electric circuit with sets of relays or cam-operated switches which are wired up in such a way as to give the required sequence. Such hard-wired circuits are now more likely to have been replaced by a microprocessor-controlled system, with the sequencing being controlled by means of a software program.

As an illustration of sequential control, consider the domestic washing machine. A number of operations have to be carried out in the correct sequence. These may involve a pre-wash cycle when the clothes in the drum are given a wash in cold water, followed by a main wash cycle when they are washed in hot water, then a rinse cycle when they are rinsed with cold water a number of times, followed by spinning to remove water from the clothes. Each of these operations involves a number of steps. For example, a prewash cycle involves opening a valve to fill the machine drum to the required level, closing the valve, switching on the drum motor to rotate the drum for a specific time and operating the pump to empty the water from the drum. The operating sequence is called a **program,** the sequence of instructions in each program being predefined and 'built' into the controller used.

Figure 1.17 shows the basic washing machine system and gives a rough idea of its constituent elements. The system that used to be used for the washing machine controller was a mechanical system which involved a set of cam-operated switches, i.e. mechanical switches, a system which is readily adjustable to give a greater variety of programs.

Figure 1.18 shows the basic principle of one such switch. When the machine is switched on, a small electric motor slowly rotates its shaft, giving an amount of rotation proportional to time. Its rotation turns the controller cams so that each in turn operates electrical switches and so switches on circuits in the correct sequence. The contour of a cam determines the time at which it operates a switch. Thus the contours of the cams are the means by which the program is specified and stored in the machine. The sequence of instructions and the instructions used in a particular washing program are

Figure 1.17 Washing machine system.

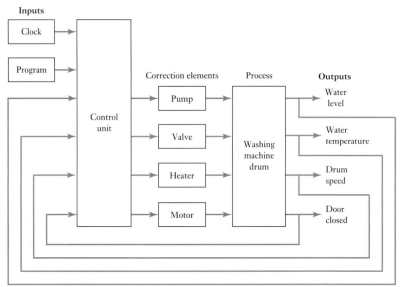

Feedback from outputs of water level, water temperature, drum speed and door closed

Figure 1.18 Cam-operated switch.

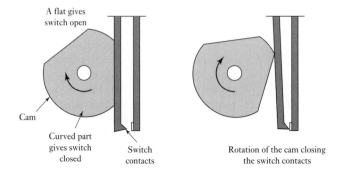

determined by the set of cams chosen. With modern washing machines the controller is a microprocessor and the program is not supplied by the mechanical arrangement of cams but by a software program. The microprocessor-controlled washing machine can be considered an example of a mechatronics approach in that a mechanical system has become integrated with electronic controls. As a consequence, a bulky mechanical system is replaced by a much more compact microprocessor.

For the pre-wash cycle an electrically operated valve is opened when a current is supplied and switched off when it ceases. This valve allows cold water into the drum for a period of time determined by the profile of the cam or the output from the microprocessor used to operate its switch. However, since the requirement is a specific level of water in the washing machine drum, there needs to be another mechanism which will stop the water going into the tank, during the permitted time, when it reaches the required level. A sensor is used to give a signal when the water level has reached the preset level and give an output from the microprocessor which is used to switch off the current to the valve. In the case of a cam-controlled valve, the sensor actuates a switch which closes the valve admitting water to the washing machine drum. When this event is completed, the microprocessor, or the rotation of the cams, initiates a pump to empty the drum.

For the main wash cycle, the microprocessor gives an output which starts when the pre-wash part of the program is completed; in the case of the cam-operated system the cam has a profile such that it starts in operation when the pre-wash cycle is completed. It switches a current into a circuit to open a valve to allow cold water into the drum. This level is sensed and the water shut off when the required level is reached. The microprocessor or cams then supply a current to activate a switch which applies a larger current to an electric heater to heat the water. A temperature sensor is used to switch off the current when the water temperature reaches the preset value. The microprocessor or cams then switch on the drum motor to rotate the drum. This will continue for the time determined by the microprocessor or cam profile before switching off. Then the microprocessor or a cam switches on the current to a discharge pump to empty the water from the drum.

The rinse part of the operation is now switched as a sequence of signals to open valves which allow cold water into the machine, switch it off, operate the motor to rotate the drum, operate a pump to empty the water from the drum, and repeat this sequence a number of times.

The final part of the operation is when the microprocessor or a cam switches on just the motor, at a higher speed than for the rinsing, to spin the clothes.

1.6 Programmable logic controller

In many simple systems there might be just an embedded microcontroller, this being a microprocessor with memory all integrated on one chip, which has been specifically programmed for the task concerned. A more adaptable form is the **programmable logic controller (PLC)**. This is a microprocessor-based controller which uses programmable memory to store instructions and to implement functions such as logic, sequence, timing, counting and arithmetic to control events and can be readily reprogrammed for different tasks. Figure 1.19 shows the control action of a programmable logic controller, the inputs being signals from, say, switches being closed

Figure 1.19   Programmable logic controller.

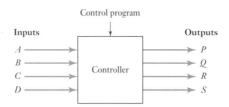

and the program used to determine how the controller should respond to the inputs and the output it should then give.

Programmable logic controllers are widely used in industry where on/off control is required. For example, they might be used in process control where a tank of liquid is to be filled and then heated to a specific temperature before being emptied. The control sequence might thus be as follows.

1 Switch on pump to move liquid into the tank.
2 Switch off pump when a level detector gives the on signal, so indicating that the liquid has reached the required level.
3 Switch on heater.
4 Switch off heater when a temperature sensor gives the on signal to indicate the required temperature has been reached.
5 Switch on pump to empty the liquid from the container.
6 Switch off pump when a level detector gives an on signal to indicate that the tank is empty.

See Chapter 14 for a more detailed discussion of programmable logic controllers and examples of their use.

1.7   Examples of mechatronic systems

Mechatronics brings together the technology of sensors and measurements systems, embedded microprocessor systems, actuators and engineering design. The following are examples of mechatronic systems and illustrate how microprocessor-based systems have been able not only to carry out tasks that previously were done 'mechanically' but also to do tasks that were not easily automated before.

1.7.1   The digital camera and autofocus

A digital camera is likely to have an autofocus control system. A basic system used with less expensive cameras is an open–loop system (Figure 1.20(a)). When the photographer presses the shutter button, a transducer on the front of the camera sends pulses of infrared (IR) light towards the subject of the photograph. The infrared pulses bounce off the subject and are reflected back to the camera where the same transducer picks them up. For each metre the subject is distant from the camera, the round-trip is about 6 ms. The time difference between the output and return pulses is detected and fed to a microprocessor. This has a set of values stored in its memory and so gives an output which rotates the lens housing and moves the lens to a position where the object is in focus. This type of autofocus can only be used for distances up to about 10 m as the returning infrared pulses are too weak at greater distances. Thus for greater distances the microprocessor gives an output which moves the lens to an infinity setting.

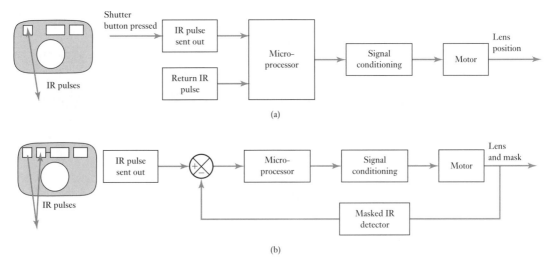

Figure 1.20 Autofocus.

A system used with more expensive cameras involves triangulation (Figure 1.20(b)). Pulses of infrared radiation are sent out and the reflected pulses are detected, not by the same transducer that was responsible for the transmission, but by another transducer. However, initially this transducer has a mask across it. The microprocessor thus gives an output which causes the lens to move and simultaneously the mask to move across the transducer. The mask contains a slot which is moved across the face of the transducer. The movement of the lens and the slot continues until the returning pulses are able to pass through the slot and impact on the transducer. There is then an output from the transducer which leads the microprocessor to stop the movement of the lens, and so give the in-focus position.

1.7.2 The engine management system

The engine management system of an automobile is responsible for managing the ignition and fuelling requirements of the engine. With a four-stroke internal combustion engine there are several cylinders, each of which has a piston connected to a common crankshaft and each of which carries out a four-stroke sequence of operations (Figure 1.21).

When the piston moves down a valve opens and the air–fuel mixture is drawn into the cylinder. When the piston moves up again the valve closes and the air–fuel mixture is compressed. When the piston is near the top of the cylinder the spark plug ignites the mixture with a resulting expansion of the hot gases. This expansion causes the piston to move back down again and so the cycle is repeated. The pistons of each cylinder are connected to a common crankshaft and their power strokes occur at different times so that there is continuous power for rotating the crankshaft.

The power and speed of the engine are controlled by varying the ignition timing and the air–fuel mixture. With modern automobile engines this is done by a microprocessor. Figure 1.22 shows the basic elements of a microprocessor control system. For ignition timing, the crankshaft drives a

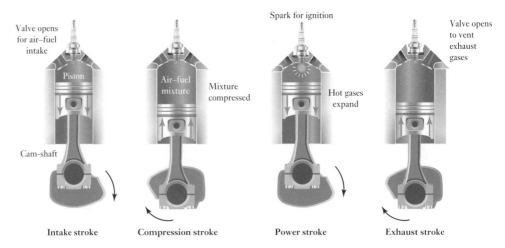

Figure 1.21   Four-stroke sequence.

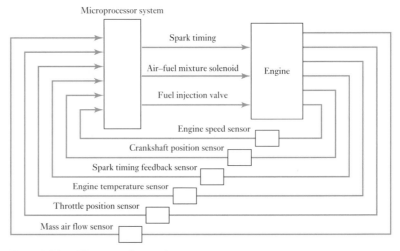

Figure 1.22   Elements of an engine management system.

distributor which makes electrical contacts for each spark plug in turn and a timing wheel. This timing wheel generates pulses to indicate the crankshaft position. The microprocessor then adjusts the timing at which high-voltage pulses are sent to the distributor so they occur at the 'right' moments of time. To control the amount of air–fuel mixture entering a cylinder during the intake strokes, the microprocessor varies the time for which a solenoid is activated to open the intake valve on the basis of inputs received of the engine temperature and the throttle position. The amount of fuel to be injected into the air stream can be determined by an input from a sensor of the mass rate of air flow, or computed from other measurements, and the microprocessor then gives an output to control a fuel injection valve. Note that the above is a very simplistic indication of engine management. See Chapter 24 for a more detailed discussion.

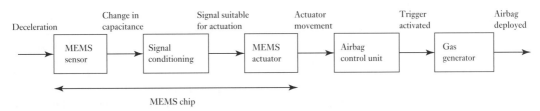

Figure 1.23   Airbag control system.

1.7.3   MEMS and the automobile airbag

Microelectromechanical systems (MEMS) are mechanical devices that are built onto semiconductor chips, generally ranging in size from about 20 micrometres to a millimetre and made up of components 0.001 to 0.1 mm in size. They usually consist of a microprocessor and components such as microsensors and microactuators. MEMS can sense, control and activate mechanical processes on the micro scale. Such MEMS chips are becoming increasingly widely used, and the following is an illustration.

Airbags in automobiles are designed to inflate in the event of a crash and so cushion the impact effects on the vehicle occupant. The airbag sensor is a MEMS accelerometer with an integrated micromechanical element which moves in response to rapid deceleration. See Figure 2.9 for basic details of the ADXL-50 device which is widely used. The rapid deceleration causes a change in capacitance in the MEMS accelerometer, which is detected by the electronics on the MEMS chip and actuates the airbag control unit to fire the airbag. The airbag control unit then triggers the ignition of a gas generator propellant to rapidly inflate a nylon fabric bag (Figure 1.23). As the vehicle occupant's body collides with and squeezes the inflated bag, the gas escapes in a controlled manner through small vent holes and so cushions the impact. From the onset of the crash, the entire deployment and inflation process of the airbag is about 60 to 80 milliseconds.

Summary

Mechatronics is a co-ordinated, and concurrently developed, integration of mechanical engineering with electronics and intelligent computer control in the design and manufacture of products and processes. It involves the bringing together of a number of technologies: mechanical engineering, electronic engineering, electrical engineering, computer technology and control engineering. Mechatronics provides an opportunity to take a new look at problems, with engineers not just seeing a problem in terms of mechanical principles but having to see it in terms of a range of technologies. The electronics, etc., should not be seen as a bolt-on item to existing mechanical hardware. A mechatronics approach needs to be adopted right from the design phase.

Microprocessors are generally involved in mechatronics systems and these are **embedded**. An embedded system is one that is designed to control a range of functions and is not designed to be programmed by the end user in the same way that a computer is. Thus, with an embedded system, the user cannot change what the system does by adding or replacing software.

A **system** can be thought of as a box or block diagram which has an input and an output and where we are concerned not with what goes on inside the box but with only the relationship between the output and the input.

In order to predict how systems behave when there are inputs to them, we need to devise **models** which relate the output to the input so that we can work out, for a given input, how the output will vary with time.

Measurement systems can, in general, be considered to be made up of three basic elements: sensor, signal conditioner and display.

There are two basic forms of **control system: open loop** and **closed loop**. With closed loop there is feedback, a system containing a comparison element, a control element, correction element, process element and the feedback involving a measurement element.

Problems

1.1   Identify the sensor, signal conditioner and display elements in the measurement systems of (a) a mercury-in-glass thermometer, (b) a Bourdon pressure gauge.

1.2   Explain the difference between open- and closed-loop control.

1.3   Identify the various elements that might be present in a control system involving a thermostatically controlled electric heater.

1.4   The automatic control system for the temperature of a bath of liquid consists of a reference voltage fed into a differential amplifier. This is connected to a relay which then switches on or off the electrical power to a heater in the liquid. Negative feedback is provided by a measurement system which feeds a voltage into the differential amplifier. Sketch a block diagram of the system and explain how the error signal is produced.

1.5   Explain the function of a programmable logic controller.

1.6   Explain what is meant by sequential control and illustrate your answer by an example.

1.7   State steps that might be present in the sequential control of a dishwasher.

1.8   Compare and contrast the traditional design of a watch with that of the mechatronics-designed product involving a microprocessor.

1.9   Compare and contrast the control system for the domestic central heating system involving a bimetallic thermostat and that involving a microprocessor.

Part II
Sensors and signal conditioning

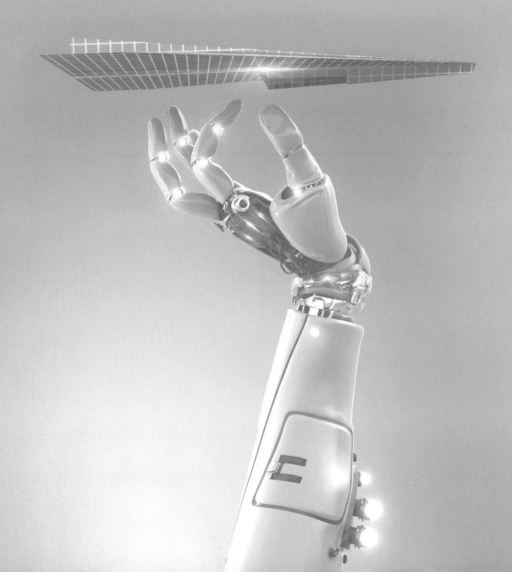

Chapter two Sensors and transducers

Objectives

The objectives of this chapter are that, after studying it, the reader should be able to:
- Describe the performance of commonly used sensors using terms such as range, span, error, accuracy, sensitivity, hysteresis and non-linearity error, repeatability, stability, dead band, resolution, output impedance, response time, time constant, rise time and settling time.
- Evaluate sensors used in the measurement of displacement, position and proximity, velocity and motion, force, fluid pressure, liquid flow, liquid level, temperature, and light intensity.
- Explain the problem of bouncing when mechanical switches are used for inputting data and how it might be overcome.

2.1 Sensors and transducers

The term **sensor** is used for an element which produces a signal relating to the quantity being measured. Thus in the case of, say, an electrical resistance temperature element, the quantity being measured is temperature and the sensor transforms an input of temperature into a change in resistance. The term **transducer** is often used in place of the term sensor. Transducers are defined as elements that when subject to some physical change experience a related change. Thus sensors are transducers. However, a measurement system may use transducers, in addition to the sensor, in other parts of the system to convert signals in one form to another form. A sensor/transducer is said to be **analogue** if it gives an output which is analogue and so changes in a continuous way and typically has an output whose size is proportional to the size of the variable being measured. The term **digital** is used if the systems give outputs which are digital in nature, i.e. a sequence of essentially on/off signals which spell out a number whose value is related to the size of the variable being measured.

This chapter is about transducers and in particular those used as sensors. The terminology that is used to specify the performance characteristics of transducers is defined and examples of transducers commonly used in engineering are discussed.

2.1.1 Smart sensors

Some sensors come combined with their signal conditioning all in the same package. Such an integrated sensor does still, however, require further data processing. However, it is possible to have the sensor and signal conditioning combined with a microprocessor all in the same package. Such an arrangement is termed a **smart sensor**. A smart sensor is able to have such functions as the ability to compensate for random errors, to adapt to changes in the environment, give an automatic calculation of measurement accuracy, adjust for non-linearities to give a linear output, self-calibrate and give self-diagnosis of faults.

Such sensors have their own standard, IEEE 1451, so that smart sensors conforming to this standard can be used in a 'plug-and-play' manner, holding and communicating data in a standard way. Information is stored in the form of a TEDS (Transducer Electronic Datasheet), generally in EEPROM, and identifies each device and gives calibration data.

2.2   Performance terminology

The following terms are used to define the performance of transducers, and often measurement systems as a whole.

1 *Range and span*
   The range of a transducer defines the limits between which the input can vary. The span is the maximum value of the input minus the minimum value. Thus, for example, a load cell for the measurement of forces might have a range of 0 to 50 kN and a span of 50 kN.

2 *Error*
   Error is the difference between the result of the measurement and the true value of the quantity being measured:

   error $=$ measured value $-$ true value

   Thus if a measurement system gives a temperature reading of 25°C when the actual temperature is 24°C, then the error is $+1$°C. If the actual temperature had been 26°C then the error would have been -1°C. A sensor might give a resistance change of $10.2\,\Omega$ when the true change should have been $10.5\,\Omega$. The error is $-0.3\,\Omega$.

3 *Accuracy*
   Accuracy is the extent to which the value indicated by a measurement system might be wrong. It is thus the summation of all the possible errors that are likely to occur, as well as the accuracy to which the transducer has been calibrated. A temperature-measuring instrument might, for example, be specified as having an accuracy of ±2°C. This would mean that the reading given by the instrument can be expected to lie within plus or minus 2°C of the true value. Accuracy is often expressed as a percentage of the full range output or full-scale deflection. The percentage of full-scale deflection term results from when the outputs of measuring systems were displayed almost exclusively on a circular or linear scale. A sensor might, for example, be specified as having an accuracy of $\pm5\%$ of full range output. Thus if the range of the sensor was, say, 0 to 200°C, then the reading given can be expected to be within plus or minus 10°C of the true reading.

4 *Sensitivity*
   The sensitivity is the relationship indicating how much output there is per unit input, i.e. output/input. For example, a resistance thermometer may have a sensitivity of $0.5\,\Omega/$°C. This term is also frequently used to indicate the sensitivity to inputs other than that being measured,

i.e. environmental changes. Thus there can be the sensitivity of the transducer to temperature changes in the environment or perhaps fluctuations in the mains voltage supply. A transducer for the measurement of pressure might be quoted as having a temperature sensitivity of $\pm 0.1\%$ of the reading per °C change in temperature.

5 *Hysteresis error*
Transducers can give different outputs from the same value of quantity being measured according to whether that value has been reached by a continuously increasing change or a continuously decreasing change. This effect is called hysteresis. Figure 2.1 shows such an output with the hysteresis error as the maximum difference in output for increasing and decreasing values.

Figure 2.1 Hysteresis.

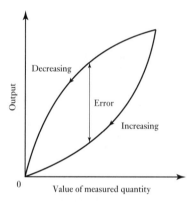

6 *Non-linearity error*
For many transducers a linear relationship between the input and output is assumed over the working range, i.e. a graph of output plotted against input is assumed to give a straight line. Few transducers, however, have a truly linear relationship and thus errors occur as a result of the assumption of linearity. The error is defined as the maximum difference from the straight line. Various methods are used for the numerical expression of the non-linearity error. The differences occur in determining the straight line relationship against which the error is specified. One method is to draw the straight line joining the output values at the end points of the range; another is to find the straight line by using the method of least squares to determine the best fit line when all data values are considered equally likely to be in error; another is to find the straight line by using the method of least squares to determine the best fit line which passes through the zero point. Figure 2.2 illustrates these three methods and how they can affect the non-linearity error quoted. The error is generally quoted as a percentage of the full range output. For example, a transducer for the measurement of pressure might be quoted as having a non-linearity error of $\pm 0.5\%$ of the full range.

7 *Repeatability / reproducibility*
The terms repeatability and reproducibility of a transducer are used to describe its ability to give the same output for repeated applications of the same input value. The error resulting from the same output not being

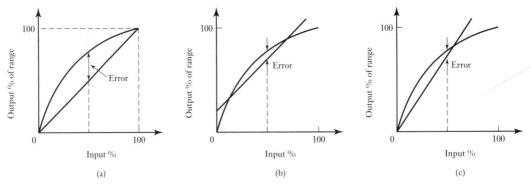

Figure 2.2   Non-linearity error using: (a) end-range values, (b) best straight line for all values, (c) best straight line through the zero point.

given with repeated applications is usually expressed as a percentage of the full range output:

$$\text{repeatability} = \frac{\text{max.} - \text{min. values given}}{\text{full range}} \times 100$$

A transducer for the measurement of angular velocity typically might be quoted as having a repeatability of $\pm0.01\%$ of the full range at a particular angular velocity.

8 *Stability*

The stability of a transducer is its ability to give the same output when used to measure a constant input over a period of time. The term **drift** is often used to describe the change in output that occurs over time. The drift may be expressed as a percentage of the full range output. The term **zero drift** is used for the changes that occur in output when there is zero input.

9 *Dead band/time*

The dead band or dead space of a transducer is the range of input values for which there is no output. For example, bearing friction in a flow-meter using a rotor might mean that there is no output until the input has reached a particular velocity threshold. The dead time is the length of time from the application of an input until the output begins to respond and change.

10 *Resolution*

When the input varies continuously over the range, the output signals for some sensors may change in small steps. A wire-wound potentiometer is an example of such a sensor, the output going up in steps as the potentiometer slider moves from one wire turn to the next. The resolution is the smallest change in the input value that will produce an observable change in the output. For a wire-wound potentiometer the resolution might be specified as, say, $0.5°$ or perhaps a percentage of the full-scale deflection. For a sensor giving a digital output the smallest change in output signal is 1 bit. Thus for a sensor giving a data word of N bits, i.e. a total of 2^N bits, the resolution is generally expressed as $1/2^N$.

11 *Output impedance*

When a sensor giving an electrical output is interfaced with an electronic circuit it is necessary to know the output impedance since this impedance is being connected in either series or parallel with that circuit. The inclusion of the sensor can thus significantly modify the behaviour of the system to which it is connected. See Section 6.1.1 for a discussion of loading.

To illustrate the above, consider the significance of the terms in the following specification of a strain gauge pressure transducer:

Ranges: 70 to 1000 kPa, 2000 to 70 000 kPa
Supply voltage: 10 V d.c. or a.c. r.m.s.
Full range output: 40 mV
Non-linearity and hysteresis: ±0.5% full range output
Temperature range: −54°C to +120°C when operating
Thermal zero shift: 0.030% full range output/°C

The range indicates that the transducer can be used to measure pressures between 70 and 1000 kPa or 2000 and 70 000 kPa. It requires a supply of 10 V d.c. or a.c. r.m.s. for its operation and will give an output of 40 mV when the pressure on the lower range is 1000 kPa and on the upper range 70 000 kPa. Non-linearity and hysteresis will lead to errors of ±0.5% of 1000, i.e. ±5 kPa on the lower range and ±0.5% of 70 000, namely ±350 kPa, on the upper range. The transducer can be used between the temperatures of −54 and +120°C. When the temperature changes by 1°C the output of the transducer for zero input will change by 0.030% of 1000 = 0.3 kPa on the lower range and 0.030% of 70 000 = 21 kPa on the upper range.

2.2.1 Static and dynamic characteristics

The **static characteristics** are the values given when steady-state conditions occur, i.e. the values given when the transducer has settled down after having received some input. The terminology defined above refers to such a state. The **dynamic characteristics** refer to the behaviour between the time that the input value changes and the time that the value given by the transducer settles down to the steady-state value. Dynamic characteristics are stated in terms of the response of the transducer to inputs in particular forms. For example, this might be a step input when the input is suddenly changed from zero to a constant value, or a ramp input when the input is changed at a steady rate, or a sinusoidal input of a specified frequency. Thus we might find the following terms (see Chapter 19 for a more detailed discussion of dynamic systems).

1 *Response time*

This is the time which elapses after a constant input, a step input, is applied to the transducer up to the point at which the transducer gives an output corresponding to some specified percentage, e.g. 95%, of the value of the input (Figure 2.3). For example, if a mercury-in-glass thermometer is put into a hot liquid there can be quite an appreciable time lapse, perhaps as much as 100 s or more, before the thermometer indicates 95% of the actual temperature of the liquid.

Figure 2.3 Response to a step input.

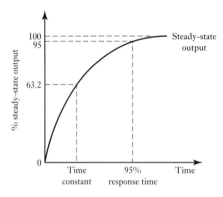

2 *Time constant*

This is the 63.2% response time. A thermocouple in air might have a time constant of perhaps 40 to 100 s. The time constant is a measure of the inertia of the sensor and so how fast it will react to changes in its input: the bigger the time constant, the slower the reaction to a changing input signal. See Section 12.3.4 for a mathematical discussion of the time constant in terms of the behaviour of a system when subject to a step input.

3 *Rise time*

This is the time taken for the output to rise to some specified percentage of the steady-state output. Often the rise time refers to the time taken for the output to rise from 10% of the steady-state value to 90 or 95% of the steady-state value.

4 *Settling time*

This is the time taken for the output to settle to within some percentage, e.g. 2%, of the steady-state value.

To illustrate the above, consider the graph in Figure 2.4 which indicates how an instrument reading changed with time, being obtained from a thermometer plunged into a liquid at time $t = 0$. The steady-state value is 55°C and so, since 95% of 55°C is 52.25°C, the 95% response time is about 228 s.

Figure 2.4 Thermometer in liquid.

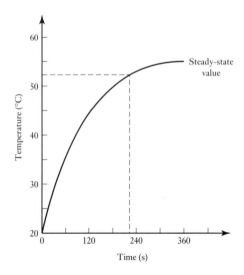

The following sections give examples of transducers grouped according to what they are being used to measure. The measurements considered are those frequently encountered in mechanical engineering, namely: displacement, proximity, velocity, force, pressure, fluid flow, liquid level, temperature and light intensity.

2.3 Displacement, position and proximity

Displacement sensors are concerned with the measurement of the amount by which some object has been moved; **position** sensors are concerned with the determination of the position of some object in relation to some reference point. **Proximity** sensors are a form of position sensor and are used to determine when an object has moved to within some particular critical distance of the sensor. They are essentially devices which give on/off outputs.

Displacement and position sensors can be grouped into two basic types: contact sensors in which the measured object comes into mechanical contact with the sensor, or non-contacting where there is no physical contact between the measured object and the sensor. For those linear displacement methods involving contact, there is usually a sensing shaft which is in direct contact with the object being monitored. The displacement of this shaft is then monitored by a sensor. The movement of the shaft may be used to cause changes in electrical voltage, resistance, capacitance or mutual inductance. For angular displacement methods involving mechanical connection, the rotation of a shaft might directly drive, through gears, the rotation of the transducer element. Non-contacting sensors might involve the presence in the vicinity of the measured object causing a change in the air pressure in the sensor, or perhaps a change in inductance or capacitance. The following are examples of commonly used displacement sensors.

2.3.1 Potentiometer sensor

A **potentiometer** consists of a resistance element with a sliding contact which can be moved over the length of the element. Such elements can be used for linear or rotary displacements, the displacement being converted into a potential difference. The rotary potentiometer consists of a circular wire-wound track or a film of conductive plastic over which a rotatable sliding contact can be rotated (Figure 2.5). The track may be a single turn or helical. With a constant input voltage V_s, between terminals 1 and 3, the output voltage V_o between terminals 2 and 3 is a fraction of the input voltage,

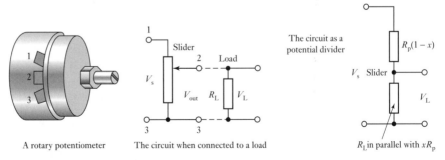

A rotary potentiometer   The circuit when connected to a load R_L in parallel with xR_p

Figure 2.5   Rotary potentiometer.

the fraction depending on the ratio of the resistance R_{23} between terminals 2 and 3 compared with the total resistance R_{13} between terminals 1 and 3, i.e. $V_o/V_s = R_{23}/R_{13}$. If the track has a constant resistance per unit length, i.e. per unit angle, then the output is proportional to the angle through which the slider has rotated. Hence an angular displacement can be converted into a potential difference.

With a wire-wound track the slider in moving from one turn to the other will change the voltage output in steps, each step being a movement of one turn. If the potentiometer has N turns then the resolution, as a percentage, is $100/N$. Thus the resolution of a wire track is limited by the diameter of the wire used and typically ranges from about 1.5 mm for a coarsely wound track to 0.5 mm for a finely wound one. Errors due to non-linearity of the track tend to range from less than 0.1% to about 1%. The track resistance tends to range from about 20 Ω to 200 kΩ. Conductive plastic has ideally infinite resolution, errors due to non-linearity of the track of the order of 0.05% and resistance values from about 500 to 80 kΩ. The conductive plastic has a higher temperature coefficient of resistance than the wire and so temperature changes have a greater effect on accuracy.

An important effect to be considered with a potentiometer is the effect of a load R_L connected across the output. The potential difference across the load V_L is only directly proportional to V_o if the load resistance is infinite. For finite loads, however, the effect of the load is to transform what was a linear relationship between output voltage and angle into a non-linear relationship. The resistance R_L is in parallel with the fraction x of the potentiometer resistance R_p. This combined resistance is $R_L x R_p/(R_L + x R_p)$. The total resistance across the source voltage is thus

$$\text{total resistance} = R_p(1 - x) + R_L x R_p/(R_L + x R_p)$$

The circuit is a potential divider circuit and thus the voltage across the load is the fraction that the resistance across the load is of the total resistance across which the applied voltage is connected:

$$\frac{V_L}{V_s} = \frac{x R_L R_p/(R_L + x R_p)}{R_p(1 - x) + x R_L R_p/(R_L + x R_p)}$$

$$= \frac{x}{(R_p/R_L)x(1 - x) + 1}$$

If the load is of infinite resistance then we have $V_L = x V_s$. Thus the error introduced by the load having a finite resistance is

$$\text{error} = x V_s - V_L = x V_s - \frac{x V_s}{(R_p/R_L)x(1 - x) + 1}$$

$$= V_s \frac{R_p}{R_L}(x^2 - x^3)$$

To illustrate the above, consider the non-linearity error with a potentiometer of resistance 500 Ω, when at a displacement of half its maximum slider travel, which results from there being a load of resistance 10 kΩ. The supply voltage is 4 V. Using the equation derived above

$$\text{error} = 4 \times \frac{500}{10\,000}(0.5^2 - 0.5^3) = 0.025 \text{ V}$$

As a percentage of the full range reading, this is 0.625%.

Potentiometers are used as sensors with the electronic systems in cars, being used for such things as the accelerator pedal position and throttle position.

2.3.2 Strain-gauged element

The electrical resistance strain gauge (Figure 2.6) is a metal wire, metal foil strip or a strip of semiconductor material which is wafer-like and can be stuck onto surfaces like a postage stamp. When subject to strain, its resistance R changes, the fractional change in resistance $\Delta R/R$ being proportional to the strain ε, i.e.

$$\frac{\Delta R}{R} = G\varepsilon$$

where G, the constant of proportionality, is termed the gauge factor.

Figure 2.6 Strain gauges: (a) metal wire, (b) metal foil, (c) semiconductor.

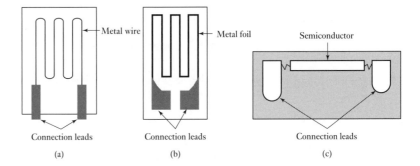

(a) (b) (c)

Since strain is the ratio (change in length/original length) then the resistance change of a strain gauge is a measurement of the change in length of the element to which the strain gauge is attached. The gauge factor of metal wire or foil strain gauges with the metals generally used is about 2.0 and resistances are generally of the order of about 100 Ω. Silicon p- and n-type semiconductor strain gauges have gauge factors of about +100 or more for p-type silicon and −100 or more for n-type silicon and resistances of the order of 1000 to 5000 Ω. The gauge factor is normally supplied by the manufacturer of the strain gauges from a calibration made of a sample of strain gauges taken from a batch. The calibration involves subjecting the sample gauges to known strains and measuring their changes in resistance. A problem with all strain gauges is that their resistance changes not only with strain but also with temperature. Ways of eliminating the temperature effect have to be used and are discussed in Chapter 3. Semiconductor strain gauges have a much greater sensitivity to temperature than metal strain gauges.

To illustrate the above, consider an electrical resistance strain gauge with a resistance of 100 Ω and a gauge factor of 2.0. What is the change in resistance of the gauge when it is subject to a strain of 0.001? The fractional change in resistance is equal to the gauge factor multiplied by the strain, thus

$$\text{change in resistance} = 2.0 \times 0.001 \times 100 = 0.2\ \Omega$$

One form of displacement sensor has strain gauges attached to flexible elements in the form of cantilevers (Figure 2.7(a)), rings (Figure 2.7(b)) or

Figure 2.7   Strain-gauged element.

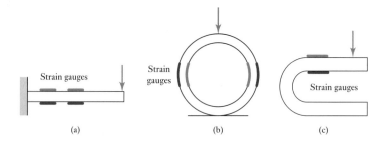

U-shapes (Figure 2.7(c)). When the flexible element is bent or deformed as a result of forces being applied by a contact point being displaced, then the electrical resistance strain gauges mounted on the element are strained and so give a resistance change which can be monitored. The change in resistance is thus a measure of the displacement or deformation of the flexible element. Such arrangements are typically used for linear displacements of the order of 1 to 30 mm and have a non-linearity error of about ±1% of full range.

2.3.3   Capacitive element

The capacitance C of a parallel plate capacitor is given by

$$C = \frac{\varepsilon_r \varepsilon_0 A}{d}$$

where ε_r is the relative permittivity of the dielectric between the plates, ε_0 a constant called the permittivity of free space, A the area of overlap between the two plates and d the plate separation. Capacitive sensors for the monitoring of linear displacements might thus take the forms shown in Figure 2.8. In (a) one of the plates is moved by the displacement so that the plate separation changes; in (b) the displacement causes the area of overlap to change; in (c) the displacement causes the dielectric between the plates to change.

For the displacement changing the plate separation (Figure 2.8(a)), if the separation d is increased by a displacement x then the capacitance becomes

$$C - \Delta C = \frac{\varepsilon_0 \varepsilon_r A}{d + x}$$

Hence the change in capacitance ΔC as a fraction of the initial capacitance is given by

$$\frac{\Delta C}{C} = -\frac{d}{d + x} - 1 = -\frac{x/d}{1 + (x/d)}$$

Figure 2.8   Forms of capacitive sensing element.

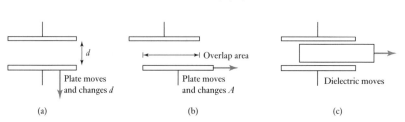

Figure 2.9 (a) Push–pull
sensor, (b) such a sensor used
as an element in the ADXL-50
MEMS accelerometer. The
Analog Devices ADXL50
consists of a mass spring system
as well as a system to measure
displacement and the appropriate
signal conditioning circuitry.

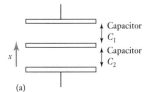

(a)

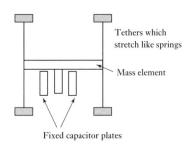

Tethers which
stretch like springs

Mass element

Fixed capacitor plates

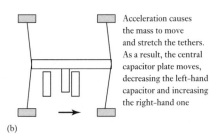

Acceleration causes
the mass to move
and stretch the tethers.
As a result, the central
capacitor plate moves,
decreasing the left-hand
capacitor and increasing
the right-hand one

(b)

Figure 2.9 (a) Push–pull
sensor, (b) such a sensor used
as an element in the ADXL-50
MEMS accelerometer. The
Analog Devices ADXL50
consists of a mass spring system
as well as a system to measure
displacement and the appropriate
signal conditioning circuitry.

There is thus a non-linear relationship between the change in capacitance
ΔC and the displacement x. This non-linearity can be overcome by using
what is termed a **push–pull displacement sensor** (Figure 2.9(a)).
Figure 2.9(b) shows how this can be realised in practice. This has three
plates with the upper pair forming one capacitor and the lower pair another
capacitor. The displacement moves the central plate between the two other
plates. The result of, for example, the central plate moving downwards
is to increase the plate separation of the upper capacitor and decrease the
separation of the lower capacitor. We thus have

$$C_1 = \frac{\varepsilon_0 \varepsilon_r A}{d + x}$$

$$C_2 = \frac{\varepsilon_0 \varepsilon_r A}{d - x}$$

When C_1 is in one arm of an a.c. bridge and C_2 in the other, then the resulting
out-of-balance voltage is proportional to x. Such a sensor is typically used
for monitoring displacements from a few millimetres to hundreds of
millimetres. Non-linearity and hysteresis are about $\pm 0.01\%$ of full range.

One form of capacitive proximity sensor consists of a single capacitor
plate probe with the other plate being formed by the object, which has to
be metallic and earthed (Figure 2.10). As the object approaches so the 'plate
separation' of the capacitor changes, becoming significant and detectable
when the object is close to the probe.

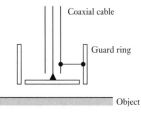

Coaxial cable

Guard ring

Object

Figure 2.10 Capacitive
proximity sensor.

2.3.4 Differential transformers

The linear variable differential transformer, generally referred to by the acronym LVDT, consists of three coils symmetrically spaced along an insulated tube (Figure 2.11). The central coil is the primary coil and the other two are identical secondary coils which are connected in series in such a way that their outputs oppose each other. A magnetic core is moved through the central tube as a result of the displacement being monitored.

Figure 2.11 LVDT.

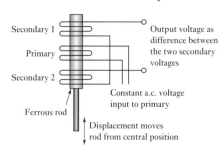

Secondary 1

Primary

Secondary 2

Ferrous rod

Output voltage as difference between the two secondary voltages

Constant a.c. voltage input to primary

Displacement moves rod from central position

When there is an alternating voltage input to the primary coil, alternating e.m.f.s are induced in the secondary coils. With the magnetic core central, the amount of magnetic material in each of the secondary coils is the same. Thus the e.m.f.s induced in each coil are the same. Since they are so connected that their outputs oppose each other, the net result is zero output.

However, when the core is displaced from the central position there is a greater amount of magnetic core in one coil than the other, e.g. more in secondary coil 2 than coil 1. The result is that a greater e.m.f. is induced in one coil than the other. There is then a net output from the two coils. Since a greater displacement means even more core in one coil than the other, the output, the difference between the two e.m.f.s increases the greater the displacement being monitored (Figure 2.12).

The e.m.f. induced in a secondary coil by a changing current i in the primary coil is given by

$$e = M\frac{\mathrm{d}i}{\mathrm{d}t}$$

where M is the mutual inductance, its value depending on the number of turns on the coils and the ferromagnetic core. Thus, for a sinusoidal input current of $i = I \sin \omega t$ to the primary coil, the e.m.f.s induced in the two secondary coils 1 and 2 can be represented by

$$v_1 = k_1 \sin(\omega t - \phi) \text{ and } v_2 = k_2 \sin(\omega t - \phi)$$

where the values of k_1, k_2 and ϕ depend on the degree of coupling between the primary and secondary coils for a particular core position. ϕ is the phase difference between the primary alternating voltage and the secondary alternating voltages. Because the two outputs are in series, their difference is the output

$$\text{output voltage} = v_1 - v_2 = (k_1 - k_2) \sin(\omega t - \phi)$$

When the core is equally in both coils, k_1 equals k_2 and so the output voltage is zero. When the core is more in 1 than in 2 we have $k_1 > k_2$ and

$$\text{output voltage} = (k_1 - k_2) \sin(\omega t - \phi)$$

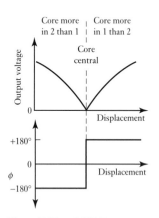

Core more in 2 than 1

Core more in 1 than 2

Core central

Output voltage

Displacement

$+180°$

0

$-180°$

ϕ

Displacement

Figure 2.12 LVDT output.

When the core is more in 2 than in 1 we have $k_1 < k_2$. A consequence of k_1 being less than k_2 is that there is a phase change of 180° in the output when the core moves from more in 1 to more in 2. Thus

$$\text{output voltage} = -(k_1 - k_2)\sin(\omega t - \phi)$$
$$= (k_2 - k_1)\sin[\omega t + (\pi - \phi)]$$

Figure 2.12 shows how the size and phase of the output change with the displacement of the core.

Figure 2.13   LVDT d.c. output.

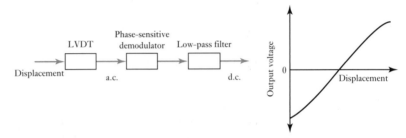

With this form of output, the same amplitude output voltage is produced for two different displacements. To give an output voltage which is unique to each value of displacement we need to distinguish between where the amplitudes are the same but there is a phase difference of 180°. A phase-sensitive demodulator, with a low-pass filter, is used to convert the output into a d.c. voltage which gives a unique value for each displacement (Figure 2.13). Such circuits are available as integrated circuits.

Typically, LVDTs have operating ranges from about ±2 to ±400 mm with non-linearity errors of about ±0.25%. LVDTs are very widely used as primary transducers for monitoring displacements. The free end of the core may be spring loaded for contact with the surface being monitored, or threaded for mechanical connection. They are also used as secondary transducers in the measurement of force, weight and pressure; these variables are transformed into displacements which can then be monitored by LVDTs.

A rotary variable differential transformer (RVDT) can be used for the measurement of rotation (Figure 2.14); it operates on the same principle as the LVDT. The core is a cardioid-shaped piece of magnetic material and rotation causes more of it to pass into one secondary coil than the other. The range of operation is typically ±40° with a linearity error of about ±0.5% of the range.

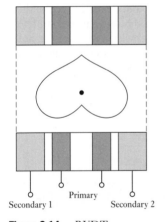

Figure 2.14   RVDT.

2.3.5 Eddy current proximity sensors

If a coil is supplied with an alternating current, an alternating magnetic field is produced. If there is a metal object in close proximity to this alternating magnetic field, then eddy currents are induced in it. The eddy currents themselves produce a magnetic field. This distorts the magnetic field responsible for their production. As a result, the impedance of the coil changes and so does the amplitude of the alternating current. At some preset level, this change can be used to trigger a switch. Figure 2.15 shows the basic form of such a sensor; it is used for the detection of non-magnetic but conductive materials. They have the advantages of being relatively inexpensive, small in size, with high reliability, and can have high sensitivity to small displacements.

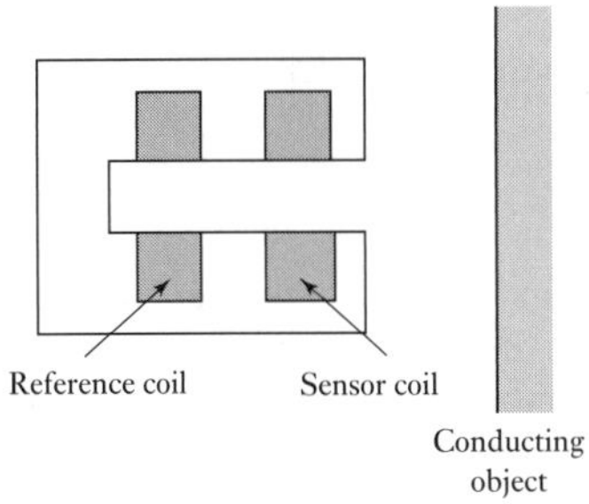

Figure 2.15   Eddy current sensor.

2.3.6   Inductive proximity switch

This consists of a coil wound round a core. When the end of the coil is close to a metal object its inductance changes. This change can be monitored by its effect on a resonant circuit and the change used to trigger a switch. It can only be used for the detection of metal objects and is best with ferrous metals.

2.3.7   Optical encoders

An **encoder** is a device that provides a digital output as a result of a linear or angular displacement. Position encoders can be grouped into two categories: **incremental encoders**, which detect changes in rotation from some datum position; and **absolute encoders**, which give the actual angular position.

Figure 2.16(a) shows the basic form of an incremental encoder for the measurement of angular displacement. A beam of light passes through slots in a disc and is detected by a suitable light sensor. When the disc is rotated, a pulsed output is produced by the sensor with the number of pulses being proportional to the angle through which the disc rotates. Thus the angular position of the disc, and hence the shaft rotating it, can be determined by the number of pulses produced since some datum position. In practice three concentric tracks with three sensors are used (Figure 2.16(b)). The inner track has just one hole and is used to locate the 'home' position of the disc. The other two tracks have a series of equally spaced holes that go completely round the disc but with the holes in the middle track offset from the holes in the outer track by one-half the width of a hole. This offset enables the direction of rotation to be determined. In a clockwise direction the pulses in the outer track lead those in the inner; in the anti-clockwise direction they lag. The resolution is determined by the number of slots on the disc. With 60 slots in 1 revolution then, since 1 revolution is a rotation of $360°$, the resolution is $360/60 = 6°$.

Figure 2.16   Incremental encoder: (a) the basic principle, (b) concentric tracks.

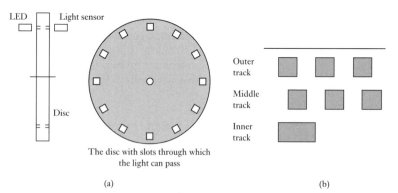

(a)

The disc with slots through which the light can pass

(b)

Figure 2.17 shows the basic form of an absolute encoder for the measurement of angular displacement. This gives an output in the form of a binary number of several digits, each such number representing a particular angular position. The rotating disc has three concentric circles of slots and three sensors to detect the light pulses. The slots are arranged in such a way that the sequential output from the sensors is a number in the binary code. Typical encoders tend to have up to 10 or 12 tracks. The number of bits in

Figure 2.17 A 3-bit absolute encoder.

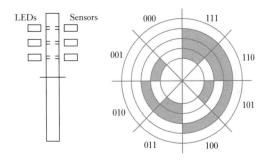

the binary number will be equal to the number of tracks. Thus with 10 tracks there will be 10 bits and so the number of positions that can be detected is 2^{10}, i.e. 1024, a resolution of $360/1024 = 0.35°$.

The normal form of binary code is generally not used because changing from one binary number to the next can result in more than one bit changing, e.g. in changing from 001 to 110 we have two bits changing, and if, through some misalignment, one of the bits changes fractionally before the others then an intermediate binary number is momentarily indicated and so can lead to false counting. To overcome this the **Gray code** is generally used (see Appendix B). With this code only one bit changes in moving from one number to the next. Figure 2.18 shows the track arrangements with normal binary code and the Gray code.

Optical encoders, e.g. HEDS-5000 from Hewlett Packard, are supplied for mounting on shafts and contain a light-emitting diode (LED) light source and a code wheel. Interface integrated circuits are also available to decode the encoder and convert from Gray code to give a binary output suitable for a microprocessor. For an absolute encoder with seven tracks on its code disc, each track will give one of the bits in the binary number and thus we have 2^7 positions specified, i.e. 128. With eight tracks we have 2^8 positions, i.e. 256.

	Normal binary	Gray code
0	0000	0000
1	0001	0001
2	0010	0011
3	0011	0010
4	0100	0110
5	0101	0111
6	0110	0101
7	0111	0100
8	1000	1100
9	1001	1101
10	1010	1111

Figure 2.18 Binary and Gray codes.

2.3.8 Pneumatic sensors

Pneumatic sensors involve the use of compressed air, displacement or the proximity of an object being transformed into a change in air pressure. Figure 2.19 shows the basic form of such a sensor. Low-pressure air is allowed

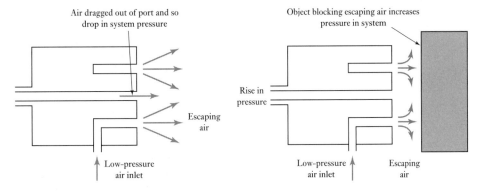

Figure 2.19 Pneumatic proximity sensor.

to escape through a port in the front of the sensor. This escaping air, in the absence of any close-by object, escapes and in doing so also reduces the pressure in the nearby sensor output port. However, if there is a close-by object, the air cannot so readily escape and the result is that the pressure increases in the sensor output port. The output pressure from the sensor thus depends on the proximity of objects.

Such sensors are used for the measurement of displacements of fractions of millimetres in ranges which typically are about 3 to 12 mm.

2.3.9   Proximity switches

There are a number of forms of switch which can be activated by the presence of an object in order to give a proximity sensor with an output which is either on or off.

The **microswitch** is a small electrical switch which requires physical contact and a small operating force to close the contacts. For example, in the case of determining the presence of an item on a conveyor belt, this might be actuated by the weight of the item on the belt depressing the belt and hence a spring-loaded platform under it, with the movement of this platform then closing the switch. Figure 2.20 shows examples of ways such switches can be actuated.

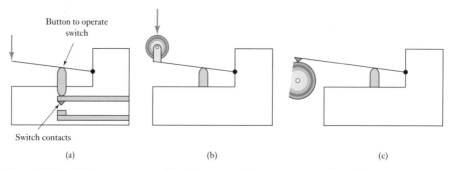

Figure 2.20   (a) Lever-operated, (b) roller-operated, (c) cam-operated switches.

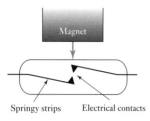

Figure 2.21   Reed switch.

Figure 2.21 shows the basic form of a **reed switch**. It consists of two magnetic switch contacts sealed in a glass tube. When a magnet is brought close to the switch, the magnetic reeds are attracted to each other and close the switch contacts. It is a non-contact proximity switch. Such a switch is very widely used for checking the closure of doors. It is also used with such devices as tachometers, which involve the rotation of a toothed wheel past the reed switch. If one of the teeth has a magnet attached to it, then every time it passes the switch it will momentarily close the contacts and hence produce a current/voltage pulse in the associated electrical circuit.

Photosensitive devices can be used to detect the presence of an opaque object by its breaking a beam of light, or infrared radiation, falling on such a device or by detecting the light reflected back by the object (Figure 2.22).

Figure 2.22   Using photoelectric sensors to detect objects by (a) the object breaking the beam, (b) the object reflecting light.

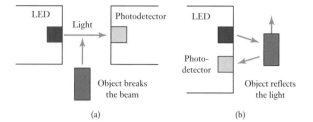

2.3.10   Hall effect sensors

When a beam of charged particles passes through a magnetic field, forces act on the particles and the beam is deflected from its straight line path. A current flowing in a conductor is like a beam of moving charges and thus can be deflected by a magnetic field. This effect was discovered by E.R. Hall in 1879 and is called the **Hall effect**. Consider electrons moving in a conductive plate with a magnetic field applied at right angles to the plane of the plate (Figure 2.23). As a consequence of the magnetic field, the moving electrons are deflected to one side of the plate and thus that side becomes negatively charged, while the opposite side becomes positively charged since the electrons are directed away from it. This charge separation produces an electric field in the material. The charge separation continues until the forces on the charged particles from the electric field just balance the forces produced by the magnetic field. The result is a transverse potential difference V given by

$$V = K_{\mathrm{H}}\frac{BI}{t}$$

where B is the magnetic flux density at right angles to the plate, I the current through it, t the plate thickness and K_{H} a constant called the **Hall coefficient**. Thus if a constant current source is used with a particular sensor, the Hall voltage is a measure of the magnetic flux density.

Figure 2.23   Hall effect.

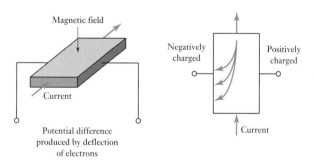

Hall effect sensors are generally supplied as an integrated circuit with the necessary signal processing circuitry. There are two basic forms of such sensor: linear, where the output varies in a reasonably linear manner with the magnetic flux density (Figure 2.24(a)); and threshold, where the output shows a sharp drop at a particular magnetic flux density (Figure 2.24(b)). The linear output Hall effect sensor 634SS2 gives an output which is fairly

Figure 2.24 Hall effect sensors: (a) linear, (b) threshold.

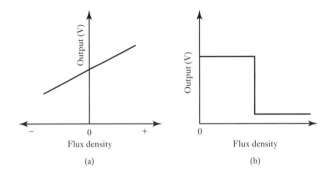

linear over the range -40 to $+40$ mT (-400 to $+400$ gauss) at about 10 mV per mT (1 mV per gauss) when there is a supply voltage of 5 V. The threshold Hall effect sensor Allegro UGN3132U gives an output which switches from virtually zero to about 145 mV when the magnetic flux density is about 3 mT (30 gauss). Hall effect sensors have the advantages of being able to operate as switches which can operate up to 100 kHz repetition rate, cost less than electromechanical switches and do not suffer from the problems associated with such switches of contact bounce occurring and hence a sequence of contacts rather than a single clear contact. The Hall effect sensor is immune to environmental contaminants and can be used under severe service conditions.

Such sensors can be used as position, displacement and proximity sensors if the object being sensed is fitted with a small permanent magnet. As an illustration, such a sensor can be used to determine the level of fuel in an automobile fuel tank. A magnet is attached to a float and as the level of fuel changes so the float distance from the Hall sensor changes (Figure 2.25). The result is a Hall voltage output which is a measure of the distance of the float from the sensor and hence the level of fuel in the tank.

Another application of Hall effect sensors is in brushless d.c. motors. With such motors it is necessary to determine when the permanent magnet rotor is correctly aligned with the windings on the stator so that the current through the windings can be switched on at the right instant to maintain the rotor rotation. Hall effect sensors are used to detect when the alignment is right.

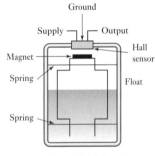

Figure 2.25 Fluid-level detector.

2.4   Velocity and motion

The following are examples of sensors that can be used to monitor linear and angular velocities and detect motion. The application of motion detectors includes security systems used to detect intruders and interactive toys and appliances, e.g. the cash machine screen which becomes active when you get near to it.

2.4.1   Incremental encoder

The incremental encoder described in Section 2.3.7 can be used for the measurement of angular velocity, the number of pulses produced per second being determined.

2.4.2 Tachogenerator

The tachogenerator is used to measure angular velocity. One form, the **variable reluctance tachogenerator**, consists of a toothed wheel of ferromagnetic material which is attached to the rotating shaft (Figure 2.26). A pick-up coil is wound on a permanent magnet. As the wheel rotates, so the teeth move past the coil and the air gap between the coil and the ferromagnetic material changes. We have a magnetic circuit with an air gap which periodically changes. Thus the flux linked by a pick-up coil changes. The resulting cyclic change in the flux produces an alternating e.m.f. in the coil.

Figure 2.26 Variable reluctance tachogenerator.

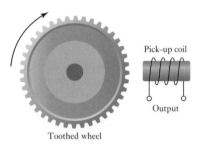

Pick-up coil

Output

Toothed wheel

If the wheel contains n teeth and rotates with an angular velocity ω, then the flux change with time for the coil can be considered to be of the form

$$\Phi = \Phi_0 + \Phi_a \cos n\omega t$$

where Φ_0 is the mean value of the flux and Φ_a the amplitude of the flux variation. The induced e.m.f. e in the N turns of the pick-up coil is $-N\,d\Phi/dt$ and thus

$$e = N\Phi_a n\omega \sin \omega t$$

and so we can write

$$e = E_{max} \sin \omega t$$

where the maximum value of the induced e.m.f. E_{max} is $N\Phi_a n\omega$ and so is a measure of the angular velocity.

Instead of using the maximum value of the e.m.f. as a measure of the angular velocity, a pulse-shaping signal conditioner can be used to transform the output into a sequence of pulses which can be counted by a counter, the number counted in a particular time interval being a measure of the angular velocity.

Another form of tachogenerator is essentially an **a.c. generator**. It consists of a coil, termed the rotor, which rotates with the rotating shaft. This coil rotates in the magnetic field produced by a stationary permanent magnet or electromagnet (Figure 2.27) and so an alternating e.m.f. is induced in it. The amplitude or frequency of this alternating e.m.f. can be used as a measure of the angular velocity of the rotor. The output may be rectified to give a d.c. voltage with a size which is proportional to the angular velocity. Non-linearity for such sensors is typically of the order of ±0.15% of the full range and the sensors are typically used for rotations up to about 10 000 rev/min.

Figure 2.27   a.c. generator form of tachogenerator.

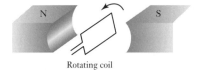

2.4.3   Pyroelectric sensors

Pyroelectric materials, e.g. lithium tantalate, are crystalline materials which generate charge in response to heat flow. When such a material is heated to a temperature just below the Curie temperature, this being about 610°C for lithium tantalate, in an electric field and the material cooled while remaining in the field, electric dipoles within the material line up and it becomes polarised (Figure 2.28, (a) leading to (b)). When the field is then removed, the material retains its polarisation; the effect is rather like magnetising a piece of iron by exposing it to a magnetic field. When the pyroelectric material is exposed to infrared radiation, its temperature rises and this reduces the amount of polarisation in the material, the dipoles being shaken up more and losing their alignment (Figure 2.28(c)).

Figure 2.28   (a), (b) Polarising a pyroelectric material, (c) the effect of temperature on the amount of polarisation.

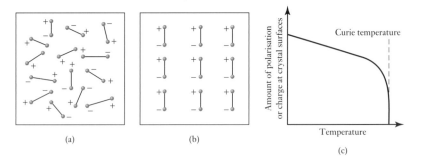

A pyroelectric sensor consists of a polarised pyroelectric crystal with thin metal film electrodes on opposite faces. Because the crystal is polarised with charged surfaces, ions are drawn from the surrounding air and electrons from any measurement circuit connected to the sensor to balance the surface charge (Figure 2.29(a)). If infrared radiation is incident on the crystal and changes its temperature, the polarisation in the crystal is reduced and consequently that is a reduction in the charge at the surfaces of the crystal. There is then an excess of charge on the metal electrodes over that needed to balance the charge on the crystal surfaces (Figure 2.29(b)). This charge leaks away through the measurement circuit until the charge on the crystal

Figure 2.29   Pyroelectric sensor.

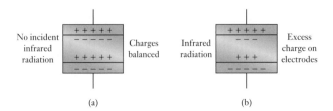

once again is balanced by the charge on the electrodes. The pyroelectric sensor thus behaves as a charge generator which generates charge when there is a change in its temperature as a result of the incidence of infrared radiation. For the linear part of the graph in Figure 2.28(c), when there is a temperature change the change in charge Δq is proportional to the change in temperature Δt:

$$\Delta q = k_p \Delta t$$

where k_p is a sensitivity constant for the crystal. Figure 2.30 shows the equivalent circuit of a pyroelectric sensor, it effectively being a capacitor charged by the excess charge with a resistance R to represent either internal leakage resistance or that combined with the input resistance of an external circuit.

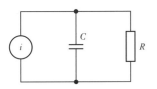

Figure 2.30   Equivalent circuit.

To detect the motion of a human or other heat source, the sensing element has to distinguish between general background heat radiation and that given by a moving heat source. A single pyroelectric sensor would not be capable of this and so a dual element is used (Figure 2.31). One form has the sensing element with a single front electrode but two, separated, back electrodes. The result is two sensors which can be connected so that when both receive the same heat signal their outputs cancel. When a heat source moves so that the heat radiation moves from one of the sensing elements to the other, then the resulting current through the resistor alternates from being first in one direction and then reversed to the other direction. Typically a moving human gives an alternating current of the order of 10^{-12} A. The resistance R has thus to be very high to give a significant voltage. For example, 50 GΩ with such a current gives 50 mV. For this reason a transistor is included in the circuit as a voltage follower to bring the output impedance down to a few kilo-ohms.

A focusing device is needed to direct the infrared radiation onto the sensor. While parabolic mirrors can be used, a more commonly used method is a Fresnel plastic lens. Such a lens also protects the front surface of the sensor and is the form commonly used for sensors to trigger intruder alarms or switch on a light when someone approaches.

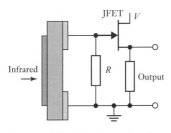

Figure 2.31   Dual pyroelectric sensor.

2.5   Force

A spring balance is an example of a force sensor in which a force, a weight, is applied to the scale pan and causes a displacement, i.e. the spring stretches. The displacement is then a measure of the force. Forces are commonly measured by the measurement of displacements, the following method illustrating this.

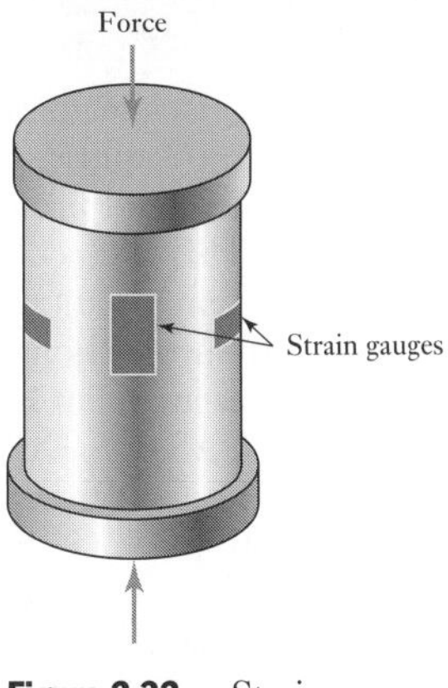

2.5.1   Strain gauge load cell

A very commonly used form of force-measuring transducer is based on the use of electrical resistance strain gauges to monitor the strain produced in some member when stretched, compressed or bent by the application of the force. The arrangement is generally referred to as a **load cell**. Figure 2.32 shows an example of such a cell. This is a cylindrical tube to which strain gauges have been attached. When forces are applied to the cylinder to compress it, then the strain gauges give a resistance change which is a measure of the strain and hence the applied forces. Since temperature also

Figure 2.32   Strain gauge load cell.

produces a resistance change, the signal conditioning circuit used has to be able to eliminate the effects due to temperature (see Section 3.5.1). Typically such load cells are used for forces up to about 10 MN, the non-linearity error being about ±0.03% of full range, hysteresis error ±0.02% of full range and repeatability error ±0.02% of full range. Strain gauge load cells based on the bending of a strain-gauged metal element tend to be used for smaller forces, e.g. with ranges varying from 0 to 5 N up to 0 to 50 kN. Errors are typically a non-linearity error of about ±0.03% of full range, hysteresis error ±0.02% of full range and repeatability error ±0.02% of full range.

2.6   Fluid pressure

Many of the devices used to monitor fluid pressure in industrial processes involve the monitoring of the elastic deformation of diaphragms, capsules, bellows and tubes. The types of pressure measurements that can be required are: absolute pressure where the pressure is measured relative to zero pressure, i.e. a vacuum, differential pressure where a pressure difference is measured and gauge pressure where the pressure is measured relative to the barometric pressure.

For a diaphragm (Figure 2.33(a) and (b)), when there is a difference in pressure between the two sides then the centre of the diaphragm becomes displaced. Corrugations in the diaphragm result in a greater sensitivity. This movement can be monitored by some form of displacement sensor, e.g. a strain gauge, as illustrated in Figure 2.34. A specially designed strain gauge is often used, consisting of four strain gauges with two measuring the strain in a circumferential direction while two measure strain in a radial direction. The four strain gauges are then connected to form the arms of a Wheatstone bridge (see Chapter 3). While strain gauges can be stuck on a diaphragm, an alternative is to create a silicon diaphragm with the strain gauges as specially doped areas of the diaphragm. Such an arrangement is used with the electronic systems for cars to monitor the inlet manifold pressure.

Figure 2.33   Diaphragms: (a) flat, (b) corrugated.

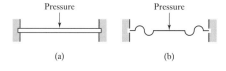

(a)   (b)

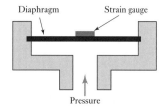

Figure 2.34   Diaphragm pressure gauge.

With the Motorola MPX pressure sensors, the strain gauge element is integrated, together with a resistive network, in a single silicon diaphragm chip. When a current is passed through the strain gauge element and pressure applied at right angles to it, a voltage is produced. This element, together with signal conditioning and temperature compensation circuitry, is packaged as the MPX sensor. The output voltage is directly proportional to the pressure. Such sensors are available for use for the measurement of absolute pressure (the MX numbering system ends with A, AP, AS or ASX), differential pressure (the MX numbering system ends with D or DP) and gauge pressure (the MX numbering system ends with GP, GVP, GS, GVS, GSV or GVSX). For example, the MPX2100 series has a pressure range of 100 kPa and with a supply voltage of 16 V d.c. gives in the absolute pressure and differential pressure forms a voltage output over the full range of 40 mV. The response time, 10 to 90%, for a step change from 0 to 100 kPa is about 1.0 ms and the output impedance is of the order of 1.4 to 3.0 kΩ. The absolute pressure sensors are used for such applications as altimeters and barometers, the differential pressure sensors for air flow measurements and the gauge pressure sensors for engine pressure and tyre pressure.

Capsules (Figure 2.35(a)) can be considered to be just two corrugated diaphragms combined and give even greater sensitivity. A stack of capsules is just a bellows (Figure 2.35(b)) and even more sensitive. Figure 2.36 shows how a bellows can be combined with an LVDT to give a pressure sensor with an electrical output. Diaphragms, capsules and bellows are made from such materials as stainless steel, phosphor bronze and nickel, with rubber and nylon also being used for some diaphragms. Pressures in the range of about 10^3 to 10^8 Pa can be monitored with such sensors.

Figure 2.35 (a) Capsule, (b) bellows.

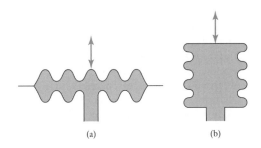

(a)　　　　　(b)

Figure 2.36 LVDT with bellows.

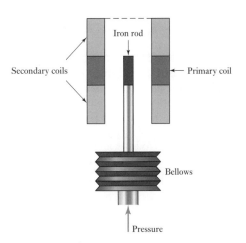

A different form of deformation is obtained using a tube with an elliptical cross-section (Figure 2.37(a)). Increasing the pressure in such a tube causes it to tend to a more circular cross-section. When such a tube is in the form of a C-shaped tube (Figure 2.37(b)), this being generally known as a **Bourdon tube**, the C opens up to some extent when the pressure in the tube increases. A helical form of such a tube (Figure 2.37(c)) gives a greater sensitivity. The tubes are made from such materials as stainless steel and phosphor bronze and are used for pressures in the range 10^3 to 10^8 Pa.

Figure 2.37   Tube pressure
sensors.

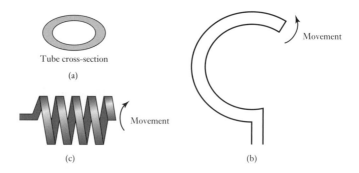

Tube cross-section

(a)

Movement

(c)

Movement

(b)

2.6.1   Piezoelectric sensors

Piezoelectric materials when stretched or compressed generate electric
charges with one face of the material becoming positively charged and
the opposite face negatively charged (Figure 2.38(a)). As a result, a voltage is produced. Piezoelectric materials are ionic crystals, which when
stretched or compressed result in the charge distribution in the crystal
changing so that there is a net displacement of charge with one face of the
material becoming positively charged and the other negatively charged.
The net charge q on a surface is proportional to the amount x by which
the charges have been displaced, and since the displacement is proportional to the applied force F:

$$q = kx = SF$$

where k is a constant and S a constant termed the **charge sensitivity**. The
charge sensitivity depends on the material concerned and the orientation
of its crystals. Quartz has a charge sensitivity of 2.2 pC/N when the
crystal is cut in one particular direction and the forces applied in a specific
direction; barium titanate has a much higher charge sensitivity of the order
of 130 pC/N and lead zirconate–titanate about 265 pC/N.

Metal electrodes are deposited on opposite faces of the piezoelectric
crystal (Figure 2.38(b)). The capacitance C of the piezoelectric material
between the plates is

$$C = \frac{\varepsilon_0 \varepsilon_r A}{t}$$

Figure 2.38   (a) Piezoelectricity,
(b) piezoelectric capacitor.

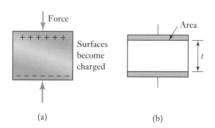

Force

+ + + + + +

Surfaces
become
charged

Area

t

(a)

(b)

where ε_r is the relative permittivity of the material, A is area and t its thickness. Since the charge $q = Cv$, where v is the potential difference produced across a capacitor, then

$$v = \frac{St}{\varepsilon_0 \varepsilon_r A} F$$

The force F is applied over an area A and so the applied pressure p is F/A and if we write $S_v = (S/\varepsilon_0 \varepsilon_r)$, this being termed the **voltage sensitivity factor**, then

$$v = S_v t p$$

The voltage is proportional to the applied pressure. The voltage sensitivity for quartz is about 0.055 V/m Pa. For barium titanate it is about 0.011 V/m Pa.

Piezoelectric sensors are used for the measurement of pressure, force and acceleration. The applications have, however, to be such that the charge produced by the pressure does not have much time to leak off and thus tends to be used mainly for transient rather than steady pressures.

The equivalent electric circuit for a piezoelectric sensor is a charge generator in parallel with capacitance C_s and in parallel with the resistance R_s arising from leakage through the dielectric (Figure 2.39(a)). When the sensor is connected via a cable, of capacitance C_c, to an amplifier of input capacitance C_A and resistance R_A, we have effectively the circuit shown in Figure 2.39(b) and a total circuit capacitance of $C_s + C_c + C_A$ in parallel with a resistance of $R_A R_s / (R_A + R_s)$. When the sensor is subject to pressure it becomes charged, but because of the resistance the capacitor will discharge with time. The time taken for the discharge will depend on the time constant of the circuit.

Figure 2.39 (a) Sensor equivalent circuit, (b) sensor connected to charge amplifier.

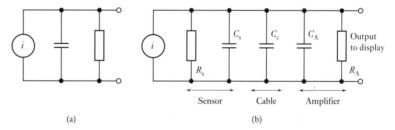

(a) (b)

2.6.2 Tactile sensor

A tactile sensor is a particular form of pressure sensor. Such a sensor is used on the 'fingertips' of robotic 'hands' to determine when a 'hand' has come into contact with an object. They are also used for 'touch display' screens where a physical contact has to be sensed. One form of tactile sensor uses piezoelectric polyvinylidene fluoride (PVDF) film. Two layers of the film are used and are separated by a soft film which transmits vibrations (Figure 2.40). The lower PVDF film has an alternating voltage applied to it and this results in mechanical oscillations of the film (the piezoelectric effect described above in reverse). The intermediate film transmits these vibrations to the upper PVDF film. As a consequence of the piezoelectric effect, these

Figure 2.40 PVDF tactile sensor.

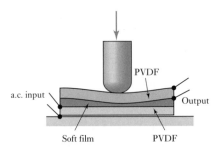

vibrations cause an alternating voltage to be produced across the upper film. When pressure is applied to the upper PVDF film its vibrations are affected and the output alternating voltage is changed.

2.7 Liquid flow

The traditional method of measuring the flow rate of liquids involves devices based on the measurement of the pressure drop occurring when the fluid flows through a constriction (Figure 2.41). For a horizontal tube, where v_1 is the fluid velocity, p_1 the pressure and A_1 the cross-sectional area of the tube prior to the constriction, v_2 the velocity, p_2 the pressure and A_2 the cross-sectional area at the constriction, with ρ the fluid density, then Bernoulli's equation gives

$$\frac{v_1^2}{2g} + \frac{p_1}{\rho g} = \frac{v_2^2}{2g} + \frac{p_2}{\rho g}$$

Figure 2.41 Fluid flow through a constriction.

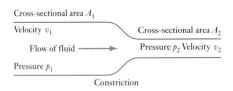

Since the mass of liquid passing per second through the tube prior to the constriction must equal that passing through the tube at the constriction, we have $A_1 v_1 \rho = A_2 v_2 \rho$. But the quantity Q of liquid passing through the tube per second is $A_1 v_1 = A_2 v_2$. Hence

$$Q = \frac{A}{\sqrt{1 - (A_2/A_1)^2}} \sqrt{\frac{2(p_1 - p_2)}{\rho}}$$

Thus the quantity of fluid flowing through the pipe per second is proportional to $\sqrt{}$(pressure difference). Measurements of the pressure difference can thus be used to give a measure of the rate of flow. There are many devices based on this principle, and the following example of the orifice plate is probably one of the commonest.

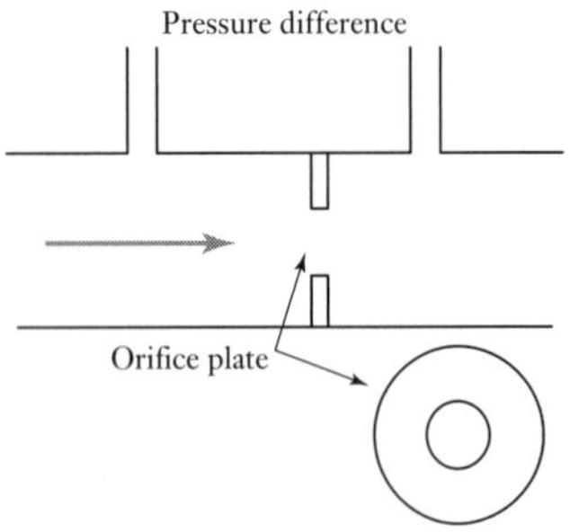

2.7.1 Orifice plate

The orifice plate (Figure 2.42) is simply a disc, with a central hole, which is placed in the tube through which the fluid is flowing. The pressure difference

Figure 2.42 Orifice plate.

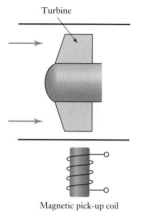

Turbine

Magnetic pick-up coil

Figure 2.43 Turbine flowmeter.

is measured between a point equal to the diameter of the tube upstream and a point equal to half the diameter downstream. The orifice plate is simple, cheap, with no moving parts, and is widely used. It does not, however, work well with slurries. The accuracy is typically about ±1.5% of full range, it is non-linear, and it does produce quite an appreciable pressure loss in the system to which it is connected.

2.7.2 Turbine meter

The turbine flowmeter (Figure 2.43) consists of a multi-bladed rotor that is supported centrally in the pipe along which the flow occurs. The fluid flow results in rotation of the rotor, the angular velocity being approximately proportional to the flow rate. The rate of revolution of the rotor can be determined using a magnetic pick-up. The pulses are counted and so the number of revolutions of the rotor can be determined. The meter is expensive with an accuracy of typically about ±0.3%.

2.8 Liquid level

The level of liquid in a vessel can be measured directly by monitoring the position of the liquid surface or indirectly by measuring some variable related to the height. Direct methods can involve floats; indirect methods include the monitoring of the weight of the vessel by, perhaps, load cells. The weight of the liquid is $Ah\rho g$, where A is the cross-sectional area of the vessel, h the height of liquid, ρ its density and g the acceleration due to gravity. Thus changes in the height of liquid give weight changes. More commonly, indirect methods involve the measurement of the pressure at some point in the liquid, the pressure due to a column of liquid of height h being $h\rho g$, where ρ is the liquid density.

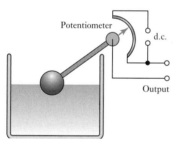

Potentiometer

d.c.

Output

Figure 2.44 Float system.

2.8.1 Floats

A direct method of monitoring the level of liquid in a vessel is by monitoring the movement of a float. Figure 2.44 illustrates this with a simple float system. The displacement of the float causes a lever arm to rotate and so move a slider across a potentiometer. The result is an output of a voltage related to the height of liquid. Other forms of this involve the lever causing the core in an LVDT to become displaced, or stretch or compress a strain-gauged element.

2.8.2 Differential pressure

Figure 2.45 shows two forms of level measurement based on the measurement of differential pressure. In Figure 2.45(a), the differential pressure cell determines the pressure difference between the liquid at the base of the vessel and atmospheric pressure, the vessel being open to atmospheric pressure. With a closed or open vessel the system illustrated in (b) can be used. The differential pressure cell monitors the difference in pressure between the base of the vessel and the air or gas above the surface of the liquid.

Figure 2.45   Using a differential pressure sensor.

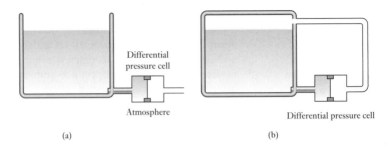

Differential pressure cell

Atmosphere

(a)

Differential pressure cell

(b)

Changes that are commonly used to monitor temperature are the expansion or contraction of solids, liquids or gases, the change in electrical resistance of conductors and semiconductors and thermoelectric e.m.f.s. The following are some of the methods that are commonly used with temperature control systems.

2.9.1   Bimetallic strips

This device consists of two different metal strips bonded together. The metals have different coefficients of expansion and when the temperature changes the composite strip bends into a curved strip, with the higher coefficient metal on the outside of the curve. This deformation may be used as a temperature-controlled switch, as in the simple thermostat which was commonly used with domestic heating systems (Figure 2.46). The small magnet enables the sensor to exhibit hysteresis, meaning that the switch contacts close at a different temperature from that at which they open.

Figure 2.46   Bimetallic thermostat.

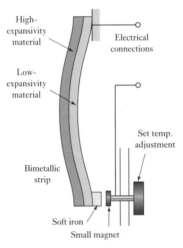

High-expansivity material

Electrical connections

Low-expansivity material

Set temp. adjustment

Bimetallic strip

Soft iron

Small magnet

2.9.2   Resistance temperature detectors (RTDs)

The resistance of most metals increases, over a limited temperature range, in a reasonably linear way with temperature (Figure 2.47). For such a linear relationship:

$$R_t = R_0(1 + \alpha t)$$

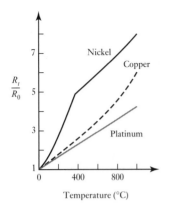

Figure 2.47 Variation of resistance with temperature for metals.

Figure 2.48 Thermistors: (a) common forms, (b) typical variation of resistance with temperature.

where R_t is the resistance at a temperature $t(^{\circ}C)$, R_0 the resistance at $0^{\circ}C$ and α a constant for the metal termed the temperature coefficient of resistance. Resistance temperature detectors (RTDs) are simple resistive elements in the form of coils of wire of such metals as platinum, nickel or nickel–copper alloys; platinum is the most widely used. Thin-film platinum elements are often made by depositing the metal on a suitable substrate, wire-wound elements involving a platinum wire held by a high-temperature glass adhesive inside a ceramic tube. Such detectors are highly stable and give reproducible responses over long periods of time. They tend to have response times of the order of 0.5 to 5 s or more.

2.9.3 Thermistors

Thermistors are small pieces of material made from mixtures of metal oxides, such as those of chromium, cobalt, iron, manganese and nickel. These oxides are semiconductors. The material is formed into various forms of element, such as beads, discs and rods (Figure 2.48(a)).

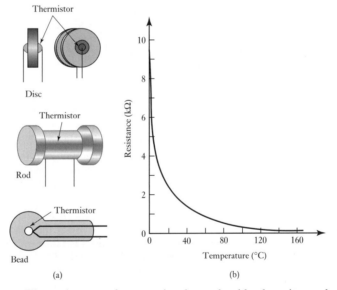

(a)   (b)

The resistance of conventional metal-oxide thermistors decreases in a very non-linear manner with an increase in temperature, as illustrated in Figure 2.48(b). Such thermistors have negative temperature coefficients (NTCs). Positive temperature coefficient (PTC) thermistors are, however, available. The change in resistance per degree change in temperature is considerably larger than that which occurs with metals. The resistance–temperature relationship for a thermistor can be described by an equation of the form

$$R_t = Ke^{\beta/t}$$

where R_t is the resistance at temperature t, with K and β being constants. Thermistors have many advantages when compared with other temperature sensors. They are rugged and can be very small, so enabling temperatures to be monitored at virtually a point. Because of their small size they respond very rapidly to changes in temperature. They give very large changes in resistance per degree change in temperature. Their main disadvantage is their non-linearity. Thermistors are used with the electronic systems for cars to monitor such variables as air temperature and coolant air temperature.

2.9.4 Thermodiodes and transistors

A junction semiconductor diode is widely used as a temperature sensor. When the temperature of doped semiconductors changes, the mobility of their charge carriers changes and this affects the rate at which electrons and holes can diffuse across a p–n junction. Thus when a p–n junction has a potential difference V across it, the current I through the junction is a function of the temperature, being given by

$$I = I_0\left(e^{eV/kT} - 1\right)$$

where T is the temperature on the Kelvin scale, e the charge on an electron, and k and I_0 are constants. By taking logarithms we can write the equation in terms of the voltage as

$$V = \left(\frac{kT}{e}\right)\ln\left(\frac{I}{I_0} + 1\right)$$

Thus, for a constant current, we have V proportional to the temperature on the Kelvin scale and so a measurement of the potential difference across a diode at constant current can be used as a measure of the temperature. Such a sensor is compact like a thermistor but has the great advantage of giving a response which is a linear function of temperature. Diodes for use as temperature sensors, together with the necessary signal conditioning, are supplied as integrated circuits, e.g. LM3911, and give a very small compact sensor. The output voltage from LM3911 is proportional to the temperature at the rate of 10 mV/°C.

In a similar manner to the thermodiode, for a thermotransistor the voltage across the junction between the base and the emitter depends on the temperature and can be used as a measure of temperature. A common method is to use two transistors with different collector currents and determine the difference in the base–emitter voltages between them, this difference being directly proportional to the temperature on the Kelvin scale. Such transistors can be combined with other circuit components on a single chip to give a temperature sensor with its associated signal conditioning, e.g. LM35 (Figure 2.49). This sensor can be used in the range −40 to 110°C and gives an output of 10 mV/°C.

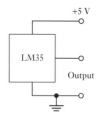

Figure 2.49   LM35.

2.9.5 Thermocouples

If two different metals are joined together, a potential difference occurs across the junction. The potential difference depends on the metals used and the temperature of the junction. A thermocouple is a complete circuit involving two such junctions (Figure 2.50(a)).

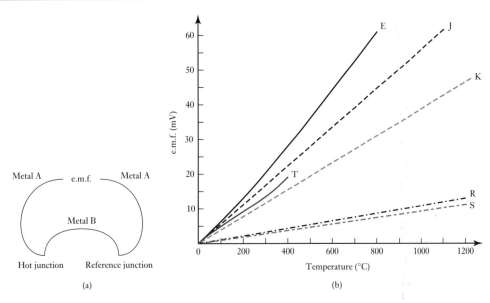

Figure 2.50 (a) A thermocouple, (b) thermoelectric e.m.f.–temperature graphs.

If both junctions are at the same temperature there is no net e.m.f. If, however, there is a difference in temperature between the two junctions, there is an e.m.f. The value of this e.m.f. E depends on the two metals concerned and the temperatures t of both junctions. Usually one junction is held at 0°C and then, to a reasonable extent, the following relationship holds:

$$E = at + bt^2$$

where a and b are constants for the metals concerned. Commonly used thermocouples are shown in Table 2.1, with the temperature ranges over which they are generally used and typical sensitivities. These commonly used thermocouples are given reference letters. For example, the iron–constantan thermocouple is called a type J thermocouple. Figure 2.50(b) shows how the e.m.f. varies with temperature for a number of commonly used pairs of metals.

Table 2.1 Thermocouples.

Ref.	Materials	Range (°C)	(μV/°C)
B	Platinum 30% rhodium/platinum 6% rhodium	0 to 1800	3
E	Chromel/constantan	−200 to 1000	63
J	Iron/constantan	−200 to 900	53
K	Chromel/alumel	−200 to 1300	41
N	Nirosil/nisil	−200 to 1300	28
R	Platinum/platinum 13% rhodium	0 to 1400	6
S	Platinum/platinum 10% rhodium	0 to 1400	6
T	Copper/constantan	−200 to 400	43

A thermocouple circuit can have other metals in the circuit and they will have no effect on the thermoelectric e.m.f. provided all their junctions are at the same temperature. This is known as the **law of intermediate metals.**

A thermocouple can be used with the reference junction at a temperature other than 0°C. The standard tables, however, assume a 0°C junction and hence a correction has to be applied before the tables can be used. The correction is applied using what is known as the **law of intermediate temperatures,** namely

$$E_{t,0} = E_{t,I} + E_{I,0}$$

The e.m.f. $E_{t,0}$ at temperature t when the cold junction is at 0°C equals the e.m.f. $E_{t,I}$ at the intermediate temperature I plus the e.m.f. $E_{I,0}$ at temperature I when the cold junction is at 0°C. To illustrate this, consider a type E thermocouple which is to be used for the measurement of temperature with a cold junction at 20°C. What will be the thermoelectric e.m.f. at 200°C? The following is data from standard tables:

Temp. (°C)	0	20	200
e.m.f. (mV)	0	1.192	13.419

Using the law of intermediate temperatures

$$E_{200,0} = E_{200,20} + E_{20,0} = 13.419 - 1.192 = 12.227 \, \text{mV}$$

Note that this is not the e.m.f. given by the tables for a temperature of 180°C with a cold junction at 0°C, namely 11.949 mV.

To maintain one junction of a thermocouple at 0°C, i.e. have it immersed in a mixture of ice and water, is often not convenient. A compensation circuit can, however, be used to provide an e.m.f. which varies with the temperature of the cold junction in such a way that when it is added to the thermocouple e.m.f. it generates a combined e.m.f. which is the same as would have been generated if the cold junction had been at 0°C (Figure 2.51). The compensating e.m.f. can be provided by the voltage drop across a resistance thermometer element.

Figure 2.51 Cold junction compensation.

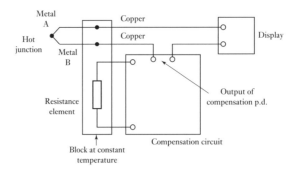

The base-metal thermocouples, E, J, K and T, are relatively cheap but deteriorate with age. They have accuracies which are typically about ±1 to 3%. Noble-metal thermocouples, e.g. R, are more expensive but are more stable with longer life. They have accuracies of the order of ±1% or better.

Thermocouples are generally mounted in a sheath to give them mechanical and chemical protection. The type of sheath used depends on the temperatures at which the thermocouple is to be used. In some cases the sheath is packed with a mineral which is a good conductor of heat and a good electrical insulator. The response time of an unsheathed thermocouple is very fast. With a sheath this may be increased to as much as a few seconds if a large sheath is used. In some instances a group of thermocouples are connected in series so that there are perhaps 10 or more hot junctions sensing the temperature. The e.m.f. produced by each is added together. Such an arrangement is known as a **thermopile**.

2.10   Light sensors

Photodiodes are semiconductor junction diodes (see Section 9.3.1 for a discussion of diodes) which are connected into a circuit in reverse bias, so giving a very high resistance (Figure 2.52(a)). With no incident light, the reverse current is almost negligible and is termed the dark current. When light falls on the junction, extra hole–electron pairs are produced and there is an increase in the reverse current and the diode resistance drops (Figure 2.52(b)). The reverse current is very nearly proportional to the intensity of the light. For example, the current in the absence of light with a reverse bias of 3 V might be 25 μA and when illuminated by 25 000 lumens/m² the current rises to 375 μA. The resistance of the device with no light is $3/(25 \times 10^{-6}) = 120\ k\Omega$ and with light is $3/(375 \times 10^{-6}) = 8\ k\Omega$. A photodiode can thus be used as a variable resistance device controlled by the light incident on it. Photodiodes have a very fast response to light.

Figure 2.52   Photodiode.

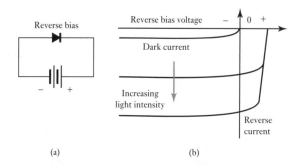

(a)   (b)

The **phototransistors** (see Section 9.3.3 for a discussion of transistors) have a light-sensitive collector–base p–n junction. When there is no incident light there is a very small collector-to-emitter current. When light is incident, a base current is produced that is directly proportional to the light intensity. This leads to the production of a collector current which is then a measure of the light intensity. Phototransistors are often available as integrated packages with the phototransistor connected in a Darlington arrangement with a conventional transistor (Figure 2.53). Because this arrangement gives a higher current gain, the device gives a much greater collector current for a given light intensity.

A **photoresistor** has a resistance which depends on the intensity of the light falling on it, decreasing linearly as the intensity increases. The cadmium sulphide photoresistor is most responsive to light having wavelengths shorter than about 515 nm and the cadmium selinide photoresistor for wavelengths less than about 700 nm.

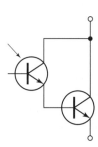

Figure 2.53   Photo Darlington.

An array of light sensors is often required in a small space in order to determine the variations of light intensity across that space. An example of this is in the digital camera to capture the image being photographed and convert it into a digital form. For this purpose a **charge-coupled device (CCD)** is often used. A CCD is a light-sensitive arrangement of many small light-sensitive cells termed pixels. These cells are basically a p-layer of silicon, separated by a depletion layer from an n-type silicon layer. When exposed to light, a cell becomes electrically charged and this charge is then converted by electronic circuitry into an 8-bit digital number. In taking a photograph the digital camera electronic circuitry discharges the light-sensitive cells, activates an electromechanical shutter to expose the cells to the image, then reads the 8-bit charge value for each cell and so captures the image. Since the pn cells are colour blind and we need colour photographs, the light passes through a colour filter matrix before striking the cells. This allows just green light to fall on some cells, blue on others and red light on others. Then, by later taking account of the output from neighbouring cells, a colour image can be created.

2.11 Selection of sensors

In selecting a sensor for a particular application there are a number of factors that need to be considered.

1 The nature of the measurement required, e.g. the variable to be measured, its nominal value, the range of values, the accuracy required, the required speed of measurement, the reliability required, the environmental conditions under which the measurement is to be made.
2 The nature of the output required from the sensor, this determining the signal conditioning requirements in order to give suitable output signals from the measurement.
3 Then possible sensors can be identified, taking into account such factors as their range, accuracy, linearity, speed of response, reliability, maintainability, life, power supply requirements, ruggedness, availability, cost.

The selection of sensors cannot be taken in isolation from a consideration of the form of output that is required from the system after signal conditioning, and thus there has to be a suitable marriage between sensor and signal conditioner.

To illustrate the above, consider the selection of a sensor for the measurement of the level of a corrosive acid in a vessel. The level can vary from 0 to 2 m in a circular vessel which has a diameter of 1 m. The empty vessel has a weight of 100 kg. The minimum variation in level to be detected is 10 cm. The acid has a density of 1050 kg/m^3. The output from the sensor is to be electrical.

Because of the corrosive nature of the acid an indirect method of determining the level seems appropriate. Thus it is possible to use a load cell, or load cells, to monitor the weight of the vessel. Such cells would give an electrical output. The weight of the liquid changes from 0 when empty to, when full, $1050 \times 2 \times \pi(1^2/4) \times 9.8 = 16.2$ kN. Adding this to the weight of the empty vessel gives a weight that varies from about 1 to 17 kN. The resolution required is for a change of level of 10 cm, i.e. a change in weight of $0.10 \times 1050 \times \pi(1^2/4) \times 9.8 = 0.8$ kN. If three load cells are used to support the tank then each will require a range of about 0 to 6 kN with a resolution of 0.27 kN. Manufacturers' catalogues can then be consulted to see if such load cells can be obtained.

2.12 Inputting data by switches

Mechanical switches consist of one or more pairs of contacts which can be mechanically closed or opened and in doing so make or break electrical circuits. Thus 0 or 1 signals can be transmitted by the act of opening or closing a switch. The term **limit switch** is used when the switches are opened or closed by the displacement of an object and used to indicate the limit of its displacement before action has to be initiated.

Mechanical switches are specified in terms of their number of poles and throws. **Poles** are the number of separate circuits that can be completed by the same switching action and **throws** are the number of individual contacts for each pole. Figure 2.54(a) shows a single pole–single throw (SPST) switch, Figure 2.54(b) a single pole–double throw (SPDT) switch and Figure 2.54(c) a double pole–double throw (DPDT) switch.

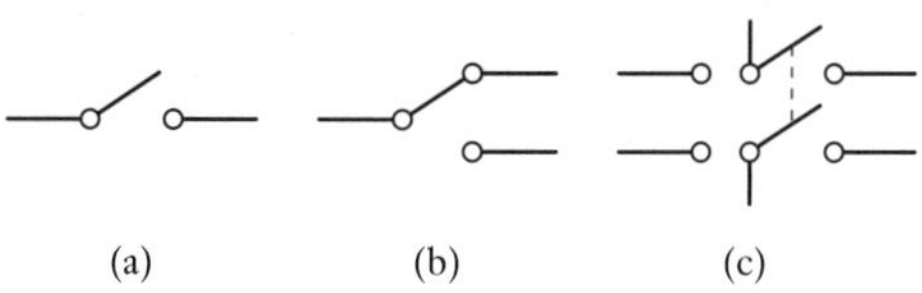

Figure 2.54 Switches: (a) SPST, (b) SPDT, (c) DPDT.

(a)   (b)   (c)

2.12.1 Debouncing

A problem that occurs with mechanical switches is **switch bounce**. When a mechanical switch is switched to close the contacts, we have one contact being moved towards the other. It hits the other and, because the contacting elements are elastic, bounces. It may bounce a number of times (Figure 2.55(a)) before finally settling to its closed state after, typically, some 20 ms. Each of the contacts during this bouncing time can register as a separate contact. Thus, to a microprocessor, it might appear that perhaps two or more separate switch actions have occurred. Similarly, when a mechanical switch is opened, bouncing can occur. To overcome this problem either hardware or software can be used.

Figure 2.55 (a) Switch bounce on closing a switch, (b) debouncing using an SR flip-flop, (c) debouncing using a D flip-flop.

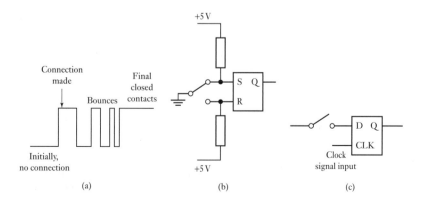

(a)   (b)   (c)

With software, the microprocessor is programmed to detect if the switch is closed and then wait, say, 20 ms. After checking that bouncing has ceased and the switch is in the same closed position, the next part of the program can take place. The hardware solution to the bounce problem is based on the use of a flip-flop. Figure 2.55(b) shows a circuit for debouncing an SPDT switch which is based on the use of an SR flip-flop (see Section 5.4.1). As shown, we have S at 0 and R at 1 with an output of 0. When the switch is moved to its lower position, initially S becomes 1 and R becomes 0. This gives an output of 1. Bouncing in changing S from 1 to 0 to 1 to 0, etc., gives no change in the output. Such a flip-flop can be derived from two NOR or two NAND gates. An SPDT switch can be debounced by the use of a D flip-flop (see Section 5.4.4). Figure 2.55(c) shows the circuit. The output from such a flip-flop only changes when the clock signal changes. Thus by choosing a clock period which is greater than the time for which the bounces last, say, 20 ms, the bounce signals will be ignored.

An alternative method of debouncing using hardware is to use a **Schmitt trigger**. This device has the 'hysteresis' characteristic shown in Figure 2.56(a). When the input voltage is beyond an upper switching threshold and giving a low output, then the input voltage needs to fall below the lower threshold before the output can switch to high. Conversely, when the input voltage is below the lower switching threshold and giving a high, then the input needs to rise above the upper threshold before the output can switch to low. Such a device can be used to sharpen slowly changing signals: when the signal passes the switching threshold it becomes a sharply defined edge between two well-defined logic levels. The circuit shown in Figure 2.56(b) can be used for debouncing; note the circuit symbol for a Schmitt trigger. With the switch open, the capacitor becomes charged and the voltage applied to the Schmitt trigger becomes high and so it gives a low output. When the switch is closed, the capacitor discharges very rapidly and so the first bounce discharges the capacitor; the Schmitt trigger thus switches to give a high output. Successive switch bounces do not have time to recharge the capacitor to the required threshold value and so further bounces do not switch the Schmitt trigger.

Figure 2.56   Schmitt trigger: (a) characteristic, (b) used for debouncing a switch.

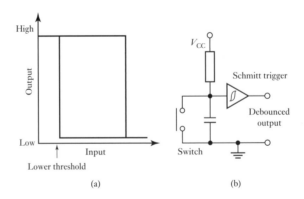

2.12.2   Keypads

A keypad is an array of switches, perhaps the keyboard of a computer or the touch input membrane pad for some device such as a microwave oven. A

Figure 2.57 (a) Contact key, (b) membrane key, (c) 16-way keypad.

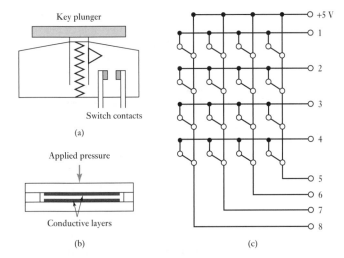

contact-type key of the form generally used with a keyboard is shown in Figure 2.57(a). Depressing the key plunger forces the contacts together with the spring returning the key to the off position when the key is released. A typical membrane switch (Figure 2.57(b)) is built up from two wafer-thin plastic films on which conductive layers have been printed. These layers are separated by a spacer layer. When the switch area of the membrane is pressed, the top contact layer closes with the bottom one to make the connection and then opens when the pressure is released.

While each switch in such arrays could be connected individually to give signals when closed, a more economical method is to connect them in an array where an individual output is not needed for each key but each key gives a unique row–column combination. Figure 2.57(c) shows the connections for a 16-way keypad.

Summary

A **sensor** is an element which produces a signal relating to the quantity being measured. A **transducer** is an element that, when subject to some physical change, experiences a related change. Thus sensors are transducers. However, a measurement system may use transducers, in addition to the sensor, in other parts of the system to convert signals in one form to another form.

The **range** of a transducer defines the limits between which the input can vary. **Span** is the maximum value of the input minus the minimum value. The **error** is the difference between the result of a measurement and its true value. **Accuracy** is the extent to which the measured value might be wrong. **Sensitivity** indicates how much output there is per unit input. **Hysteresis error** is the difference between the values obtained when reached by a continuously increasing and a continuously decreasing change. **Non-linearity error** is the error given by assuming a linear relationship. **Repeatability/reproducibility** is a measure of the ability to give the same output for repeated applications of the same input. **Stability** is the ability to give the same output for a constant input. **Dead band** is the range of input values for which there is no output. **Resolution** is the smallest change in input that will produce an observable change in output. **Response time** is the time that elapses after a step input before the output reaches a specified percentage, e.g. 95%, of the input. The **time constant** is 63.2% of the response time. **Rise time** is the time taken for the output to rise to some specified percentage of its steady-state output. **Settling time** is the time taken for the output to settle down to within some percentage, e.g. 2%, of the steady-state value.

Problems

2.1 Explain the significance of the following information given in the specification of transducers.

(a) A piezoelectric accelerometer.
 Non-linearity: ±0.5% of full range.
(b) A capacitive linear displacement transducer.
 Non-linearity and hysteresis: ±0.01% of full range.
(c) A resistance strain gauge force measurement transducer.
 Temperature sensitivity: ±1% of full range over normal environmental temperatures.
(d) A capacitance fluid pressure transducer.
 Accuracy: ±1% of displayed reading.
(e) Thermocouple.
 Sensitivity: nickel chromium/nickel aluminium thermocouple: 0.039 mV/°C when the cold junction is at 0°C.
(f) Gyroscope for angular velocity measurement.
 Repeatability: ±0.01% of full range.
(g) Inductive displacement transducer.
 Linearity: ±1% of rated load.
(h) Load cell.
 Total error due to non-linearity, hysteresis and non-repeatability: ±0.1%.

2.2 A copper–constantan thermocouple is to be used to measure temperatures between 0 and 200°C. The e.m.f. at 0°C is 0 mV, at 100°C it is 4.277 mV and at 200°C it is 9.286 mV. What will be the non-linearity error at 100°C as a percentage of the full range output if a linear relationship is assumed between e.m.f. and temperature over the full range?

2.3 A thermocouple element when taken from a liquid at 50°C and plunged into a liquid at 100°C at time $t = 0$ gave the following e.m.f. values. Determine the 95% response time.

Time (s)	0	20	40	60	80	100	120
e.m.f. (mV)	2.5	3.8	4.5	4.8	4.9	5.0	5.0

2.4 What is the non-linearity error, as a percentage of full range, produced when a 1 kΩ potentiometer has a load of 10 kΩ and is at one-third of its maximum displacement?

2.5 What will be the change in resistance of an electrical resistance strain gauge with a gauge factor of 2.1 and resistance 50 Ω if it is subject to a strain of 0.001?

2.6   You are offered a choice of an incremental shaft encoder or an absolute shaft encoder for the measurement of an angular displacement. What is the principal difference between the results that can be obtained by these methods?

2.7   A shaft encoder is to be used with a 50 mm radius tracking wheel to monitor linear displacement. If the encoder produces 256 pulses per revolution, what will be the number of pulses produced by a linear displacement of 200 mm?

2.8   A rotary variable differential transformer has a specification which includes the following information:

Ranges: $\pm 30°$, linearity error $\pm 0.5\%$ full range
            $\pm 60°$, linearity error $\pm 2.0\%$ full range
Sensitivity: 1.1 (mV/V input)/degree
Impedance: primary 750 Ω, secondary 2000 Ω

What will be (a) the error in a reading of $40°$ due to non-linearity when the RDVT is used on the $\pm 60°$ range, and (b) the output voltage change that occurs per degree if there is an input voltage of 3 V?

2.9   What are the advantages and disadvantages of the plastic film type of potentiometer when compared with the wire-wound potentiometer?

2.10   A pressure sensor consisting of a diaphragm with strain gauges bonded to its surface has the following information in its specification:

Ranges: 0 to 1400 kPa, 0 to 35 000 kPa
Non-linearity error: $\pm 0.15\%$ of full range
Hysteresis error: $\pm 0.05\%$ of full range

What is the total error due to non-linearity and hysteresis for a reading of 1000 kPa on the 0 to 1400 kPa range?

2.11   The water level in an open vessel is to be monitored by a differential pressure cell responding to the difference in pressure between that at the base of the vessel and the atmosphere. Determine the range of differential pressures the cell will have to respond to if the water level can vary between zero height above the cell measurement point and 2 m above it.

2.12   An iron–constantan thermocouple is to be used to measure temperatures between 0 and $400°C$. What will be the non-linearity error as a percentage of the full-scale reading at $100°C$ if a linear relationship is assumed between e.m.f. and temperature?

        e.m.f. at $100°C = 5.268$ mV; e.m.f. at $400°C = 21.846$ mV

2.13   A platinum resistance temperature detector has a resistance of 100.00 Ω at $0°C$, 138.50 Ω at $100°C$ and 175.83 Ω at $200°C$. What will be the non-linearity error in $°C$ at $100°C$ if the detector is assumed to have a linear relationship between 0 and $200°C$?

2.14   A strain gauge pressure sensor has the following specification. Will it be suitable for the measurement of pressure of the order of 100 kPa to an accuracy of ± 5 kPa in an environment where the temperature is reasonably constant at about $20°C$?

Ranges: 2 to 70 MPa, 70 kPa to 1 MPa
Excitation: 10 V d.c. or a.c. (r.m.s.)
Full range output: 40 mV
Non-linearity and hysteresis errors: $\pm 0.5\%$

Temperature range: -54 to $+120°C$

Thermal shift zero: 0.030% full range output/°C
Thermal shift sensitivity: 0.030% full range output/°C

2.15   A float sensor for the determination of the level of water in a vessel has a cylindrical float of mass 2.0 kg, cross-sectional area 20 cm^2 and a length of 1.5 m. It floats vertically in the water and presses upwards against a beam attached to its upward end. What will be the minimum and maximum upthrust forces exerted by the float on the beam? Suggest a means by which the deformation of the beam under the action of the upthrust force could be monitored.

2.16   Suggest a sensor that could be used as part of the control system for a furnace to monitor the rate at which the heating oil flows along a pipe. The output from the measurement system is to be an electrical signal which can be used to adjust the speed of the oil pump. The system must be capable of operating continuously and automatically, without adjustment, for long periods of time.

2.17   Suggest a sensor that could be used, as part of a control system, to determine the difference in levels between liquids in two containers. The output is to provide an electrical signal for the control system.

2.18   Suggest a sensor that could be used as part of a system to control the thickness of rolled sheet by monitoring its thickness as it emerges from rollers. The sheet metal is in continuous motion and the measurement needs to be made quickly to enable corrective action to be taken quickly. The measurement system has to supply an electrical signal.

Chapter three Signal conditioning

Objectives

The objectives of this chapter are that, after studying it, the reader should be able to:
- Explain the requirements for signal conditioning.
- Explain how operational amplifiers can be used.
- Explain the requirements for protection and filtering.
- Explain the principles of the Wheatstone bridge and, in particular, how it is used with strain gauges.
- Explain the principle of pulse modulation.
- Explain the problems that can occur with ground loops and interference and suggest possible solutions to these problems.
- State the requirements for maximum power transfer between electrical components.

3.1 Signal conditioning

The output signal from the sensor of a measurement system has generally to be processed in some way to make it suitable for the next stage of the operation. The signal may be, for example, too small and have to be amplified, contain interference which has to be removed, be non-linear and require linearisation, be analogue and have to be made digital, be digital and have to be made analogue, be a resistance change and have to be made into a current change, be a voltage change and have to be made into a suitable size current change, etc. All these changes can be referred to as **signal conditioning**. For example, the output from a thermocouple is a small voltage, a few millivolts. A signal conditioning module might then be used to convert this into a suitable size current signal, provide noise rejection, linearisation and cold junction compensation (i.e. compensating for the cold junction not being at 0°C).

Chapter 4 continues with a discussion of signal conditioning involving digital signals.

3.1.1 Signal conditioning processes

Some of the processes that can occur in conditioning a signal are outlined below.

1 *Protection* to prevent damage to the next element, e.g. a microprocessor, as a result of high current or voltage. Thus there can be series current-limiting resistors, fuses to break if the current is too high, polarity protection and voltage limitation circuits (see Section 3.3).

2 Getting the signal into the *right type of signal*. This can mean making the signal into a d.c. voltage or current. Thus, for example, the resistance change

of a strain gauge has to be converted into a voltage change. This can be done by the use of a Wheatstone bridge and using the out-of-balance voltage (see Section 3.5). It can mean making the signal digital or analogue (see Section 4.3 for analogue-to-digital and analogue-to-digital converters).

3 Getting the *level* of the signal right. The signal from a thermocouple might be just a few millivolts. If the signal is to be fed into an analogue-to-digital converter for inputting to a microprocessor then it needs to be made much larger, volts rather than millivolts. Operational amplifiers are widely used for amplification (see Section 3.2).

4 Eliminating or reducing *noise*. For example, filters might be used to eliminate mains noise from a signal (see Section 3.4).

5 Signal *manipulation*, e.g. making it a linear function of some variable. The signals from some sensors, e.g. a flowmeter, are non-linear and thus a signal conditioner might be used so that the signal fed on to the next element is linear (see Section 3.2.6).

The following sections outline some of the elements that might be used in signal conditioning.

3.2   The operational amplifier

An amplifier can be considered to be essentially a system which has an input and an output (Figure 3.1), the **voltage gain** of the amplifier being the ratio of the output and input voltages when each is measured relative to the earth. The **input impedance** of an amplifier is defined as the input voltage divided by the input current, the **output impedance** being the output voltage divided by the output current.

The basis of many signal conditioning modules is the **operational amplifier**. The operational amplifier is a high-gain d.c. amplifier, the gain typically being of the order of 100 000 or more, that is supplied as an integrated circuit on a silicon chip. It has two inputs, known as the inverting input ($-$) and the non-inverting input ($+$). The output depends on the connections made to these inputs. There are other inputs to the operational amplifier, namely a negative voltage supply, a positive voltage supply and two inputs termed offset null, these being to enable corrections to be made for the non-ideal behaviour of the amplifier (see Section 3.2.8). Figure 3.2 shows the pin connections for a 741-type operational amplifier.

An ideal model for an operational amplifier is as an amplifier with an infinite gain, infinite input impedance and zero output impedance, i.e. the output voltage is independent of the load.

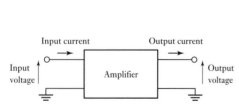

Figure 3.1   Amplifier.

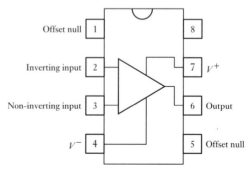

Figure 3.2   Pin connections for a 741 operational amplifier.

Figure 3.3 Inverting amplifier.

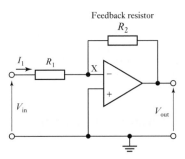

Feedback resistor

The following indicates the types of circuits that might be used with operational amplifiers when used as signal conditioners.

3.2.1 Inverting amplifier

Figure 3.3 shows the connections made to the amplifier when used as an **inverting amplifier**. The input is taken to the inverting input through a resistor R_1 with the non-inverting input being connected to ground. A feedback path is provided from the output, via the resistor R_2 to the inverting input. The operational amplifier has a voltage gain of about 100 000 and the change in output voltage is typically limited to about $\pm 10\,\text{V}$. The input voltage must then be between $+0.0001$ and -0.0001 V. This is virtually zero and so point X is at virtually earth potential. For this reason it is called a **virtual earth**. The potential difference across R_1 is $(V_{in} - V_X)$. Hence, for an ideal operational amplifier with an infinite gain, and hence $V_X = 0$, the input potential V_{in} can be considered to be across R_1. Thus

$$V_{in} = I_1 R_1$$

The operational amplifier has a very high impedance between its input terminals; for a 741 about $2\,\text{M}\Omega$. Thus virtually no current flows through X into it. For an ideal operational amplifier the input impedance is taken to be infinite and so there is no current flow through X. Hence the current I_1 through R_1 must be the current through R_2. The potential difference across R_2 is $(V_X - V_{out})$ and thus, since V_X is zero for the ideal amplifier, potential difference across R_2 is $-V_{out}$. Thus

$$-V_{out} = I_1 R_2$$

Dividing these two equations,

$$\text{voltage gain of circuit} = \frac{V_{out}}{V_{in}} = -\frac{R_2}{R_1}$$

Thus the voltage gain of the circuit is determined solely by the relative values of R_2 and R_1. The negative sign indicates that the output is inverted, i.e. 180° out of phase, with respect to the input.

To illustrate the above, consider an inverting operational amplifier circuit which has a resistance of $1\,\text{M}\Omega$ in the inverting input line and a feedback resistance of 10 MΩ. What is the voltage gain of the circuit?

$$\text{Voltage gain of circuit} = \frac{V_{out}}{V_{in}} = -\frac{R_2}{R_1} = -\frac{10}{1} = -10$$

As an example of the use of the inverting amplifier circuit, photodiodes are widely used sensors (see Section 2.10) and give small currents on exposure to light. The inverting amplifier circuit can be used with such a sensor to give a current to voltage converter, the photodiode being reverse bias connected in place of resistor R_1, and so enable the output to be used as input to a microcontroller.

3.2.2   Non-inverting amplifier

Figure 3.4(a) shows the operational amplifier connected as a non-inverting amplifier. The output can be considered to be taken from across a potential divider circuit consisting of R_1 in series with R_2. The voltage V_X is then the fraction $R_1/(R_1 + R_2)$ of the output voltage, i.e.

$$V_X = \frac{R_1}{R_1 + R_2}V_{out}$$

Since there is virtually no current through the operational amplifier between the two inputs there can be virtually no potential difference between them. Thus, with the ideal operational amplifier, we must have $V_X = V_{in}$. Hence

$$\text{voltage gain of circuit} = \frac{V_{out}}{V_{in}} = \frac{R_1 + R_2}{R_1} = 1 + \frac{R_2}{R_1}$$

A particular form of this amplifier is when the feedback loop is a short circuit, i.e. $R_2 = 0$. Then the voltage gain is 1. The input to the circuit is into a large resistance, the input resistance typically being 2 MΩ. The output resistance, i.e. the resistance between the output terminal and the ground line, is, however, much smaller, e.g. 75 Ω. Thus the resistance in the circuit that follows is a relatively small one and is less likely to load that circuit. Such an amplifier is referred to as a **voltage follower**; Figure 3.4(b) shows the basic circuit.

Figure 3.4   (a) Non-inverting amplifier, (b) voltage follower.

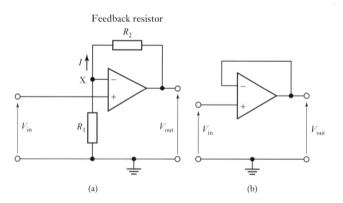

(a)   (b)

3.2.3   Summing amplifier

Figure 3.5 shows the circuit of a summing amplifier. As with the inverting amplifier (Section 3.2.1), X is a virtual earth. Thus the sum of the currents entering X must equal that leaving it. Hence

$$I = I_A + I_B + I_C$$

Figure 3.5 Summing amplifier.

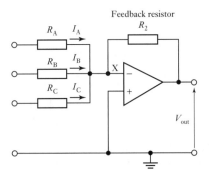

But $I_A = V_A/R_A$, $I_B = V_B/R_B$ and $I_C = V_C/R_C$. Also we must have the same current I passing through the feedback resistor. The potential difference across R_2 is $(V_X - V_{out})$. Hence, since V_X can be assumed to be zero, it is $-V_{out}$ and so $I = -V_{out}/R_2$. Thus

$$-\frac{V_{out}}{R_2} = \frac{V_A}{R_A} + \frac{V_B}{R_B} + \frac{V_C}{R_C}$$

The output is thus the scaled sum of the inputs, i.e.

$$V_{out} = -\left(\frac{R_2}{R_A}V_A + \frac{R_2}{R_B}V_B + \frac{R_2}{R_C}V_C\right)$$

If $R_A = R_B = R_C = R_1$ then

$$V_{out} = -\frac{R_2}{R_1}(V_A + V_B + V_C)$$

To illustrate the above, consider the design of a circuit that can be used to produce an output voltage which is the average of the input voltages from three sensors. Assuming that an inverted output is acceptable, a circuit of the form shown in Figure 3.5 can be used. Each of the three inputs must be scaled to 1/3 to give an output of the average. Thus a voltage gain of the circuit of 1/3 for each of the input signals is required. Hence, if the feedback resistance is 4 kΩ the resistors in each input arm will be 12 kΩ.

3.2.4 Integrating and differentiating amplifiers

Consider an inverting operational amplifier circuit with the feedback being via a capacitor, as illustrated in Figure 3.6(a).

The current is the rate of movement of charge q and since for a capacitor the charge $q = Cv$, where v is the voltage across it, then the current through the capacitor $i = dq/dt = C\,dv/dt$. The potential difference across C is $(v_X - v_{out})$ and since v_X is effectively zero, being the virtual earth, it is $-v_{out}$. Thus the current through the capacitor is $-C\,dv_{out}/dt$. But this is also the current through the input resistance R. Hence

$$\frac{v_{in}}{R} = -C\frac{dv_{out}}{dt}$$

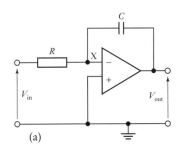

(a)

(b)

Figure 3.6 (a) Integrating amplifier, (b) differentiator amplifier.

Rearranging this gives

$$dv_{out} = -\left(\frac{1}{RC}\right)v_{in}\, dt$$

Integrating both sides gives

$$v_{out}(t_2) - v_{out}(t_1) = -\frac{1}{RC}\int_{t_1}^{t_2} v_{in}\, dt$$

$v_{out}(t_2)$ is the output voltage at time t_2 and $v_{out}(t_1)$ is the output voltage at time t_1. The output is proportional to the integral of the input voltage, i.e. the area under a graph of input voltage with time.

A differentiation circuit can be produced if the capacitor and resistor are interchanged in the circuit for the integrating amplifier. Figure 3.6(b) shows the circuit. The input current i_{in} to capacitor C is $dq/dt = C\, dv/dt$. With the ideal case of zero op-amp current, this is also the current through the feedback resistor R, i.e. $-v_{out}/R$ and so

$$\frac{v_{out}}{R} = -C\frac{dv_{in}}{dt}$$

$$v_{out} = -RC\frac{dv_{in}}{dt}$$

At high frequencies the differentiator circuit is susceptible to stability and noise problems. A solution is to add an input resistor R_{in} to limit the gain at high frequencies and so reduce the problem.

Figure 3.7 Difference amplifier.

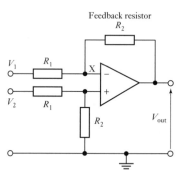

3.2.5   Difference amplifier

A difference amplifier is one that amplifies the difference between two input voltages. Figure 3.7 shows the circuit. Since there is virtually no current through the high resistance in the operational amplifier between the two input terminals, there is no potential drop and thus both the inputs X will be at the same potential. The voltage V_2 is across resistors R_1 and R_2 in series. Thus the potential V_X at X is

$$\frac{V_X}{V_2} = \frac{R_2}{R_1 + R_2}$$

The current through the feedback resistance must be equal to that from V_1 through R_1. Hence

$$\frac{V_1 - V_X}{R_1} = \frac{V_X - V_{out}}{R_2}$$

This can be rearranged to give

$$\frac{V_{out}}{R_2} = V_X\left(\frac{1}{R_2} + \frac{1}{R_1}\right) - \frac{V_1}{R_1}$$

Hence substituting for V_X using the earlier equation,

$$V_{out} = \frac{R_2}{R_1}(V_2 - V_1)$$

The output is thus a measure of the difference between the two input voltages.

As an illustration of the use of such a circuit with a sensor, Figure 3.8 shows it used with a thermocouple. The difference in voltage between the e.m.f.s of the two junctions of the thermocouple is being amplified. The values of R_1 and R_2 can, for example, be chosen to give a circuit with an output of 10 mV for a temperature difference between the thermocouple junctions of 10°C if such a temperature difference produces an e.m.f. difference between the junctions of 530 μV. For the circuit we have

$$V_{out} = \frac{R_2}{R_1}(V_2 - V_1)$$

$$10 \times 10^{-3} = \frac{R_2}{R_1} \times 530 \times 10^{-6}$$

Hence $R_2/R_1 = 18.9$. Thus if we take for R_1 a resistance of 10 kΩ then R_2 must be 189 kΩ.

A difference amplifier might be used with a Wheatstone bridge (see Section 3.5), perhaps one with strain gauge sensors in its arms, to amplify the out-of-balance potential difference that occurs when the resistance in one or more arms changes. When the bridge is balanced, both the output terminals of the bridge are at the same potential; there is thus no output potential difference. The output terminals from the bridge might both be at, say, 5.00 V. Thus the differential amplifier has both its inputs at 5.00 V. When

Figure 3.8 Difference amplifier with a thermocouple.

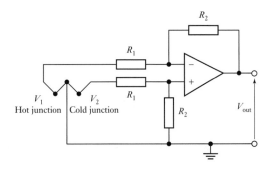

the bridge is no longer balanced we might have one output terminal at 5.01 V and the other at 4.99 V and so the inputs to the differential amplifier are 5.01 and 4.99 V. The amplifier amplifies this difference in the voltages of 0.02 V. The original 5.00 V signal which is common to both inputs is termed the **common mode voltage** V_{CM}. For the amplifier only to amplify the difference between the two signals assumes that the two input channels are perfectly matched and the operational amplifier has the same, high, gain for both of them. In practice this is not perfectly achieved and thus the output is not perfectly proportional to the difference between the two input voltages. Thus we write for the output

$$V_{out} = G_d \Delta V + G_{CM} V_{CM}$$

where G_d is the gain for the voltage difference ΔV, G_{CM} the gain for the common mode voltage V_{CM}. The smaller the value of G_{CM}, the smaller the effect of the common mode voltage on the output. The extent to which an operational amplifier deviates from the ideal situation is specified by the **common mode rejection ratio (CMRR)**:

$$CMRR = \frac{G_d}{G_{CM}}$$

To minimise the effect of the common mode voltage on the output, a high CMRR is required. CMRRs are generally specified in decibels (dB). Thus, on the decibel scale a CMRR of, say, 10 000 would be 20 lg 10 000 = 80 dB. A typical operational amplifier might have a CMRR between about 80 and 100 dB.

A common form of **instrumentation amplifier** involves three operational amplifiers (Figure 3.9), rather than just a single difference amplifier, and is available as a single integrated circuit. Such a circuit is designed to have a high input impedance, typically about 300 MΩ, a high voltage gain and excellent CMRR, typically more than 100 dB. The first stage involves the amplifiers A_1 and A_2, one being connected as an inverting amplifier and the other as a non-inverting amplifier. Amplifier A_3 is a differential amplifier with inputs from A_1 and A_2.

Figure 3.9 Instrumentation amplifier.

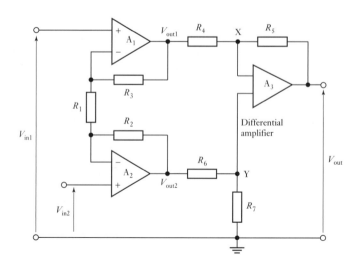

Because virtually no current passes through A_3, the current through R_4 will be the same as that through R_5. Hence

$$\frac{V_{out1} - V_X}{R_4} = \frac{V_X - V_{out}}{R_5}$$

The differential input to A_3 is virtually zero, so $V_Y = V_X$. Hence the above equation can be written as

$$V_{out} = \left(1 + \frac{R_5}{R_4}\right)V_Y - \frac{R_5}{R_4}V_{out1}$$

R_6 and R_7 form a potential divider for the voltage V_{out2} so that

$$V_Y = \frac{R_6}{R_6 + R_7}V_{out2}$$

Hence we can write

$$V_{out} = \frac{1 + \dfrac{R_5}{R_4}}{1 + \dfrac{R_7}{R_6}}V_{out2} - \frac{R_5}{R_4}V_{out1}$$

Hence by suitable choice of resistance values we obtain equal multiplying factors for the two inputs to the difference amplifier. This requires

$$1 + \frac{R_5}{R_4} = \left(1 + \frac{R_7}{R_6}\right)\frac{R_5}{R_4}$$

and hence $R_4/R_5 = R_6/R_7$.

We can apply the **principle of superposition**, i.e. we can consider the output produced by each source acting alone and then add them to obtain the overall response. Amplifier A_1 has an input of the difference signal V_{in1} on its non-inverting input and amplifies this with a gain of $1 + R_3/R_1$. It also has an input of V_{in2} on its inverting input and this is amplified to give a gain of $-R_3/R_1$. Also the common mode voltage V_{cm} on the non-inverting input is amplified by A_1. Thus the output of A_1 is

$$V_{out1} = \left(1 + \frac{R_3}{R_1}\right)V_{in1} - \left(\frac{R_3}{R_1}\right)V_{in2} + \left(1 + \frac{R_3}{R_1}\right)V_{cm}$$

Amplifier A_2 likewise gives

$$V_{out2} = \left(1 + \frac{R_2}{R_1}\right)V_{in2} - \left(\frac{R_2}{R_1}\right)V_{in1} + \left(1 + \frac{R_2}{R_1}\right)V_{cm}$$

The difference input to A_3 is $V_{out1} - V_{out2}$ and so

$$V_{out2} - V_{out1} = \left(1 + \frac{R_3}{R_1} + \frac{R_2}{R_1}\right)V_{in1} - \left(1 + \frac{R_2}{R_3} + \frac{R_3}{R_1}\right)V_{in2}$$

$$+ \left(\frac{R_3}{R_1} - \frac{R_2}{R_1}\right)V_{cm}$$

Figure 3.10   INA114.

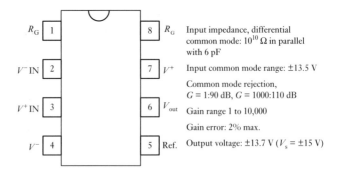

With $R_2 = R_3$ the common mode voltage term disappears, thus

$$V_{out2} - V_{out1} = \left(1 + \frac{2R_2}{R_1}\right)(V_{in1} - V_{in2})$$

The overall gain is thus $(1 + 2R_2/R_1)$ and is generally set by varying R_1.

Figure 3.10 shows the pin connections and some specification details for a low-cost, general-purpose instrumentation amplifier (Burr-Brown INA114) using this three op-amps form of design. The gain is set by connecting an external resistor R_G between pins 1 and 8, the gain then being $1 + 50/R_G$ when R_G is in kΩ. The 50 kΩ term arises from the sum of the two internal feedback resistors.

3.2.6   Logarithmic amplifier

Some sensors have outputs which are non-linear. For example, the output from a thermocouple is not a perfectly linear function of the temperature difference between its junctions. A signal conditioner might then be used to linearise the output from such a sensor. This can be done using an operational amplifier circuit which is designed to have a non-linear relationship between its input and output so that when its input is non-linear, the output is linear. This is achieved by a suitable choice of component for the feedback loop.

The logarithmic amplifier shown in Figure 3.11 is an example of such a signal conditioner. The feedback loop contains a diode (or a transistor with a grounded base). The diode has a non-linear characteristic. It might be represented by $V = C \ln I$, where C is a constant. Then, since the current through the feedback loop is the same as the current through the input resistance and the potential difference across the diode is $-V_{out}$, we have

$$V_{out} = -C \ln(V_{in}/R) = K \ln V_{in}$$

where K is some constant. However, if the input V_{in} is provided by a sensor with an input t, where $V_{in} = A\,e^{at}$, with A and a being constants, then

$$V_{out} = K \ln V_{in} = K \ln(A\,e^{at}) = K \ln A + Kat$$

The result is a linear relationship between V_{out} and t.

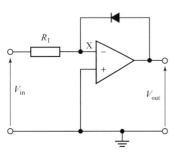

Figure 3.11   Logarithmic amplifier.

3.2.7 Comparator

A comparator indicates which of two voltages is the larger. An operational amplifier used with no feedback or other components can be used as a comparator. One of the voltages is applied to the inverting input and the other to the non-inverting input (Figure 3.12(a)). Figure 3.12(b) shows the relationship between the output voltage and the difference between the two input voltages. When the two inputs are equal there is no output. However, when the non-inverting input is greater than the inverting input by more than a small fraction of a volt then the output jumps to a steady positive saturation voltage of typically $+10$ V. When the inverting input is greater than the non-inverting input then the output jumps to a steady negative saturation voltage of typically -10 V. Such a circuit can be used to determine when a voltage exceeds a certain level, the output then being used to perhaps initiate some action.

Figure 3.12 Comparator.

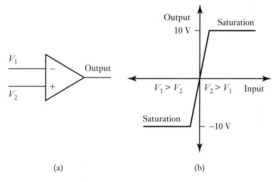

As an illustration of such a use, consider the circuit shown in Figure 3.13. This is designed so that when a critical temperature is reached a relay is activated and initiates some response. The circuit has a Wheatstone bridge with a thermistor in one arm. The resistors in the bridge have their resistances selected so that at the critical temperature the bridge will be balanced. When the temperature is below this value the thermistor resistance R_1 is more than R_2 and the bridge is out of balance. As a consequence there is a voltage difference between the inputs to the operational amplifier and it gives an output at its lower saturated level. This keeps the transistor off, i.e. both the base–emitter and base–collector junctions are reverse biased, and so no current passes through the relay coil. When the temperature rises and the resistance of the thermistor falls, the bridge becomes balanced

Figure 3.13 Temperature switch circuit.

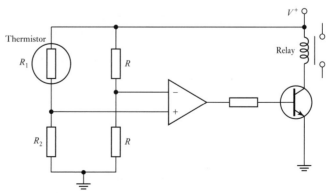

Figure 3.14 Focusing system
for a CD player.

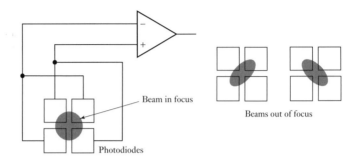

Beam in focus

Beams out of focus

Photodiodes

and the operational amplifier then switches to its upper saturation level.
Consequently the transistor is switched on, i.e. its junctions become forward
biased, and the relay energised.

As another illustration of the use of a comparator, consider the system
used to ensure that in a compact disc player the laser beam is focused on the
disc surface. With a CD player, lenses are used to focus a laser beam onto a
CD, this having the audio information stored as a sequence of microscopic
pits and flats. The light is reflected back from the disc to an array of four
photodiodes (Figure 3.14). The output from these photodiodes is then used
to reproduce the sound. The reason for having four photodiodes is that the
array can also be used to determine whether the beam of laser light is in
focus. When the beam is in focus on the disc then a circular spot of light
falls on the photodiode array with equal amounts of light falling on each
photodiode. As a result the output from the operational amplifier, which
is connected as a comparator, is zero. When the beam is out of focus an
elliptical spot of light is produced. This results in different amounts of light
falling on each of the photocells. The outputs from the two diagonal sets
of cells are compared and, because they are different, the comparator gives
an output which indicates that the beam is out of focus and in which direction
it is out of focus. The output can then be used to initiate correcting action
by adjusting the lenses focusing the beam onto the disc.

3.2.8 Real amplifiers

Operational amplifiers are not in the real world the perfect (ideal) element
discussed in the previous sections of this chapter. A particularly significant
problem is that of the **offset voltage**.

An operational amplifier is a high-gain amplifier which amplifies the
difference between its two inputs. Thus if the two inputs are shorted we
might expect to obtain no output. However, in practice this does not occur
and quite a large output voltage might be detected. This effect is produced
by imbalances in the internal circuitry in the operational amplifier. The
output voltage can be made zero by applying a suitable voltage between the
input terminals. This is known as the **offset voltage**. Many operational
amplifiers are provided with arrangements for applying such an offset voltage via a potentiometer. With the 741 this is done by connecting a 10 kΩ potentiometer between pins 1 and 5 (see Figure 3.2) and connecting the sliding
contact of the potentiometer to a negative voltage supply (Figure 3.15). The
imbalances within the operational amplifier are corrected by adjusting the

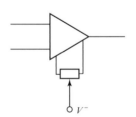

V^-

Figure 3.15 Correcting the
offset voltage.

position of the slider until with no input to the amplifier there is no output. Typically a general-purpose amplifier will have an offset voltage between 1 and 5 mV.

Operational amplifiers draw small currents at the input terminals in order to bias the input transistors. The bias current flowing through the source resistance at each terminal generates a voltage in series with the input. Ideally, the bias currents at the two inputs will be equal; however, in practice this will not be the case. Thus the effect of these bias currents is to produce an output voltage when there is no input signal and the output should be zero. This is particularly a problem when the amplifier is operating with d.c. voltages. The average value of the two bias currents is termed the **input bias current**. For a general-purpose operational amplifier, a typical value is about 100 nA. The difference between the two bias currents is termed the **input-offset current**. Ideally this would be zero, but for a typical general-purpose amplifier it is likely to be about 10 nA, about 10 to 25% of the input bias current.

An important parameter which affects the use of an operational amplifier with alternating current applications is the **slew rate**. This is the maximum rate of change at which the output voltage can change with time in response to a perfect step-function input. Typical values range from 0.2 V/μs to over 20 V/μs. With high frequencies, the large-signal operation of an amplifier is determined by how fast the output can swing from one voltage to another. Thus for use with high-frequency inputs a high value of slew rate is required.

As an illustration of the above, the general-purpose amplifier LM348 with an open-loop voltage gain of 96 dB has an input bias current of 30 nA and a slew rate of 0.6 V/μs. The wide-band amplifier AD711 with an open-loop gain of 100 has an input bias current of 25 pA and a slew rate of 20 V/μs.

| 3.3 | **Protection** |

There are many situations where the connection of a sensor to the next unit, e.g. a microprocessor, can lead to the possibility of damage as a result of perhaps a high current or high voltage. A high current can be protected against by the incorporation in the input line of a series resistor to limit the current to an acceptable level and a fuse to break if the current does exceed a safe level. High voltages, and wrong polarity, may be protected against by the use of a Zener diode circuit (Figure 3.16). Zener diodes behave like ordinary diodes up to some breakdown voltage when they become conducting. Thus to allow a maximum voltage of 5 V but stop voltages above 5.1 V getting through, a Zener diode with a voltage rating of 5.1 V might be chosen. When the voltage rises to 5.1 V the Zener diode breakdown and its resistance drop to a very low value. The result is that the voltage across the diode, and hence that outputted to the next circuit, drops. Because the Zener diode is a diode with a low resistance for current in one direction through it and a high resistance for the opposite direction, it also provides protection against wrong polarity. It is connected with the correct polarity to give a high resistance across the output and so a high voltage drop. When the supply polarity is reversed, the diode has low resistance and so little voltage drop occurs across the output.

In some situations it is desirable to isolate circuits completely and remove all electrical connections between them. This can be done using

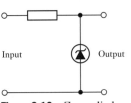

Input   Output

Figure 3.16   Zener diode protection circuit.

an **optoisolator**. Thus we might have the output from a microprocessor applied to a light-emitting diode (LED) which emits infrared radiation. This radiation is detected by a phototransistor or triac and gives rise to a current which replicates the changes occurring in the voltage applied to the LED. Figure 3.17 shows a number of forms of optoisolator. The term **transfer ratio** is used to specify the ratio of the output current to the input current. Typically, a simple transistor optoisolator (Figure 3.17(a)) gives an output current which is smaller than the input current and a transfer ratio of perhaps 30% with a maximum value of 7 mA. However, the Darlington form (Figure 3.17(b)) gives an output current larger than the input current, e.g. the Siemens 6N139 gives a transfer ratio of 800% with a maximum output value of 60 mA. Another form of optoisolator (Figure 3.17(c)) uses the triac and so can be used with alternating current, a typical triac optoisolator being able to operate with the mains voltage. Yet another form (Figure 3.17(d)) uses a triac with a zero-crossing unit, e.g. Motorola MOC3011, to reduce transients and electromagnetic interference.

Figure 3.17 Optoisolators: (a) transistor, (b) Darlington, (c) triac, (d) triac with zero-crossing unit.

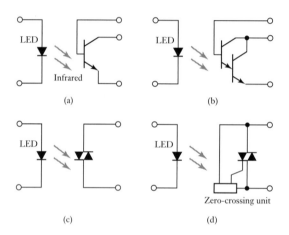

Optoisolator outputs can be used directly to switch low-power load circuits. Thus a Darlington optoisolator might be used as the interface between a microprocessor and lamps or relays. To switch a high-power circuit, an optocoupler might be used to operate a relay and so use the relay to switch the high-power device.

A protection circuit for a microprocessor input is thus likely to be like that shown in Figure 3.18; to prevent the LED having the wrong polarity or too high an applied voltage, it is also likely to be protected by the Zener diode circuit shown in Figure 3.16 and if there is alternating signal in the input a diode would be put in the input line to rectify it.

Figure 3.18 Protection circuit.

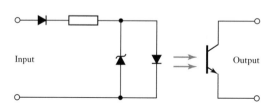

3.4 Filtering

The term **filtering** is used to describe the process of removing a certain band of frequencies from a signal and permitting others to be transmitted. The range of frequencies passed by a filter is known as the **pass band**, the range not passed as the **stop band** and the boundary between stopping and passing as the **cut-off frequency**. Filters are classified according to the frequency ranges they transmit or reject. A **low-pass filter** (Figure 3.19(a)) has a pass band which allows all frequencies from zero up to some frequency to be transmitted. A **high-pass filter** (Figure 3.19(b)) has a pass band which allows all frequencies from some value up to infinity to be transmitted. A **band-pass filter** (Figure 3.19(c)) allows all the frequencies within a specified band to be transmitted. A **band-stop filter** (Figure 3.19(d)) stops all frequencies with a particular band from being transmitted. In all cases the cut-off frequency is defined as being that at which the output voltage is 70.7% of that in the pass band. The term **attenuation** is used for the ratio of input and output powers, this being written as the ratio of the logarithm of the ratio and so gives the attenuation in units of bels. Since this is a rather large unit, decibels (dB) are used and then attenuation in dB = 10 lg(input power/output power). Since the power through an impedance is proportional to the square of the voltage, the attenuation in dB = 20 lg(input voltage/ output voltage). The output voltage of 70.7% of that in the pass band is thus an attenuation of 3 dB.

Figure 3.19 Characteristics of ideal filters: (a) low-pass, (b) high-pass, (c) band-pass, (d) band-stop.

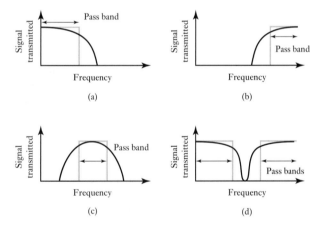

The term **passive** is used to describe a filter made up using only resistors, capacitors and inductors, the term **active** being used when the filter also involves an operational amplifier. Passive filters have the disadvantage that the current that is drawn by the item that follows can change the frequency characteristic of the filter. This problem does not occur with an active filter.

Low-pass filters are very commonly used as part of signal conditioning. This is because most of the useful information being transmitted is low frequency. Since noise tends to occur at higher frequencies, a low-pass filter can be used to block it off. Thus a low-pass filter might be selected with a cut-off frequency of 40 Hz, thus blocking off any interference signals from the a.c. mains supply and noise in general. Figure 3.20 shows the basic form that can be used for a low-pass filter.

Figure 3.20 Low-pass filter: (a) passive, (b) active using an operational amplifier.

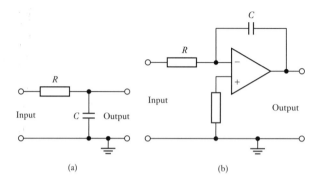

(a)   (b)

3.5 Wheatstone bridge

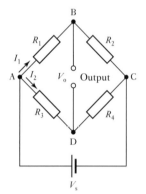

Figure 3.21 Wheatstone bridge.

The **Wheatstone bridge** can be used to convert a resistance change to a voltage change. Figure 3.21 shows the basic form of such a bridge. When the output voltage V_o is zero, then the potential at B must equal that at D. The potential difference across R_1, i.e. V_{AB}, must then equal that across R_3, i.e. V_{AD}. Thus $I_1 R_1 = I_2 R_2$. It also means that the potential difference across R_2, i.e. V_{BC}, must equal that across R_4, i.e. V_{DC}. Since there is no current through BD, then the current through R_2 must be the same as that through R_1 and the current through R_4 the same as that through R_3. Thus $I_1 R_2 = I_2 R_4$. Dividing these two equations gives

$$\frac{R_1}{R_2} = \frac{R_3}{R_4}$$

The bridge is said to be **balanced**.

Now consider what happens when one of the resistances changes from this balanced condition. The supply voltage V_s is connected between points A and C and thus the potential drop across the resistor R_1 is the fraction $R_1/(R_1 + R_2)$ of the supply voltage. Hence

$$V_{AB} = \frac{V_s R_1}{R_1 + R_2}$$

Similarly, the potential difference across R_3 is

$$V_{AD} = \frac{V_s R_3}{R_3 + R_4}$$

Thus the difference in potential between B and D, i.e. the output potential difference V_o, is

$$V_o = V_{AB} - V_{AD} = V_s \left(\frac{R_1}{R_1 + R_2} - \frac{R_3}{R_3 + R_4} \right)$$

This equation gives the balanced condition when $V_o = 0$.

Consider resistance R_1 to be a sensor which has a resistance change. A change in resistance from R_1 to $R_1 + \delta R_1$ gives a change in output from V_o to $V_o + \delta V_o$, where

$$V_o + \delta V_o = V_s \left(\frac{R_1 + \delta R_1}{R_1 + \delta R_1 + R_2} - \frac{R_3}{R_3 + R_4} \right)$$

Hence

$$(V_o + \delta V_o) - V_o = V_s\left(\frac{R_1 + \delta R_1}{R_1 + \delta R_1 + R_2} - \frac{R_1}{R_1 + R_2}\right)$$

If δR_1 is much smaller than R_1 then the above equation approximates to

$$\delta V_o \approx V_s\left(\frac{\delta R_1}{R_1 + R_2}\right)$$

With this approximation, the change in output voltage is thus proportional to the change in the resistance of the sensor. This gives the output voltage when there is no load resistance across the output. If there is such a resistance then the loading effect has to be considered.

To illustrate the above, consider a platinum resistance temperature sensor which has a resistance at 0°C of 100 Ω and forms one arm of a Wheatstone bridge. The bridge is balanced, at this temperature, with each of the other arms also being 100 Ω. If the temperature coefficient of resistance of platinum is 0.0039/K, what will be the output voltage from the bridge per degree change in temperature if the load across the output can be assumed to be infinite? The supply voltage, with negligible internal resistance, is 6.0 V. The variation of the resistance of the platinum with temperature can be represented by

$$R_t = R_0(1 + \alpha t)$$

where R_t is the resistance at t (°C), R_0 the resistance at 0°C and α the temperature coefficient of resistance. Thus

$$\text{change in resistance} = R_t - R_0 = R_0\alpha t$$
$$= 100 \times 0.0039 \times 1 = 0.39 \ \Omega/\text{K}$$

Since this resistance change is small compared with the 100 Ω, the approximate equation can be used. Hence

$$\delta V_o \approx V_s\left(\frac{\delta R_1}{R_1 + R_2}\right) = \frac{6.0 \times 0.39}{100 + 100} = 0.012 \ \text{V}$$

3.5.1 Temperature compensation

In many measurements involving a resistive sensor the actual sensing element may have to be at the end of long leads. Not only the sensor but the resistance of these leads will be affected by changes in temperature. For example, a platinum resistance temperature sensor consists of a platinum coil at the ends of leads. When the temperature changes, not only will the resistance of the coil change but so also will the resistance of the leads. What is required is just the resistance of the coil and so some means has to be employed to compensate for the resistance of the leads to the coil. One method of doing this is to use three leads to the coil, as shown in Figure 3.22. The coil is connected into the Wheatstone bridge in such a way that lead 1 is in series with the R_3 resistor while lead 3 is in series with the platinum resistance coil R_1. Lead 2 is the connection to the power supply. Any change in lead resistance is likely to affect all three leads equally, since they are of the

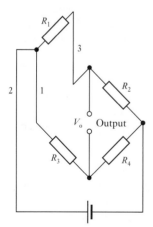

Figure 3.22 Compensation for leads.

same material, diameter and length and held close together. The result is that changes in lead resistance occur equally in two arms of the bridge and cancel out if R_1 and R_3 are the same resistance.

The electrical resistance strain gauge is another sensor where compensation has to be made for temperature effects. The strain gauge changes resistance when the strain applied to it changes. Unfortunately, it also changes if the temperature changes. One way of eliminating the temperature effect is to use a **dummy strain gauge**. This is a strain gauge which is identical to the one under strain, the active gauge, and is mounted on the same material but is not subject to the strain. It is positioned close to the active gauge so that it suffers the same temperature changes. Thus a temperature change will cause both gauges to change resistance by the same amount. The active gauge is mounted in one arm of a Wheatstone bridge (Figure 3.23(a)) and the dummy gauge in another arm so that the effects of temperature-induced resistance changes cancel out.

Figure 3.23 Compensation with strain gauges: (a) use of dummy gauge, (b) four active arm bridge.

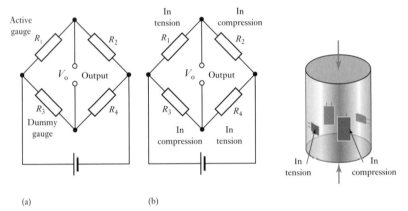

(a)   (b)

Strain gauges are often used with other sensors such as load cells or diaphragm pressure gauges to give measures of the amount of displacement occurring. In such situations, temperature compensation is still required. While dummy gauges could be used, a better solution is to use four strain gauges. Two of them are attached so that when forces are applied they are in tension and the other two in compression. The load cell in Figure 3.23(b) shows such a mounting. The gauges that are in tension will increase in resistance while those in compression will decrease in resistance. As the gauges are connected as the four arms of a Wheatstone bridge (Figure 3.23(b)), then since all will be equally affected by any temperature changes, the arrangement is temperature compensated. It also gives a much greater output voltage than would occur with just a single active gauge.

To illustrate this, consider a load cell with four strain gauges arranged as shown in Figure 3.23 which is to be used with a four active arm strain gauge bridge. The gauges have a gauge factor of 2.1 and a resistance of 100 Ω. When the load cell is subject to a compressive force, the vertical gauges show compression and, because when an item is squashed there is a consequential sideways extension, the horizontal gauges are subject to tensile strain (the ratio of the transverse to longitudinal strains is called Poisson's ratio and is usually about 0.3). Thus, if the compressive gauges suffer a strain of -1.0×10^{-5} and the tensile gauges $+0.3 \times 10^{-5}$, the supply voltage for the bridge is 6 V and the output voltage from the bridge is amplified by a differential operational amplifier circuit, what will be the ratio of the feedback resistance to that of the input resistances in the two inputs of the amplifier if the load is to produce an output of 1 mV?

The change in resistance of a gauge subject to the compressive strain is given by $\Delta R / R = G\varepsilon$:

$$\text{change in resistance} = G\varepsilon R = -2.1 \times 1.0 \times 10^{-5} \times 100$$
$$= -2.1 \times 10^{-3}\,\Omega$$

For a gauge subject to tension we have

$$\text{change in resistance} = G\varepsilon R = 2.1 \times 0.3 \times 10^{-5} \times 100$$
$$= 6.3 \times 10^{-4}\,\Omega$$

The out-of-balance potential difference is given by (see earlier in Section 3.5)

$$V_o = V_s\left(\frac{R_1}{R_1 + R_2} - \frac{R_3}{R_3 + R_4}\right)$$

$$= V_s\left(\frac{R_1(R_3 + R_4) - R_3(R_1 + R_2)}{(R_1 + R_2)(R_3 + R_4)}\right)$$

$$= V_s\left(\frac{R_1 R_4 - R_2 R_3}{(R_1 + R_2)(R_3 + R_4)}\right)$$

We now have each of the resistors changing. We can, however, neglect the changes in relation to the denominators where the effect of the changes on the sum of the two resistances is insignificant. Thus

$$V_o = V_s\left(\frac{(R_1 + \delta R_1)(R_4 + \delta R_4) - (R_2 + \delta R_2)(R_3 + \delta R_3)}{(R_1 + R_2)(R_3 + R_4)}\right)$$

Neglecting products of δ terms and since we have an initially balanced bridge with $R_1 R_4 = R_2 R_3$, then

$$V_o = \frac{V_s R_1 R_4}{(R_1 + R_2)(R_3 + R_4)}\left(\frac{\delta R_1}{R_1} - \frac{\delta R_2}{R_2} - \frac{\delta R_3}{R_3} + \frac{\delta R_4}{R_4}\right)$$

Hence

$$V_o = \frac{6 \times 100 \times 100}{200 \times 200}\left(\frac{2 \times 6.3 \times 10^{-4} + 2 \times 2.1 \times 10^{-3}}{100}\right)$$

The output is thus 8.19×10^{-5} V. This becomes the input to the differential amplifier; hence, using the equation developed in Section 3.2.5,

$$V_o = \frac{R_2}{R_1}(V_2 - V_1)$$

$$1.0 \times 10^{-3} = \frac{R_2}{R_1} \times 8.19 \times 10^{-5}$$

Thus $R_2 / R_1 = 12.2$.

3.5.2 Thermocouple compensation

A thermocouple gives an e.m.f. which depends on the temperature of its two junctions (see Section 2.9.5). Ideally, if one junction is kept at 0°C, then the temperature relating to the e.m.f. can be directly read from tables. However, this is not always feasible and the cold junction is often allowed to be at the ambient temperature. To compensate for this a potential difference has to be added to the thermocouple. This must be the same as the e.m.f. that would be generated by the thermocouple with one junction at 0°C and the other at the ambient temperature. Such a potential difference can be produced by using a resistance temperature sensor in a Wheatstone bridge. The bridge is balanced at 0°C and the output voltage from the bridge provides the correction potential difference at other temperatures.

The resistance of a metal resistance temperature sensor can be described by the relationship

$$R_t = R_0(1 + \alpha t)$$

where R_t is the resistance at t (°C), R_0 the resistance at 0°C and α the temperature coefficient of resistance. Thus

$$\text{change in resistance} = R_t - R_0 = R_0 \alpha t$$

The output voltage for the bridge, taking R_1 to be the resistance temperature sensor, is given by

$$\delta V_o \approx V_s\left(\frac{\delta R_1}{R_1 + R_2}\right) = \frac{V_s R_0 \alpha t}{R_0 + R_2}$$

The thermocouple e.m.f. e is likely to vary with temperature t in a reasonably linear manner over the small temperature range being considered — from 0°C to the ambient temperature. Thus $e = kt$, where k is a constant, i.e. the e.m.f. produced per degree change in temperature. Hence for compensation we must have

$$kt = \frac{V_s R_0 \alpha t}{R_0 + R_2}$$

and so

$$kR_2 = R_0(V_s \alpha - k)$$

For an iron–constantan thermocouple giving 51 μV/°C, compensation can be provided by a nickel resistance element with a resistance of 10 Ω at 0°C and a temperature coefficient of resistance of 0.0067/K, a supply voltage for the bridge of 1.0 V and R_2 as 1304 Ω.

3.6   Pulse modulation

A problem that is often encountered with dealing with the transmission of low-level d.c. signals from sensors is that the gain of an operational amplifier used to amplify them may drift and so the output drifts. This problem can be overcome if the signal is a sequence of pulses rather than a continuous–time signal.

One way this conversion can be achieved is by chopping the d.c. signal in the way suggested in Figure 3.24. The output from the chopper is a chain of

Figure 3.24 Pulse amplitude modulation.

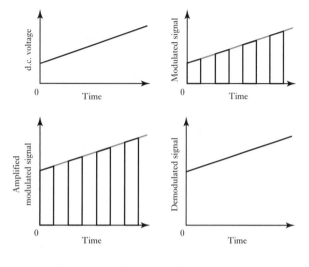

Figure 3.25 PWM for voltage control: (a) duty cycle 50%, (b) duty cycle 25%.

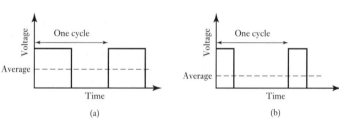

pulses, the heights of which are related to the d.c. level of the input signal. This process is called **pulse amplitude modulation**. After amplification and any other signal conditioning, the modulated signal can be demodulated to give a d.c. output. With pulse amplitude modulation, the height of the pulses is related to the size of the d.c. voltage.

Pulse width modulation (PWM) is widely used with control systems as a means of controlling the average value of a d.c. voltage. Thus if there is a constant analogue voltage and it is chopped into pulses, by varying the width of the pulses so the average value of the voltage can be changed. Figure 3.25 illustrates this. The term **duty cycle** is used for the fraction of each cycle for which the voltage is high. Thus for a PWM signal where the signal is high for half of each cycle, the duty cycle is ½ or 50%. If it is only on for a quarter of each cycle then the duty cycle is ¼ or 25%.

3.7 Problems with signals

When connecting sensors to signal conditioning equipment and controllers, problems can occur with signals as a result of grounding and electromagnetic interference.

3.7.1 Grounding

Generally the signals from sensors and signal conditioning equipment are transmitted as voltages to the controller. Such voltages are the potential differences between two points. If one of the points is earthed it is said to be a **grounded signal source**. If neither point is grounded then it is said to be a **floating signal source**. With a grounded source the voltage output is the potential difference between the system ground and the positive signal lead of the source. If it is a floating source, the signal source is not referenced to any absolute value and each of the voltage lines may have a potential relative to the ground.

Differential systems, e.g. a differential amplifier, are concerned with the potential difference between two input lines. If each has a voltage referred to a common ground, V_A and V_B, then the **common mode voltage** is the average of the two, i.e. $\frac{1}{2} (V_A + V_B)$. Thus if we have one input line at 10 V and the other at 12 V, the potential difference will be 2 V and the common mode voltage 11 V. The differential measurement system is concerned with the difference between the two inputs $(V_A - V_B)$ and not the common mode voltage. Unfortunately, the common mode voltage can have an effect on the indicated potential difference value, and the extent to which it affects the difference is described by the **common mode rejection ratio** (CMRR) (see Section 3.2.5). This is the ratio of differential gain of the system to the common mode gain or, when expressed in decibels, 20 lg (differential gain/ common mode gain). The higher the CMRR the greater the differential gain when compared with the common mode gain and the less significance is attached to the common mode voltage. A CMRR of 10 000, or 80 dB, for a differential amplifier would mean that, if the desired difference signal was the same size as the common mode voltage, it will appear at the output 10 000 times greater in size that the common mode.

Problems can arise with systems when a circuit has several grounding points. For example, it might be that both the sensor and the signal conditioner are grounded. In a large system, multiple grounding is largely inevitable. Unfortunately, there may be a potential difference between the two grounding points and thus significant currents can flow between the grounding points through the low but finite ground resistance (Figure 3.26). These are termed **ground-loop currents**. This potential difference between two grounding points is not necessarily just d.c. but can also be a.c., e.g. a.c. mains hum. There is also the problem that we have a loop in which currents can be induced by magnetic coupling with other nearby circuits. Thus a consequence of having a ground loop can be to make remote measurements difficult.

Ground loops from multiple point grounding can be minimised if the multiple earth connections are made close together and the common ground has

Figure 3.26   A ground loop.

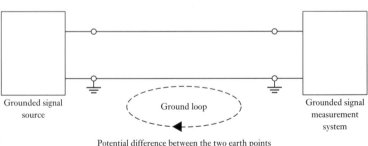

Grounded signal source

Ground loop

Grounded signal measurement system

Potential difference between the two earth points gives rise to the ground loop current

Figure 3.27 Isolation using (a) an optoisolator, (b) a transformer.

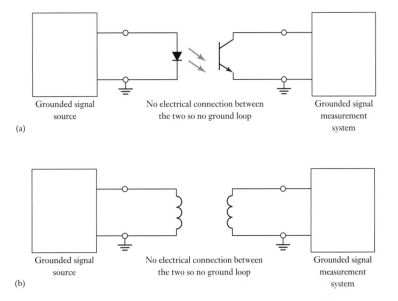

(a)
Grounded signal
source

No electrical connection between
the two so no ground loop

Grounded signal
measurement
system

(b)
Grounded signal
source

No electrical connection between
the two so no ground loop

Grounded signal
measurement
system

a resistance small enough to make the voltage drops between the earthing points negligible. Ground loops can be eliminated if there is electrical isolation of the signal source system from the measurement system. This can be achieved by using an optoisolator (see Section 3.3) or a transformer (Figure 3.27).

3.7.2 Electromagnetic interference

Electromagnetic interference is an undesirable effect on circuits resulting from time-varying electric and magnetic fields. Common sources of such interference are fluorescent lamps, d.c. motors, relay coils, household appliances and the electrics of motor cars.

Electrostatic interference occurs as a result of mutual capacitance between neighbouring conductors. Electric shielding can guard against this. This is a shield of electrically conductive material, e.g. copper or aluminium, that is used to enclose a conductor or circuit. Thus a screened cable might be used to connect a sensor to its measurement system. If the sensor is earthed then the screen should be connected to the same point where the sensor is earthed, thus minimising the ground loop (Figure 3.28).

Figure 3.28 Use of a shielded cable to minimise electrostatic interference.

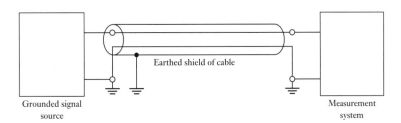

Grounded signal
source

Earthed shield of cable

Measurement
system

Figure 3.29 Twisted pair
of cables to minimise
electromagnetic interference.

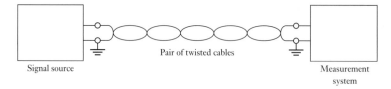

Signal source   Pair of twisted cables   Measurement
                                                                                            system

Interference also occurs when there is a changing magnetic field which
induces voltages in the measurement system. Protection against this can be
achieved by such methods as placing components as far as possible from
sources of interference and minimising the area of any loops in the system
and by the use of twisted pairs of wires for interconnections (Figure 3.29).
With twisted wire, the coupling alternates in phase between adjacent twists
and so leads to cancellation of the effect.

3.8 Power transfer

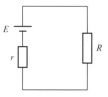

Figure 3.30 A d.c. source
supplying a load.

There are many control system situations when components are interconnected. Thus with electrical components we might have a sensor system
connected to an amplifier. With a mechanical system we might have a motor
rotating a load. A concern is the condition for maximum power transfer
between the two elements.

As an introduction, consider a d.c. source of e.m.f. E and internal
resistance r supplying a load of resistance R (Figure 3.30). The current
supplied to the load is $I = E/(R + r)$ and so the power supplied to the load is

$$P = I^2 R = \frac{E^2 R}{(R + r)^2}$$

The maximum power supplied to the load will when $dP/dt = 0$.

$$\frac{dP}{dt} = \frac{(R + r)^2 E^2 - E^2 R^2 (R + r)}{(R + r)^3}$$

When this is zero then $(R + r) = 2R$ and so the condition for maximum power
transfer is $R = r$, i.e. when the source and load resistances are matched.

With an alternating current source having an internal impedance supplying a load impedance, the condition for maximum power transfer can similarly be derived and is when the source and load impedances are matched.
With, for example, a high impedance sensor being matched to an electronic
system, an impedance matching amplifier might be used between the source
and the load in order to achieve this maximum power transmission. Such an
amplifier is typically a high gain amplifier with a high input impedance and
a low output impedance.

Summary

Signal conditioning can involve protection to prevent damage to the next
element in a system, getting a signal into the form required, getting the level
of a signal right, reducing noise, manipulating a signal to perhaps make it
linear.

Commonly used signal conditioning elements are **operational amplifiers**, these being high-gain d.c. amplifiers with gains of the order of 100 000 or more.

Protection against perhaps a high voltage or current can involve the use of resistors and fuses; Zener diodes can be used to protect against wrong polarity and high voltages. Optoisolators are used to isolate circuits completely, removing all electrical connections between them.

Filters can be used to remove a particular band of frequencies from a signal and permit others to be transmitted.

The **Wheatstone bridge** can be used to convert an electrical resistance change to a voltage change.

When connecting sensors to signal conditioning equipment and controllers, problems can occur with signals when a circuit has several **grounding** points and **electromagnetic interference** as a result of time-varying electric and magnetic fields.

For **maximum power transfer** between electrical components the impedances must match.

Problems

3.1 Design an operational amplifier circuit that can be used to produce an output that ranges from 0 to −5 V when the input goes from 0 to 100 mV.

3.2 An inverting amplifier has an input resistance of 2 kΩ. Determine the feedback resistance needed to give a voltage gain of 100.

3.3 Design a summing amplifier circuit that can be used to produce an output that ranges from −1 to −5 V when the input goes from 0 to 100 mV.

3.4 A differential amplifier is used with a thermocouple sensor in the way shown in Figure 3.8. What values of R_1 and R_2 would give a circuit which has an output of 10 mV for a temperature difference between the thermocouple junctions of 100°C with a copper–constantan thermocouple if the thermocouple is assumed to have a constant sensitivity of 43 μV/°C?

3.5 The output from the differential pressure sensor used with an orifice plate for the measurement of flow rate is non-linear, the output voltage being proportional to the square of the flow rate. Determine the form of characteristic required for the element in the feedback loop of an operational amplifier signal conditioner circuit in order to linearise this output.

3.6 A differential amplifier is to have a voltage gain of 100. What will be the feedback resistance required if the input resistances are both 1 kΩ?

3.7 A differential amplifier has a differential voltage gain of 2000 and a common mode gain of 0.2. What is the common mode rejection ratio in dB?

3.8 Digital signals from a sensor are polluted by noise and mains interference and are typically of the order of 100 V or more. Explain how protection can be afforded for a microprocessor to which these signals are to be inputted.

3.9 A platinum resistance temperature sensor has a resistance of 120 Ω at 0°C and forms one arm of a Wheatstone bridge. At this temperature the bridge is balanced with each of the other arms being 120Ω. The temperature coefficient of resistance of the platinum is 0.0039/K. What will be the output voltage from the bridge for a change in temperature of 20°C? The loading across the output is effectively open circuit and the supply voltage to the bridge is from a source of 6.0 V with negligible internal resistance.

3.10   A diaphragm pressure gauge employs four strain gauges to monitor the displacement of the diaphragm. The four active gauges form the arms of a Wheatstone bridge, in the way shown in Figure 3.23. The gauges have a gauge factor of 2.1 and resistance 120 Ω. A differential pressure applied to the diaphragm results in two of the gauges on one side of the diaphragm being subject to a tensile strain of 1.0×10^{-5} and the two on the other side a compressive strain of 1.0×10^{-5}. The supply voltage for the bridge is 10 V. What will be the voltage output from the bridge?

3.11   A Wheatstone bridge has a single strain gauge in one arm and the other arms are resistors with each having the same resistance as the unstrained gauge. Show that the output voltage from the bridge is given by $\frac{1}{4}V_s G\varepsilon$, where V_s is the supply voltage to the bridge, G the gauge factor of the strain gauge and ε the strain acting on it.

Chapter four Digital signals

Objectives

The objectives of this chapter are that, after studying it, the reader should be able to:
- Explain the principles and main methods of analogue-to-digital and digital-to-analogue converters.
- Explain the principles and uses of multiplexers.
- Explain the principles of digital signal processing.

4.1 Digital signals

The output from most sensors tends to be in analogue form, the size of the output being related to the size of the input. Where a microprocessor is used as part of the measurement or control system, the analogue output from the sensor has to be converted into a *digital* form before it can be used as an input to the microprocessor. Likewise, most actuators operate with analogue inputs and so the digital output from a microprocessor has to be converted into an analogue form before it can be used as an input by the actuator.

4.1.1 Binary numbers

The **binary system** is based on just the two symbols or states 0 and 1, these possibly being 0 V and 5 V signals. These are termed *bi*nary dig*its* or **bits**. When a number is represented by this system, the digit position in the number indicates the weight attached to each digit, the weight increasing by a factor of 2 as we proceed from right to left:

$$\ldots \qquad 2^3 \qquad 2^2 \qquad 2^1 \qquad 2^0$$
$$\text{bit 3} \qquad \text{bit 2} \qquad \text{bit 1} \qquad \text{bit 0}$$

For example, the decimal number 15 is $2^0 + 2^1 + 2^2 + 2^3 = 1111$ in the binary system. In a binary number the bit 0 is termed the **least significant bit** (LSB) and the highest bit the **most significant bit** (MSB). The combination of bits to represent a number is termed a **word**. Thus 1111 is a 4-bit word. Such a word could be used to represent the size of a signal. The term **byte** is used for a group of 8 bits. See Appendix B for more discussion of binary numbers.

4.2 Analogue and digital signals

Analogue-to-digital conversion involves converting analogue signals into binary words. Figure 4.1(a) shows the basic elements of analogue-to-digital conversion.

Figure 4.1   (a) Analogue-to-digital conversion, (b) analogue input, (c) clock signal, (d) sampled signal, (e) sampled and held signal.

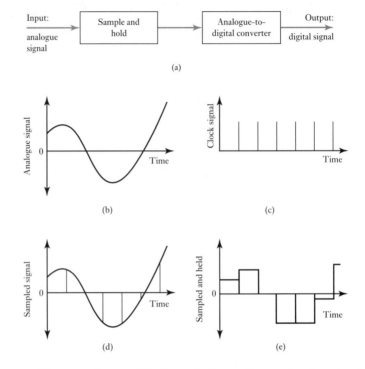

The procedure used is that a clock supplies regular time signal pulses to the analogue-to-digital converter (ADC) and every time it receives a pulse it samples the analogue signal. Figure 4.1 illustrates this analogue-to-digital conversion by showing the types of signals involved at the various stages. Figure 4.1(b) shows the analogue signal and Figure 4.1(c) the clock signal which supplies the time signals at which the sampling occurs. The result of the sampling is a series of narrow pulses (Figure 4.1(d)). A **sample and hold** unit is then used to hold each sampled value until the next pulse occurs, with the result shown in Figure 4.1(e). The sample and hold unit is necessary because the ADC requires a finite amount of time, termed the **conversion time**, to convert the analogue signal into a digital one.

The relationship between the sampled and held input and the output for an ADC is illustrated by the graph shown in Figure 4.2 for a digital output

Figure 4.2   Input/output for an ADC.

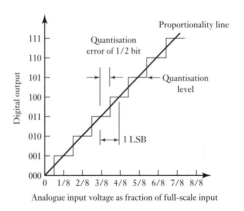

which is restricted to 3 bits. With 3 bits there are $2^3 = 8$ possible output levels. Thus, since the output of the ADC to represent the analogue input can be only one of these eight possible levels, there is a range of inputs for which the output does not change. The eight possible output levels are termed **quantisation levels** and the difference in analogue voltage between two adjacent levels is termed the **quantisation interval**. Thus for the ADC given in Figure 4.2, the quantisation interval is 1 V. Because of the step-like nature of the relationship, the digital output is not always proportional to the analogue input and thus there will be error, this being termed the **quantisation error**. When the input is centred over the interval, the quantisation error is zero, the maximum error being equal to one-half of the interval or $\pm\frac{1}{2}$ bit.

The word length possible determines the **resolution** of the element, i.e. the smallest change in input which will result in a change in the digital output. The smallest change in digital output is 1 bit in the least significant bit position in the word, i.e. the far right bit. Thus with a word length of n bits the full-scale analogue input V_{FS} is divided into 2^n pieces and so the minimum change in input that can be detected, i.e. the resolution, is $V_{FS}/2^n$.

Thus if we have an ADC with a word length of 10 bits and the analogue signal input range is 10 V, then the number of levels with a 10-bit word is $2^{10} = 1024$ and thus the resolution is $10/1024 = 9.8$ mV.

Consider a thermocouple giving an output of 0.5 mV/°C. What will be the word length required when its output passes through an ADC if temperatures from 0 to 200°C are to be measured with a resolution of 0.5°C? The full-scale output from the sensor is $200 \times 0.5 = 100$ mV. With a word length n, this voltage will be divided into $100/2^n$ mV steps. For a resolution of 0.5°C we must be able to detect a signal from the sensor of $0.5 \times 0.5 = 0.25$ mV. Thus we require

$$0.25 = \frac{100}{2^n}$$

Hence $n = 8.6$. Thus a 9-bit word length is required.

4.2.1 Sampling theorem

ADCs sample analogue signals at regular intervals and convert these values to binary words. How often should an analogue signal be sampled in order to give an output which is representative of the analogue signal?

Figure 4.3 illustrates the problem with different sampling rates being used for the same analogue signal. When the signal is reconstructed from the samples, it is only when the sampling rate is at least twice that of the highest frequency in the analogue signal that the sample gives the original form of signal. This criterion is known as the **Nyquist criterion** or **Shannon's sampling theorem**. When the sampling rate is less than twice the highest frequency, the reconstruction can represent some other analogue signal and we obtain a false image of the real signal. This is termed **aliasing**. In Figure 4.3(c) this could be an analogue signal with a much smaller frequency than that of the analogue signal that was sampled.

Whenever a signal is sampled too slowly, there can be a false interpretation of high-frequency components as arising from lower frequency aliases.

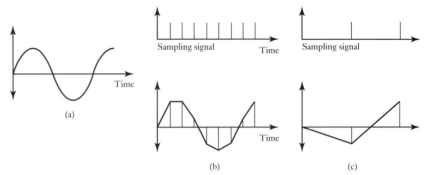

Figure 4.3   Effect of sampling frequency: (a) analogue signal, (b) sampled signal, (c) sampled signal.

High-frequency noise can also create errors in the conversion process. To minimise errors due to both aliasing and high-frequency noise, a low-pass filter is used to precede the ADC, the filter having a bandwidth such that it passes only low frequencies for which the sampling rate will not give aliasing errors. Such a filter is termed an **anti-aliasing filter.**

4.2.2   Digital-to-analogue conversion

The input to a digital-to-analogue converter (DAC) is a binary word; the output is an analogue signal that represents the weighted sum of the non-zero bits represented by the word. Thus, for example, an input of 0010 must give an analogue output which is twice that given by an input of 0001. Figure 4.4 illustrates this for an input to a DAC with a resolution of 1 V for unsigned binary words. Each additional bit increases the output voltage by 1 V.

Consider the situation where a microprocessor gives an output of an 8-bit word. This is fed through an 8-bit DAC to a control valve. The control valve requires 6.0 V to be fully open. If the fully open state is indicated by 11111111 what will be the output to the valve for a change of 1 bit?

The full-scale output voltage of 6.0 V will be divided into 2^8 intervals. A change of 1 bit is thus a change in the output voltage of $6.0/2^8 = 0.023$ V.

Figure 4.4   Input/output for a DAC.

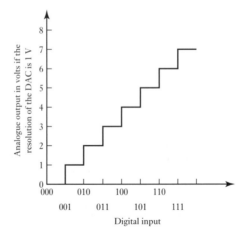

Digital-to-analogue and analogue-to-digital converters

The following are commonly encountered forms of DACs and ADCs.

4.3.1 DACs

A simple form of DAC uses a summing amplifier (see Section 3.2.3) to form the weighted sum of all the non–zero bits in the input word (Figure 4.5). The reference voltage is connected to the resistors by means of electronic switches which respond to binary 1. The values of the input resistances depend on which bit in the word a switch is responding to, the value of the resistor for successive bits from the LSB being halved. Hence the sum of the voltages is a weighted sum of the digits in the word. Such a system is referred to as a **weighted-resistor network.** The function of the op-amp circuit is to act as a buffer to ensure that the current out of the resistor network is not affected by the output load and also so that the gain can be adjusted to give an output range of voltages appropriate to a particular application.

Figure 4.5 Weighted–resistor DAC.

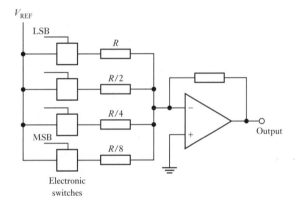

A problem with the weighted-resistor network is that accurate resistances have to be used for each of the resistors and it is difficult to obtain such resistors over the wide range needed. As a result this form of DAC tends be limited to 4-bit conversions.

Another, more commonly used, version uses an **R–2R ladder network** (Figure 4.6). This overcomes the problem of obtaining accurate resistances

Figure 4.6 *R–2R* ladder DAC.

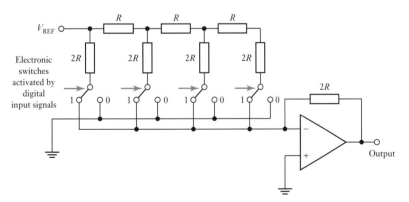

Figure 4.7 ZN558D DAC.

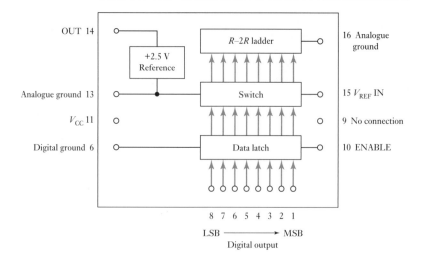

over a wide range of values, only two values being required. The output voltage is generated by switching sections of the ladder to either the reference voltage or 0 V according to whether there is a 1 or 0 in the digital input.

Figure 4.7 shows details of the GEC Plessey ZN558D 8-bit latched input DAC using a R-$2R$ ladder network. After the conversion is complete, the 8-bit result is placed in an internal latch until the next conversion is complete. Data is held in the latch when ENABLE is high, the latch being said to be transparent when ENABLE is low. A **latch** is just a device to retain the output until a new one replaces it. When a DAC has a latch it may be interfaced directly to the data bus of a microprocessor and treated by it as just an address to send data. A DAC without a latch would be connected via a peripheral interface adapter (PIA), such a device providing latching (see Section 13.4). Figure 4.8 shows how the ZN558D might be used with a microprocessor when the output is required to be a voltage which varies between zero and the reference voltage, this being termed **unipolar operation**. With $V_{\text{ref in}} = 2.5$ V, the output range is $+5$ V when $R_1 = 8$ kΩ and $R_2 = 8$ kΩ and the range is $+10$ V when $R_1 = 16$ kΩ and $R_2 = 5.33$ kΩ.

Figure 4.8 Unipolar operation.

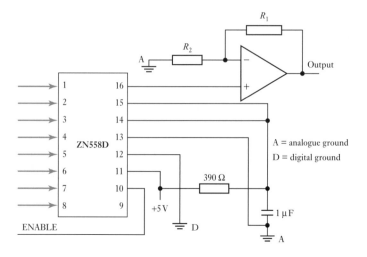

The specifications of DACs include such terms as the following.

1 The *full-scale output*, i.e. the output when the input word is all ones. For the ZN558D this is typically 2.550 V.
2 The *resolution*, 8-bit DACS generally being suitable for most microprocessor control systems. The ZN558D is 8-bit.
3 The *settling time*, this being the time taken by the DAC to reach within $\frac{1}{2}$ LSB of its new voltage after a binary change. This is 800 ns for the ZN558D.
4 The *linearity*, this being the maximum deviation from the straight line through zero and the full range of the output. This is a maximum of ± 0.5 LSB for the ZN558D.

4.3.2 ADCs

The input to an ADC is an analogue signal and the output is a binary word that represents the level of the input signal. There are a number of forms of ADC, the commonest being successive approximations, ramp, dual ramp and flash.

Successive approximations is probably the most commonly used method. Figure 4.9 illustrates the subsystems involved. A voltage is generated by a clock emitting a regular sequence of pulses which are counted, in a binary manner, and the resulting binary word converted into an analogue voltage by a DAC. This voltage rises in steps and is compared with the analogue input voltage from the sensor. When the clock-generated voltage passes the input analogue voltage, the pulses from the clock are stopped from being counted by a gate being closed. The output from the counter at that time is then a digital representation of the analogue voltage. While the comparison could be accomplished by starting the count at 1, the LSB, and then proceeding bit by bit upwards, a faster method is by successive approximations. This involves selecting the MSB that is less than the analogue value, then adding successive lesser bits for which the total does not exceed the analogue value. For example, we might start the comparison with 1000. If this is too large we try 0100. If this is too small we then try 0110. If this is too large we try 0101. Because each of the bits in the word is tried in sequence, with an n-bit word it only takes n steps to make the comparison. Thus if the clock has a frequency f, the time between pulses is $1/f$. Hence the time taken to generate the word, i.e. the conversion time, is n/f.

Figure 4.10 shows the typical form of an 8-bit ADC (GEC Plessey ZN439) designed for use with microprocessors and using the successive approximations method. Figure 4.11 shows how it can be connected so that it is controlled by a microprocessor and sends its digital output to the

Figure 4.9 Successive approximations ADC.

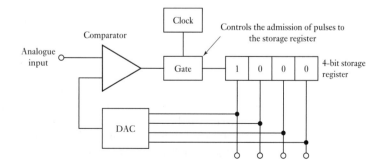

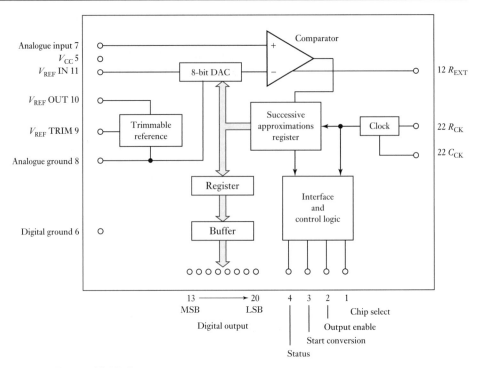

Figure 4.10   ZN439 ADC.

Figure 4.11   ZN439 connected to a microprocessor.

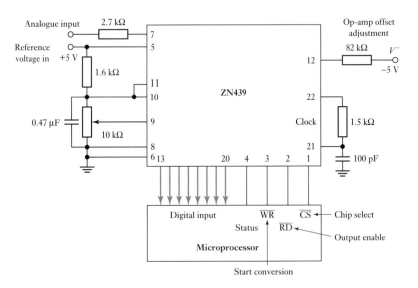

microprocessor. All the active circuitry, including the clock, is contained on a single chip. The ADC is first selected by taking the chip select pin low. When the start conversion pin receives a negative-going pulse the conversion starts. At the end of the conversion the status pin goes low. The digital output is sent to an internal buffer where it is held until read as a result of the output enable pin being taken low.

Figure 4.12 Ramp ADC.

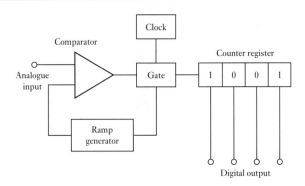

The **ramp** form of ADC involves an analogue voltage which is increased at a constant rate, a so-called ramp voltage, and applied to a comparator where it is compared with the analogue voltage from the sensor. The time taken for the ramp voltage to increase to the value of the sensor voltage will depend on the size of the sampled analogue voltage. When the ramp voltage starts, a gate is opened which starts a binary counter counting the regular pulses from a clock. When the two voltages are equal, the gate closes and the word indicated by the counter is the digital representation of the sampled analogue voltage. Figure 4.12 indicates the subsystems involved in the ramp form of ADC.

The **dual ramp converter** is more common than the single ramp. Figure 4.13 shows the basic circuit. The analogue voltage is applied to an integrator which drives a comparator. The output from the comparator goes high as soon as the integrator output is more than a few millivolts. When the comparator output is high, an AND gate passes pulses to a binary counter. The counter counts pulses until it overflows. The counter then resets to zero, sends a signal to a switch which disconnects the unknown voltage and connects a reference voltage, and starts counting again. The polarity of the reference voltage is opposite to that of the input voltage. The integrator voltage then decreases at a rate proportional to the reference voltage. When the integrator output reaches zero, the comparator goes low, bringing the AND gate low and so switching the clock off. The count is then a measure of the analogue input voltage. Dual ramp ADCs have excellent noise rejection

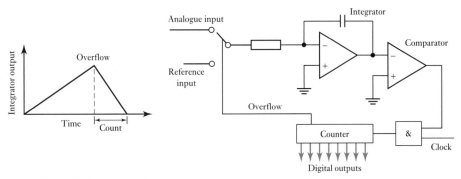

Figure 4.13 Dual ramp ADC.

Figure 4.14   Flash ADC.

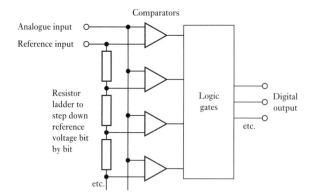

because the integral action averages out random negative and positive contributions over the sampling period. They are, however, very slow.

The **flash ADC** is very fast. For an *n*-bit converter, $2^n - 1$ separate voltage comparators are used in parallel, with each having the analogue input voltage as one input (Figure 4.14). A reference voltage is applied to a ladder of resistors so that the voltage applied as the other input to each comparator is 1 bit larger in size than the voltage applied to the previous comparator in the ladder. Thus when the analogue voltage is applied to the ADC, all those comparators for which the analogue voltage is greater than the reference voltage of a comparator will give a high output and those for which it is less will be low. The resulting outputs are fed in parallel to a logic gate system which translates them into a digital word.

In considering the specifications of ADCs the following terms will be encountered.

1   *Conversion time*, i.e. the time required to complete a conversion of the input signal. It establishes the upper signal frequency that can be sampled without aliasing; the maximum frequency is $1/(2 \times \text{conversion time})$.
2   *Resolution*, this being the full-scale signal divided by 2^n, where *n* is the number of bits. It is often just specified by a statement of the number of bits.
3   *Linearity error*, this being the deviation from a straight line drawn through zero and full-scale. It is a maximum of $\pm\frac{1}{2}$ LSB.

Table 4.1 shows some specification details of commonly used ADCs.

Table 4.1   ADCs.

ADC	Type	Resolution (bits)	Conversion time (ns)	Linearity error (LSB)
ZN439	SA	8	5000	$\pm 1/2$
ZN448E	SA	8	9000	$\pm 1/2$
ADS7806	SA	12	20 000	$\pm 1/2$
ADS7078C	SA	16	20 000	$\pm 1/2$
ADC302	F	8	20	$\pm 1/2$

SA = successive approximations, F = flash.

4.3.3 Sample and hold amplifiers

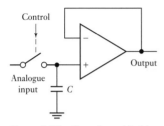

Figure 4.15 Sample and hold.

It takes a finite time for an ADC to convert an analogue signal to digital and problems can arise if the analogue signal changes during the conversion time. To overcome this, a sample and hold system is used to sample the analogue signal and hold it while the conversion takes place.

The basic circuit (Figure 4.15) consists of an electronic switch to take the sample, with a capacitor for the hold and an operational amplifier voltage follower. The electronic switch is controlled so that the sample is taken at the instant dictated by the control input. When the switch closes, the input voltage is applied across the capacitor and the output voltage becomes the same as the input voltage. If the input voltage changes while the switch is closed, the voltage across the capacitor and the output voltage change accordingly. When the switch opens, the capacitor retains its charge and the output voltage remains equal to the input voltage at the instant the switch was opened. The voltage is thus held until such time as the switch closes again. The time required for the capacitor to charge to a new sample of the input analogue voltage is called the **acquisition time** and depends on the value of the capacitance and the circuit resistance when the switch is on. Typical values are of the order of 4 µs.

4.4   Multiplexers

A **multiplexer** is a circuit that is able to have inputs of data from a number of sources and then, by selecting an input channel, give an output from just one of them. In applications where there is a need for measurements to be made at a number of different locations, rather than use a separate ADC and microprocessor for each measurement, a multiplexer can be used to select each input in turn and switch it through a single ADC and microprocessor (Figure 4.16). The multiplexer is essentially an electronic switching device which enables each of the inputs to be sampled in turn.

Figure 4.16 Multiplexer.

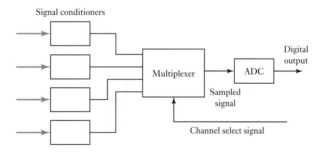

As an illustration of the types of analogue multiplexers available, the DG508ACJ has eight input channels with each channel having a 3-bit binary address for selection purposes. The transition time between taking samples is 0.6 µs.

4.4.1 Digital multiplexer

Figure 4.17 shows the basic principle of a multiplexer which can be used to select digital data inputs; for simplicity only a two-input channel system is

Figure 4.17   Two-channel multiplexer.

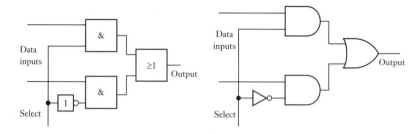

shown. The logic level applied to the select input determines which AND gate is enabled so that its data input passes through the OR gate to the output (see Chapter 5 for a discussion of such gates). A number of forms of multiplexers are available in integrated packages. The 151 types enable one line from eight to be selected, the 153 types one line from four inputs which are supplied as data on two lines each, the 157 types one line from two inputs which are supplied as data on four lines.

4.4.2   Time division multiplexing

Often there is a need for a number of peripheral devices to share the same input/output lines from a microprocessor. So that each peripheral can be supplied with different data it is necessary to allocate each a particular time slot during which data is transmitted. This is termed **time division multiplexing**. Figure 4.18 illustrates how this can be used to drive two display devices. In Figure 4.18(a) the system is not time multiplexed, in (b) it is.

Figure 4.18   Time division multiplexing.

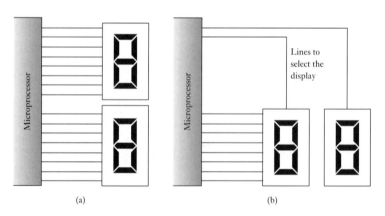

(a)   (b)

4.5   Data acquisition

The term **data acquisition**, or **DAQ**, is used for the process of taking data from sensors and inputting that data into a computer for processing. The sensors are connected, generally via some signal conditioning, to a data acquisition board which is plugged into the back of a computer (Figure 4.19(a)). The DAQ board is a printed circuit board that, for analogue inputs, basically provides a multiplexer, amplification, analogue-to-digital conversion, registers and control circuitry so that sampled digital signals are applied to the computer system. Figure 4.19(b) shows the basic elements of such a board.

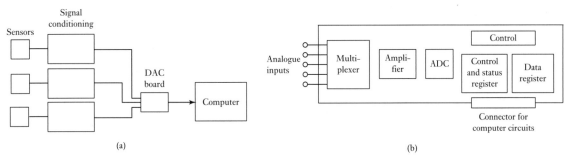

Figure 4.19 DAQ system.

Computer software is used to control the acquisition of data via the DAQ board. When the program requires an input from a particular sensor, it activates the board by sending a control word to the control and status register. Such a word indicates the type of operation that the board has to carry out. As a consequence the board switches the multiplexer to the appropriate input channel. The input from the sensor connected to that input channel is then passed via an amplifier to the ADC. After conversion the resulting digital signal is passed to the data register and the word in the control and status register changes to indicate that the signal has arrived. Following that signal, the computer then issues a signal for the data to be read and taken into the computer for processing. This signal is necessary to ensure that the computer does not wait doing nothing while the board carries out its acquisition of data, but uses this to signal to the computer when the acquisition is complete and then the computer can interrupt any program it is implementing, read the data from the DAQ and then continue with its program. A faster system does not involve the computer in the transfer of the data into memory but transfers the acquired data directly from the board to memory without involving the computer, this being termed **direct memory address** (DMA).

The specifications for a DAQ board include the sampling rate for analogue inputs, which might be 100 kS/s (100 000 samples per second). The Nyquist criteria for sampling indicate that the maximum frequency of analogue signal that can be sampled with such a board is 50 kHz, the sample rate having to be twice the maximum frequency component. In addition to the above basic functions of a DAQ board, it may also supply analogue outputs, timers and counters which can be used to provide triggers for the sensor system.

As an example of a low-cost multifunction board for use with an IBM computer, Figure 4.20 shows the basic structure of the National Instruments DAQ board PC–LPM–16. This board has 16 analogue input channels, a sampling rate of 50 kS/s, an 8-bit digital input and an 8-bit digital output, and a counter/timer which can give an output. Channels can be scanned in sequence, taking one reading from each channel in turn, or there can be continuous scanning of a single channel.

4.5.1 Data accuracy

An advantage of digital signal processing is that two voltage ranges are used rather than two exact voltage levels to distinguish between the two binary states for each bit. Thus data accuracy is less affected by noise, drift,

Figure 4.20   PC-LPM-16 DAQ
board.

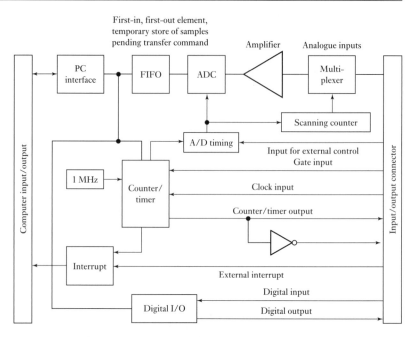

component tolerances and other factors causing fluctuations in voltages
which would be critical for transmission as analogue voltages. For example,
in a 5 V system, the difference between the two binary states is typically a
minimum of 3 V. So two signals could be 0 and 5 V or 1 V and 4 V and still
be distinguished as 0 and 1.

4.5.2   Parity method for error detection

The movement of digital data from one location to another can result in
transmission errors, the receiver not receiving the same signal as transmitted
by the transmitter as a result of electrical noise in the transmission process.
Sometimes a noise pulse may be large enough at some point to alter the logic
level of the signal. For example, the sequence 1001 may be transmitted and be
received as though 1101. In order to detect such errors a **parity bit** is often used.
A parity bit is an extra 0 or 1 bit attached to a code group at transmission. In **even
parity** the value of the bit is chosen so that the total number of ones in the code
group, including the parity bit, is an even number. For example, in transmitting
1001 the parity bit used would be 0 to give 01001 and so an even number of
ones. In transmitting 1101 the parity bit used would be 1 to give 11101 and so
an even number of ones. With **odd parity** the parity bit is chosen so that the
total number of ones, including the parity bit, is odd. Thus if at the receiver the
number of ones in a code group does not give the required parity, the receiver
will know that there is an error and can request the code group be retransmitted.

An extension of the parity check is the **sum check** in which blocks of code
may be checked by sending a series of bits representing their binary sum.
Parity and sum checks can only detect single errors in blocks of code; double
errors go undetected. Also the error is not located so that correction by the
receiver can be made. Multiple error detection techniques and methods to
pinpoint errors have been devised.

4.6   **Digital signal processing**

The term **digital signal processing or discrete-time signal processing** is used for the processing applied to a signal by a microprocessor. Digital signals are discrete-time signals in that they are not continuous functions of time but exist at only discrete times. Whereas signal conditioning of analogue signals requires components such as amplifiers and filter circuits, digital signal conditioning can be carried out by a program applied to a microprocessor, i.e. processing the signal. To change the characteristics of a filter used with analogue signals it is necessary to change hardware components, whereas to change the characteristics of a digital filter all that is necessary is to change the software, i.e. the program of instructions given to a microprocessor.

With a digital signal processing system there is an input of a word representing the size of a pulse and an output of another word. The output pulse at a particular instant is computed by the system as a result of processing the present input pulse, together with previous pulse inputs and possibly previous system outputs.

For example, the program used by the microprocessor might read the value of the present input and add to it the previous output value to give the new output. If we consider the present input to be the kth pulse in the input sequence of pulses we can represent this pulse as $x[k]$. The kth output of a sequence of pulses can be represented by $y[k]$. The previous output, i.e. the $(k-1)$th pulse, can be represented by $y[k-1]$. Thus we can describe the program which gives an output obtained by adding to the value of the present input the previous output by

$$y[k] = x[k] + y[k-1]$$

Such an equation is called a **difference equation**. It gives the relationship between the output and input for a discrete-time system and is comparable with a differential equation which is used to describe the relationship between the output and input for a system having inputs and outputs which vary continuously with time.

For the above difference equation, suppose we have an input of a sampled sine wave signal which gives a sequence of pulses of

$$0.5, 1.0, 0.5, -0.5, -1.0, -0.5, 0.5, 1.0, \ldots$$

The $k=1$ input pulse has a size of 0.5. If we assume that previously the output was zero then $y[k-1] = 0$ and so $y[1] = 0.5 + 0 = 0.5$. The $k=2$ input pulse has a size of 1.0 and so $y[2] = x[2] + y[2-1] = 1.0 + 0.5 = 1.5$. The $k=3$ input pulse has a size of 0.5 and so $y[3] = x[3] + y[3-1] = 0.5 + 1.5 = 2.0$. The $k=4$ input pulse has a size of -0.5 and so $y[4] = x[4] + y[4-1] = -0.5 + 2.0 = 1.5$. The $k=5$ input pulse has a size of -1.0 and so $y[5] = x[5] + y[5-1] = -1.0 + 1.5 = 0.5$. The output is thus the pulses

$$0.5, 1.5, 2.0, 1.5, 0.5, \ldots$$

We can continue in this way to obtain the output for all the pulses.

As another example of a difference equation we might have

$$y[k] = x[k] + ay[k-1] - by[k-2]$$

The output is the value of the current input plus a times the previous output and minus b times the last but one output. If we have $a = 1$ and $b = 0.5$ and consider the input to be the sampled sine wave signal considered above, then the output now becomes

$$0.5, 1.5, 1.75, 0.5, -1.37, \ldots$$

We can have a difference equation which produces an output which is similar to that which would have been obtained by integrating a continuous-time signal. Integration of a continuous-time signal between two times can be considered to be the area under the continuous-time function between those times. Thus if we consider two discrete-time signals $x[k]$ and $x[k-1]$ occurring with a time interval of T between them (Figure 4.21), the change in area is $\frac{1}{2}T(x[k] + x[k-1])$. Thus if the output is to be the sum of the previous area and this change in area, the difference equation is

$$y[k] = y[k-1] + \tfrac{1}{2}T(x[k] + x[k-1])$$

This is known as *Tustin's approximation* for integration.

Differentiation can be approximated by determining the rate at which the input changes. Thus when the input changes from $x[k-1]$ to $x[k]$ in time T the output is

$$y[k] = (x[k] - x[k-1])/T$$

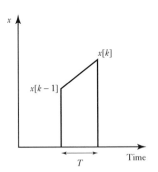

Figure 4.21 Integration.

Summary

Analogue-to-digital conversion involves converting analogue signals into binary words. A clock supplies a regular time signal to the analogue-to-digital converter (ADC) and it samples the analogue signal at each clock pulse. A sample and hold unit then holds each sampled value until the next pulse occurs. Forms of ADC are successive approximations, ramp, dual ramp and flash.

Digital-to-analogue conversion involves converting a binary word into an analogue signal. Forms of digital-to-analogue converters (DACs) are weighted-resistor and R–$2R$ ladder.

A **multiplexer** is a circuit that is able to have inputs of data from a number of sources and then, by selecting an input channel, give an output for just one of them.

The term **data acquisition**, or DAQ, is used for the process of taking data from sensors and inputting that data into a computer for processing.

The term **digital signal processing** or **discrete-time signal processing** is used for the processing applied to a signal by a microprocessor.

Problems

4.1   What is the resolution of an ADC with a word length of 12 bits and an analogue signal input range of 100 V?

4.2   A sensor gives a maximum analogue output of 5 V. What word length is required for an ADC if there is to be a resolution of 10 mV?

4.3   An R–$2R$ DAC ladder of resistors has its output fed through an inverting operational amplifier with a feedback resistance of $2R$. If the DAC is 3 bit and the reference voltage is 5 V, determine the resolution of the converter.

4.4   For a binary weighted-resistor DAC how should the values of the input resistances be weighted for a 4-bit DAC?

4.5   What is the conversion time for a 12-bit ADC with a clock frequency of 1 MHz?

4.6   In monitoring the inputs from a number of thermocouples the following sequence of modules is used for each thermocouple in its interface with a microprocessor

Protection, cold junction compensation, amplification, linearisation, sample and hold, analogue-to-digital converter, buffer, multiplexer.

Explain the function of each of the modules.

4.7   Suggest the modules that might be needed to interface the output of a microprocessor with an actuator.

4.8   For the 4-bit weighted-resistor DAC shown in Figure 4.5, determine the output from the resistors to the amplifier for inputs of 0001, 0010, 0100 and 1000 if the inputs are 0 V for a logic 0 and 5 V for a logic 1.

4.9   If the smallest resistor in a 16-bit weighted-resistor DAC is R, how big would the largest resistor need to be?

4.10   A 10-bit ramp ADC has a full-scale input of 10 V. How long will it take to convert such a full-scale input if the clock period is 15 μs?

4.11   For a 12-bit ADC with full-scale input, how much faster will a successive approximations ADC be than a ramp ADC?

Chapter five Digital logic

Objectives

The objectives of this chapter are that, after studying it, the reader should be able to:
- Recognise the symbols used for the logic gates AND, OR, NOT, NAND, NOR and XOR, and use such gates in applications, recognising the significance of logic families.
- Explain how SR, JK and D flip-flops can be used in control systems.
- Explain the operation of decoders and the 555 timer.

5.1   Digital logic

Many control systems are concerned with setting events in motion or stopping them when certain conditions are met. For example, with the domestic washing machine, the heater is only switched on when there is water in the drum and it is to the prescribed level. Such control involves *digital* signals where there are only two possible signal levels. Digital circuitry is the basis of digital computers and microprocessor controlled systems.

With **digital control** we might, for example, have the water input to the domestic washing machine switched on if we have both the door to the machine closed and a particular time in the operating cycle has been reached. There are two input signals which are either yes or no signals and an output signal which is a yes or no signal. The controller is here programmed to only give a yes output if both the input signals are yes, i.e. if input A and input B are both 1 then there is an output of 1. Such an operation is said to be controlled by a **logic gate**, in this example an AND gate. There are many machines and processes which are controlled in this way. The term **combinational logic** is used for the combining of two or more basic logic gates to form a required function. For example, a requirement might be that a buzzer sounds in a car if the key is in the ignition and a door is opened or if the headlights are on and the driver's door is opened. Combinational logic depends only on the values of the inputs at a particular instant of time.

In addition to a discussion of combinational logic, this chapter also includes a discussion of **sequential logic**. Such digital circuitry is used to exercise control in a specific sequence dictated by a control clock or enable–disable control signals. These are combinational logic circuits with memory. Thus the timing or sequencing history of the input signals plays a part in determining the output.

5.2   Logic gates

Logic gates are the basic building blocks for digital electronic circuits.

5.2.1   AND gate

Suppose we have a gate giving a high output only when both input A and input B are high; for all other conditions it gives a low output. This is an AND logic gate. We can visualise the AND gate as an electric circuit involving two switches in series (Figure 5.1(a)). Only when switch A and switch B are closed is there a current. Different sets of standard circuit symbols for logic gates have been used, with the main form being that originated in the United States. An international standard form (IEEE/ANSI), however, has now been developed; this removes the distinctive shape and uses a rectangle with the logic function written inside it. Figure 5.1(b) shows the US form of symbol used for an AND gate and (c) shows the new standardised form, the & symbol indicating AND. Both forms will be used in this book. As illustrated in the figure, we can express the relationship between the inputs and the outputs of an AND gate in the form of an equation, termed a **Boolean equation** (see Appendix C). The Boolean equation for the AND gate is written as

$$A \cdot B = Q$$

Figure 5.1 AND gate:
(a) represented by switches,
(b) US symbols, (c) new
standardised symbols.

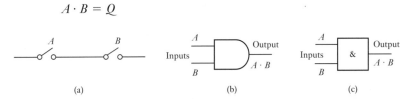

An example of an AND gate is an interlock control system for a machine tool such that if the safety guard is in place and gives a 1 signal and the power is on, giving a 1 signal, then there can be an output, a 1 signal, and the machine operates. Another example is a burglar alarm in which it gives an output, the alarm sounding, when the alarm is switched on and when a door is opened to activate a sensor.

The relationships between inputs to a logic gate and the outputs can be tabulated in a form known as a **truth table**. This specifies the relationships between the inputs and outputs. Thus for an AND gate with inputs A and B and a single output Q, we will have a 1 output when, and only when, $A = 1$ and $B = 1$. All other combinations of A and B will generate a 0 output. We can thus write the truth table as

Inputs		Output
A	B	Q
0	0	0
0	1	0
1	0	0
1	1	1

Consider what happens when we have two digital inputs which are functions of time, as in Figure 5.2. Such a figure is termed an AND gate timing diagram. There will only be an output from the AND gate when each of the inputs is high and thus the output is as shown in the figure.

Figure 5.2 AND gate.

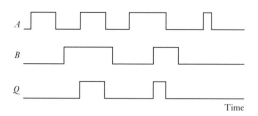

5.2.2 OR gate

An OR gate with inputs A and B gives an output of a 1 when A or B is 1. We can visualise such a gate as an electric circuit involving two switches in parallel (Figure 5.3(a)). When switch A or B is closed, then there is a current. OR gates can also have more than two inputs. The truth table for the gate is

Inputs		Output
A	B	Q
0	0	0
0	1	1
1	0	1
1	1	1

Figure 5.3 OR gate:
(a) representation by switches,
(b) symbols, (c) timing diagram.

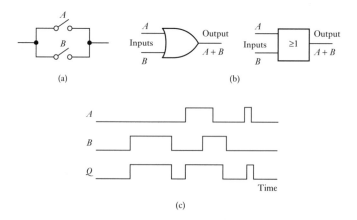

We can write the Boolean equation for an OR gate as

$$A + B = Q$$

The symbols used for an OR gate are shown in Figure 5.3(b); the use of a greater than or equal to 1 sign to depict OR arises from the OR function being true if at least more than one input is true. Figure 5.3(c) shows a timing diagram.

5.2.3 NOT gate

A NOT gate has just one input and one output, giving a 1 output when the input is 0 and a 0 output when the input is 1. The NOT gate gives an output which is the inversion of the input and is called an **inverter**. Figure 5.4(a) shows the symbols used for a NOT gate. The 1 representing NOT actually symbolises logic identity, i.e. no operation, and the inversion is depicted by the circle on the output. Thus if we have a digital input which varies with time, as in Figure 5.4(b), the out variation with time is the inverse.

Figure 5.4 NOT gate.

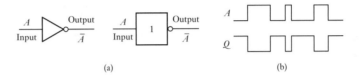

(a)   (b)

The following is the truth table for the NOT gate:

Input A	Output Q
0	1
1	0

The Boolean equation describing the NOT gate is

$$\overline{A} = Q$$

A bar over a symbol is used to indicate that the inverse, or complement, is being taken; thus the bar over the A indicates that the output Q is the inverse value of A.

5.2.4 NAND gate

The NAND gate can be considered as a combination of an AND gate followed by a NOT gate (Figure 5.5(a)). Thus when input A is 1 and input B is 1, there is an output of 0, all other inputs giving an output of 1.

The NAND gate is just the AND gate truth table with the outputs inverted. An alternative way of considering the gate is as an AND gate with a NOT gate applied to invert both the inputs before they reach the AND gate. Figure 5.5(b) shows the symbols used for the NAND gate, being the

Figure 5.5   NAND gate.

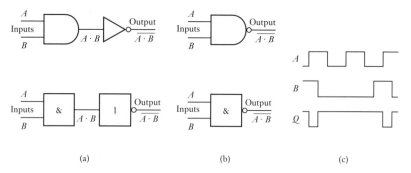

AND symbol followed by the circle to indicate inversion. The following is the truth table:

Inputs		Output
A	B	Q
0	0	1
0	1	1
1	0	1
1	1	0

The Boolean equation describing the NAND gate is

$$\overline{A \cdot B} = Q$$

Figure 5.5(c) shows the output that occurs for a NAND gate when its two inputs are digital signals which vary with time. There is only a low output when both the inputs are high.

5.2.5   NOR gate

The NOR gate can be considered as a combination of an OR gate followed by a NOT gate (Figure 5.6(a)). Thus when input A or input B is 1 there is an output of 0. It is just the OR gate with the outputs inverted. An alternative way of considering the gate is as an OR gate with a NOT gate applied to invert both the inputs before they reach the OR gate. Figure 5.6(b) shows the symbols used for the NOR gate; it is the OR symbol followed by the circle to

Figure 5.6   NOR gate.

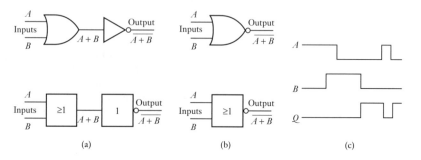

indicate inversion. The Boolean equation for the NOR gate is

$$\overline{A + B} = Q$$

The following is the truth table for the NOR gate and Figure 5.6(c) shows its timing diagram:

Inputs		Output
A	B	Q
0	0	1
0	1	0
1	0	0
1	1	0

5.2.6 XOR gate

The EXCLUSIVE-OR gate (XOR) can be considered to be an OR gate with a NOT gate applied to one of the inputs to invert it before the inputs reach the OR gate (Figure 5.7(a)). Alternatively it can be considered as an AND gate with a NOT gate applied to one of the inputs to invert it before the inputs reach the AND gate. The symbols are shown in Figure 5.7(b); the $=1$ depicts that the output is true if only one input is true. The following is the truth table and Figure 5.7(c) shows a timing diagram:

Figure 5.7 XOR gate.

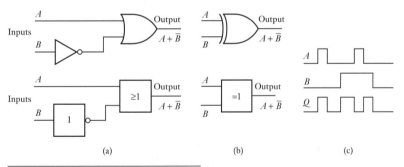

(a) (b) (c)

Inputs		Output
A	B	Q
0	0	0
0	1	1
1	0	1
1	1	0

5.2.7 Combining gates

It might seem that to make logic systems we require a range of gates. However, as the following shows, we can make up all the gates from just one. Consider the combination of three NOR gates shown in Figure 5.8.

Figure 5.8   Three NOR gates.

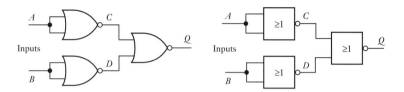

The truth table, with the intermediate and final outputs, is as follows:

A	B	C	D	Q
0	0	1	1	0
0	1	1	0	0
1	0	0	1	0
1	1	0	0	1

The result is the same as an AND gate. If we followed this assembly of gates by a NOT gate then we would obtain a truth table the same as a NAND gate.

A combination of three NAND gates is shown in Figure 5.9. The truth table, with the intermediate and final outputs, is as follows:

A	B	C	D	Q
0	0	1	1	0
0	1	1	0	1
1	0	0	1	1
1	1	0	0	1

Figure 5.9   Three NAND gates.

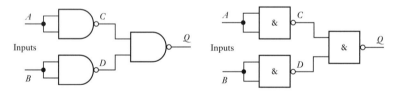

The result is the same as an OR gate. If we followed this assembly of gates by a NOT gate then we would obtain a truth table the same as a NOR gate.

The above two illustrations of gate combinations show how one type of gate, a NOR or a NAND, can be used to substitute for other gates, provided we use more than one gate. Gates can also be combined to make complex gating circuits and sequential circuits.

Logic gates are available as integrated circuits. The different manufacturers have standardised their numbering schemes so that the basic part numbers are the same regardless of the manufacturer. For example, Figure 5.10(a) shows the gate systems available in integrated circuit 7408; it has four two-input AND gates and is supplied in a 14-pin package. Power supply connections are made to pins 7 and 14, these supplying the operating voltage for all the four AND gates. In order to indicate at which end of the package pin 1 starts, a notch is cut between pins 1 and 14. Integrated circuit

Figure 5.10 Integrated circuits: (a) 7408, (b) 7402.

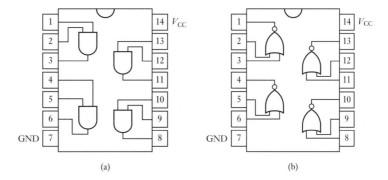

(a)   (b)

7411 has three AND gates with each having three inputs; integrated circuit 7421 has two AND gates with each having four inputs. Figure 5.10(b) shows the gate systems available in integrated circuit 7402. This has four two–input NOR gates in a 14-pin package, power connections being to pins 7 and 14. Integrated circuit 7427 has three gates with each having three inputs; integrated circuit 7425 has two gates with each having four inputs.

For a discussion of how Boolean algebra and techniques such as De Morgan's law and Karnaugh maps can be used to generate the required logic functions from logic gates, see Appendix C.

5.2.8 Logic families and integrated circuits

In order to implement digital logic designs, it is necessary to understand the significance of logic families and their different operating principles. Integrated circuits made with the same technology and electrical characteristics comprise a **logic family**. Commonly encountered families are transistor–transistor logic (TTL), complementary metal-oxide semiconductor (CMOS) and emitter-coupled logic (ECL). The general parameters are outlined below.

1 **Logic level**, i.e. the range of voltage levels that can correspond to the binary 1 and 0 states. For the standard 74XX TTL series, the typical voltage guaranteed to register as binary 0 is between 0 and 0.4 V and for binary 1 between 2.4 V and 5.0 V. For CMOS, the levels depend on the supply voltage V_{DD} used. This can be from +3 V to +15 V and the maximum voltage for logic 1 is $0.3V_{DD}$ while the minimum for logic 1 is $0.7V_{DD}$.

2 **Noise immunity** or **noise margin**, i.e. the circuit's ability to tolerate noise without causing spurious changes in the output voltage. For the standard 74XX TTL series, the noise margin is 0.4 V. Thus 0.4 V is the leeway that can be accepted on the logic 0 and logic 1 inputs and they still register as 0 and 1. For CMOS the noise margin depends on the supply voltage and is $0.3V_{DD}$.

3 **Fan-out**, i.e. the number of gate inputs that can be driven by a standard gate output while maintaining the desired LOW or HIGH levels. This is determined by how much current a gate can supply and how much is needed to drive a gate. For a standard TTL gate, the fan-out is 10, for CMOS it is 50 and for ECL 25. If more gates are connected to the driver gate then it will not supply enough current to drive them.

4 **Current-sourcing** or **current-sinking action**, i.e. how the current flows between the output of one logic gate and the input of another. For one gate driving another, with current-sourcing the driving gate when high supplies a current to the input of the next gate. With current-sinking, the driver gate when low receives a current back to it from the driven gate. TTL gates operate as current sinking.

5 **Propagation delay time**, i.e. how fast a digital circuit responds to a change in the input level. Typically TTL gates have delay times of 2 to 40 ns, this being generally about 5 to 10 times faster than CMOS gates but slower than ECL gates which typically have propagation delays of 2 ns.

6 **Power consumption**, i.e. the amount of power the logic gate will drain from the power supply. TTL consumes about 10 mW per gate while CSMOS draws no power unless it is in the act of switching. ECL consumes about 25 to 60 mW per gate.

The main criteria generally involved in determining which logic family to use are propagation delay and power consumption. The principal advantage of CMOS over TTL is the low power consumption which makes it ideal for battery-operated equipment. It is possible for integrated circuits from different logic families to be connected together, but special interfacing techniques must be used.

The TTL family is widely used, the family being identified as the 74XX series. There are a number of forms. Typically, the standard TTL is 7400 with a power dissipation of 10 mW and a propagation delay of 10 ns. The low-power Schottky TTL (LS) is 74LS00 with a power dissipation of 2 mW and the same propagation delay. The advanced low power Schottky TTL (ALS) is 74ALS00 and is faster and dissipates even lower power, the propagation delay being 4 ns and the power dissipation 1 mW. The fast TTL(F) is 74F00 and has a propagation delay of 3 ns and power dissipation of 6 mW.

The CMOS family includes the 4000 series which had the low power dissipation advantage over the TTL series, but unfortunately was much slower. The 40H00 series was faster but still slower than TTL (LS). The 74C00 series was developed to be pin-compatible with the TTL family, using the same numbering system but beginning with 74C. While it has a power advantage over the TTL family, it is still slower. The 74HC00 and 74HCT00 are faster with speeds comparable with the TTL (LS) series.

5.3   Applications of logic gates

The following are some examples of the uses of logic gates for a number of simple applications.

5.3.1   Parity generators

In the previous chapter the use of parity bits as an error detection method was discussed. A single bit is added to each code block to force the number of ones in the block, including the parity bit, to be an odd number if odd parity is being used or an even number if even parity is being used.

Figure 5.11 shows a logic gate circuit that could be used to determine and add the appropriate parity bit. The system employs XOR gates; with an XOR gate if all the inputs are 0 or all are 1 the output is 0, and if the inputs are not equal the output is a 1. Pairs of bits are checked and an output of 1 given if they are not equal. If odd parity is required the bias bit is 0; if

Figure 5.11 Parity bit generator.

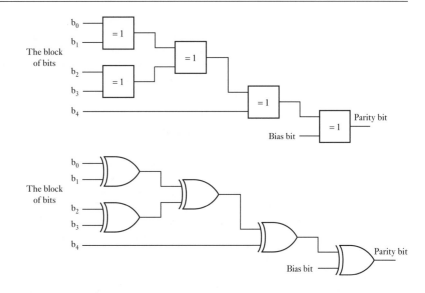

even parity it is 1. The appropriate bias bit can then be added to the signal for transmission. The same circuit can be used to check the parity at the receiver, with the final output being a 1 when there is an error. Such circuits are available as integrated circuits.

5.3.2 Digital comparator

A digital comparator is used to compare two digital words to determine if they are exactly equal. The two words are compared bit by bit and a 1 output given if the words are equal. To compare the equality of two bits, an XOR gate can be used; if the bits are both 0 or both 1 the output is 0, and if they are not equal the output is a 1. To obtain a 1 output when the bits are the same we need to add a NOT gate, this combination of XOR and NOT being termed an XNOR gate. To compare each of the pairs of bits in two words we need an XNOR gate for each pair. If the pairs are made up of the same bits then the output from each XNOR gate is a 1. We can then use an AND gate to give a 1 output when all the XNOR outputs are ones. Figure 5.12 shows the system.

Figure 5.12 Comparator.

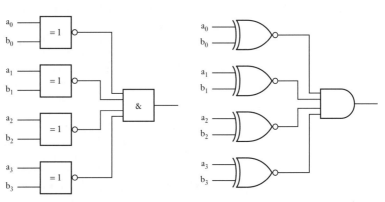

Digital comparators are available as integrated circuits and can generally determine not only if two words are equal but which one is greater than the other. For example, the 7485 4-bit magnitude comparator compares two 4-bit words A and B, giving a 1 output from pin 5 if A is greater than B, a 1 output from pin 6 if A equals B and a 1 output from pin 7 if A is less than B.

5.3.3   Coder

Figure 5.13 shows a simple system by which a controller can send a coded digital signal to a set of traffic lights so that the code determines which light, red, amber or green, will be turned on. To illuminate the red light we might use the transmitted signal $A = 0$, $B = 0$, for the amber light $A = 0$, $B = 1$ and for the green light $A = 1$, $B = 0$. We can switch on the lights using these codes by using three AND gates and two NOT gates.

Figure 5.13   Traffic lights.

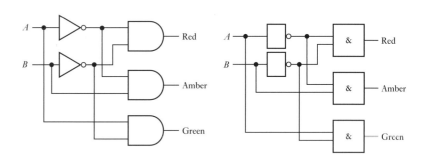

5.3.4   Code converter

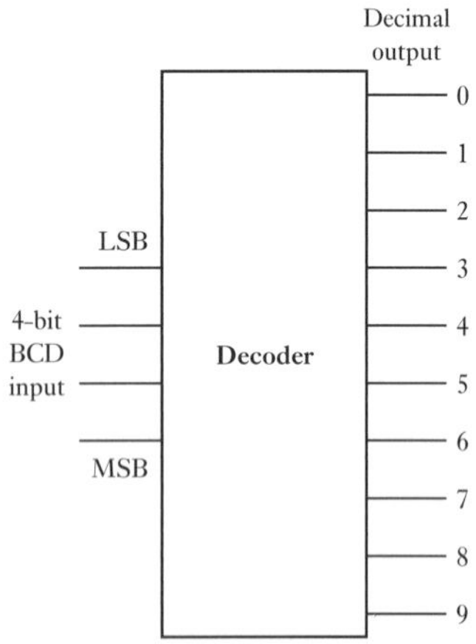

In many applications there is a need to change data from one type of code to another. For example, the output from a microprocessor system might be BCD (binary-coded decimal) and need to be transformed into a suitable code to drive a seven-segment display. The term **data decoding** is used for the process of converting some code group, e.g. BCD, binary, hex, into an individual active output representing that group. A decoder has n binary input lines for the coded input of an n-bit word and gives m output lines such that only one line is activated for one possible combination of inputs, i.e. only one output line gives an output for a particular word input code. For example, a BCD-to-decimal decoder has a 4-bit input code and 10 output lines so that a particular BCD input will give rise to just one of the output lines being activated and so indicate a particular decimal number with each output line corresponding to a decimal number (Figure 5.14).

Thus, in general, a **decoder** is a logic circuit that looks at its inputs, determines which number is there, and activates the one output that corresponds to that number. Decoders are widely used in microprocessor circuits.

Decoders can have the active output high and the inactive ones low or the active output low and the inactive ones high. For active-high output a decoder can be assembled from AND gates, while for active-low output NAND gates can be used. Figure 5.15 shows how a BCD-to-decimal

Figure 5.14   Decoder.

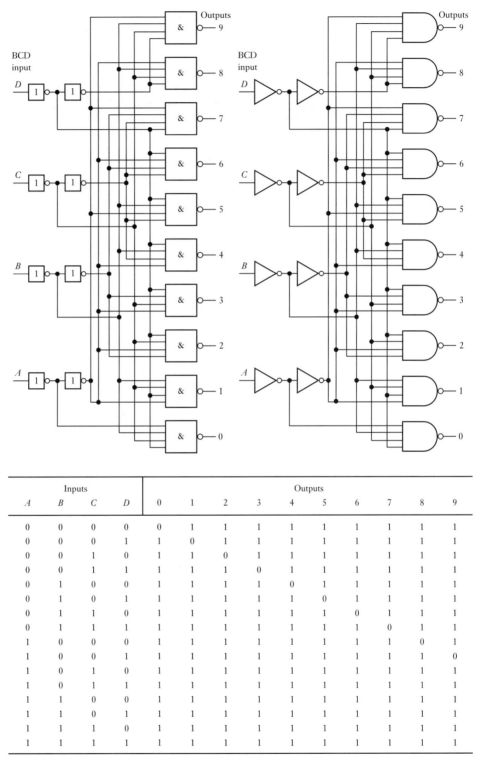

| | Inputs | | | | | | | Outputs | | | | | |
A	B	C	D	0	1	2	3	4	5	6	7	8	9
0	0	0	0	0	1	1	1	1	1	1	1	1	1
0	0	0	1	1	0	1	1	1	1	1	1	1	1
0	0	1	0	1	1	0	1	1	1	1	1	1	1
0	0	1	1	1	1	1	0	1	1	1	1	1	1
0	1	0	0	1	1	1	1	0	1	1	1	1	1
0	1	0	1	1	1	1	1	1	0	1	1	1	1
0	1	1	0	1	1	1	1	1	1	0	1	1	1
0	1	1	1	1	1	1	1	1	1	1	0	1	1
1	0	0	0	1	1	1	1	1	1	1	1	0	1
1	0	0	1	1	1	1	1	1	1	1	1	1	0
1	0	1	0	1	1	1	1	1	1	1	1	1	1
1	0	1	1	1	1	1	1	1	1	1	1	1	1
1	1	0	0	1	1	1	1	1	1	1	1	1	1
1	1	0	1	1	1	1	1	1	1	1	1	1	1
1	1	1	0	1	1	1	1	1	1	1	1	1	1
1	1	1	1	1	1	1	1	1	1	1	1	1	1

Figure 5.15 BCD–to–decimal decoder: 1 = HIGH, 0 = LOW.

decoder for active-low output can be assembled and the resulting truth table. Such a decoder is readily available as an integrated circuit, e.g. 74LS145.

A decoder that is widely used is BCD-to-seven, e.g. 74LS244, for taking a 4-bit BCD input and giving an output to drive the seven segments of a display.

The term **3-line-to-8-line decoder** is used where a decoder has three input lines and eight output lines. It takes a 3-bit binary number and activates the one of the eight outputs corresponding to that number. Figure 5.16 shows how such a decoder can be realised from logic gates and its truth table.

Some decoders have one or more ENABLE inputs that are used to control the operation of the decoder. Thus with the ENABLE line HIGH the decoder will function in its normal way and the inputs will determine

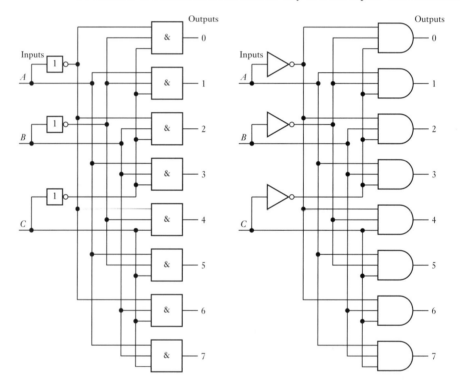

Inputs			Outputs							
C	B	A	0	1	2	3	4	5	6	7
0	0	0	1	0	0	0	0	0	0	0
0	0	1	0	1	0	0	0	0	0	0
0	1	0	0	0	1	0	0	0	0	0
0	1	1	0	0	0	1	0	0	0	0
1	0	0	0	0	0	0	1	0	0	0
1	0	1	0	0	0	0	0	1	0	0
1	1	0	0	0	0	0	0	0	1	0
1	1	1	0	0	0	0	0	0	0	1

Figure 5.16 The 3-line-to-8-line decoder.

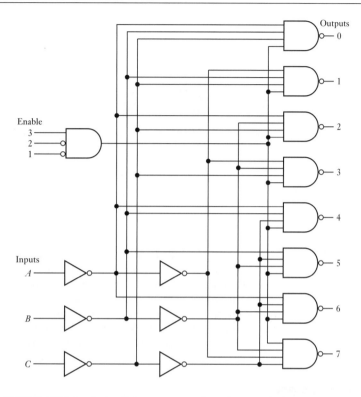

Enable			Inputs			Outputs							
E1	E2	E3	C	B	A	0	1	2	3	4	5	6	7
1	X	X	X	X	X	1	1	1	1	1	1	1	1
X	1	X	X	X	X	1	1	1	1	1	1	1	1
X	X	0	X	X	X	1	1	1	1	1	1	1	1
0	0	1	0	0	0	0	1	1	1	1	1	1	1
0	0	1	0	0	1	1	0	1	1	1	1	1	1
0	0	1	0	1	0	1	1	0	1	1	1	1	1
0	0	1	0	1	1	1	1	1	0	1	1	1	1
0	0	1	1	0	0	1	1	1	1	0	1	1	1
0	0	1	1	0	1	1	1	1	1	1	0	1	1
0	0	1	1	1	0	1	1	1	1	1	1	0	1
0	0	1	1	1	1	1	1	1	1	1	1	1	0

Figure 5.17 The 74LS138: 1 = HIGH, 0 = LOW, X = does not matter.

which output is HIGH; with the ENABLE line LOW all the outputs are held LOW regardless of the inputs. Figure 5.17 shows a commonly used 3-line-to-8-line decoder with this facility, the 74LS138. Note that the outputs are active-LOW rather than the active-HIGH of Figure 5.16, and that the decoder has three ENABLE lines with the requirement for normal functioning that E1 and E3 are LOW and E3 is HIGH. All other variations result in the decoder being disabled and just a HIGH output.

Figure 5.18 illustrates the type of response we can get from a 74LS138 decoder for different inputs.

Figure 5.18　The 74LS138.

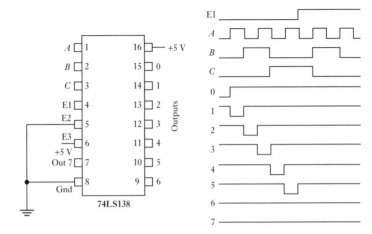

A 74LS138 decoder might be used with a microprocessor with the ENABLE used to switch on the decoder and then depending on the output from three output lines from the microprocessor so one of the eight decoder outputs receives the LOW output with all the others remaining HIGH. Thus, we can consider each output device to have an address, i.e. a unique binary output number, so that when a microprocessor sends an address to the decoder it activates the device which has been allocated that address. The 74LS138 can then be referred to as an address decoder.

5.4　Sequential logic

The logic circuits considered in earlier sections of this chapter are all examples of combinational logic systems. With such systems the output is determined by the combination of the input variables at a particular instant of time. For example, if input A and input B occur at the same time then an AND gate gives an output. The output does not depend on what the inputs previously were. Where a system requires an output which depends on earlier values of the inputs, a **sequential logic** system is required. The main difference between a combinational logic system and a sequential logic system is that the sequential logic system must have some form of memory.

Figure 5.19 shows the basic form of a sequential logic system. The combinational part of the system accepts logic signals from external inputs and from outputs from the memory. The combinational system then operates on these inputs to produce its outputs. The outputs are thus a function of both its external inputs and the information stored in its memory.

Figure 5.19　Sequential logic system.

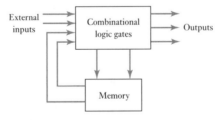

5.4.1　The flip-flop

The **flip-flop** is a basic memory element which is made up of an assembly of logic gates and is a sequential logic device. There are a number of forms of flip-flops. Figure 5.20(a) shows one form, the SR (set–reset) flip-flop, involving NOR gates. If initially we have both outputs 0 and $S = 0$ and $R = 0$, then when we set and have S change from 0 to 1, the output from NOR gate 2 will become 0. This will then result in both the inputs to NOR gate 1 becoming 0 and so its output becomes 1. This feedback acts as an input to NOR gate 2 which then has both its inputs at 1 and results in no further change.

Figure 5.20 SR flip-flop.

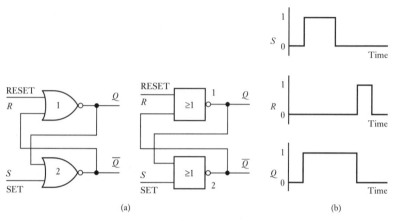

(a)   (b)

Now if S changes from 1 to 0, the output from NOR gate 1 remains at 1 and the output from NOR gate 2 remains at 0. There is no change in the outputs when the input S changes from 1 to 0. It will remain in this state indefinitely if the only changes are to S. It 'remembers' the state it was set to. Figure 5.20(b) illustrates this with a timing diagram in which a rectangular pulse is used as the input S.

If we change R from 0 to 1 when S is 0, the output from NOR gate 1 changes to 0 and hence the output from NOR gate 2 changes to 1. The flip-flop has been reset. A change then of R to 0 will have no effect on these outputs.

Thus when S is set to 1 and R made 0, the output Q will change to 1 if it was previously 0, remaining at 1 it was previously 1. This is the set condition and it will remain in this condition even when S changes to 0. When S is 0 and R is made 1, the output Q is reset to 0 if it was previously 1, remaining at 0 if it was previously 0. This is the rest condition. The output Q that occurs at a particular time will depend on the inputs S and R and also the last value of the output. The following state table illustrates this.

S	R	$Q_t \rightarrow Q_{t+1}$		$\overline{Q}_t \rightarrow \overline{Q}_{t+1}$
0	0	$0 \rightarrow 0$		$1 \rightarrow 1$
0	0	$1 \rightarrow 1$		$0 \rightarrow 0$
0	1	$0 \rightarrow 0$		$1 \rightarrow 1$
0	1	$1 \rightarrow 0$		$0 \rightarrow 0$
1	0	$0 \rightarrow 1$		$1 \rightarrow 0$
1	0	$1 \rightarrow 1$		$0 \rightarrow 0$
1	1	Not allowed		
1	1	Not allowed		

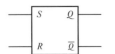

Figure 5.21   SR flip-flop.

Note that if S and R are simultaneously made equal to 1, no stable state can occur and so this input condition is not allowed. Figure 5.21 shows the simplified block symbol used for the SR flip-flop.

As a simple illustration of the use of a flip-flop, consider a simple alarm system in which an alarm is to sound when a beam of light is interrupted and remain sounding even when the beam is no longer interrupted. Figure 5.22 shows a possible system. A phototransistor might be used as the sensor and so connected that when it is illuminated it gives a virtually 0 V input to S, but when the illumination ceases it gives about 5 V input to S. When the light beam is interrupted, S becomes 1 and the output from the flip-flop becomes 1 and the alarm sounds. The output will remain as 1 even when S changes to 0. The alarm can only be stopped if the reset switch is momentarily opened to produce a 5 V input to R.

Figure 5.22   Alarm circuit.

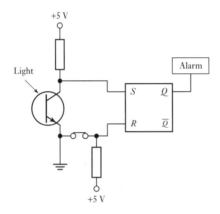

5.4.2   Synchronous systems

It is often necessary for set and reset operations to occur at particular times. With an unclocked or **asynchronous system** the outputs of logic gates can change state at any time when any one or more of the inputs change. With a clocked or **synchronous system** the exact times at which any output can change state are determined by a signal termed the clock signal. This is generally a rectangular pulse train and when the same clock signal is used for all parts of the system, outputs are synchronised. Figure 5.23(a) shows the principle of a **gated SR flip-flop**. The set and clock signal are supplied through an AND gate to the S input of the flip-flop. Thus the set signal only arrives at the flip-flop when both it and the clock signal are 1. Likewise the reset signal is supplied with the clock signal to the R input via another AND gate. As a consequence, setting and resetting can only occur at the time determined by the clock. Figure 5.23(b) shows the timing diagram.

5.4.3   JK flip-flop

For many applications the indeterminate state that occurs with the SR flip-flop when $S = 1$ and $R = 1$ is not acceptable and another form of flip-flop

Figure 5.23 Clocked SR flip-flop.

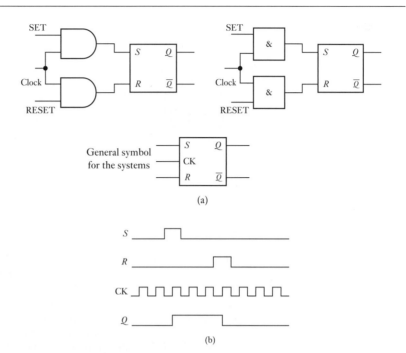

(a)

(b)

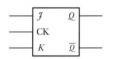

Figure 5.24 JK flip-flop.

is used, the **JK flip-flop** (Figure 5.24). This has become a very widely used flip-flop device.

The following is the truth table for this flip-flop; note that the only changes from the state table for the SR flip-flop are the entries when both inputs are 1:

J	K	$Q_t \rightarrow Q_{t+1}$	$\overline{Q}_t \rightarrow \overline{Q}_{t+1}$
0	0	$0 \rightarrow 0$	$1 \rightarrow 1$
0	0	$1 \rightarrow 1$	$0 \rightarrow 0$
0	1	$0 \rightarrow 0$	$1 \rightarrow 1$
0	1	$1 \rightarrow 0$	$0 \rightarrow 0$
1	0	$0 \rightarrow 1$	$1 \rightarrow 0$
1	0	$1 \rightarrow 1$	$0 \rightarrow 0$
1	1	$0 \rightarrow 1$	$1 \rightarrow 0$
1	1	$1 \rightarrow 0$	$0 \rightarrow 1$

As an illustration of the use of such a flip-flop, consider the requirement for a high output when input A goes high and then some time later B goes high. An AND gate can be used to determine whether two inputs are both high, but its output will be high regardless of which input goes high first. However, if the inputs A and B are used with a JK flip-flop, then A must be high first in order for the output to go high when B subsequently goes high.

5.4.4 D flip-flop

The data or **D flip-flop** is basically a clocked SR flip-flop or a JK flip-flop with the D input being connected directly to the S or J inputs and via a NOT

Figure 5.25 D flip-flop.

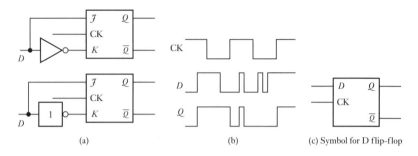

gate to the R or K inputs (Figure 5.25(a)); in the symbol for the D flip-flop this joined R and K input is labelled D. This arrangement means that a 0 or a 1 input will then switch the outputs to follow the input when the clock pulse is 1 (Figure 5.25(b)). A particular use of the D flip-flop is to ensure that the output will only take on the value of the D input at precisely defined times. Figure 5.25(c) shows the symbol used for a D flip-flop.

With the above form of D flip-flop, when the clock or enable input goes high, the output follows the data presented at input D. The flip-flop is said to be transparent. When there is a high-to-low transition at the enable input, output Q is held at the data level just prior to the transition. The data at transition is said to be **latched**. D flip-flops are available as integrated circuits. The 7475 is an example; it contains four transparent D latches.

The 7474 D flip-flop differs from the 7475 in being an edge-triggered device; there are two such flip-flops in the package. With an edge-triggered D flip-flop, transitions in Q only occur at the edge of the input clock pulse and with the 7474 it is the positive edge, i.e. low-to-high transition. Figure 5.26(a) illustrates this. The basic symbol for an edge-triggered D flip-flop differs from that of a D flip-flop by a small triangle being included on the CK input (Figure 5.26(b)). There are also two other inputs called preset and clear. A low on the preset sets the output Q to 1 while a low on clear clears the output, setting Q to 0.

Figure 5.26 (a) Positive edge-triggered, (b) symbol for edge-triggered D flip-flop.

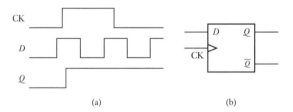

As an illustration of a simple application for such a flip-flop, Figure 5.27 shows a system that could be used to show a green light when the sensor input is low and a red light when it goes high and sound an alarm. The red light is to remain on as long as the sensor input is high but the alarm can be switched off. This might be a monitoring system for the temperature in some process, the sensor and signal conditioning giving a low signal when the temperature is below the safe level and a high signal when it is above. The flip-flop has a high input. When a low input is applied to the CK input and the sensor input is low, the green light is on. When the sensor input changes to high, the green light goes out, the red light on and the alarm

Figure 5.27 Alarm system.

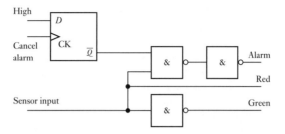

sounds. The alarm can be cancelled by applying a high signal to the CK input, but the red light remains on as long as the sensor input is high. Such a system could be constructed using a 7474 and an integrated circuit or circuits giving three NAND gates.

5.4.5 Registers

A **register** is a set of memory elements and is used to hold information until it is needed. It can be implemented by a set of flip-flops. Each flip-flop stores a binary signal, i.e. a 0 or a 1. Figure 5.28 shows the form a 4-bit register can take when using D flip-flops.

Figure 5.28 Register.

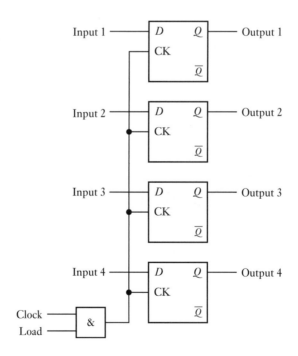

When the load signal is 0, no clock input occurs to the D flip-flops and so no change occurs to the states of the flip-flops. When the load signal is 1, then the inputs can change the states of the flip-flops. As long as the load signal is 0, the flip-flops will hold their old state values.

5.4.6   The 555 timer

The 555 timer chip is very widely used in digital circuits as it can provide a wide variety of timing tasks. It consists of an SR flip–flop with inputs fed by two comparators (Figure 5.29). The comparators each have an input voltage derived from a potentiometric chain of equal size resistors. So comparator A has an non-inverting voltage input of $V_{CC}/3$ and comparator B has an inverting input of $2V_{CC}/3$.

One use of a 555 timer is as a **monostable multivibrator**, this being a circuit which will generate a single pulse of the desired time duration when it receives a trigger signal. Figure 5.30(a) shows how the time is connected for such a use. Initially, the output will be low with the transistor shorting the capacitor and the outputs of both comparators low (Figure 5.30(b)).

Figure 5.29   The 555 timer.

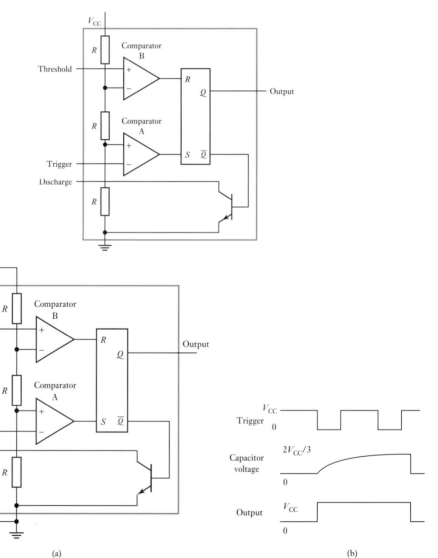

(a)   (b)

Figure 5.30   Monostable multivibrator.

When the trigger pulse goes below $V_{CC}/3$, the trigger comparator goes high and sets the flip-flop. The output is then high and the transistor cuts off and the capacitor begins to charge. When the capacitor reaches $2V_{CC}/3$, the threshold comparator resets the flip-flop and thus resets the output to low and discharges the capacitor. If the trigger is pulsed while the output is high it has no effect. The length of the pulse is thus the time taken for the capacitor to charge up to $2V_{CC}/3$ and this depends on its time constant, i.e. its value of $R_t C$, and is given by the normal relationship for the charging up of a capacitor through a resistance as $1.1R_t C$. As an illustration, consider the situation where a burglar alarm is to sound if a door is opened and the rightful householder does not enter the requisite number on a keypad within 30 s. If the circuit of Figure 5.30 is used with a capacitor of 1 μF then R_t would need to have a value of $30/(1.1 \times 1 \times 10^{-6}) = 27.3$ MΩ.

Summary

With **combinational logic systems,** the output is determined by the combination of the input variables at a particular instant of time. The output does not depend on what the inputs previously were. Where a system requires an output which depends on earlier values of the inputs, a **sequential logic system** is required. The main difference between a combinational logic system and a sequential logic system is that the sequential logic system must have some form of memory.

Commonly encountered logic families are **transistor–transistor logic (TTL), complementary metal-oxide semiconductor (CMOS)** and **emitter-coupled logic (ECL),** being distinguished by their logic levels, noise immunity, fan-out, current-sourcing or current-sinking action, propagation delay time and power dissipation.

A **decoder** is a logic circuit that looks at its inputs, determines which number is there, and activates the one output that corresponds to that number.

The **flip-flop** is a basic memory element which is made up of an assembly of logic gates and is a sequential logic device.

A **register** is a set of memory elements and is used to hold information until it is needed.

The **555 timer** chip consists of an SR flip-flop with inputs fed by two comparators.

Problems

5.1 Explain what logic gates might be used to control the following situations.

(a) The issue of tickets at an automatic ticket machine at a railway station.

(b) A safety lock system for the operation of a machine tool.

(c) A boiler shut-down switch when the temperature reaches, say, 60°C and the circulating pump is off.

(d) A signal to start a lift moving when the lift door is closed and a button has been pressed to select the floor.

5.2   For the time signals shown as A and B in Figure 5.31, which will be the output signal if A and B are inputs to (a) an AND gate, (b) an OR gate?

Figure 5.31 Problem 5.2.

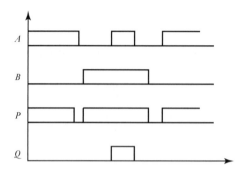

5.3   A clock signal as a continuous sequence of pulses is applied to a logic gate and is to be outputted only when an enable signal is also applied to the gate. What logic gate can be used?

5.4   Input A is applied directly to a two–input AND gate. Input B is applied to a NOT gate and then to the AND gate. What condition of inputs A and B will result in a 1 output from the AND gate?

5.5   Figure 5.32(a) shows the input signals A and B applied to the gate system shown in Figure 5.32(b). Draw the resulting output waveforms P and Q.

Figure 5.32 Problem 5.5.

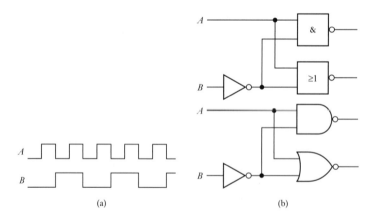

5.6   Figure 5.33 shows the timing diagram for the S and R inputs for an SR flip-flop. Complete the diagram by adding the Q output.

Figure 5.33 Problem 5.6.

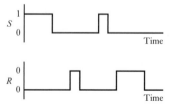

5.7   Explain how the arrangement of gates shown in Figure 5.34 gives an SR flip-flop.

Figure 5.34 Problem 5.7.

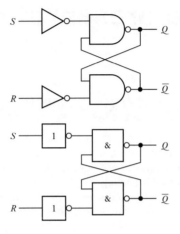

Chapter six Data presentation systems

Objectives

The objectives of this chapter are that, after studying it, the reader should be able to:
- Explain the problem of loading in measurement systems.
- Identify commonly used data presentation elements and describe their characteristics.
- Explain the principles of magnetic and optical recording.
- Explain the principles of displays, in particular the LED seven-segment and dot matrix displays.
- Describe the basic elements of data acquisition systems and virtual instruments.

6.1 Displays

This chapter is about how data can be displayed, e.g. as digits on an light-emitting diode (LED) display or as a display on a computer screen, and stored, e.g. on a computer hard disk or a CD.

Measurement systems consist of three elements: sensor, signal conditioner and display or data presentation element (see Section 1.4). There are a very wide range of elements that can be used for the presentation of data. Traditionally they have been classified into two groups: indicators and recorders. **Indicators** give an instant visual indication of the sensed variable, while **recorders** record the output signal over a period of time and give automatically a permanent record.

This chapter can also be considered as the completion of the group of chapters concerned with measurement systems, i.e. sensors, signal conditioning and now display, and so the chapter is used to bring the items together in a consideration of examples of complete measurement systems.

6.1.1 Loading

A general point that has to be taken account of when putting together any measurement system is **loading**, i.e. the effect of connecting a load across the output terminals of any element of a measurement system.

Connecting an ammeter into a circuit to make a measurement of the current changes the resistance of the circuit and so changes the current. The act of attempting to make such a measurement has modified the current that was being measured. When a voltmeter is connected across a resistor then we effectively have put two resistances in parallel, and if the resistance of the voltmeter is not considerably higher than that of the resistor the current through the resistor is markedly changed and so the voltage being measured

Figure 6.1 Measurement system loading.

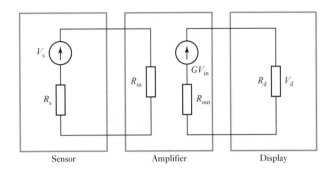

Sensor   Amplifier   Display

is changed. The act of attempting to make the measurement has modified the voltage that was being measured. Such acts are termed loading.

Loading can also occur within a measurement system when the connection of one element to another modifies the characteristics of the preceding element. Consider, for example, a measurement system consisting of a sensor, an amplifier and a display element (Figure 6.1). The sensor has an open-circuit output voltage of V_s and a resistance R_s. The amplifier has an input resistance of R_{in}. This is thus the load across the sensor. Hence the input voltage from the sensor is divided so that the potential difference across this load, and thus the input voltage V_{in} to the amplifier, is

$$V_{in} = \frac{V_s R_{in}}{R_s + R_{in}}$$

If the amplifier has a voltage gain of G then the open-circuit voltage output from it will be GV_{in}. If the amplifier has an output resistance of R_{out} then the output voltage from the amplifier is divided so that the potential difference V_d across the display element, resistance R_d, is

$$V_d = \frac{GV_{in}R_d}{R_{out} + R_d} = \frac{GV_s R_{in} R_d}{(R_{out} + R_d)(R_s + R_{in})}$$

$$= \frac{GV_s}{\left(\dfrac{R_{out}}{R_d} + 1\right)\left(\dfrac{R_s}{R_{in}} + 1\right)}$$

Thus if loading effects are to be negligible we require $R_{out} \gg R_d$ and $R_s \gg R_{in}$.

6.2 Data presentation elements

This section is a brief overview of commonly used examples of data presentation elements.

6.2.1 Analogue and digital meters

The **moving-coil meter** is an analogue indicator with a pointer moving across a scale. The basic instrument movement is a d.c. microammeter with shunts, multipliers and rectifiers being used to convert it to other ranges of direct current and measurement of alternating current, direct voltage and alternating voltage. With alternating current and voltages, the instrument is restricted to between about 50 Hz and 10 kHz. The accuracy of such a meter depends on a number of factors, among which are temperature, the presence nearby of magnetic fields or ferrous materials, the way the meter is mounted, bearing friction, inaccuracies in scale marking during manufacture, etc. In addition there are errors involved in reading the meter, e.g. parallax errors when the position of the pointer against the scale is read from an angle other than directly at right angles to the scale and errors arising from estimating the position of the pointer between scale markings. The overall accuracy is generally of the order of ± 0.1 to $\pm 5\%$. The time taken for a moving-coil meter to reach a steady deflection is typically in the region of a few seconds. The low resistance of the meter can present loading problems.

The **digital voltmeter** gives its reading in the form of a sequence of digits. Such a form of display eliminates parallax and interpolation errors and can give accuracies as high as $\pm 0.005\%$. The digital voltmeter is essentially just a sample and hold unit feeding an analogue-to-digital converter (ADC) with its output counted by a counter (Figure 6.2). It has a high resistance, of the order of 10 MΩ, and so loading effects are less likely than with the moving-coil meter with its lower resistance. Thus, if a digital voltmeter specification includes the statement 'sample rate approximately 5 readings per second' then this means that every 0.2 s the input voltage is sampled. It is the time taken for the instrument to process the signal and give a reading. Thus, if the input voltage is changing at a rate which results in significant changes during 0.2 s then the voltmeter reading can be in error. A low-cost digital voltmeter has typically a sample rate of 3 per second and an input impedance of 100 MΩ.

Figure 6.2 Principle of digital voltmeter.

6.2.2 Analogue chart recorders

Analogue chart recorders have data recorded on paper by fibre-tipped ink pens, by the impact of a pointer pressing a carbon ribbon against the paper, by the use of thermally sensitive paper which changes colour when a heated pointer moves across it, by a beam of ultraviolet light falling on paper sensitive to it and by a tungsten-wire stylus moving across the surface of specially coated paper, a thin layer of aluminium over coloured dye, and the electrical discharge removing the aluminium and exposing the dye. In many applications they have been superseded by virtual instruments (see Section 6.6.1).

6.2.3 Cathode-ray oscilloscope

The cathode-ray oscilloscope is a voltage-measuring instrument which is capable of displaying extremely high-frequency signals. A general-purpose instrument can respond to signals up to about 10 MHz, while more specialist instruments can respond up to about 1 GHz. Double-beam oscilloscopes enable two separate traces to be observed simultaneously on the screen while storage oscilloscopes enable the trace to remain on the screen after the input signal has ceased, only being removed by a deliberate action of erasure. Digital storage oscilloscopes digitise the input signal and store the digital signal in a memory. The signal can then be analysed and manipulated and the analogue display on the oscilloscope screen obtained from reconstructing the analogue signal. Permanent records of traces can be made with special-purpose cameras that attach directly to the oscilloscope.

A general-purpose oscilloscope is likely to have vertical deflection, i.e. Y-deflection, sensitivities which vary between 5 mV per scale division and 20 V per scale division. In order that a.c. components can be viewed in the presence of high d.c. voltages, a blocking capacitor can be switched into the input line. When the amplifier is in its a.c. mode, its bandwidth typically extends from about 2 Hz to 10 MHz and when in the d.c. mode, from d.c. to 10 MHz. The Y-input impedance is typically about 1 MΩ shunted with about 20 pF capacitance. When an external circuit is connected to the Y-input, problems due to loading and interference can distort the input signal. While interference can be reduced by the use of coaxial cable, the capacitance of the coaxial cable and any probe attached to it can be enough, particularly at low frequencies, to introduce a relatively low impedance across the input impedance of the oscilloscope and so introduce significant loading. A number of probes exist for connection to the input cable and which are designed to increase the input impedance and avoid this loading problem. A passive voltage probe that is often used is a 10-to-1 attenuator (Figure 6.3). This has a 9 MΩ resistor and variable capacitor in the probe tip. However, this reduces not only the capacitive loading but also the voltage sensitivity and so an active voltage probe using an FET is often used.

Figure 6.3 Passive voltage probe.

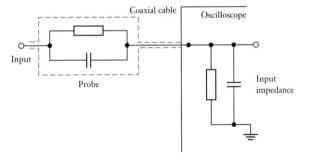

6.2.4 Visual display unit

Output data is increasingly being presented using a visual display unit (VDU). With a cathode-ray tube screen, the picture on the screen is built up by moving the spot formed by an electron beam in a series of horizontal scan lines, one after another down the screen. The image is built up by varying the intensity of the spot on the screen as each line is scanned. This raster form of display is termed **non-interlaced** (Figure 6.4(a)). To reduce the effects of flicker two scans down the screen are used to trace a complete picture. On the first scan all the odd-numbered lines are traced out and on the second the even-numbered lines are traced. This technique is called **interlaced scanning** (Figure 6.4(b)).

Figure 6.4 (a) Non-interlaced, (b) interlaced displays.

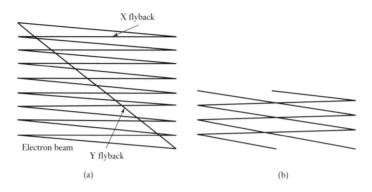

(a)   (b)

Figure 6.5 Character build-up by selective lighting.

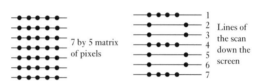

The screen of the VDU is coated with a large number of phosphor dots, these dots forming the **pixels**. The term pixel is used for the smallest addressable dot on any display device. A text character or a diagram is produced on the screen by selectively lighting these dots. Figure 6.5 shows how, for a 7 by 5 matrix, characters are built up by the electron beam moving in its zigzag path down the screen. The input data to the VDU is usually in digital **ASCII (American Standard Code for Information Interchange)** format. This is a 7-bit code and so can be used to represent $2^7 = 128$ characters. It enables all the standard keyboard characters to be covered, as well as some control functions such as RETURN which is used to indicate the return from the end of a line to the start of the next line. Table 6.1 gives an abridged list of the code.

Table 6.1 ASCII code.

Character	ASCII	Character	ASCII	Character	ASCII
A	100 0001	N	100 1110	0	011 0000
B	100 0010	O	100 1111	1	011 0001
C	100 0011	P	101 0000	2	011 0010
D	100 0100	Q	101 0001	3	011 0011
E	100 0101	R	101 0010	4	011 0100
F	100 0110	S	101 0011	5	011 0101
G	100 0111	T	101 0100	6	011 0110
H	100 1000	U	101 0101	7	011 0111
I	100 1001	V	101 0110	8	011 1000
J	100 1010	W	101 0111	9	011 1001
K	100 1011	X	101 1000		
L	100 1100	Y	101 1001		
M	100 1101	Z	101 1010		

Figure 6.6 Dot matrix print head mechanism.

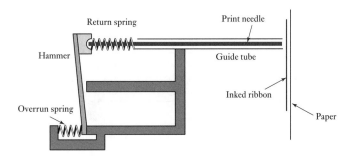

6.2.5   Printers

Printers provide a record of data on paper. There are a number of versions of such printers: the dot matrix printer, the ink/bubble jet printer and the laser printer.

The **dot matrix printer** has a print head (Figure 6.6) which consists of either 9 or 24 pins in a vertical line. Each pin is controlled by an electromagnet which when turned on propels the pin onto the inking ribbon. This transfers a small blob of ink onto the paper behind the ribbon. A character is formed by moving the print head in horizontal lines back and forth across the paper and firing the appropriate pins.

The **ink jet printer** uses a conductive ink which is forced through a small nozzle to produce a jet of very small drops of ink of constant diameter at a constant frequency. With one form a constant stream of ink passes along a tube and is pulsed to form fine drops by a piezoelectric crystal which vibrates at a frequency of about 100 kHz (Figure 6.7). Another form uses a small heater in the print head with vaporised ink in a capillary tube, so producing gas bubbles which push out drops of ink (Figure 6.8). In one printer version each drop of ink is given a charge as a result of passing through a charging

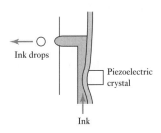

Figure 6.7 Producing a stream of drops.

Figure 6.8 Principle of the bubble jet.

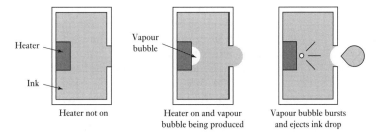

electrode and the charged drops are deflected by passing between plates between which an electric field is maintained; in another version a vertical stack of nozzles is used and each jet is just switched on or off on demand. Ink jet printers can give colour prints by the use of three different colour ink jet systems. The fineness of the drops is such that prints can be produced with more than 600 dots per inch.

The **laser printer** has a photosensitive drum which is coated with a selenium-based light-sensitive material (Figure 6.9). In the dark the selenium has a high resistance and consequently becomes charged as it passes close to the charging wire; this is a wire at a high voltage and off which charge leaks.

Figure 6.9 Basic elements of a laser printer.

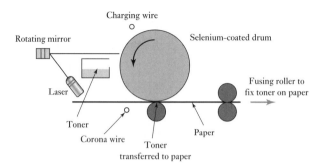

A light beam is made to scan along the length of the drum by a small rotating eight-sided mirror. When light strikes the selenium its resistance drops and it can no longer remain charged. By controlling the brightness of the beam of light, so points on the drum can be discharged or left charged. As the drum passes the toner reservoir, the charged areas attract particles of toner, which thus stick to the areas that have not been exposed to light and do not stick on the areas that have been exposed to light. The paper is given a charge as it passes another charging wire, the so-called corona wire, so that as it passes close to the drum, it attracts the toner off the drum. A hot fusing roller is then used to melt the toner particles so that, after passing between rollers, they firmly adhere to the paper. General-use laser printers are currently able to produce 1200 dots per inch.

6.3 Magnetic recording

Magnetic recording is used for the storage of data on the floppy disks and hard disks of computers. The basic principles are that a recording head, which responds to the input signal, produces corresponding magnetic patterns on a thin layer of magnetic material and a read head gives an output by converting the magnetic patterns on the magnetic material to electrical signals. In addition to these heads the systems require a transport system which moves the magnetic material in a controlled way under the heads.

Figure 6.10(a) shows the basic elements of the recording head; it consists of a core of ferromagnetic material which has a non-magnetic gap. When electrical signals are fed to the coil which is wound round the core, magnetic flux is produced in the core. The proximity of the magnetic coated plastic to the non-magnetic gap means that the magnetic flux readily follows a path through the core and that part of the magnetic coating in the region of the gap. When there is magnetic flux passing through a region of the magnetic coating, it becomes permanently magnetised. Hence a magnetic record is produced of the electrical input signal. Reversing the direction of the current reverses the flux direction.

The replay head (Figure 6.10(b)) has a similar construction to that of the recording head. When a piece of magnetised coating bridges the non-magnetised gap, then magnetic flux is induced in the core. Flux changes in the core induce e.m.f.s in the coil wound round the core. Thus the output from the coil is an electrical signal which is related to the magnetic record on the coating.

Figure 6.10 Basis of magnetic (a) recording, (b) replay head.

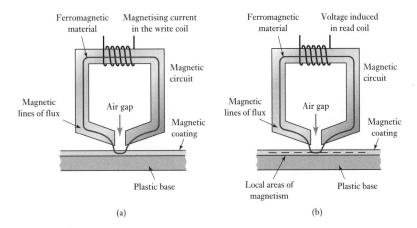

(a)   (b)

6.3.1 **Magnetic recording codes**

Digital recording involves the recording of signals as a coded combination of bits. A bit cell is the element of the magnetic coating where the magnetism is either completely saturated in one direction or completely saturated in the reverse direction. Saturation is when the magnetising field has been increased to such an extent that the magnetic material has reached its maximum amount of magnetic flux and further increases in magnetising current produce no further change.

The bit cells on the magnetic surface might then appear in the form shown in Figure 6.11. An obvious method of putting data on the magnetic material might seem to be to use the magnetic flux in one direction to represent a 0 and in the reverse direction a 1. However, it is necessary to read each cell and thus accurate timing points are needed in order to indicate clearly when sampling should take place. Problems can arise if some external clock is used to give the timing signals, as a small mismatch between the timing signals and the rate at which the magnetic surface is moving under the read head can result in perhaps a cell being missed or even read twice. Synchronisation is essential. Such synchronisation is achieved by using the bit cells themselves to generate the signals for taking samples. One method is to use transitions on the magnetic surface from saturation in one direction to saturation in the other, i.e. where the demarcation between two bits is clearly evident, to give feedback to the timing signal generation in order to adjust it so that it is in synchronisation with the bit cells.

If the flux reversals do not occur sufficiently frequently, this method of synchronisation can still result in errors occurring. One way of overcoming this problem is to use a form of encoding. Some of the methods commonly used are outlined below.

1 *Non-return-to-zero* (NRZ)
   With this system the flux is recorded on the tape such that no change in flux represents 0 and a change in flux 1 (Figure 6.12(a)). It is, however, not self-clocking.

2 *Phase encoding* (PE)
   Phase encoding has the advantage of being self-clocking with no external clock signals being required. Each cell is split in two with one half having

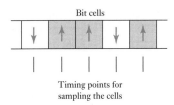

Bit cells

Timing points for sampling the cells

Figure 6.11 Bit cells.

Figure 6.12 (a) Non-return-to-zero, (b) phase, (c) frequency, (d) modified frequency modulation.

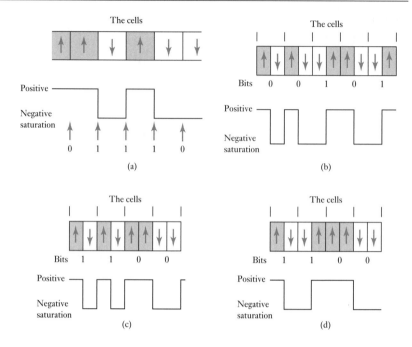

positive saturation flux and the other a negative saturation flux. A digit 0 is then recorded as a half-bit positive saturation followed by a half-bit negative saturation; a 1 digit is represented by a half-bit negative saturation followed by a half-bit positive saturation. The mid-cell transition of positive to negative thus indicates a 0 and a negative to positive transition a 1 (Figure 6.12(b)).

3 *Frequency modulation* (FM)
   This is self-clocking and similar to phase encoding but there is always a flux direction reversal at the beginning of each cell (Figure 6.12(c)). For a 0 bit there is then no additional flux reversal during the cell, but for a 1 there is an additional flux reversal during the cell.

4 *Modified frequency modulation* (MFM)
   This is a modification of the frequency modulation code, the difference being that the flux reversal at the beginning of each bit code is only present if the current and previous bit were 0 (Figure 6.12(d)). This means that only one flux reversal is required for each bit. This and the run length limited code are the codes generally used for magnetic discs.

5 *Run length limited* (RLL)
   This is a group of self-clocking codes which specify a minimum and maximum distance, i.e. run, between flux reversals. The maximum run is short enough to ensure that the flux reversals are sufficiently frequent for the code to be self-clocking. A commonly used form of this code is $RLL_{2,7}$, the 2,7 indicating that the minimum distance between flux reversals is to be 2 bits and the maximum is to be 7. The sequence of codes is described as a sequence of S-codes and R-codes. An S-code, a space code, has no reversals of flux while an R-code, a reversal code, has a reversal during the bit. Two S/R-codes are used to represent each bit.

The bits are grouped into sequences of 2, 3 and 4 bits and a code assigned to each group, the codes being:

Bit sequence	Code sequence
10	SRSS
11	RSSS
000	SSSRSS
010	RSSRSS
011	SSRSSS
0010	SSRSSRSS
0011	SSSSRSSS

Figure 6.13 shows the coding for the sequence 0110010, it being broken into groups 011 and 0010 and so represented by SSRSSSSRSSRSS. There are at least two S-codes between R-codes and there can be no more than seven S-codes between R-codes.

Figure 6.13 RLL code.

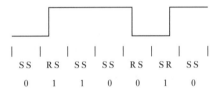

SS	RS	SS	SS	RS	SR	SS
0	1	1	0	0	1	0

The optimum code is the one that allows the bits to be packed as close as possible and which can be read without error. The read heads can locate reversals quite easily but they must not be too close together. The RLL code has the advantage of being more compact than the other codes, PE and FM taking up the most space. MFM and NRZ take up the same amount of space. NRZ has the disadvantage of, unlike the other codes, not being self-clocking.

6.3.2 Magnetic disks

Digital recording is very frequently to a hard disk. The digital data is stored on the disk surface along concentric circles called tracks, a single disk having many such tracks. A single read/write head is used for each disk surface and the heads are moved, by means of a mechanical actuator, backwards and forwards to access different tracks. The disk is spun by the drive and the read/write heads read or write data into a track. Hard disks (Figure 6.14(a)) are sealed units with data stored on the disk surface along concentric circles. A hard disk assembly has more than one such disk and the data is stored on magnetic coatings on both sides of the disks. The disks are rotated at high speeds and the tracks accessed by moving the read/write heads. Large amounts of data can be stored on such assemblies of disks; storages of hundreds of gigabytes are now common.

The disk surface is divided into sectors (Figure 6.14(b)) and so a unit of information on a disk has an address consisting of a track number and

Figure 6.14 Hard disk:
(a) arrangement of disks,
(b) tracks and sectors.

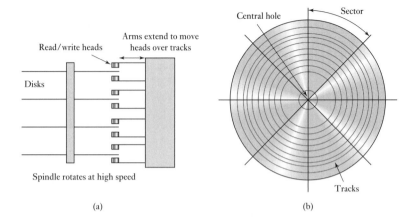

(a)   (b)

a sector number. A floppy disk has normally between 8 and 18 sectors and about 100 tracks. A hard disk might have about 2000 tracks per surface and 32 sectors. To seek data the head must be moved to over the required track, the time this takes being termed the **seek time**, and then wait there until the required segment moves under it, this time being termed the **latency**. In order that an address can be identified, it is necessary for information to have been recorded on the disk to identify segments and tracks. The writing of this information is called **formatting** and has to be carried out before data can be stored on a disk. The technique usually used is to store this location information on the tracks so that when data is stored the sequence of information on a track becomes:

index marker,
sector 0 header, sector 0 data, sector 0 trailer,
sector 1 header, sector 1 data, sector 1 trailer,
sector 2 header, sector 2 data, sector 2 trailer,
etc.

The index marker contains the track number with the sector header identifying the sector. The sector trailer contains information, e.g. a cyclic redundancy check, which can be used to check that a sector was read correctly.

6.4   Optical recording

Like magnetic disks, CD-ROMs store the data along their tracks. Unlike a magnetic disk which has a series of concentric tracks, a CD-ROM has a spiral track. The recording surface is coated with aluminium and is highly reflective. The information is then stored in a track about 0.6 μm wide as a series of pits, etched into the surface by focused light from a laser in a beam about 1 μm in diameter, and these result in light being strongly reflected or not reflected according to whether it strikes a pit or a non-depressed area. Data is thus read as a sequence of reflected and non-reflected light pulses.

Optical recording uses similar coding methods to that used with magnetic recording, the RLL form of coding being generally used. Because optical recordings can be very easily corrupted by scratches or dust obstructing the laser beams used to read them, methods have to be used to detect and correct for errors. One method is **parity checking**. With this method, groups of bits are augmented with an extra, parity, bit which is either set to 0 or 1 so that the total number of 1 bits in a group is either an odd or an even number. When the information is read, if one of the bits has been corrupted, then the number of bits will have changed and this will be detected as an error.

6.5 Displays

Many display systems use light indicators to indicate on/off status or give alphanumeric displays. The term **alphanumeric** is a contraction of the terms alphabetic and numeric and describes displays of the letters of the alphabet and numbers 0 to 9 with decimal points. One form of such a display involves seven 'light' segments to generate the alphabetic and numeric characters. Figure 6.15 shows the segments and Table 6.2 shows how a 4-bit binary code input can be used to generate inputs to switch on the various segments.

Another format involves a 7 by 5 or 9 by 7 dot matrix (Figure 6.16). The characters are then generated by the excitation of appropriate dots.

The light indicators for such displays might be neon lamps, incandescent lamps, **light-emitting diodes** (LEDs) or **liquid crystal displays** (LCDs). **Neon lamps** need high voltages and low currents and can be powered directly from the mains voltage but can only be used to give a red light. **Incandescent lamps** can be used with a wide range of voltages but need a comparatively high current. They emit white light and so use lenses to generate any required colour. Their main advantage is their brightness.

Figure 6.15 Seven-segment display.

Table 6.2 Seven-segment display.

Binary input				Segments activated							Number displayed
				a	b	c	d	e	f	g	
0	0	0	0	1	1	1	1	1	1	0	0
0	0	0	1	0	1	1	0	0	0	0	1
0	0	1	0	1	1	0	1	1	0	1	2
0	0	1	0	1	1	1	1	0	0	1	3
0	1	0	0	0	1	1	0	0	1	1	4
0	1	0	1	1	0	1	1	0	1	1	5
0	1	1	0	0	0	1	1	1	1	1	6
0	1	1	1	1	1	1	0	0	0	0	7
1	0	0	0	1	1	1	1	1	1	1	8
1	0	0	1	1	1	1	0	0	1	1	9

Figure 6.16 A 7 by 5 dot matrix display.

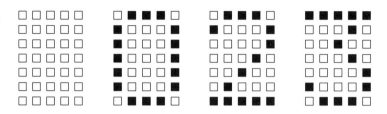

6.5.1 Light-emitting diodes

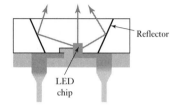

Figure 6.17 LED.

LEDs require low voltages and low currents and are cheap. These diodes when forward biased emit light over a certain band of wavelengths. Figure 6.17 shows the basic form of a LED, the light emitted from the diode being enhanced in one direction by means of reflectors. Commonly used LED materials are gallium arsenide, gallium phosphide and alloys of gallium arsenide with gallium phosphide. The most commonly used LEDs can give red, yellow or green colours. With microprocessor-based systems, LEDs are the most common form of indicator used.

A current-limiting resistor is generally required with an LED in order to limit the current to below the maximum rated current of about 10 to 30 mA. Typically an LED might give a voltage drop across it of 2.1 V when the current is limited to 20 mA. Thus when, say, a 5 V output is applied, 2.9 V has to be dropped across a series resistor. This means that a resistance of $2.9/0.020 = 145\ \Omega$ is required and so a standard resistor of 150 Ω is likely to be used. Some LEDs are supplied with built-in resistors so they can be directly connected to microprocessor systems.

LEDs are available as single light displays, seven- and sixteen-segment alphanumeric displays, in dot matrix format and bar graph form.

Figure 6.18(a) shows how seven LEDs, to give the seven segments of a display of the form shown in Figure 6.16, might be connected to a driver so that when a line is driven low, a voltage is applied and the LED in that line is switched on. The voltage has to be above a 'turn-on' value before the LED emits significant light; typical turn-on voltages are about 1.5 V. Such

Figure 6.18 (a) Common anode connection for LEDs, (b) common cathode.

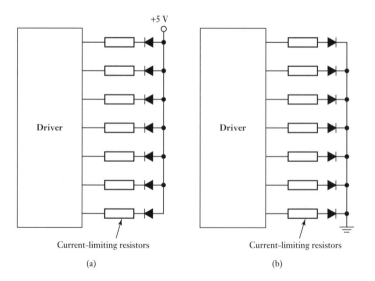

Current-limiting resistors   Current-limiting resistors

(a)   (b)

an arrangement is known as the **common anode** form of connection since all the LED anodes are connected together. An alternative arrangement is the **common cathode** (Figure 6.18(b)). The elements in common anode form are made active by the input going low, in the common cathode type by going high. Common anode is the usual choice since the direction of current flow and the size of current involved are usually most appropriate.

Examples of such types of display are the seven-segment 7.6 mm and 10.9 mm high-intensity displays of Hewlett Packard which are available as either common anode or common cathode form. In addition to the seven segments to form the characters, there is either a left-hand or right-hand decimal point. By illuminating different segments of the display, the full range of numbers and a small range of alphabetical characters can be formed.

Often the output from the driver is not in the normal binary form but in **Binary-Coded Decimal** (BCD) (see Appendix B). With BCD, each decimal digit is coded separately in binary. For example, the decimal number 15 has the 1 coded as 0001 and the 5 as 0101 to give the BCD code of 0001 0101. The driver output has then to be decoded into the required format for the LED display. The 7447 is a commonly used decoder for driving displays (Figure 6.19).

Figure 6.19 Decoder with seven-segment display.

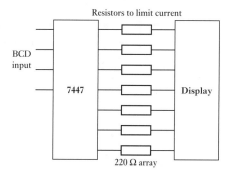

Figure 6.20 shows the basic form used for a 5 by 7 dot matrix LED display. The array consists of five column connectors, each connecting the anodes of seven LEDs. Each row connects to the cathodes of five LEDs. To turn on a particular LED, power is applied to its column and its row is grounded. Such a display enables all the ASCII characters to be produced.

6.5.2 Liquid crystal displays

Liquid crystal displays do not produce any light of their own but rely on reflected light or transmitted light. The liquid crystal material is a compound with long rod-shaped molecules which is sandwiched between two sheets of polymer containing microscopic grooves. The upper and lower sheets are grooved in directions at 90° to each other. The molecules of the liquid crystal material align with the grooves in the polymer and adopt a smooth 90° twist between them (Figure 6.21).

When plane polarised light is incident on the liquid crystal material its plane of polarisation is rotated as it passes through the material. Thus if it is sandwiched between two sheets of polariser with their transmission directions at right angles, the rotation allows the light to be transmitted and so the material appears light.

Figure 6.20 Dot matrix display.

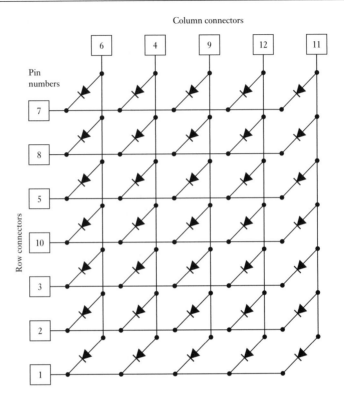

Figure 6.21 Liquid crystal:
(a) no electric field, (b) electric
field.

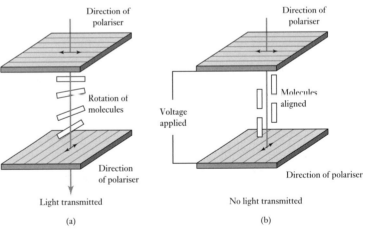

However, if an electric field is applied across the material, the molecules
become aligned with the field and the light passing through the top polariser
is not rotated and cannot pass through the lower polariser but becomes
absorbed. The material then appears dark.

The arrangement is put between two sheets of glass on which are
transparent electrodes in the form of the required display. An LED display
can be transmissive or reflective. With the transmissive display, the
display is back-lit. When the rotation of the plane of polarisation allows light
to be transmitted, the display is light, otherwise it is dark. With the reflective

light display, there is a reflective surface behind the crystals. Thus when the incident light passes through the display, it is reflected back and so the display appears light. When the incident light cannot pass through the display it appears dark.

LCDs are available in many segment layouts, including a seven-segment display similar to the seven-segment LED display. The application of voltages to the various display elements results in them appearing black against the lighter display where there is no electric field. To turn on a segment, an a.c. electric field of about 3 to 12 V is used. The drive voltage must not be d.c. but a.c. since d.c. voltages generate reactions which destroy the crystals. LCDs have a relatively slow response time, typically about 100 to 150 ms. Power consumption is low.

LCDs are also available as dot matrix displays. LCD modules are also available with displays with one or more rows of characters. For example, a two row 40 character display is available.

Integrated circuit drivers are available to drive LEDs. Thus the MC14543B can be used for a seven-segment LCD display. Drivers are available for when the input is in BCD code. A 5 × 8-dot matrix display can be driven by the MC145000 driver. Displays are available combined with drivers. For example, the Hitachi LM018L is a 40 character × 2 line reflective-type LCD module with a built-in driver HD44780 which provides a range of features, including 192 5 × 7-dot characters plus 8 user-defined characters and thus can be directly interfaced to a 4-bit or 8-bit microprocessor.

LEDs are the form of display used in battery-operated devices such as cell phones, watches and calculators.

6.6 Data acquisition systems

The term **data acquisition** (DAQ) tends to be frequently used for systems in which inputs from sensors are converted into a digital form for processing, analysis and display by a computer. The systems thus contain: sensors, wiring to connect the sensors to signal conditioning to carry out perhaps filtering and amplification, DAQ hardware to carry out such functions as conversion of input to digital format and conversion of output signals to analogue format for control systems, the computer and DAQ software. The software carries out analysis of the digital input signals. Such systems are also often designed to exercise control functions as well.

6.6.1 Computer with plug-in boards

Figure 6.22 shows the basic elements of a DAQ system using plug-in boards with a computer for the DAQ hardware. The signal conditioning prior to the inputs to the board depends on the sensors concerned, e.g. it might be for thermocouples: amplification, cold junction compensation and linearisation; for strain gauges:

Figure 6.22 Data acquisition system.

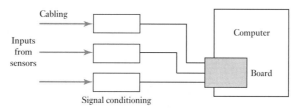

Wheatstone bridge, voltage supply for bridge and linearisation; for resistance temperature detectors (RTDs): current supply, circuitry and linearisation.

In selecting the DAQ board to be used the following criteria have to be considered:

1 What type of computer software system is being used, e.g. Windows, MacOS?
2 What type of connector is the board to be plugged into, e.g. PCMCIA for laptops, NuBus for MacOS, PCI?
3 How many analogue inputs will be required and what are their ranges?
4 How many digital inputs will be required?
5 What resolution will be required?
6 What is the minimum sampling rate required?
7 Are any timing or counting signals required?

Figure 6.23 shows the basic elements of a DAQ board. Some boards will be designed only to handle analogue inputs/outputs and others digital inputs/outputs.

Figure 6.23 DAQ board elements.

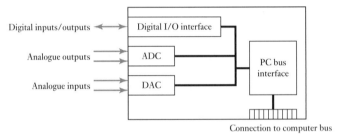

All DAQ boards use **drivers**, software generally supplied by the board manufacturer with a board, to communicate with the computer and tell it what has been inserted and how the computer can communicate with the board. Before a board can be used, three parameters have to be set. These are the addresses of the input and output channels, the interrupt level and the channel to be used for direct memory access. With 'plug-and-play' boards for use with Windows software, these parameters are set by the software; otherwise microswitches have to be set on the card in accordance with the instructions in the manual supplied with the board.

Application software can be used to assist in the designing of measurement systems and the analysis of the data. As an illustration of the type of application software available, LabVIEW is a graphical programming software package that has been developed by National Instruments for DAQ and instrument control. LabVIEW programs are called **virtual instruments** because in appearance and operation they imitate actual instruments. A virtual instrument has three parts: a front panel which is the interactive user interface and simulates the front panel of an instrument by containing control knobs, push-buttons and graphical displays; a block diagram which is the source code for the program with the programming being done graphically by drawing lines between connection points on selected icons on the computer screen; and representation as an icon and connector which can provide a graphical representation of the virtual instrument if it is wanted for use in other block diagrams.

Figure 6.24(a) shows the icon selected for a virtual instrument where one analogue sample is obtained from a specified input channel, the icon having been selected from the Analog Input palette. The 'Device' is the device

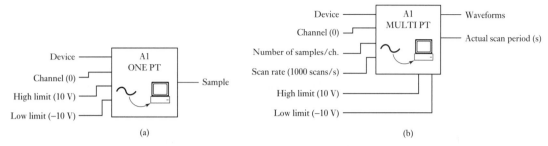

Figure 6.24 Analogue input icon: (a) single input, (b) for sampling from a number of channels.

number assigned to the DAQ board, the 'Channel' is the source of the data, a 'Sample' is one analogue-to-digital conversion, and 'High limit' and 'Low limit' are the voltage limits expected for the signal (the default is +10 V and −10 V and changing these values automatically changes the gain of the amplifier on the DAQ board).

If we want a waveform from each channel in a designated channel string then the icon shown in Figure 6.24(b) can be selected. For each input channel a set of samples is acquired over a period of time, at a specified sampling rate, and gives a waveform output showing how the analogue quantity varies with time.

By connecting other icons to, say, the above icon, a block diagram can be built up which might take the inputs from a number of analogue channels, sample them in sequence and display the results as a sequence of graphs. The type of front panel display we might have for a simple DAQ acquisition of samples and display is shown in Figure 6.25. By using the up and down arrows the parameters can be changed and the resulting display viewed.

Virtual instruments have a great advantage over traditional instruments in that the vendor of a traditional instrument determines its characteristics and interface while with a virtual instrument these can all be defined by the user and readily changed.

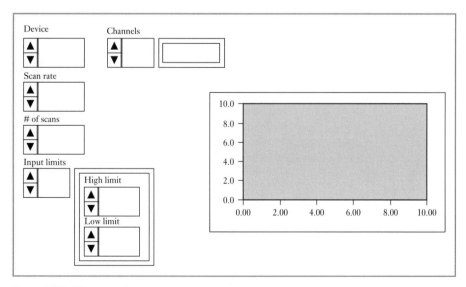

Figure 6.25 Front panel.

6.6.2 Data loggers

The term **data logger** is used for DAQ systems which are able to be used away from a computer. Once the program has been set by a computer, it can be put onto a memory card which can be inserted into the logger or have the program downloaded to it from a computer, so enabling it to carry out the required DAQ functions.

Figure 6.26 shows the basic elements of a data logger. Such a unit can monitor the inputs from a large number of sensors. Inputs from individual sensors, after suitable signal conditioning, are fed into the multiplexer. The multiplexer is used to select one signal which is then fed, after amplification, to the ADC. The digital signal is then processed by a microprocessor. The microprocessor is able to carry out simple arithmetic operations, perhaps taking the average of a number of measurements. The output from the system might be displayed on a digital meter that indicates the output and channel number, used to give a permanent record with a printer, stored on a floppy disk or transferred to perhaps a computer for analysis.

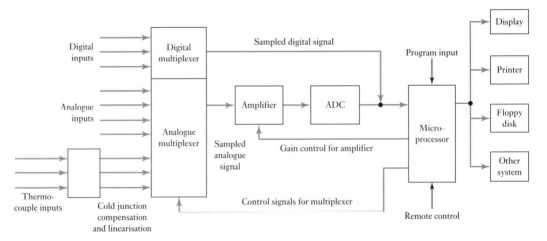

Figure 6.26 Data logger system.

Because data loggers are often used with thermocouples, there are often special inputs for thermocouples, these providing cold junction compensation and linearisation. The multiplexer can be switched to each sensor in turn and so the output consists of a sequence of samples. Scanning of the inputs can be selected by programming the microprocessor to switch the multiplexer to sample just a single channel, carry out a single scan of all channels, a continuous scan of all channels, or perhaps a periodic scan of all channels, say every 1, 5, 15, 30 or 60 minutes.

Typically a data logger may handle 20 to 100 inputs, though some may handle considerably more, perhaps 1000. It might have a sample and conversion time of 10 μs and be used to make perhaps 1000 readings per second. The accuracy is typically about 0.01% of full-scale input and linearity is about ±0.005% of full-scale input. Cross-talk is typically 0.01% of full-scale input on any one input. The term **cross-talk** is used to describe the interference that can occur when one sensor is being sampled as a result of signals from other sensors.

6.7 Measurement systems

The following examples illustrate some of the points involved in the design of measurement systems for particular applications.

6.7.1 Load cell for use as a link to detect load lifted

A link-type load cell, of the form shown in Figure 6.27, has four strain gauges attached to its surface and can be inserted in between the cable lifting a load and the load to give a measure of the load being lifted. Two of the strain gauges are in the longitudinal axis direction and two in a transverse direction. When the link is subject to tensile forces, the axial gauges will be in tension and the transverse gauges in compression. Suppose we have the design criteria for the load cell of a sensitivity such that there is an output of about 30 mV when the stress applied to the link is 500 MPa. We will assume that the strain gauges may have gauge factors of 2.0 and resistances of 100 Ω.

Figure 6.27 Load cell.

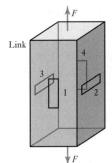

Link

Gauges 1 and 4 are on opposite faces and in tension when the link is subject to tensile forces

Gauges 2 and 3 are on opposite faces and in compression when the link is subject to tensile forces

When a load F is applied to the link then, since the elastic modulus E is stress/strain and stress is force per unit area, the longitudinal axis strain ε_1 is F/AE and the transverse strain ε_t is $-\nu F/AE$, where A is the cross-sectional area and ν is Poisson's ratio for the link material. The responses of the strain gauges (see Section 2.3.2) to these strains are

$$\frac{\delta R_1}{R_1} = \frac{\delta R_4}{R_4} = G\varepsilon_1 = \frac{GF}{AE}$$

$$\frac{\delta R_3}{R_3} = \frac{\delta R_2}{R_2} = G\varepsilon_t = -\frac{\nu GF}{AE}$$

The output voltage from the Wheatstone bridge (see Section 3.5) is given by

$$V_o = \frac{V_s R_1 R_4}{(R_1 + R_2)(R_3 + R_4)}\left(\frac{\delta R_1}{R_1} - \frac{\delta R_2}{R_2} - \frac{\delta R_3}{R_3} + \frac{\delta R_4}{R_4}\right)$$

With $R_1 = R_2 = R_3 = R_4 = R$, and with $\delta R_1 = \delta R_4$ and $\delta R_2 = \delta R_3$, then

$$V_o = \frac{V_s}{2R}(\delta R_1 - \delta R_2) = \frac{V_s GF}{2AE}(1 + \nu)$$

Suppose we consider steel for the link. Then tables give E as about 210 GPa and ν about 0.30. Thus with a stress $(= F/A)$ of 500 MPa we have, for strain gauges with a gauge factor of 2.0,

$$V_o = 3.09 \times 10^{-3} V_s$$

For a bridge voltage with a supply voltage of 10 V this would be an output voltage of 30.9 mV. No amplification is required if this is the only load value required; if, however, this is a maximum value and we want to determine loads below this level then we might use a differential amplifier. The output can be displayed on a high-resistance voltmeter – high resistance to avoid loading problems. A digital voltmeter might thus be suitable.

6.7.2 Temperature alarm system

A measurement system is required which will set off an alarm when the temperature of a liquid rises above 40°C. The liquid is normally at 30°C. The output from the system must be a 1 V signal to operate the alarm.

Since the output is to be electrical and a reasonable speed of response is likely to be required, an obvious possibility is an electrical resistance element. To generate a voltage output the resistance element could be used with a Wheatstone bridge. The output voltage will probably be less than 1 V for a change from 30 to 40°C, but a differential amplifier could be used to enable the required voltage to be obtained. A comparator can then be used to compare the value with the set value for the alarm.

Suppose a nickel element is used. Nickel has a temperature coefficient of resistance of 0.0067/K. Thus if the resistance element is taken as being 100 Ω at 0°C then its resistance at 30°C will be

$$R_{30} = R_0(1 + \alpha t) = 100(1 + 0.0067 \times 30) = 120.1 \, \Omega$$

and at 40°C

$$R_{40} = 100(1 + 0.0067 \times 40) = 126.8 \, \Omega$$

Thus there is a change in resistance of 6.7 Ω. If this element forms one arm of a Wheatstone bridge which is balanced at 30°C, then the output voltage V_o is given by (see Section 3.5)

$$\delta V_o = \frac{V_s \delta R_1}{R_1 + R_2}$$

With the bridge balanced at 30°C and, say, all the arms have the same value and a supply voltage of 4 V, then

$$\delta V_o = \frac{4 \times 6.7}{126.8 + 120.1} = 0.109 \text{ V}$$

To amplify this to 1 V we can use a difference amplifier (see Section 3.2.5)

$$V_o = \frac{R_2}{R_1}(V_2 - V_1)$$

$$1 = \frac{R_2}{R_1} \times 0.109$$

Hence $R_2/R_1 = 9.17$ and so if we use an input resistance of $1\,\mathrm{k\Omega}$ the feedback resistance must be $9.17\,\mathrm{k\Omega}$.

6.7.3 Angular position of a pulley wheel

A potentiometer is to be used to monitor the angular position of a pulley wheel. Consider the items that might be needed to enable there to be an output to a recorder of 10 mV per degree if the potentiometer has a full-scale angular rotation of 320°.

When the supply voltage V_s is connected across the potentiometer we will need to safeguard it and the wiring against possible high currents and so a resistance R_s can be put in series with the potentiometer R_p. The total voltage drop across the potentiometer is thus $V_s R_p/(R_s + R_p)$. For an angle θ with a potentiometer having a full-scale angular deflection of θ_F we will obtain an output from the potentiometer of

$$V_\theta = \frac{\theta}{\theta_F} \frac{V_s R_p}{R_s + R_p}$$

Suppose we consider a potentiometer with a resistance of $4\,\mathrm{k\Omega}$ and let R_s be $2\,\mathrm{k\Omega}$. Then for 1 mV per degree we have

$$0.01 = \frac{1}{320} \frac{4V_s}{4 + 2}$$

Hence we would need a supply voltage of 4.8 V. To prevent loading of the potentiometer by the resistance of the recorder, a voltage follower circuit can be used. Thus the circuit might be of the form shown in Figure 6.28.

Figure 6.28 Pulley wheel monitor.

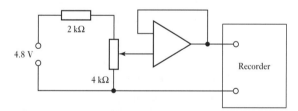

6.7.4 Temperature measurement to give a binary output

Consider the requirement for a temperature measurement system for temperatures in the range 0 to 100°C and which will give an 8-bit binary output with a change in 1 bit corresponding to a temperature change of 1°C. The output is intended for inputting to a microprocessor as part of a temperature control system.

A linear temperature sensor is required and so the thermotransistor LM35 can be used (see Section 2.9.4). LM35 gives an output of 10 mV/°C when it has a supply voltage of 5 V. If we apply the output from LM35 to an 8-bit ADC then a digital output can be obtained. We need the resolution of the ADC to be 10 mV so that each step of 10 mV will generate a change in output of 1 bit. Suppose we use a successive approximations ADC, e.g. ADC0801; then this requires an input of a reference voltage which when subdivided into $2^8 = 256$ bits gives 10 mV per bit. Thus a reference voltage of 2.56 V is required. For this to be obtained the reference voltage

Figure 6.29 Temperature sensor.

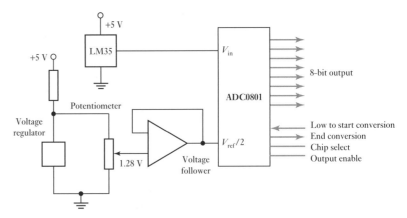

input to the ADC0801 has to be $V_{ref}/2$ and so an accurate input voltage of 1.28 V is required. Such a voltage can be obtained by using a potentiometer circuit across the 5 V supply with a voltage follower to avoid loading problems. Because the voltage has to remain steady at 1.28 V, even if the 5 V supply voltage fluctuates, a voltage regulator is likely to be used, e.g. a 2.45 V voltage regulator ZN458/B. Thus the circuit might be as in Figure 6.29.

6.8 Testing and calibration

Testing a measurement system installation falls into three stages:

1 *Pre-installation testing*
This is the testing of each instrument for correct calibration and operation prior to it being installed.

2 *Piping and cabling testing*
In the case of pneumatic lines this involves, prior to the connection of the instruments, blowing through with clean, dry air prior to connection and pressure testing to ensure they are leak free. With process piping, all the piping should be flushed through and tested prior to the connection of instruments. With instrument cables, all should be checked for continuity and insulation resistance prior to the connection of any instruments.

3 *Pre-commissioning*
This involves testing that the installation is complete, all instrument components are in full operational order when interconnected and all control room panels or displays function.

6.8.1 Calibration

Calibration consists of comparing the output of a measurement system and its subsystems against standards of known accuracy. The standards may be other instruments which are kept specially for calibration duties or some means of defining standard values. In many companies some instruments and items such as standard resistors and cells are kept in a company standards department and used solely for calibration purposes. The likely relationship between the calibration of an instrument in everyday use and national standards is outlined below.

1 National standards are used to calibrate standards for calibration centres.
2 Calibration centre standards are used to calibrate standards for instrument manufacturers.

3 Standardised instruments from instrument manufacturers are used to provide in-company standards.
4 In-company standards are used to calibrate process instruments.

There is a simple traceability chain from the instrument used in a process back to national standards. The following are some examples of calibration procedures that might be used in-company:

1 *Voltmeters*
These can be checked against standard voltmeters or standard cells giving standard e.m.f.s.

2 *Ammeters*
These can be checked against standard ammeters.

3 *Gauge factor of strain gauges*
This can be checked by taking a sample of gauges from a batch and applying measured strains to them when mounted on some test piece. The resistance changes can be measured and hence the gauge factor computed.

4 *Wheatstone bridge circuits*
The output from a Wheatstone bridge can be checked when a standard resistance is introduced into one of the arms.

5 *Load cells*
For low-capacity load cells, dead-weight loads using standard weights can be used.

6 *Pressure sensors*
Pressure sensors can be calibrated by using a dead-weight tester (Figure 6.30). The calibration pressures are generated by adding standard weights W to the piston tray. After the weights are placed on the tray, a screw-driven plunger is forced into the hydraulic oil in the chamber to lift the piston–weight assembly. The calibration pressure is then W/A, where A is the cross-sectional area of the piston. Alternatively the dead-weight tester can be used to calibrate a pressure gauge and this gauge can be used for the calibration of other gauges.

Figure 6.30 Dead-weight calibration for pressure gauges.

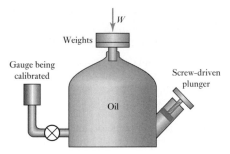

7 *Temperature sensors*
These can be calibrated by immersion in a melt of a pure metal or water. The temperature of the substance is then slowly reduced and a temperature–time record obtained. When the substance changes state from liquid to solid, the temperature remains constant. Its value can be looked up from tables and hence an accurate reference temperature for calibration obtained. Alternatively, the temperature at which a liquid boils can be

used. However, the boiling point depends on the atmospheric pressure and corrections have to be applied if it differs from the standard atmospheric pressure. Alternatively, in-company the readings given by the measurement system can be compared with those of a standard thermometer.

Summary

A general point that has to be taken account of when putting together any measurement system is **loading**, i.e. the effect of connecting a load across the output terminals of any element of a measurement system.

Indicators give an instant visual indication of the sensed variable while **recorders** record the output signal over a period of time and give automatically a permanent record.

The term **data acquisition** (DAQ) tends to be frequently used for systems in which inputs from sensors are converted into a digital form for processing, analysis and display by a computer. The term **data logger** is used for DAQ systems which are able to be used away from a computer.

Virtual instruments are software-generated instruments; in appearance and operation they imitate actual instruments.

Testing a measurement system installation falls into three stages: pre-installation testing, piping and cabling testing, pre-commissioning.

Calibration consists of comparing the output of a measurement system and its subsystems against standards of known accuracy.

Problems

6.1 Explain the significance of the following terms taken from the specifications of display systems.

(a) Recorder: dead band $\pm 0.2\%$ of span.

(b) The hard disk has two disks with four read/write heads, one for each surface of the disks. Each surface has 614 tracks and each track 32 sectors.

(c) Data logger: number of inputs 100, cross-talk on any one input 0.01% of full-scale input.

(d) Double-beam oscilloscope: vertical deflection with two identical channels, bandwidth d.c. to 15 MHz, deflection factor of 10 mV/div to 20 V/div in 11 calibrated steps, time base of 0.5 μs/div to 0.5 s/div in 19 calibrated steps.

6.2 Explain the problems of loading when a measurement system is being assembled from a sensor, signal conditioner and display.

6.3 Suggest a display unit that could be used to give:

(a) a permanent record of the output from a thermocouple;

(b) a display which enables the oil pressure in a system to be observed;

(c) a record to be kept of the digital output from a microprocessor;

(d) the transient voltages resulting from monitoring of the loads on an aircraft during simulated wind turbulence.

6.4 A cylindrical load cell, of the form shown in Figure 2.32, has four strain gauges attached to its surface. Two of the gauges are in the circumferential direction and two in the longitudinal axis direction. When the cylinder is subject to a compressive load, the axial gauges will be in compression while the circumferential ones will be in tension. If the material of the cylinder has a cross-sectional area A and an elastic modulus E, then a force F acting on the cylinder will give a strain acting on the axial gauges of $-F/AE$ and on the circumferential gauges of $+\nu F/AE$, where ν is Poisson's ratio for the material. Design a complete measurement system, using load cells, which could be used to monitor the mass of water in a tank. The tank itself has a mass of 20 kg and the water when at the required level 40 kg. The mass is to be monitored to an accuracy of ± 0.5 kg. The strain gauges have a gauge factor of 2.1 and are all of the same resistance of 120.0 Ω. For all other items, specify what your design requires. If you use mild steel for the load cell material, then the tensile modulus may be taken as 210 GPa and Poisson's ratio 0.30.

6.5 Design a complete measurement system involving the use of a thermocouple to determine the temperature of the water in a boiler and give a visual indication on a meter. The temperature will be in the range 0 to 100°C and is required to an accuracy of $\pm 1\%$ of full-scale reading. Specify the materials to be used for the thermocouple and all other items necessary. In advocating your design you must consider the problems of cold junction and non-linearity. You will probably need to consult thermocouple tables. The following data is taken from such tables, the cold junction being at 0°C, and may be used as a guide:

Materials	e.m.f. in mV at				
	20°C	40°C	60°C	80°C	100°C
Copper–constantan	0.789	1.611	2.467	3.357	4.277
Chromel–constantan	1.192	2.419	3.683	4.983	6.317
Iron–constantan	1.019	2.058	3.115	4.186	5.268
Chromel–alumel	0.798	1.611	2.436	3.266	4.095
Platinum–10% Rh, Pt	0.113	0.235	0.365	0.502	0.645

6.6 Design a measurement system which could be used to monitor the temperatures, of the order of 100°C, in positions scattered over a number of points in a plant and present the results on a control panel.

6.7 A suggested design for the measurement of liquid level in a vessel involves a float which in its vertical motion bends a cantilever. The degree of bending of the cantilever is then taken as a measure of the liquid level. When a force F is applied to the free end of a cantilever of length L, the strain on its surface a distance x from the clamped end is given by

$$\text{strain} = \frac{6(L - x)}{wt^2E}$$

where w is the width of the cantilever, t its thickness and E the elastic modulus of the material. Strain gauges are to be used to monitor the bending of the cantilever with two strain gauges being attached longitudinally to the upper surface and two longitudinally to the lower surface. The gauges are then to be incorporated into a four-gauge Wheatstone bridge and the output voltage, after possible amplification, then taken as a measure of the liquid level. Determine the specifications required for the components of this system if there is to be an output of 10 mV per 10 cm change in level.

6.8   Design a static pressure measurement system based on a sensor involving a 40 mm diameter diaphragm across which there is to be a maximum pressure difference of 500 MPa. For a diaphragm where the central deflection y is much smaller than the thickness t of the diaphragm,

$$y \approx \frac{3r^2P(1 - v^2)}{16Et^3}$$

where r is the radius of the diaphragm, P the pressure difference, E the modulus of elasticity and v Poisson's ratio. Explain how the deflection y will be converted into a signal that can be displayed on a meter.

6.9   Suggest the elements that might be considered for the measurement systems to be used to:

(a) Monitor the pressure in an air pressure line and present the result on a dial, no great accuracy being required.

(b) Monitor continuously and record the temperature of a room with an accuracy of $\pm 1°C$.

(c) Monitor the weight of lorries passing over a weighing platform.

(d) Monitor the angular speed of rotation of a shaft.

Part III
Actuation

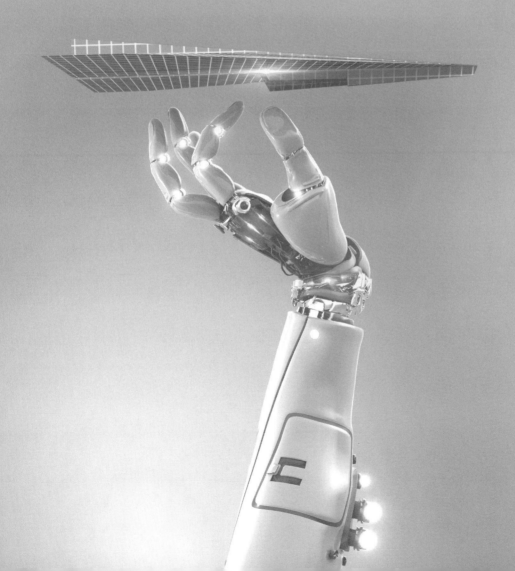

Chapter seven Pneumatic and hydraulic actuation systems

Objectives

The objectives of this chapter are that, after studying it, the reader should be able to:
- Interpret system drawings, and design simple systems, for sequential control systems involving hydraulic/pneumatic directional control valves and cylinders.
- Explain the principles of process control valves, their characteristics and sizing.

7.1 Actuation systems

Actuation systems are the elements of control systems which are responsible for transforming the output of a microprocessor or control system into a controlling action on a machine or device. Thus, for example, we might have an electrical output from the controller which has to be transformed into a linear motion to move a load. Another example might be where an electrical output from the controller has to be transformed into an action which controls the amount of liquid passing along a pipe.

In this chapter fluid power systems, namely pneumatic and hydraulic actuation systems, are discussed. **Pneumatics** is the term used when compressed air is used and **hydraulics** when a liquid, typically oil. In Chapter 8 mechanical actuator systems are discussed and in Chapter 9 electrical actuation systems.

7.2 Pneumatic and hydraulic systems

Pneumatic signals are often used to control final control elements, even when the control system is otherwise electrical. This is because such signals can be used to actuate large valves and other high-power control devices and so move significant loads. The main drawback with pneumatic systems is, however, the compressibility of air. Hydraulic systems can be used for even higher power control devices but are more expensive than pneumatic systems and there are hazards associated with oil leaks which do not occur with air leaks.

The atmospheric pressure varies with both location and time but in pneumatics is generally taken to be 10^5 Pa, such a pressure being termed 1 bar.

7.2.1 Hydraulic systems

With a hydraulic system, pressurised oil is provided by a pump driven by an electric motor. The pump pumps oil from a sump through a non-return

Figure 7.1   (a) Hydraulic power
supply, (b) accumulator.

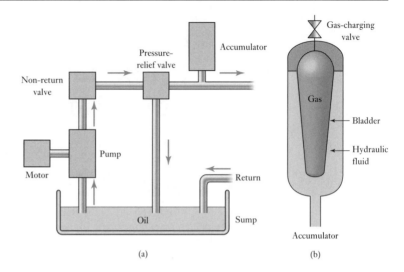

Figure 7.1   (a) Hydraulic power supply, (b) accumulator.

valve and an accumulator to the system, from which it returns to the
sump. Figure 7.1(a) illustrates the arrangement. A pressure-relief valve is
included, this being to release the pressure if it rises above a safe level, the
non-return valve is to prevent the oil being back driven to the pump and
the accumulator is to smooth out any short-term fluctuations in the output
oil pressure. Essentially the accumulator is just a container in which the oil
is held under pressure against an external force, Figure 7.1(b) showing the
most commonly used form which is gas pressurised and involves gas within a
bladder in the chamber containing the hydraulic fluid; an older type involved
a spring-loaded piston. If the oil pressure rises then the bladder contracts,
increases the volume the oil can occupy and so reduces the pressure. If the oil
pressure falls, the bladder expands to reduce the volume occupied by the oil
and so increases its pressure.

Commonly used hydraulic pumps are the gear pump, the vane pump
and the piston pump. The **gear pump** consists of two close-meshing gear
wheels which rotate in opposite directions (Figure 7.2(a)). Fluid is forced
through the pump as it becomes trapped between the rotating gear teeth and
the housing and so is transferred from the inlet port to be discharged at the
outlet port. Such pumps are widely used, being low cost and robust. They
generally operate at pressures below about 15 MPa and at 2400 rotations per
minute. The maximum flow capacity is about 0.5 m^3/min. However, leakage
occurs between the teeth and the casing and between the interlocking teeth,
and this limits the efficiency. The **vane pump** has spring-loaded sliding
vanes slotted in a driven rotor (Figure 7.2(b)). As the rotor rotates, the vanes
follow the contours of the casing. This results in fluid becoming trapped
between successive vanes and the casing and transported round from the
inlet port to outlet port. The leakage is less than with the gear pump. **Piston
pumps** used in hydraulics can take a number of forms. With the **radial
piston pump** (Figure 7.2(c)), a cylinder block rotates round the stationary
cam and this causes hollow pistons, with spring return, to move in and out.
The result is that fluid is drawn in from the inlet port and transported round
for ejection from the discharge port. The **axial piston pump** (Figure 7.2(d))
has pistons which move axially rather than radially. The pistons are arranged

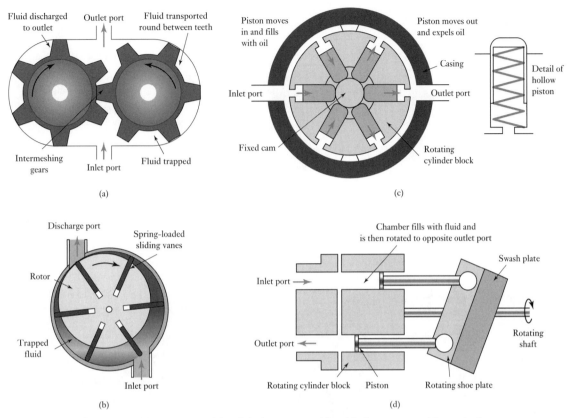

Figure 7.2 (a) Gear pump, (b) vane pump, (c) radial piston pump, (d) axial piston pump with swash plate.

axially in a rotating cylinder block and made to move by contact with the swash plate. This plate is at an angle to the drive shaft and thus as the shaft rotates they move the pistons so that air is sucked in when a piston is opposite the inlet port and expelled when it is opposite the discharge port. Piston pumps have a high efficiency and can be used at higher hydraulic pressures than gear or vane pumps.

7.2.2 Pneumatic systems

With a **pneumatic** power supply (Figure 7.3) an electric motor drives an air compressor. The air inlet to the compressor is likely to be filtered and via a silencer to reduce the noise level. A pressure-relief valve provides protection against the pressure in the system rising above a safe level. Since the air compressor increases the temperature of the air, there is likely to be a cooling system and to remove contamination and water from the air a filter with a water trap. An air receiver increases the volume of air in the system and smoothes out any short-term pressure fluctuations.

Commonly used air compressors are ones in which successive volumes of air are isolated and then compressed. Figure 7.4(a) shows the basic form of

Figure 7.3   Pneumatic power supply.

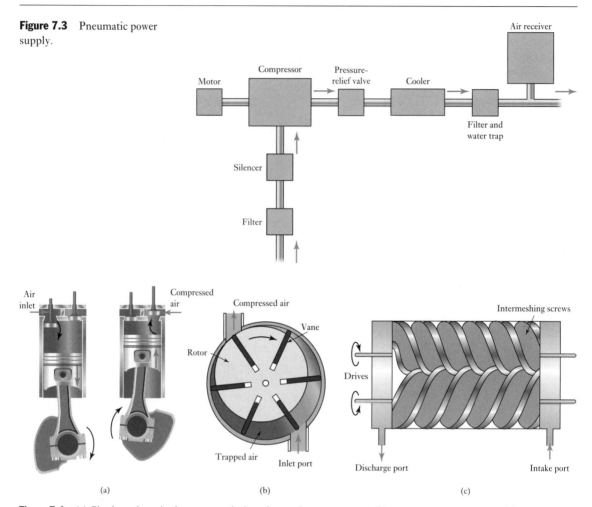

Figure 7.4   (a) Single-acting, single stage, vertical, reciprocating compressor, (b) rotary vane compressor, (c) screw compressor.

a single-acting, single-stage, vertical, reciprocating compressor. On the air intake stroke, the descending piston causes air to be sucked into the chamber through the spring-loaded inlet valve and when the piston starts to rise again, the trapped air forces the inlet valve to close and so becomes compressed. When the air pressure has risen sufficiently, the spring-loaded outlet valve opens and the trapped air flows into the compressed-air system. After the piston has reached the top dead centre it then begins to descend and the cycle repeats itself. Such a compressor is termed **single-acting** because one pulse of air is produced per piston stroke; **double-acting** compressors are designed to produce pulses of air on both the up and down strokes of the piston. It is also termed **single-stage** because the compressor goes directly from atmospheric pressure to the required pressure in a single operation. For the production of compressed air at more than a few bars, two or more stages are generally used. Normally two stages are used for pressures up to about 10 to 15 bar and more stages for higher pressures. Thus with a two-stage compressor we might have the first stage taking air at atmospheric pressure and compressing it to, say, 2 bar and then the second stage compressing this air to, say, 7 bar. Reciprocating piston compressors can be used as a single-stage compressor to produce air pressures up to about 12 bar and as a multistage compressor up to about 140 bar. Typically, air flow deliveries tend to range from about $0.02 \, \text{m}^3/\text{min}$ free air delivery to about $600 \, \text{m}^3/\text{min}$ free air delivery; free air is the term used for air at normal atmospheric pressure. Another form of compressor is the **rotary vane compressor**. This has a rotor mounted eccentrically in a cylindrical chamber (Figure 7.4(b)). The rotor has blades, the vanes, which are free to slide in radial slots with rotation causing the vanes to be driven outwards against the walls of the cylinder. As the rotor rotates, air is trapped in pockets formed by the vanes and as the rotor rotates so the pockets become smaller and the air is compressed. Compressed packets of air are thus discharged from the discharge port. Single-stage, rotary vane compressors typically can be used for pressures up to about 800 kPa with flow rates of the order of $0.3 \, \text{m}^3/\text{min}$ to $30 \, \text{m}^3/\text{min}$ free air delivery. Another form of compressor is the **rotary screw compressor** (Figure 7.4(c)). This has two intermeshing rotary screws which rotate in opposite directions. As the screws rotate, air is drawn into the casing through the inlet port and into the space between the screws. Then this trapped air is moved along the length of the screws and compressed as the space becomes progressively smaller, emerging from the discharge port. Typically, single-stage, rotary screw compressors can be used for pressures up to about 1000 kPa with flow rates of between $1.4 \, \text{m}^3/\text{min}$ and $60 \, \text{m}^3/\text{min}$ free air delivery.

7.2.3 Valves

Valves are used with hydraulic and pneumatic systems to direct and regulate the fluid flow. There are basically just two forms of valve, the **finite position** and the **infinite position** valves. The finite position valves are ones where the action is just to allow or block fluid flow and so can be used to switch actuators on or off. They can be used for directional control to switch the flow from one path to another and so from one actuator to another. The infinite position valves are able to control flow anywhere between fully on and fully off and so are used to control varying actuator forces or the rate of fluid flow for a process control situation.

7.3 Directional control valves

Pneumatic and hydraulic systems use directional control valves to direct the flow of fluid through a system. They are not intended to vary the rate of flow of fluid but are either completely open or completely closed, i.e. on/off devices. Such on/off valves are widely used to develop sequenced control systems (see Section 7.5.1). They might be activated to switch the fluid flow direction by means of mechanical, electrical or fluid pressure signals.

A common type of directional control valve is the **spool valve**. A spool moves horizontally within the valve body to control the flow. Figure 7.5

Figure 7.5 Spool valve.

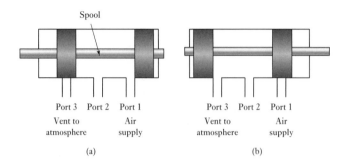

shows a particular form. In (a) the air supply is connected to port 1 and port 3 is closed. Thus the device connected to port 2 can be pressurised. When the spool is moved to the left (Figure 7.5(b)) the air supply is cut off and port 2 is connected to port 3. Port 3 is a vent to the atmosphere and so the air pressure in the system attached to port 2 is vented. Thus the movement of the spool has allowed the air firstly to flow into the system and then be reversed and flow out of the system. **Rotary spool valves** have a rotating spool which, when it rotates, opens and closes ports in a similar way.

Another common form of directional control valve is the **poppet valve**. Figure 7.6 shows one form. This valve is normally in the closed condition, there being no connection between port 1 to which the pressure supply is connected and port 2 to which the system is connected. In poppet valves, balls, discs or cones are used in conjunction with valve seats to control the flow. In the figure a ball is shown. When the push-button is depressed, the ball is pushed out of its seat and flow occurs as a result of port 1 being connected to port 2. When the button is released, the spring forces the ball back up against its seat and so closes off the flow.

Figure 7.6 Poppet valve.

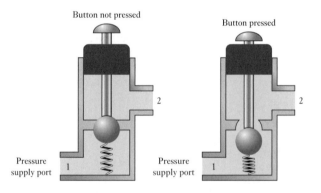

7.3.1   Valve symbols

The symbol used for a control valve consists of a square for each of its switching positions. Thus for the poppet valve shown in Figure 7.6, there are two positions: one with the button not pressed and one with it pressed. Thus a two-position valve will have two squares, a three-position valve three squares. Arrow-headed lines (Figure 7.7(a)) are used to indicate the directions of flow in each of the positions, with blocked-off lines indicating closed flow lines (Figure 7.7(b)). The initial position of the valve has the connections (Figure 7.7(c)) to the ports shown; in Figure 7.7(c) the valve has four ports. Ports are labelled by a number or a letter according to their function. The ports are labelled 1 (or P) for pressure supply, 3 (or T) for hydraulic return port, 3 or 5 (or R or S) for pneumatic exhaust ports, and 2 or 5 (or B or A) for output ports.

Figure 7.7 (a) Flow path, (b) flow shut-off, (c) initial connections.

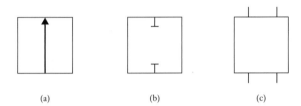

(a)　　　　　(b)　　　　　(c)

Figure 7.8(a) shows examples of some of the symbols which are used to indicate the various ways the valves can be actuated. More than one of these symbols might be used with the valve symbol. As an illustration, Figure 7.8(b) shows the symbol for the two-port, two-position poppet valve of Figure 7.6. Note that a two-port, two-position valve would be described as a 2/2 valve, the first number indicating the number of ports and the second number the number of positions. The valve actuation is by a push-button and a spring.

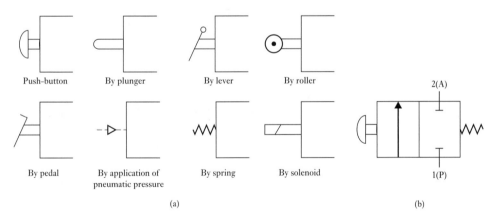

Push-button　　By plunger　　By lever　　By roller

By pedal　　By application of pneumatic pressure　　By spring　　By solenoid

(a)　　　　　　　　　　　　　　　　　　　　(b)

Figure 7.8 Valve actuation symbols.

As a further illustration, Figure 7.9 shows a solenoid-operated spool valve and its symbol. The valve is actuated by a current passing through a solenoid and returned to its original position by a spring.

Figure 7.10 shows the symbol for a 4/2 valve. The connections are shown for the initial state, i.e. 1(P) is connected to 2(A) and 3(R) closed. When the

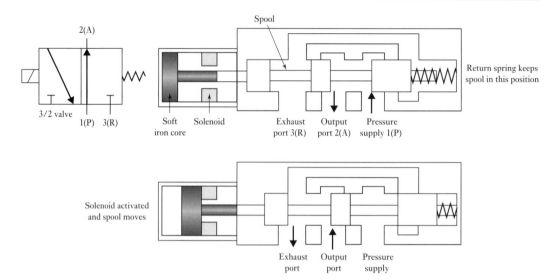

Figure 7.9   Single-solenoid valve.

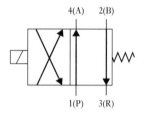

Figure 7.10   A 4/2 valve.

Figure 7.11   Lift system.

solenoid is activated it gives the state indicated by the symbols used in the square to which it is attached, i.e. we now have 1(P) closed and 2(A) connected to 3(R). When the current through the solenoid ceases, the spring pushes the valve back to its initial position. The spring movement gives the state indicated by the symbols used in the square to which it is attached.

Figure 7.11 shows a simple example of an application of valves in a pneumatic lift system. Two push-button 2/2 valves are used. When the button on the up valve is pressed, the load is lifted. When the button on the down valve is pressed, the load is lowered. Note that with pneumatic systems an open arrow is used to indicate a vent to the atmosphere.

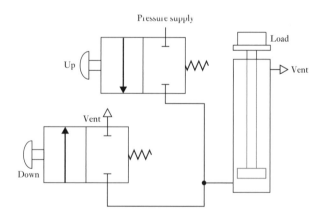

7.3.2   Pilot-operated valves

The force required to move the ball or shuttle in a valve can often be too large for manual or solenoid operation. To overcome this problem a **pilot-operated**

Figure 7.12 Pilot-operated system.

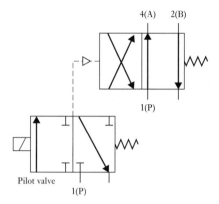

system is used where one valve is used to control a second valve. Figure 7.12 illustrates this. The pilot valve is small capacity and can be operated manually or by a solenoid. It is used to allow the main valve to be operated by the system pressure. The pilot pressure line is indicated by dashes. The pilot and main valves can be operated by two separate valves but they are often combined in a single housing.

7.3.3 Directional valves

Figure 7.13 shows a simple **directional valve** and its symbol. Free flow can only occur in one direction through the valve: that which results in the ball being pressed against the spring. Flow in the other direction is blocked by the spring forcing the ball against its seat.

Figure 7.13 Directional valve.

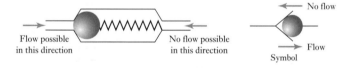

7.4 Pressure control valves

There are three main types of pressure control valves.

1 *Pressure-regulating valves*
   These are used to control the operating pressure in a circuit and maintain it at a constant value.

2 *Pressure-limiting valves*
   These are used as safety devices to limit the pressure in a circuit to below some safe value. The valve opens and vents to the atmosphere, or back to the sump, if the pressure rises above the set safe value. Figure 7.14 shows a **pressure-limiting/relief valve** which has one orifice which is normally closed. When the inlet pressure overcomes the force exerted by the spring, the valve opens and vents to the atmosphere, or back to the sump.

Figure 7.14   Pressure-limiting valve.

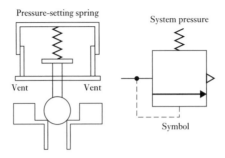

3   *Pressure sequence valves*

These valves are used to sense the pressure of an external line and give a signal when it reaches some preset value. With the pressure-limiting valve of Figure 7.15, the limiting pressure is set by the pressure at the inlet to the valve. We can adapt such a valve to give a sequence valve. This can be used to allow flow to occur to some part of the system when the pressure has risen to the required level. For example, in an automatic machine we might require some operation to start when the clamping pressure applied to a workpiece is at some particular value. Figure 7.15(a) shows the symbol for a sequence valve, the valve switching on when the inlet pressure reaches a particular value and allowing the pressure to be applied to the system that follows. Figure 7.15(b) shows a system where such a sequential valve is used. When the 4/3 valve first operates, the pressure is applied to cylinder 1 and its ram moves to the right. While this is happening the pressure is too low to operate the sequence valve and so no pressure is applied to cylinder 2. When the ram of cylinder 1 reaches the end stop, then the pressure in the system rises and, at an appropriate level, triggers the sequence valve to open and so apply pressure to cylinder 2 to start its ram in motion.

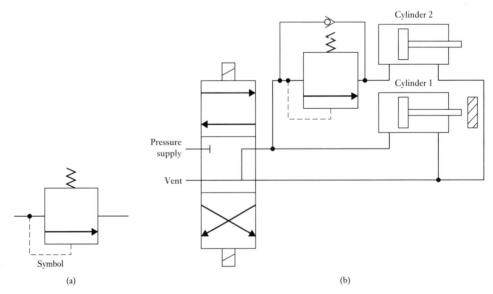

Figure 7.15   (a) Pressure sequence valve symbol, (b) a sequential system.

7.5 Cylinders

The **hydraulic** or **pneumatic cylinder** is an example of a linear actuator. The principles and form are the same for both hydraulic and pneumatic versions, differences being purely a matter of size as a consequence of the higher pressures used with hydraulics. The cylinder consists of a cylindrical tube along which a piston/ram can slide. There are two basic types, single-acting cylinders and double-acting cylinders.

The term single acting is used when the control pressure is applied to just one side of the piston, a spring often being used to provide the opposition to the movement of the piston. The other side of the piston is open to the atmosphere. Figure 7.16 shows one such cylinder with a spring return. The fluid is applied to one side of the piston at a gauge pressure p with the other side being at atmospheric pressure and so produces a force on the piston of pA, where A is the area of the piston. The actual force acting on the piston rod will be less than this because of friction.

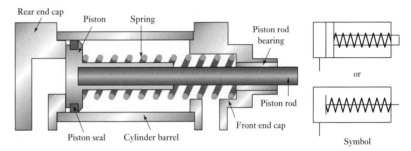

Figure 7.16 Single-acting cylinder.

For the single-acting cylinder shown in Figure 7.17, when a current passes through the solenoid, the valve switches position and pressure is applied to move the piston along the cylinder. When the current through the solenoid ceases, the valve reverts to its initial position and the air is vented from the cylinder. As a consequence the spring returns the piston back along the cylinder.

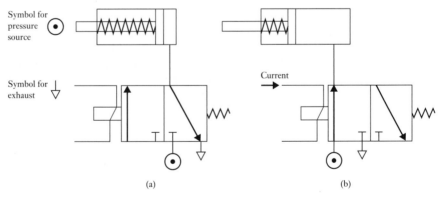

Figure 7.17 Control of a single-acting cylinder with (a) no current through solenoid, (b) a current through the solenoid.

The term 'double acting' is used when the control pressures are applied to each side of the piston (Figure 7.18). A difference in pressure between the two sides then results in motion of the piston, the piston being able to move in either direction along the cylinder as a result of high-pressure signals. For

Figure 7.18 Double-acting cylinder.

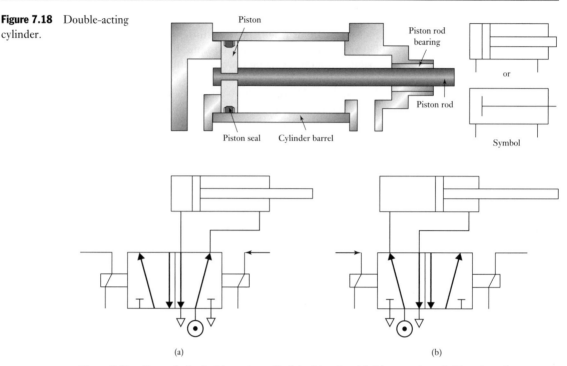

(a)   (b)

Figure 7.19   Control of a double-acting cylinder with solenoid, (a) not activated, (b) activated.

the double-acting cylinder shown in Figure 7.19, current through one solenoid causes the piston to move in one direction with current through the other solenoid reversing the direction of motion.

The choice of cylinder is determined by the force required to move the load and the speed required. Hydraulic cylinders are capable of much larger forces than pneumatic cylinders. However, pneumatic cylinders are capable of greater speeds. The force produced by a cylinder is equal to the cross-sectional area of the cylinder multiplied by the working pressure, i.e. the pressure difference between the two sides of the piston, in the cylinder. A cylinder for use with a working pneumatic pressure of 500 kPa and having a diameter of 50 mm will thus give a force of 982 N. A hydraulic cylinder with the same diameter and a working pressure of 15 000 kPa will give a force of 29.5 kN.

If the flow rate of hydraulic liquid into a cylinder is a volume of Q per second, then the volume swept out by the piston in a time of 1 s must be Q. But for a piston of cross-sectional area A this is a movement through a distance of v in 1 s, where we have $Q = Av$. Thus the speed v of a hydraulic cylinder is equal to the flow rate of liquid Q through the cylinder divided by the cross-sectional area A of the cylinder. Thus for a hydraulic cylinder of diameter 50 mm and a hydraulic fluid flow of 7.5×10^{-3} m³/s, the speed is 3.8 m/s. The speed of a pneumatic cylinder cannot be calculated in this way since its speed depends on the rate at which air can be vented ahead of the advancing piston. A valve to adjust this can be used to regulate the speed.

To illustrate the above, consider the problem of a hydraulic cylinder to be used to move a workpiece in a manufacturing operation through a distance of 250 mm in 15 s. If a force of 50 kN is required to move the workpiece, what is the required working pressure and hydraulic liquid flow rate if a cylinder with a piston diameter of 150 mm is available? The cross-sectional area of the piston is $\frac{1}{4}\pi \times 0.150^2 = 0.0177$ m^2. The force produced by the cylinder is equal to the product of the cross-sectional area of the cylinder and the working pressure. Thus the working pressure is $50 \times 10^3/0.0177 = 2.8$ MPa. The speed of a hydraulic cylinder is equal to the flow rate of liquid through the cylinder divided by the cross-sectional area of the cylinder. Thus the required flow rate is $(0.250/15) \times 0.0177 = 2.95 \times 10^{-4}$ m^3/s.

7.5.1 Cylinder sequencing

Many control systems employ pneumatic or hydraulic cylinders as the actuating elements and require a sequence of extensions and retractions of the cylinders to occur. For example, we might have two cylinders A and B and require that when the start button is pressed, the piston of cylinder A extends and then, when it is fully extended, the piston of cylinder B extends. When this has happened and both are extended, we might need the piston of cylinder A to retract, and when it is fully retracted we might then have the piston of B retract. In discussions of sequential control with cylinders it is common practice to give each cylinder a reference letter A, B, C, D, etc., and to indicate the state of each cylinder by using a $+$ sign if it is extended or a $-$ sign if retracted. Thus the above required sequence of operations is $A+, B+, A-, B-$. Figure 7.20 shows a circuit that could be used to generate this sequence.

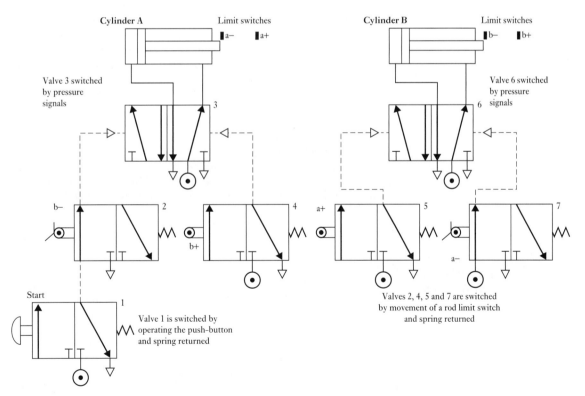

Figure 7.20 Two-actuator sequential operation.

The sequence of operations is outlined below.

1 Initially both the cylinders have retracted pistons. Start push-button on valve 1 is pressed. This applies pressure to valve 2, as initially limit switch b− is activated, hence valve 3 is switched to apply pressure to cylinder A for extension.

2 Cylinder A extends, releasing limit switch a−. When cylinder A is fully extended, limit switch a+ operates. This switches valve 5 and causes pressure to be applied to valve 6 to switch it and so apply pressure to cylinder B to cause its piston to extend.

3 Cylinder B extends, releasing limit switch b−. When cylinder B is fully extended, limit switch b+ operates. This switches valve 4 and causes pressure to be applied to valve 3 and so applies pressure to cylinder A to start its piston retracting.

4 Cylinder A retracts, releasing limit switch a+. When cylinder A is fully retracted, limit switch a− operates. This switches valve 7 and causes pressure to be applied to valve 5 and so applies pressure to cylinder B to start its piston retracting.

5 Cylinder B retracts, releasing limit switch b+. When cylinder B is fully retracted, limit switch b− operates to complete the cycle.

The cycle can be started again by pushing the start button. If we wanted the system to run continuously then the last movement in the sequence would have to trigger the first movement.

An alternative way of realising the above sequence involves the air supply being switched on and off to valves in groups and is termed **cascade control**. This avoids a problem that can occur with circuits, formed in the way shown in Figure 7.20, of air becoming trapped in the pressure line to control a valve and so preventing the valve from switching. With cascade control, the sequence of operations is divided into groups with no cylinder letter appearing more than once in each group. Thus for the sequence A+, B+, B−, A− we can have the groups A+, B+ and A−, B−. A valve is then used to switch the air supply between the two groups, i.e. air to the group A+B+ and then air switched to the group with A−B−. A start/stop valve is included in the line that selects the first group, and if the sequence is to be continuously repeated, the last operation has to supply a signal to start the sequence over again. The first function in each group is initiated by that group supply being switched on; further actions within the group are controlled by switch-operated valves, and the last valve operation initiates the next group to be selected. Figure 7.21 shows the pneumatic circuit.

7.6 Servo and proportional control valves

Servo and **proportional control valves** are both infinite position valves which give a valve spool displacement proportional to the current supplied to a solenoid. Basically, servo valves have a torque motor to move the spool within a valve (Figure 7.22). By varying the current supplied to the torque motor, an armature is deflected and this moves the spool in the valve and hence gives a flow related to the current. Servo valves are high precision and costly and generally used in a closed-loop control system.

Proportional control valves are less expensive and basically have the spool position directly controlled by the size of the current to the valve solenoid. They are often used in open-loop control systems.

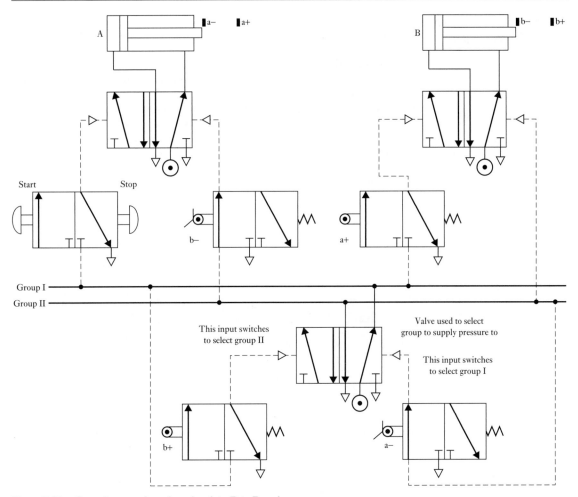

Figure 7.21 Cascade control used to give $A+$, $B+$, $B-$, $A-$.

Figure 7.22 The basic form of a servo valve.

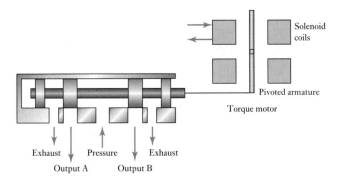

7.7   Process control valves

Process control valves are used to control the rate of fluid flow and are used where, perhaps, the rate of flow of a liquid into a tank has to be controlled. The basis of such valves is an actuator being used to move a plug into the flow pipe and so alter the cross-section of the pipe through which the fluid can flow.

A common form of pneumatic actuator used with process control valves is the **diaphragm actuator**. Essentially it consists of a diaphragm with the input pressure signal from the controller on one side and atmospheric pressure on the other, this difference in pressure being termed the **gauge pressure**. The diaphragm is made of rubber which is sandwiched in its centre between two circular steel discs. The effect of changes in the input pressure is thus to move the central part of the diaphragm, as illustrated in Figure 7.23(a). This movement is communicated to the final control element by a shaft which is attached to the diaphragm, e.g. as in Figure 7.23(b).

Figure 7.23   (a) Pneumatic diaphragm actuator, (b) control valve.

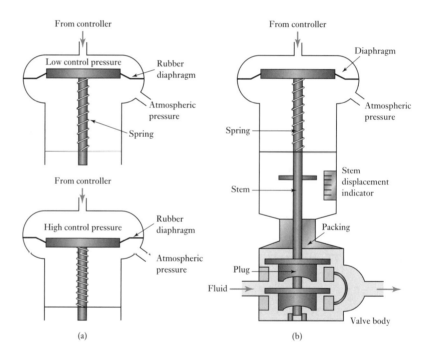

The force F acting on the shaft is the force that is acting on the diaphragm and is thus the gauge pressure P multiplied by the diaphragm area A. A restoring force is provided by a spring. Thus if the shaft moves through a distance x, and assuming the compression of the spring is proportional to the force, i.e. $F = kx$ with k being a constant, then $kx = PA$ and thus the displacement of the shaft is proportional to the gauge pressure.

To illustrate the above, consider the problem of a diaphragm actuator to be used to open a control valve if a force of 500 N must be applied to the valve. What diaphragm area is required for a control gauge pressure of 100 kPa? The force F applied to the diaphragm of area A by a pressure P is given by $P = F/A$. Hence $A = 500/(100 \times 10^3) = 0.005 \text{ m}^2$.

7.7.1 Valve bodies and plugs

Figure 7.23(b) shows a cross-section of a valve for the control of rate of flow of a fluid. The pressure change in the actuator causes the diaphragm to move and so consequently the valve stem. The result of this is a movement of the inner-valve plug within the valve body. The plug restricts the fluid flow and so its position determines the flow rate.

There are many forms of valve body and plug. Figure 7.24 shows some forms of valve bodies. The term **single-seated** is used for a valve where there is just one path for the fluid through the valve and so just one plug is needed to control the flow. The term **double-seated** is used for a valve where the fluid on entering the valve splits into two streams, as in Figure 7.23, with each stream passing through an orifice controlled by a plug. There are thus two plugs with such a valve.

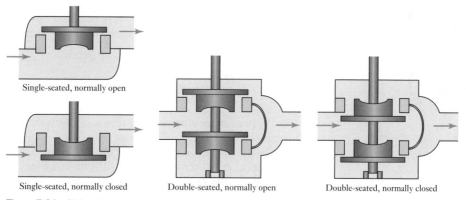

Single-seated, normally open

Single-seated, normally closed   Double-seated, normally open   Double-seated, normally closed

Figure 7.24 Valve bodies.

A single-seated valve has the advantage that it can be closed more tightly than a double-seated one, but the disadvantage that the force on the plug due to the flow is much higher and so the diaphragm in the actuator has to exert considerably higher forces on the stem. This can result in problems in accurately positioning the plug. Double-seated valves thus have an advantage here. The form of the body also determines whether an increasing air pressure will result in the valve opening or closing.

The shape of the plug determines the relationship between the stem movement and the effect on the flow rate. Figure 7.25(a) shows three commonly used types and Figure 7.25(b) how the percentage by which the volumetric rate of flow is related to the percentage displacement of the valve stem.

With the **quick-opening** type a large change in flow rate occurs for a small movement of the valve stem. Such a plug is used where on/off control of flow rate is required.

With the **linear-contoured** type, the change in flow rate is proportional to the change in displacement of the valve stem, i.e.

change in flow rate $=$ k(change in stem displacement)

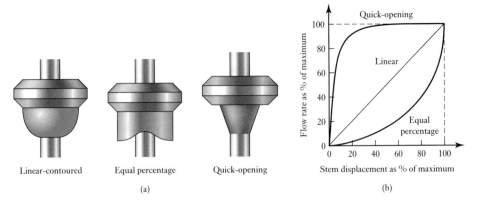

Linear-contoured   Equal percentage   Quick-opening

(a)

(b)

Figure 7.25   (a) Plug shapes, (b) flow characteristics.

where k is a constant. If Q is the flow rate at a valve stem displacement S and Q_{max} is the maximum flow rate at the maximum stem displacement S_{max}, then we have

$$\frac{Q}{Q_{max}} = \frac{S}{S_{max}}$$

or the percentage change in the flow rate equals the percentage change in the stem displacement.

To illustrate the above, consider the problem of an actuator which has a stem movement at full travel of 30 mm. It is mounted on a linear plug valve which has a minimum flow rate of 0 and a maximum flow rate of 40 m³/s. What will be the flow rate when the stem movement is (a) 10 mm, (b) 20 mm? Since the percentage flow rate is the same as the percentage stem displacement, then: (a) a percentage stem displacement of 33% gives a percentage flow rate of 33%, i.e. 13 m³/s; (b) a percentage stem displacement of 67% gives a percentage flow rate of 67%, i.e. 27 m³/s.

With the **equal percentage** type of plug, equal percentage changes in flow rate occur for equal changes in the valve stem position, i.e.

$$\frac{\Delta Q}{Q} = k\Delta S$$

where ΔQ is the change in flow rate at a flow rate of Q and ΔS the change in valve position resulting from this change. If we write this expression for small changes and then integrate it we obtain

$$\int_{Q_{min}}^{Q} \frac{1}{Q} dQ = k \int_{S_{min}}^{S} dS$$

$$\ln Q - \ln Q_{min} = k(S - S_{min})$$

If we consider the flow rate Q_{max} which is given by S_{max} then

$$\ln Q_{max} - \ln Q_{min} = k(S_{max} - S_{min})$$

Eliminating k from these two equations gives

$$\frac{\ln Q - \ln Q_{\min}}{\ln Q_{\max} - \ln Q_{\min}} = \frac{S - S_{\min}}{S_{\max} - S_{\min}}$$

$$\ln \frac{Q}{Q_{\min}} = \frac{S - S_{\min}}{S_{\max} - S_{\min}} \ln \frac{Q_{\max}}{Q_{\min}}$$

and so

$$\frac{Q}{Q_{\min}} = \left(\frac{Q_{\max}}{Q_{\min}}\right)^{(S - S_{\min})/(S_{\max} - S_{\min})}$$

The term **rangeability** R is used for the ratio $Q_{\max}/Q_{\min}$.

To illustrate the above, consider the problem of an actuator which has a stem movement at full travel of 30 mm. It is mounted with a control valve having an equal percentage plug and which has a minimum flow rate of 2 m^3/s and a maximum flow rate of 24 m^3/s. What will be the flow rate when the stem movement is (a) 10 mm, (b) 20 mm? Using the equation

$$\frac{Q}{Q_{\min}} = \left(\frac{Q_{\max}}{Q_{\min}}\right)^{(S - S_{\min})/(S_{\max} - S_{\min})}$$

we have for (a) $Q = 2 \times (24/2)^{10/30} = 4.6$ m^3/s and for (b) $Q = 2 \times (24/2)^{20/30} = 10.5$ m^3/s.

The relationship between the flow rate and the stem displacement is the inherent characteristic of a valve. It is only realised in practice if the pressure losses in the rest of the pipework, etc., are negligible compared with the pressure drop across the valve itself. If there are large pressure drops in the pipework so that, for example, less than half the pressure drop occurs across the valve, then a linear characteristic might become almost a quick-opening characteristic. The linear characteristic is thus widely used when a linear response is required and most of the system pressure is dropped across the valve. The effect of large pressure drops in the pipework with an equal percentage valve is to make it more like a linear characteristic. For this reason, if a linear response is required when only a small proportion of the system pressure is dropped across the valve, then an equal percentage value might be used.

7.7.2 Control valve sizing

The term control valve sizing is used for the procedure of determining the correct size of valve body. The equation relating the rate of flow of liquid Q through a wide open valve to its size is

$$Q = A_V \sqrt{\frac{\Delta P}{\rho}}$$

where A_V is the valve flow coefficient, ΔP the pressure drop across the valve and ρ the density of the fluid. This equation is sometimes written, with the quantities in SI units, as

$$Q = 2.37 \times 10^{-5} C_V \sqrt{\frac{\Delta P}{\rho}}$$

where C_V is the valve flow coefficient. Alternatively it may be found written as

$$Q = 0.75 \times 10^{-6} C_V \sqrt{\frac{\Delta P}{G}}$$

where G is the specific gravity or relative density. These last two forms of the equation derive from its original specification in terms of US gallons. Table 7.1 shows some typical values of A_V, C_V and valve size.

Table 7.1   Flow coefficients and valve sizes.

Flow coefficients	Valve size (mm)							
	480	640	800	960	1260	1600	1920	2560
C_V	8	14	22	30	50	75	110	200
$A_V \times 10^{-5}$	19	33	52	71	119	178	261	474

To illustrate the above, consider the problem of determining the valve size for a valve that is required to control the flow of water when the maximum flow required is $0.012 \text{ m}^3/\text{s}$ and the permissible pressure drop across the valve at this flow rate is 300 kPa. Using the equation

$$Q = A_V \sqrt{\frac{\Delta P}{\rho}}$$

then, since the density of water is 1000 kg/m^3,

$$A_V = Q \sqrt{\frac{\rho}{\Delta P}} = 0.012 \sqrt{\frac{1000}{300 \times 10^3}} = 69.3 \times 10^{-5}$$

Thus, using Table 7.1, the valve size is 960 mm.

7.7.3   Example of fluid control system

Figure 7.26(a) shows the essential features of a system for the control of a variable such as the level of a liquid in a container by controlling the rate at which liquid enters it. The output from the liquid–level sensor, after signal conditioning, is transmitted to the current-to-pressure converter as a current of 4 to 20 mA. It is then converted into a gauge pressure of 20 to 100 kPa which then actuates a pneumatic control valve and so controls the rate at which liquid is allowed to flow into the container.

Figure 7.26(b) shows the basic form of a current-to-pressure converter for such a system. The input current passes through coils mounted on a core which is attracted towards a magnet, the extent of the attraction depending on the size of the current. The movement of the core causes movement of the lever about its pivot and so the movement of a flapper above the nozzle. The position of the flapper in relation to the nozzle determines the rate at which air can escape from the system and hence the air pressure in the system. Springs on the flapper are used to adjust the sensitivity of the converter so that currents of 4 to 20 mA produce gauge pressures of 20 to 100 kPa. These are the standard values that are generally used in such systems.

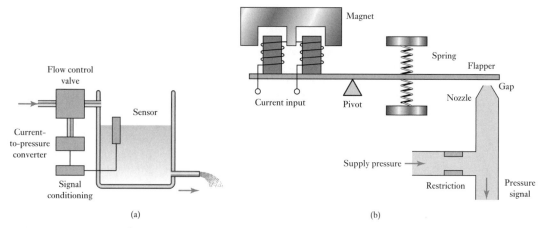

Figure 7.26 (a) Fluid control system, (b) current-to-pressure converter.

7.8 Rotary actuators

A linear cylinder can, with suitable mechanical linkages, be used to produce rotary movement through angles less than 360°, Figure 7.27(a) illustrating such an arrangement. Another alternative is a **semi-rotary actuator** involving a vane (Figure 7.27(b)). A pressure difference between the two ports causes the vane to rotate and so give a shaft rotation which is a measure of the pressure difference. Depending on the pressures, so the vane can be rotated clockwise or anti-clockwise.

For rotation through angles greater than 360° a pneumatic motor can be used; one such form is the **vane motor** (Figure 7.27(c)). An eccentric rotor has slots in which vanes are forced outwards against the walls of the cylinder by the rotation. The vanes divide the chamber into separate compartments which increase in size from the inlet port round to the exhaust port. The air entering such a compartment exerts a force on a vane and causes the rotor to rotate. The motor can be made to reverse its direction of rotation by using a different inlet port.

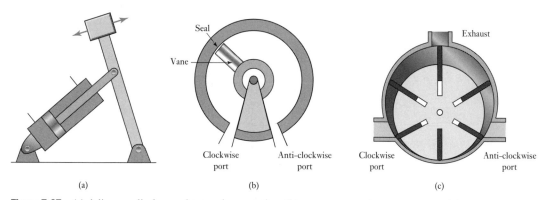

Figure 7.27 (a) A linear cylinder used to produce rotation, (b) vane-type semi-rotary actuator, (c) vane motor.

Summary

Pneumatic systems use air, hydraulic systems use oil. The main drawback with pneumatic systems is the compressibility of air. Hydraulic systems can be used for higher power control devices but are more expensive than pneumatic systems and there are hazards associated with oil leaks which do not occur with air leaks.

Pneumatic and hydraulic systems use **directional control valves** to direct the flow of fluid through a system. Such valves are on/off valves. The symbol used for such a valve is a square for each of its switching positions, the symbols used in each square indicating the connections made when that switching position is activated.

The **hydraulic** or **pneumatic cylinder** consists of a cylindrical tube along which a piston/ram can slide. There are two basic types, **single-acting cylinders** and **double-acting cylinders**. With single-acting, the control pressure is applied to just one side of the piston, a spring often being used to provide the opposition to the movement of the piston. The other side of the piston is open to the atmosphere. The term double-acting is used when the control pressures are applied to each side of the piston.

Servo and **proportional control valves** are both infinite position valves which give a valve spool displacement proportional to the current supplied to a solenoid.

Process control valves are used to control the rate of fluid flow. The basis of such valves is an actuator being used to move a plug into the flow pipe and so alter the cross-section of the pipe through which the fluid can flow. There are many forms of valve body and plug, these determining how the valve controls the fluid flow.

Problems

7.1   Describe the basic details of (a) a poppet valve, (b) a shuttle valve.

7.2   Explain the principle of a pilot-operated valve.

7.3   Explain how a sequential valve can be used to initiate an operation only when another operation has been completed.

7.4   Draw the symbols for (a) a pressure-relief valve, (b) a 2/2 valve which has actuators of a push-button and a spring, (c) a 4/2 valve, (d) a directional valve.

7.5   State the sequence of operations that will occur for the cylinders A and B in Figure 7.28 when the start button is pressed. a−, a+, b− and b+ are limit switches to detect when the cylinders are fully retracted and fully extended.

7.6   Design a pneumatic valve circuit to give the sequence A+, followed by B+ and then simultaneously followed by A− and B−.

7.7   A force of 400 N is required to open a process control valve. What area of diaphragm will be needed with a diaphragm actuator to open the valve with a control gauge pressure of 70 kPa?

7.8   A pneumatic system is operated at a pressure of 1000 kPa. What diameter cylinder will be required to move a load requiring a force of 12 kN?

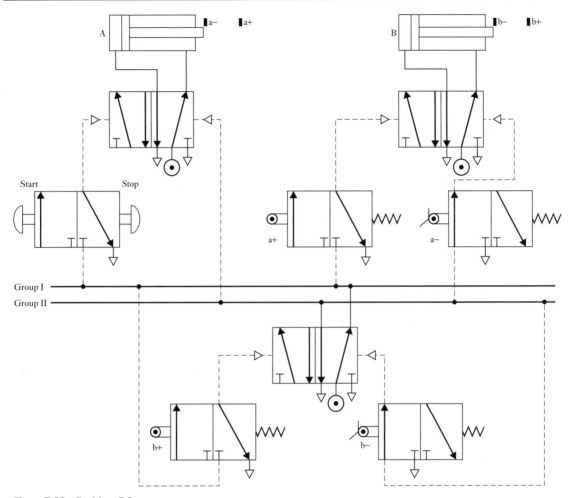

Figure 7.28 Problem 7.5.

7.9 A hydraulic cylinder is to be used to move a workpiece in a manufacturing operation through a distance of 50 mm in 10 s. A force of 10 kN is required to move the workpiece. Determine the required working pressure and hydraulic liquid flow rate if a cylinder with a piston diameter of 100 mm is available.

7.10 An actuator has a stem movement which at full travel is 40 mm. It is mounted with a linear plug process control valve which has a minimum flow rate of 0 and a maximum flow rate of 0.20 m³/s. What will be the flow rate when the stem movement is (a) 10 mm, (b) 20 mm?

7.11 An actuator has a stem movement which at full travel is 40 mm. It is mounted on a process control valve with an equal percentage plug and which has a minimum flow rate of 0.2 m³/s and a maximum flow rate of 4.0 m³/s. What will be the flow rate when the stem movement is (a) 10 mm, (b) 20 mm?

7.12 What is the process control valve size for a valve that is required to control the flow of water when the maximum flow required is 0.002 m³/s and the permissible pressure drop across the valve at this flow rate is 100 kPa? The density of water is 1000 kg/m³.

Chapter eight   Mechanical actuation systems

Objectives

The objectives of this chapter are that, after studying it, the reader should be able to:
- Determine possible mechanical actuation systems for motion transmission involving linear-to-rotary, rotary-to-rotary, rotary-to-linear and cyclic motion transmission.
- Evaluate the capabilities of linkages, cams, gears, ratchet-and-pawl, belt and chain drives and bearings for actuation systems.

8.1 Mechanical systems

This chapter is a consideration of **mechanisms**: mechanisms are devices which can be considered to be motion converters in that they transform motion from one form to some other required form. They might, for example, transform linear motion into rotational motion, or motion in one direction into a motion in a direction at right angles, or perhaps a linear reciprocating motion into rotary motion, as in the internal combustion engine where the reciprocating motion of the pistons is converted into rotation of the crank and hence the drive shaft.

Mechanical elements can include the use of linkages, cams, gears, rack-and-pinion, chains, belt drives, etc. For example, the rack-and-pinion can be used to convert rotational motion to linear motion. Parallel shaft gears might be used to reduce a shaft speed. Bevel gears might be used for the transmission of rotary motion through 90°. A toothed belt or chain drive might be used to transform rotary motion about one axis to motion about another. Cams and linkages can be used to obtain motions which are prescribed to vary in a particular manner. This chapter is a consideration of the basic characteristics of a range of such mechanisms.

Many of the actions which previously were obtained by the use of mechanisms are, however, often nowadays being obtained, as a result of a mechatronics approach, by the use of microprocessor systems. For example, cams on a rotating shaft were previously used for domestic washing machines in order to give a timed sequence of actions such as opening a valve to let water into the drum, switching the water off, switching a heater on, etc. Modern washing machines use a microprocessor-based system with the microprocessor programmed to switch on outputs in the required sequence. Another example is the hairspring balance wheel and gears and pointer of a watch which have now largely been replaced by an integrated circuit with perhaps a liquid crystal display. The mechatronics approach has resulted in a simplification, and often a reduction in cost.

Mechanisms still, however, have a role in mechatronics systems. For example, the mechatronics system in use in an automatic camera for adjusting the aperture for correct exposures involves a mechanism for adjusting the size of the diaphragm.

While electronics might now be used often for many functions that previously were fulfilled by mechanisms, mechanisms might still be used to provide such functions as:

1 force amplification, e.g. that given by levers;
2 change of speed, e.g. that given by gears;
3 transfer of rotation about one axis to rotation about another, e.g. a timing belt;
4 particular types of motion, e.g. that given by a quick-return mechanism.

The term **kinematics** is used for the study of motion without regard to forces. When we consider just the motions without any consideration of the forces or energy involved then we are carrying out a kinematic analysis of the mechanism. This chapter is an introduction to such a consideration.

8.2 Types of motion

The motion of any rigid body can be considered to be a combination of translational and rotational motions. By considering the three dimensions of space, a **translation motion** can be considered to be a movement which can be resolved into components along one or more of the three axes (Figure 8.1(a)). A **rotational motion** can be considered as a rotation which has components rotating about one or more of the axes (Figure 8.1(b)).

Figure 8.1 Types of motion: (a) translational, (b) rotational.

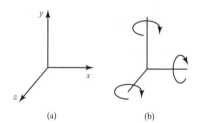

(a)   (b)

A complex motion may be a combination of translational and rotational motions. For example, think of the motion required for you to pick up a pencil from a table. This might involve your hand moving at a particular angle towards the table, rotation of the hand, and then all the movement associated with opening your fingers and moving them to the required positions to grasp the pencil. This is a sequence of quite complex motions. However, we can break down all these motions into combinations of translational and rotational motions. Such an analysis is particularly relevant if we are not moving a human hand to pick up the pencil but instructing a robot to carry out the task. Then it really is necessary to break down the motion into combinations of translational and rotational motions so that we can design mechanisms to carry out each of these components of the motion. For example, among the sequence of control signals sent to a mechanism might be such groupings of signals as those to instruct joint 1 to rotate by 20° and link 2 to be extended by 4 mm for translational motion.

8.2.1   Freedom and constraints

An important aspect in the design of mechanical elements is the orientation and arrangement of the elements and parts. A body that is free in space can move in three, independent, mutually perpendicular directions and rotate in three ways about those directions (Figure 8.1). It is said to have six degrees of freedom. The number of **degrees of freedom** is the number of components of motion that are required in order to generate the motion. If a joint is constrained to move along a line then its translational degrees of freedom are reduced to one. Figure 8.2(a) shows a joint with just this one translational degree of freedom. If a joint is constrained to move on a plane then it has two translational degrees of freedom. Figure 8.2(b) shows a joint which has one translational degree of freedom and one rotational degree of freedom.

Figure 8.2 Joints with: (a) one, (b) two degrees of freedom.

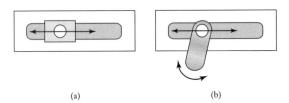

(a)   (b)

The problem in design is often to reduce the number of degrees of freedom and this then requires an appropriate number and orientation of constraints. Without any constraints a body would have six degrees of freedom. A constraint is needed for each degree of freedom that is to be prevented from occurring. Provided we have no redundant constraints then the number of degrees of freedom would be 6 minus the number of constraints. However, redundant constraints often occur and so for constraints on a single rigid body we have the basic rule

$$6 \; - \; \text{number of constraints} \; = \; \text{number of degrees of freedom}$$
$$- \; \text{number of redundancies}$$

Thus if a body is required to be fixed, i.e. have zero degrees of freedom, then if no redundant constraints are introduced the number of constraints required is 6.

A concept that is used in design is that of the **principle of least constraint**. This states that in fixing a body or guiding it to a particular type of motion, the minimum number of constraints should be used, i.e. there should be no redundancies. This is often referred to as **kinematic design**.

For example, to have a shaft which only rotates about one axis with no translational motions, we have to reduce the number of degrees of freedom to 1. Thus the minimum number of constraints to do this is 5. Any more constraints than this will give redundancies. The mounting that might be used to mount the shaft has a ball bearing at one end and a roller bearing at the other (Figure 8.3). The pair of bearings together prevent translational motion at right angles to the shaft, the y-axis, and rotations about the z-axis and the y-axis. The ball bearing prevents translational motion along the x-axis and along the z-axis. Thus there is a total of five constraints. This leaves just one degree of freedom, the required rotation about the x-axis. If

Figure 8.3 Shaft with no redundancies.

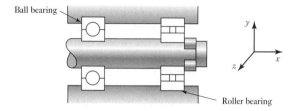

there had been a roller bearing at each end of the shaft then both the bearings could have prevented translational motion along the x-axis and the z-axis and thus there would have been redundancy. Such redundancy might cause damage. If ball bearings are used at both ends of the shaft, then in order to prevent redundancy one of the bearings would have its outer race not fixed in its housing so that it could slide to some extent in an axial direction.

8.2.2 Loading

Mechanisms are structures and as such transmit and support loads. Analysis is thus necessary to determine the loads to be carried by individual elements. Then consideration can be given to the dimensions of the element so that it might, for example, have sufficient strength and perhaps stiffness under such loading.

8.3 Kinematic chains

When we consider the movements of a mechanism without any reference to the forces involved, we can treat the mechanism as being composed of a series of individual links. Each part of a mechanism which has motion relative to some other part is termed a **link**. A link need not necessarily be a rigid body but it must be a resistant body which is capable of transmitting the required force with negligible deformation. For this reason it is usually taken as being represented by a rigid body which has two or more points of attachment to other links, these being termed **nodes**. Each link is capable of moving relative to its neighbouring links. Figure 8.4 shows examples of links with two, three and four nodes. A **joint** is a connection between two or more links at their nodes and which allows some motion between the connected links. Levers, cranks, connecting rods and pistons, sliders, pulleys, belts and shafts are all examples of links.

Figure 8.4 Links: (a) with two nodes, (b) with three nodes, (c) with four nodes.

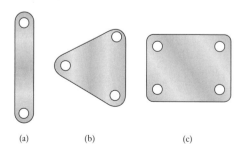

(a)   (b)   (c)

A sequence of joints and links is known as a **kinematic chain**. For a kinematic chain to transmit motion, one link must be fixed. Movement of one link will then produce predictable relative movements of the others. It is possible to obtain from one kinematic chain a number of different mechanisms by having a different link as the fixed one.

As an illustration of a kinematic chain, consider a motor car engine where the reciprocating motion of a piston is transformed into rotational motion of a crankshaft on bearings mounted in a fixed frame (Figure 8.5(a)). We can represent this as being four connected links (Figure 8.5(b)). Link 1 is the crankshaft, link 2 the connecting rod, link 3 the fixed frame and link 4 the slider, i.e. piston, which moves relative to the fixed frame (see Section 8.3.2 for further discussion).

Figure 8.5 Simple engine mechanism.

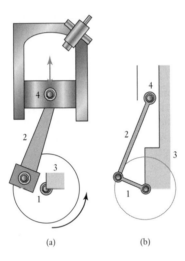

(a)   (b)

The designs of many mechanisms are based on two basic forms of kinematic chains, the four-bar chain and the slider–crank chain. The following illustrates some of the forms such chains can take.

8.3.1   The four-bar chain

The **four-bar chain** consists of four links connected to give four joints about which turning can occur. Figure 8.6 shows a number of forms of the four-bar chain produced by altering the relative lengths of the links. If the

Figure 8.6 Examples of four-bar chains.

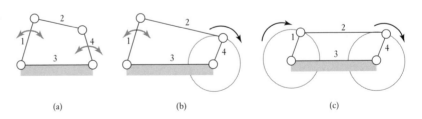

(a)   (b)   (c)

sum of the length of the shortest link plus the length of the longest link is less than or equal to the sum of the lengths of the other two links then at least one link will be capable of making a full revolution with respect to the fixed link. If this condition is not met then no link is capable of a complete revolution. This is known as the Grashof condition. In Figure 8.6(a), link 3 is fixed and the relative lengths of the links are such that links 1 and 4 can oscillate but not rotate. The result is a **double-lever mechanism**. By shortening link 4 relative to link 1, then link 4 can rotate (Figure 8.6(b)) with link 1 oscillating and the result is termed a **lever–crank mechanism**. With links 1 and 4 the same length and both able to rotate (Figure 8.6(c)), then the result is a **double-crank mechanism**. By altering which link is fixed, other forms of mechanism can be produced.

Figure 8.7 illustrates how such a mechanism can be used to advance the film in a cine camera. As link 1 rotates so the end of link 2 locks into a sprocket of the film, pulls it forward before releasing and moving up and back to lock into the next sprocket.

Figure 8.7 Cine film advance mechanism.

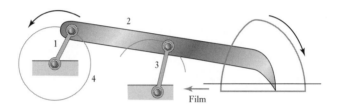

Some linkages may have **toggle positions**. These are positions where the linkage will not react to any input from one of its links. Figure 8.8 illustrates such a toggle, being the linkage used to control the movement of the tailgate of a truck so that when link 2 reaches the horizontal position no further load on link 2 will cause any further movement. There is another toggle position for the linkage and that is when links 3 and 4 are both vertical and the tailgate is vertical.

Figure 8.8 Toggle linkage.

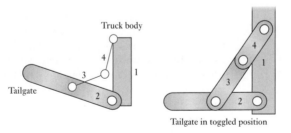

8.3.2 The slider–crank mechanism

This form of mechanism consists of a crank, a connecting rod and a slider and is the type of mechanism described in Figure 8.5 which showed the simple engine mechanism. With that configuration, link 3 is fixed, i.e. there is no relative movement between the centre of rotation of the crank and the housing in which the piston slides. Link 1 is the crank that rotates, link 2 the connecting rod and link 4 the slider which moves relative to the fixed link. When the piston moves backwards and forwards, i.e. link 4 moves

Figure 8.9 The position sequence for the links in a slider-crank mechanism.

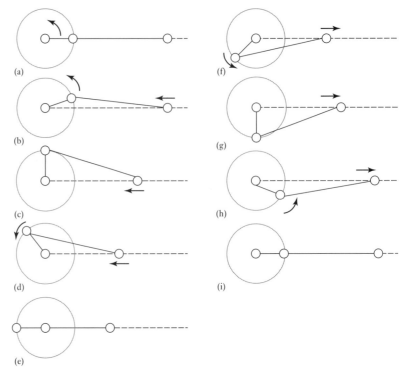

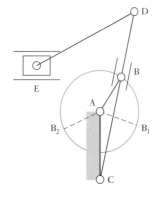

Figure 8.10 Quick-return mechanism.

backwards and forwards, then the crank, link 1, is forced to rotate. Hence the mechanism transforms an input of backwards and forwards motion into rotational motion. Figure 8.9 shows a number of stages in this motion. A useful way of seeing how any mechanism might behave is to construct, to scale, a cardboard model and move the links. Changing the length of a link then enables the changes in behaviour of the mechanism to be determined.

Figure 8.10 shows another form of this type of mechanism, a **quick-return mechanism**. It consists of a rotating crank, link AB, which rotates round a fixed centre, an oscillating lever CD, which is caused to oscillate about C by the sliding of the block at B along CD as AB rotates, and a link DE which causes E to move backwards and forwards. E might be the ram of a machine and have a cutting tool attached to it. The ram will be at the extremes of its movement when the positions of the crank are AB_1 and AB_2. Thus, as the crank moves anti-clockwise from B_1 to B_2, the ram makes a complete stroke, the cutting stroke. When the crank continues its movement from B_2 anti-clockwise to B_1 the ram again makes a complete stroke in the opposite direction, the return stroke. With the crank rotating at constant speed, because the angle of crank rotation required for the cutting stroke is greater than the angle for the return stroke, the cutting stroke takes more time than the return stroke hence the term quick-return for the mechanism. Similar diagrams, and a cardboard model, can be constructed in the same way as that shown in Figure 8.9.

8.4   Cams

A **cam** is a body which rotates or oscillates and in doing so imparts a reciprocating or oscillatory motion to a second body, called the **follower**, with which it is in contact (Figure 8.11). As the cam rotates so the follower is made to rise, dwell and fall; the lengths of times spent at each of these

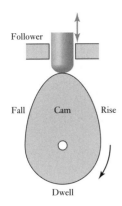

Figure 8.11 Cam and cam follower.

positions depending on the shape of the cam. The rise section of the cam is the part that drives the follower upwards, its profile determining how quickly the cam follower will be lifted. The fall section of the cam is the part that lowers the follower, its profile determining how quickly the cam follower will fall. The dwell section of the cam is the part that allows the follower to remain at the same level for a significant period of time. The dwell section of the cam is where it is circular with a radius that does not change.

The cam shape required to produce a particular motion of the follower will depend on the shape of the cam and the type of follower used. Figure 8.12 shows the type of follower displacement diagram that can be produced by an eccentric cam with a point-shaped follower. This is a circular cam with an offset centre of rotation. It produces an oscillation of the follower which is simple harmonic motion and is often used with pumps. The radial distance from the axis of rotation of the cam to the point of contact of the cam with the follower gives the displacement of the follower with reference to the axis of rotation of the cam. The figure shows how these radial distances, and hence follower displacements, vary with the angle of rotation of the cam. The vertical displacement diagram is obtained by taking the radial distance of the cam surface from the point of rotation at different angles and projecting them round to give the displacement at those angles.

Figure 8.12 Displacement diagram for an eccentric cam.

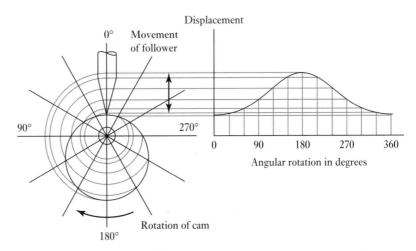

Figure 8.13 shows the types of follower displacement diagrams that can be produced with two other different shaped cams and either point or knife followers.

The heart-shaped cam (Figure 8.13(a)) gives a follower displacement which increases at a constant rate with time before decreasing at a constant rate with time, hence a uniform speed for the follower. The pear-shaped cam (Figure 8.13(b)) gives a follower motion which is stationary for about half a revolution of the cam and rises and falls symmetrically in each of the remaining quarter revolutions. Such a pear-shaped cam is used for engine valve control. The dwell holds the valve open while the petrol/air mixture passes into the cylinder. The longer the dwell, i.e. the greater the length of the cam surface with a constant radius, the more time is allowed for the cylinder to be completely charged with flammable vapour.

Figure 8.14 shows a number of examples of different types of cam followers. Roller followers are essentially ball or roller bearings. They have

Figure 8.13 Cams: (a) heart-shaped, (b) pear-shaped.

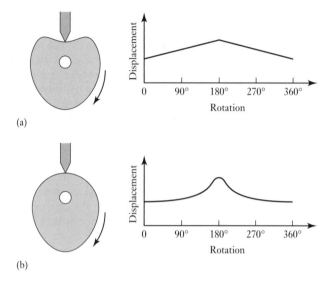

Figure 8.14 Cam followers: (a) point, (b) knife, (c) roller, (d) sliding and oscillating, (e) flat, (f) mushroom.

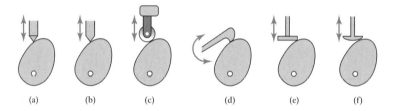

the advantage of lower friction than a sliding contact but can be more expensive. Flat-faced followers are often used because they are cheaper and can be made smaller than roller followers. Such followers are widely used with engine valve cams. While cams can be run dry, they are often used with lubrication and may be immersed in an oil bath.

8.5   Gears

Gear trains are mechanisms which are very widely used to transfer and transform rotational motion. They are used when a change in speed or torque of a rotating device is needed. For example, the car gearbox enables the driver to match the speed and torque requirements of the terrain with the engine power available.

Gears can be used for the transmission of rotary motion between parallel shafts (Figure 8.15(a)) and for shafts which have axes inclined to one another (Figure 8.15(b)). The term **bevel gears** is used when the lines of the shafts intersect, as illustrated in Figure 8.15(b). When two gears are in mesh, the larger gear wheel is often called the **spur** or **crown wheel** and the smaller one the **pinion**. Gears for use with parallel shafts may have axial teeth with the teeth cut along axial lines parallel to the axis of the shaft (Figure 8.15(c)). Such gears are then termed **spur gears**. Alternatively they may have helical teeth with the teeth being cut on a helix (Figure 8.15(d)) and are then termed

Figure 8.15 (a) Parallel gear axes, (b) axes inclined to one another, (c) axial teeth, (d) helical teeth, (e) double helical teeth.

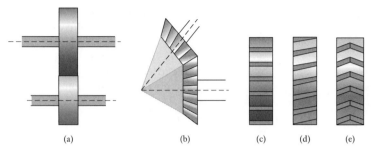

helical gears. Helical gears have the advantage that there is a gradual engagement of any individual tooth and consequently there is a smoother drive and generally prolonged life of the gears. However, the inclination of the teeth to the axis of the shaft results in an axial force component on the shaft bearing. This can be overcome by using double helical teeth (Figure 8.15(e)).

Consider two meshed gear wheels A and B (Figure 8.16). If there are 40 teeth on wheel A and 80 teeth on wheel B, then wheel A must rotate through two revolutions in the same time as wheel B rotates through one. Thus the angular velocity ω_A of wheel A must be twice that ω_B of wheel B, i.e.

$$\frac{\omega_A}{\omega_B} = \frac{\text{number of teeth on B}}{\text{number of teeth on A}} = \frac{80}{40} = 2$$

Since the number of teeth on a wheel is proportional to its diameter, we can write

$$\frac{\omega_A}{\omega_B} = \frac{\text{number of teeth on B}}{\text{number of teeth on A}} = \frac{d_B}{d_A}$$

Thus for the data we have been considering, wheel B must have twice the diameter of wheel A. The term **gear ratio** is used for the ratio of the angular speeds of a pair of intermeshed gear wheels. Thus the gear ratio for this example is 2.

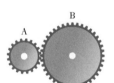

Figure 8.16 Two meshed gears.

8.5.1 Gear trains

The term **gear train** is used to describe a series of intermeshed gear wheels. The term **simple gear train** is used for a system where each shaft carries only one gear wheel, as in Figure 8.17. For such a gear train, the overall gear ratio is the ratio of the angular velocities at the input and output shafts and is thus ω_A/ω_C, i.e.

$$G = \frac{\omega_A}{\omega_C}$$

Consider a simple gear train consisting of wheels A, B and C, as in Figure 8.17, with A having 9 teeth and C having 27 teeth. Then, as the angular velocity

Figure 8.17 Simple gear train.

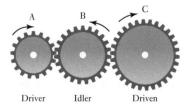

Driver   Idler   Driven

of a wheel is inversely proportional to the number of teeth on the wheel, the gear ratio is $27/9 = 3$. The effect of wheel B is purely to change the direction of rotation of the output wheel compared with what it would have been with just the two wheels A and C intermeshed. The intermediate wheel, B, is termed the **idler wheel**.

We can rewrite this equation for the overall gear ratio G as

$$G = \frac{\omega_A}{\omega_C} = \frac{\omega_A}{\omega_B} \times \frac{\omega_B}{\omega_C}$$

But ω_A/ω_B is the gear ratio for the first pair of gears and ω_B/ω_C the gear ratio for the second pair of gears. Thus the overall gear ratio for a simple gear train is the product of the gear ratios for each successive pair of gears.

The term **compound gear train** is used to describe a gear train when two wheels are mounted on a common shaft. Figure 8.18(a) and (b) shows two examples of such a compound gear train. The gear train in Figure 8.18(b) enables the input and output shafts to be in line.

Figure 8.18 Compound gear trains.

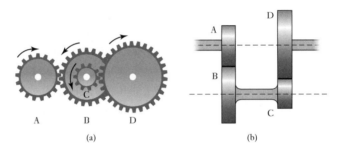

(a)   (b)

When two gear wheels are mounted on the same shaft they have the same angular velocity. Thus, for both of the compound gear trains in Figure 8.18, $\omega_B = \omega_C$. The overall gear ratio G is thus

$$G = \frac{\omega_A}{\omega_D} = \frac{\omega_A}{\omega_B} \times \frac{\omega_B}{\omega_C} \times \frac{\omega_C}{\omega_D} = \frac{\omega_A}{\omega_B} \times \frac{\omega_C}{\omega_D}$$

For the arrangement shown in Figure 8.18(b), for the input and output shafts to be in line we must also have for the radii of the gears

$$r_A + r_B = r_D + r_C$$

Consider a compound gear train of the form shown in Figure 8.18(a), with A, the first driver, having 15 teeth, B 30 teeth, C 18 teeth and D, the final driven wheel, 36 teeth. Since the angular velocity of a wheel is inversely proportional to the number of teeth on the wheel, the overall gear ratio is

$$G = \frac{30}{15} \times \frac{36}{18} = 4$$

Thus, if the input to wheel A is an angular velocity of 160 rev/min, then the output angular velocity of wheel D is $160/4 = 40$ rev/min.

A simple gear train of spur, helical or bevel gears is usually limited to an overall gear ratio of about 10. This is because of the need to keep the gear train down to a manageable size if the number of teeth on the pinion is to be kept above a minimum number, which is usually about 10 to 20. Higher gear ratios can, however, be obtained with compound gear trains. This is because the gear ratio is the product of the individual gear ratios of parallel gear sets.

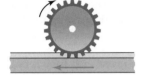

Figure 8.19 Rack-and-pinion.

8.5.2 Rotational to translational motion

The **rack-and-pinion** (Figure 8.19) is another form of gear, being essentially two intermeshed gears with one having a base circle of infinite radius. Such gears can be used to transform either linear motion to rotational motion or rotational motion to linear motion.

Another method that has been used for converting rotary to translational motion is the **screw and nut system**. With the conventional form of screw and nut, the nut is rotated and moved along the stationary screw. However, if the screw is rotated then a nut, which is attached to the part to be driven, moves along the screw thread. Such an arrangement is termed a **lead screw**. The lead L is the distance moved parallel to the screw axis when the nut is given one turn; for a single thread the lead is equal to the pitch. In n revolutions the distance moved parallel to the screw axis will be nL. If n revolutions are completed in a time t, the linear velocity v parallel to the screw axis is nL/t. As n/t is the number of revolutions per second f for the screw then:

$$v = \frac{nL}{t} = fL$$

There are, however, problems with using such an arrangement for converting rotational motion to linear motion. There are the high friction forces involved in the direct sliding contact between the screw and the nut and also the lack of rigidity. Friction can be overcome by using a **ball screw**. Such a screw is identical in principle to the lead screw, but ball bearings are located in the thread of the nut. Such an arrangement has been used with robots with the arm being driven by a ball screw powered by a geared d.c. motor (Figure 8.20). The motor rotates the screw which moved the nut up or down its thread. The movement of the nut is conveyed to the arm by means of a linkage.

Figure 8.20 Ball screw and links used to move a robot arm.

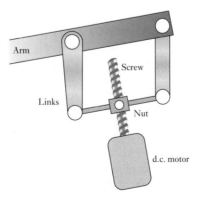

8.6   Ratchet and pawl

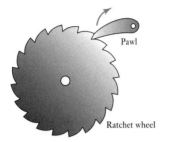

Figure 8.21   Ratchet and pawl.

Ratchets can be used to lock a mechanism when it is holding a load. Figure 8.21 shows a ratchet and pawl. The mechanism consists of a wheel, called a **ratchet**, with saw-shaped teeth which engage with an arm called **a pawl**. The arm is pivoted and can move back and forth to engage the wheel. The shape of the teeth is such that rotation can occur in only one direction. Rotation of the ratchet wheel in a clockwise direction is prevented by the pawl and can only take place when the pawl is lifted. The pawl is normally spring loaded to ensure that it automatically engages with the ratchet teeth.

Thus a winch used to wind up a cable on a drum may have a ratchet and pawl to prevent the cable unwinding from the drum when the handle is released.

8.7   Belt and chain drives

Belt drives are essentially just a pair of rolling cylinders with the motion of one cylinder being transferred to the other by a belt (Figure 8.22). Belt drives use the friction that develops between the pulleys attached to the shafts and the belt around the arc of contact in order to transmit a torque. Since the transfer relies on frictional forces then slip can occur. The transmitted torque is due to the differences in tension that occur in the belt during operation. This difference results in a tight side and a slack side for the belt. If the tension on the tight side is T_1, and that on the slack side T_2, then with pulley A in Figure 8.22 as the driver

Figure 8.22   Belt drive.

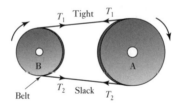

$$\text{torque on A} = (T_1 - T_2)r_A$$

where r_A is the radius of pulley A. For the driven pulley B we have

$$\text{torque on B} = (T_1 - T_2)r_B$$

where r_B is the radius of pulley B. Since the power transmitted is the product of the torque and the angular velocity, and since the angular velocity is v/r_A for pulley A and v/r_B for pulley B, where v is the belt speed, then for either pulley we have

$$\text{power} = (T_1 - T_2)v$$

As a method of transmitting power between two shafts, belt drives have the advantage that the length of the belt can easily be adjusted to suit a wide range of shaft-to-shaft distances and the system is automatically protected against overload because slipping occurs if the loading exceeds the maximum tension that can be sustained by frictional forces. If the distances between shafts is large, a belt drive is more suitable than gears, but over small distances gears are to be preferred. Different-size pulleys can be used to give a gearing effect. However, the gear ratio is limited to about 3 because of the need to maintain an adequate arc of contact between the belt and the pulleys.

The belt drive shown in Figure 8.22 gives the driven wheel rotating in the same direction as the driver wheel. Figure 8.23 shows two types of reversing drives. With both forms of drive, both sides of the belt come into contact with the wheels and so V–belts or timing belts cannot be used.

Figure 8.23 Reversed belt drives: (a) crossed belt, (b) open belt.

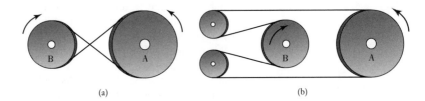

(a) (b)

8.7.1 Types of belts

The four main types of belts (Figure 8.24) are outlined below.

1 *Flat*
   The belt has a rectangular cross-section. Such a drive has an efficiency of about 98% and produces little noise. They can transmit power over long distances between pulley centres. Crowned pulleys are used to keep the belts from running off the pulleys.

2 *Round*
   The belt has a circular cross-section and is used with grooved pulleys.

3 *V*
   V–belts are used with grooved pulleys and are less efficient than flat belts but a number of them can be used on a single wheel and so give a multiple drive.

4 *Timing*
   Timing belts require toothed wheels, having teeth which fit into the grooves on the wheels. The timing belt, unlike the other belts, does not stretch or slip and consequently transmits power at a constant angular velocity ratio. The teeth make it possible for the belt to be run at slow or fast speeds.

Figure 8.24 Types of belt.

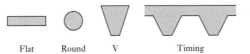

Flat   Round   V   Timing

8.7.2   Chains

Slip can be prevented by the use of chains which lock into teeth on the rotating cylinders to give the equivalent of a pair of intermeshing gear wheels. A chain drive has the same relationship for gear ratio as a simple gear train. The drive mechanism used with a bicycle is an example of a chain drive. Chains enable a number of shafts to be driven by a single wheel and so give a multiple drive. They are not as quiet as timing belts but can be used for larger torques.

8.8   Bearings

Whenever there is relative motion of one surface in contact with another, either by rotating or sliding, the resulting frictional forces generate heat which wastes energy and results in wear. The function of a **bearing** is to guide with minimum friction and maximum accuracy the movement of one part relative to another.

Of particular importance is the need to give suitable support to rotating shafts, i.e. support radial loads. The term **thrust bearing** is used for bearings that are designed to withstand forces along the axis of a shaft when the relative motion is primarily rotation. The following sections outline the characteristics of commonly used forms of bearings.

8.8.1   Plain journal bearings

Journal bearings are used to support rotating shafts which are loaded in a radial direction; the term **journal** is used for a shaft. The bearing basically consists of an insert of some suitable material which is fitted between the shaft and the support (Figure 8.25). Rotation of the shaft results in its surface sliding over that of the bearing surface. The insert may be a white metal, aluminium alloy, copper alloy, bronze or a polymer such as nylon or polytetrafluoroethylene (PTFE). The insert provides lower friction and less wear than if the shaft just rotated in a hole in the support. The bearing may be a dry rubbing bearing or lubricated. Plastics such as nylon and PTFE are generally used without lubrication, the coefficient of friction with such materials being exceptionally low. A widely used bearing material is sintered bronze; this is bronze with a porous structure which allows it to be impregnated with oil and so the bearing has a 'built-in' lubricant.

The possible lubricants are as follows.

1 *Hydrodynamic*
   The **hydrodynamic journal bearing** consists of the shaft rotating continuously in oil in such a way that it rides on oil and is not supported by metal (Figure 8.26). The load is carried by the pressure generated in the oil as a result of the shaft rotating.

2 *Hydrostatic*
   A problem with hydrodynamic lubrication is that the shaft only rides on oil when it is rotating and when at rest there is metal-to-metal contact. To avoid excessive wear at start-up and when there is only a low load, oil is pumped into the load-bearing area at a high-enough pressure to lift the shaft off the metal when at rest.

Figure 8.25   Plain journal bearing.

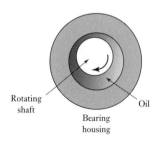

Figure 8.26   Hydrodynamic journal bearing.

3 *Solid-film*

This is a coating of a solid material such as graphite or molybdenum disulphide.

4 *Boundary layer*

This is a thin layer of lubricant which adheres to the surface of the bearing.

8.8.2 Ball and roller bearings

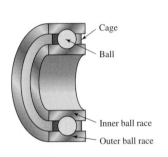

Figure 8.27 Basic elements of a ball bearing.

With this type of bearing, the main load is transferred from the rotating shaft to its support by rolling contact rather than sliding contact. A rolling element bearing consists of four main elements: an inner race, an outer race, the rolling element of either balls or rollers and a cage to keep the rolling elements apart (Figure 8.27). The inner and outer races contain hardened tracks in which the rolling elements roll.

There are a number of forms of ball bearings.

1 *Deep-groove* (Figure 8.28(a))

This is good at withstanding radial loads but is only moderately good for axial loads. It is a versatile bearing which can be used with a wide range of load and speed.

2 *Filling-slot* (Figure 8.28(b))

This is able to withstand higher radial loads than the deep-groove equivalent but cannot be used when there are axial loads.

3 *Angular contact* (Figure 8.28(c))

This is good for both radial and axial loads and is better for axial loads than the deep-groove equivalent.

4 *Double-row* (Figure 8.28(d))

Double-row ball bearings are made in a number of types and are able to withstand higher radial loads than their single-row equivalents. The figure shows a double-row deep-groove ball bearing, there being double-row versions of each of the above single-row types.

5 *Self-aligning* (Figure 8.28(e))

Single-row bearings can withstand a small amount of shaft misalignment but where there can be severe misalignment a self-aligning bearing is used. This is able to withstand only moderate radial loads and is fairly poor for axial loads.

6 *Thrust, grooved race* (Figure 8.28(f))

These are designed to withstand axial loads but are not suitable for radial loads.

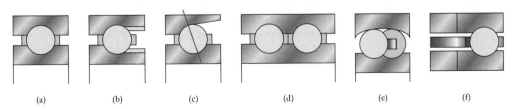

(a)   (b)   (c)   (d)   (e)   (f)

Figure 8.28 Types of ball bearings.

There are also a number of forms of roller bearing, the following being common examples.

1 *Straight roller* (Figure 8.29(a))
   This is better for radial loads than the equivalent ball bearing but is not generally suitable for axial loads. They will carry a greater load than ball bearings of the same size because of their greater contact area. However, they are not tolerant of misalignment.

2 *Taper roller* (Figure 8.29(b))
   This is good for radial loads and good in one direction for axial loads.

3 *Needle roller* (Figure 8.29(c))
   This has a roller with a high length/diameter ratio and tends to be used in situations where there is insufficient space for the equivalent ball or roller bearing.

Figure 8.29 Roller bearings.

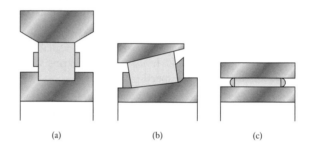

(a)   (b)   (c)

8.8.3   Selection of bearings

In general, dry sliding bearings tend to be used only for small–diameter shafts with low-load and low-speed situations, ball and roller bearings, i.e. bearings involving rolling, with a much wider range of diameter shafts and higher load and higher speed, and hydrodynamic bearings for the high loads with large-diameter shafts.

Summary

Mechanisms are devices which can be considered to be motion converters in that they transform motion from one form to some other required form.

The motion of a body can be considered to be a combination of translational and rotational motions. The number of **degrees of freedom** is the number of components to motion that are required to generate the motion.

Each part of a mechanism which has motion relative to some other part is termed a **link**. The points of attachment of a link to its neighbouring links are termed **nodes**. A **joint** is a connection between two or more links at their nodes. A sequence of joints and links is known as a **kinematic chain**. A **four-bar chain** consists of four links connected to give four joints about which turning can occur.

A **cam** is a body which rotates or oscillates and in doing so imparts a reciprocating or oscillatory motion to a second body, called the **follower**, with which it is in contact.

Gears can be used for the transmission of rotary motion between parallel shafts and for shafts which have axes inclined to one another.

The **rack-and-pinion** and the **screw-and-nut** systems can be used to convert rotational motion to translational motion.

Ratchets can be used to lock a mechanism when it is holding a load.

Belt and **chain drives** can be used to transmit rotational motion between shafts which are parallel and some distance apart.

Bearings are used to guide with minimum friction and maximum accuracy the movement of one part relative to another.

Problems

8.1　Explain the terms (a) mechanism, (b) kinematic chain.

8.2　Explain what is meant by the four-bar chain.

8.3　By examining the following mechanisms, state the number of degrees of freedom each has.

(a) A car hood hinge mechanism.

(b) An estate car tailgate mechanism.

(c) A windscreen wiper mechanism.

(d) Your knee.

(e) Your ankle.

8.4　Analyse the motions of the following mechanisms and state whether they involve pure rotation, pure translation or are a mixture of rotation and translation components.

(a) The keys on a computer keyboard.

(b) The pen in an XY plotter.

(c) The hour hand of a clock.

(d) The pointer on a moving-coil ammeter.

(e) An automatic screwdriver.

8.5　For the mechanism shown in Figure 8.30, the arm AB rotates at a constant rate. B and F are sliders moving along CD and AF. Describe the behaviour of this mechanism.

Figure 8.30　Problem 8.5.

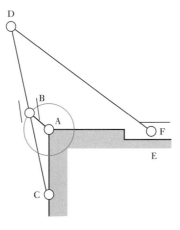

Figure 8.31
Problem 8.6.

8.6 Describe how the displacement of the cam follower shown in Figure 8.31 will vary with the angle of rotation of the cam.

8.7 A circular cam of diameter 100 mm has an eccentric axis of rotation which is offset 30 mm from the centre. When used with a knife follower with its line of action passing through the centre of rotation, what will be the difference between the maximum and minimum displacements of the follower?

8.8 Design a cam follower system to give constant follower speeds over follower displacements varying from 40 to 100 mm.

8.9 Design a mechanical system which can be used to:

(a) operate a sequence of microswitches in a timed sequence;

(b) move a tool at a steady rate in one direction and then quickly move it back to the beginning of the path;

(c) transform a rotation into a linear back-and-forth movement with simple harmonic motion;

(d) transform a rotation through some angle into a linear displacement;

(e) transform a rotation of a shaft into rotation of another, parallel shaft some distance away;

(f) transform a rotation of one shaft into rotation of another, close shaft which is at right angles to it.

8.10 A compound gear train consists of the final driven wheel with 15 teeth which meshes with a second wheel with 90 teeth. On the same shaft as the second wheel is a wheel with 15 teeth. This meshes with a fourth wheel, the first driver, with 60 teeth. What is the overall gear ratio?

Chapter nine Electrical actuation systems

Objectives

The objectives of this chapter are that, after studying it, the reader should be able to:

- Evaluate the operational characteristics of electrical actuation systems: relays, solid-state switches (thyristors, bipolar transistors and MOSFETs), solenoids, d.c. motors, a.c. motors and steppers.
- Explain the principles of d.c. motors, including the d.c. permanent magnet motor and how it can have its speed controlled.
- Explain the principle of the brushless permanent magnet d.c. motor.
- Explain the principles of the variable reluctance, permanent magnet and hybrid forms of stepper motor and how step sequences can be generated.
- Explain the requirements in selecting motors of inertia matching and torque and power requirements.

9.1 Electrical systems

In any discussion of electrical systems used as actuators for control, the discussion has to include:

1 *switching devices* such as mechanical switches, e.g. relays, and solid-state switches, e.g. diodes, thyristors, and transistors, where the control signal switches on or off some electrical device, perhaps a heater or a motor;

2 *solenoid-type devices* where a current through a solenoid is used to actuate a soft iron core, as, for example, the solenoid-operated hydraulic/pneumatic valve where a control current through a solenoid is used to actuate a hydraulic/pneumatic flow;

3 *drive systems*, such as d.c. and a.c. motors, where a current through a motor is used to produce rotation.

This chapter is an overview of such devices and their characteristics.

9.2 Mechanical switches

Mechanical switches are elements which are often used as sensors to give inputs to systems (see Section 2.12), e.g. keyboards. In this chapter we are concerned with their use as actuators to switch on perhaps electric motors or heating elements, or switch on the current to actuate solenoid valves controlling hydraulic or pneumatic cylinders. The electrical **relay** is an example of a mechanical switch used in control systems as an actuator.

9.2.1 Relays

Relays are electrically operated switches in which changing a current in one electric circuit switches a current on or off in another circuit. For the relay shown in Figure 9.1(a), when there is a current through the solenoid of the relay, a magnetic field is produced which attracts the iron armature, moves the push rod, and so closes the normally open (NO) switch contacts and opens the normally closed (NC) switch contacts.

Figure 9.1 (a) A relay and (b) a driver circuit.

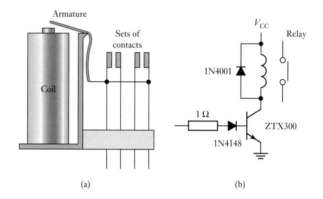

(a)   (b)

Relays are often used in control systems. The output from a controller is a relatively small current and so it is often used in conjunction with a transistor to switch on the current through the relay solenoid and hence use the relay to switch on the much larger current needed to switch on or off a final correction element such as an electric heater in a temperature control system or a motor. Figure 9.1(b) shows the type of circuit that might be used. Because relays are inductances, they can generate a back voltage when the energising current is switched off or when their input switches from a high to low signal. As a result, damage can occur in the connecting circuit. To overcome this problem, a diode is connected across the relay. When the back e.m.f. occurs, the diode conducts and shorts it out. Such a diode is termed a **free-wheeling** or **flyback** diode.

As an illustration of the ways relays can be used in control systems, Figure 9.2 shows how two relays might be used to control the operation of pneumatic valves which in turn control the movement of pistons in three cylinders A, B and C. The sequence of operation is outlined below.

1 When the start switch is closed, current is applied to the A and B solenoids and results in both A and B extending, i.e. A+ and B+.
2 The limit switches a+ and b+ are then closed; the a+ closure results in a current flowing through relay coil 1 which then closes its contacts and so supplies current to the C solenoid and results in it extending, i.e. C+.
3 Its extension causes limit switch c+ to close and so current to switch the A and B control valves and hence retraction of cylinders A and B, i.e. A− and B−.
4 Closing limit switch a− passes a current through relay coil 2; its contacts close and allow a current to valve C and cylinder C to retract, i.e. C−.

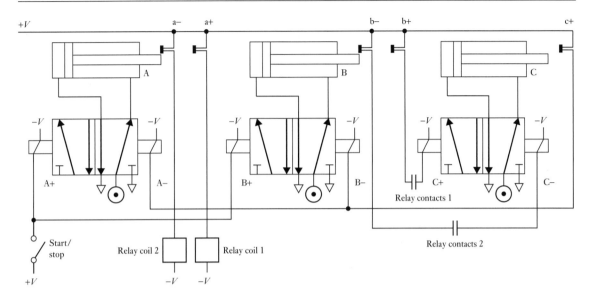

Figure 9.2 Relay-controlled system.

The sequence thus given by this system is A+ and B+ concurrently, then C+, followed by A− and B− concurrently and finally C−.

Time-delay relays are control relays that have a delayed switching action. The time delay is usually adjustable and can be initiated when a current flows through the relay coil or when it ceases to flow through the coil.

9.3 Solid-state switches

There are a number of solid-state devices which can be used electronically to switch circuits. These include:

1 diodes;
2 thyristors and triacs;
3 bipolar transistors;
4 power MOSFETs.

9.3.1 Diodes

The **diode** has the characteristic shown in Figure 9.3(a), only passing a current when forward biased, i.e. with the anode being positive with respect to the cathode. If the diode is sufficiently reverse biased, i.e. a very high voltage, it will break down. If an alternating voltage is applied across a diode, it can be regarded as only switching on when the direction of the voltage is such as to forward-bias it and being off in the reverse-biased direction. The result is that the current through the diode is half-rectified to become just the current due to the positive halves of the input voltage (Figure 9.3(b)), i.e. the circuit only 'switches on' for the positive half cycle.

Figure 9.3   (a) Diode characteristic, (b) half-wave rectification.

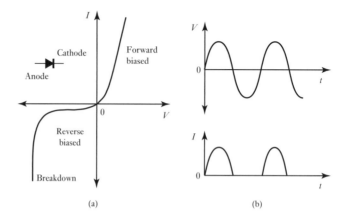

(a)   (b)

9.3.2   Thyristors and triacs

The **thyristor**, or **silicon-controlled rectifier** (SCR), can be regarded as a diode which has a gate controlling the conditions under which the diode can be switched on. Figure 9.4(a) shows the thyristor characteristic. With the gate current zero, the thyristor passes negligible current when reverse biased (unless sufficiently reverse biased, hundreds of volts, when it breaks down). When forward biased the current is also negligible until the forward breakdown voltage is exceeded. When this occurs the voltage across the diode falls to a low level, about 1 to 2 V, and the current is then only limited by the external resistance in a circuit. Thus, for example, if the forward breakdown is at 300 V then when this voltage is reached the thyristor switches on and the voltage across it drops to 1 or 2 V. If the thyristor is in series with a resistance of, say, 20 Ω (Figure 9.4(b)) then before breakdown we have a very high resistance in series with the 20 Ω and so virtually all the 300 V is across the thyristor and there is negligible current. When forward breakdown occurs, the voltage across the thyristor drops to, say, 2 V and so there is now $300 - 2 = 298$ V across the 20 Ω resistor, hence the current rises to $298/20 = 14.9$ A. Once switched on, the thyristor remains on until the forward current is reduced to below a level of a few milliamps. The voltage at which forward breakdown occurs is determined by the current entering the gate: the higher the current, the lower the breakdown voltage. The power-handling capability of a thyristor is high and thus it is widely used for switching high-power applications. As an example, the Texas Instruments

Figure 9.4   (a) Thyristor characteristic, (b) thyristor circuit.

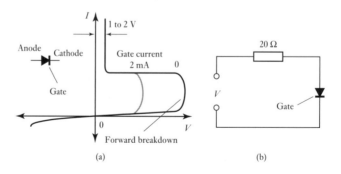

(a)   (b)

Figure 9.5 Triac characteristic.

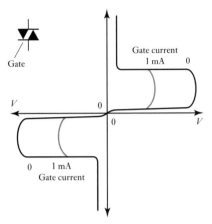

CF106D has a maximum off-state voltage of 400 V and a maximum gate trigger current of 0.2 mA.

The triac is similar to the thyristor and is equivalent to a pair of thyristors connected in reverse parallel on the same chip. The triac can be turned on in either the forward or reverse direction. Figure 9.5 shows the characteristic. As an example, the Motorola MAC212-4 triac has a maximum off-state voltage of 200 V and a maximum on-state current of 12 A r.m.s. Triacs are simple, relatively inexpensive, methods of controlling a.c. power.

Figure 9.6 shows the type of effect that occurs when a sinusoidal alternating voltage is applied across (a) a thyristor and (b) a triac. Forward breakdown occurs when the voltage reaches the breakdown value and then the voltage across the device remains low.

Figure 9.6 Voltage control: (a) thyristor, (b) triac.

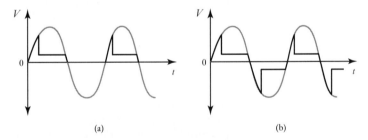

As an example of how such devices can be used for control purposes, Figure 9.7 illustrates how a thyristor could be used to control a steady d.c. voltage V. In this the thyristor is operated as a switch by using the gate to switch the device on or off. By using an alternating signal to the gate, the supply voltage can be chopped and an intermittent voltage produced. The average value of the output d.c. voltage is thus varied and hence controlled by the alternating signal to the gate.

Figure 9.7 Thyristor d.c. control.

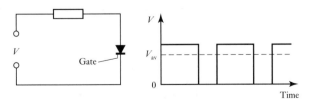

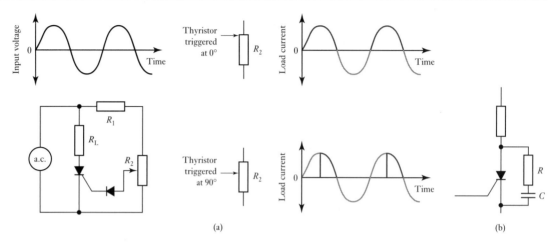

Figure 9.8 (a) Phase control, (b) snubber circuit.

Another example of control is that of alternating current for electric heaters, electric motors or lamp dimmers. Figure 9.8(a) shows a half-wave, variable resistance, phase control circuit. The alternating current is applied across the load, e.g. the lamp for the lamp dimming circuit, in series with a thyristor. R_1 is a current-limiting resistor and R_2 is a potentiometer which sets the level at which the thyristor is triggered. The diode is to prevent the negative part of the alternating voltage cycle being applied to the gate. By adjusting R_2 the thyristor can be made to trigger at any point between $0°$ and $90°$ in the positive half cycle of the applied alternating voltage. When the thyristor is triggered near the beginning of the cycle, i.e. $0°$, it conducts for the entire positive half cycle and the maximum power is delivered to the load. As the triggering of the thyristor is delayed to later in the cycle, so the power delivered to the load is reduced.

When a source voltage is suddenly applied to a thyristor, or a triac, with the gate off, the thyristor may switch from off to on. A typical rate of voltage change that would produce this effect is of the order of 50 V/μs. If the source is a d.c. voltage the thyristor can remain in this conducting state until there is a circuit interruption. In order to prevent this sudden change in source voltage producing this effect, the rate at which the voltage changes with time, i.e. dV/dt, is controlled by using a **snubber circuit**. This is a resistor in series with a capacitor and is placed in parallel with the thyristor (Figure 9.8(b)).

9.3.3 Bipolar transistors

Bipolar transistors come in two forms, the npn and the pnp. Figure 9.9(a) shows the symbol for each. For the npn transistor, the main current flows in at the collector and out at the emitter, a controlling signal being applied to the base. The pnp transistor has the main current flowing in at the emitter and out at the collector, a controlling signal being applied to the base.

For an npn transistor connected as shown in Figure 9.9(b), the so-termed common emitter circuit, the relationship between the collector current I_C and the potential difference between the collector and emitter V_{CE} is described by the series of graphs shown in Figure 9.9(c). When the base

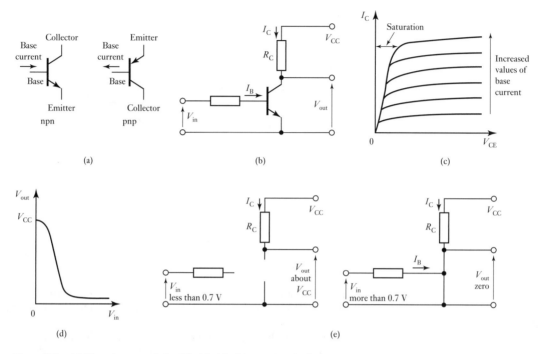

Figure 9.9 (a) Transistor symbols, (b), (c), (d), (e) transistor switch.

current I_B is zero, the transistor is cut off; in this state both the base–emitter and the base–collector junctions are reverse biased. When the base current is increased, the collector current increases and V_{CE} decreases as a result of more of the voltage being dropped across R_C. When V_{CE} reaches a value $V_{CE(sat)}$, the base–collector junction becomes forward biased and the collector current can increase no further, even if the base current is further increased. This is termed **saturation**. By switching the base current between 0 and a value that drives the transistor into saturation, bipolar transistors can be used as switches. When there is no input voltage V_{in} then virtually the entire V_{CC} voltage appears at the output. When the input voltage is made sufficiently high, the transistor switches so that very little of the V_{CC} voltage appears at the output (Figure 9.9(d)). Figure 9.9(e) summarises this switching behaviour of a typical transistor.

The relationship between collector current and the base current I_B at values below that which drives the transistor into saturation is

$$I_C = h_{FE}I_B$$

where h_{FE} is the **current gain**. At saturation the collector current $I_{C(sat)}$ is

$$I_{C(sat)} = \frac{V_{CC} - V_{CE(sat)}}{R_C}$$

To ensure that the transistor is driven into saturation the base current must thus rise to at least

$$I_{B(sat)} = \frac{I_{C(sat)}}{h_{FE}}$$

Figure 9.10   (a) Switching a
load, (b) and (c) Darlington pairs.

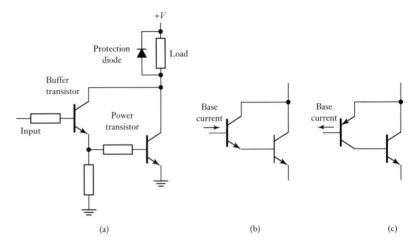

Thus for a transistor with h_{FE} of 50 and $V_{\mathrm{CE(sat)}}$ of 1 V, then for a circuit with $R_{\mathrm{C}} = 10\ \Omega$ and $V_{\mathrm{CC}} = 5$ V, the base current must rise to at least about 8 mA.

Because the base current needed to drive a bipolar power transistor is fairly large, a second transistor is often needed to enable switching to be obtained with relatively small currents, e.g. that supplied by a microprocessor. Thus the switching circuit can be of the form shown in Figure 9.10(a). Such a combination of a pair of transistors to enable a high current to be switched with a small input current is termed a **Darlington pair** and they are available as single-chip devices. A **protection diode** (free-wheeling diode) is generally connected in parallel with the load to prevent damage when the transistor is switched off since it is generally used with inductive loads and large transient voltages can occur. The integrated circuit ULN2001N from SGS-Thomson contains seven separate Darlington pairs, each pair being provided with a protection diode. Each pair is rated as 500 mA continuous and can withstand surges up to 600 mA.

Figure 9.10(b) shows the Darlington connections when a small npn transistor is combined with a large npn transistor, the result being equivalent to a large npn transistor with a large amplification factor. Figure 9.10(c) shows the Darlington connections for a small pnp transistor with a large npn transistor, the result being equivalent to a single large pnp transistor.

In using transistor-switched actuators with a microprocessor, attention has to be given to the size of the base current required and its direction. The base current required can be too high and so a **buffer** might be used. The buffer increases the drive current to the required value. It might also be used to invert. Figure 9.11 illustrates how a buffer might be used when transistor switching is used to control a d.c. motor by on/off switching. Type 240 buffer is inverting while types 241 and 244 are non-inverting. Buffer 74LS240 has a high-level maximum output current of 15 mA and a low-level maximum output current of 24 mA.

Bipolar transistor switching is implemented by base currents and higher frequencies of switching are possible than with thyristors. The power-handling capability is less than that of thyristors.

Figure 9.11 Control of d.c. motor.

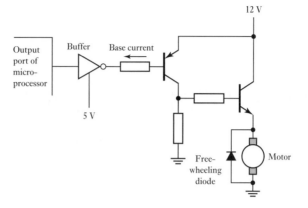

9.3.4 MOSFETs

MOSFETs (metal-oxide field-effect transistors) come in two types, the n-channel and the p-channel. Figure 9.12(a) and (b) shows the symbols. The main difference between the use of a MOSFET for switching and a bipolar transistor is that no current flows into the gate to exercise the control. The gate voltage is the controlling signal. Thus drive circuitry can be simplified in that there is no need to be concerned about the size of the current.

Figure 9.12 MOSFETs: (a) n-channel, (b) p-channel, (c) used to control a d.c. motor.

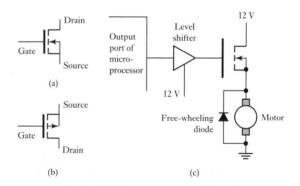

Figure 9.12(c) illustrates the use of a MOSFET as an on/off switch for a motor; compare the circuit with that in Figure 9.11 where bipolar transistors are used. A level shifter buffer is indicated, this being to raise the voltage level to that required for the MOSFET.

With MOSFETs, very high-frequency switching is possible, up to 1 MHz, and interfacing with a microprocessor is simpler than with bipolar transistors.

9.4   Solenoids

Essentially, solenoids consist of a coil of electrical wire with an armature which is attracted to the coil when a current passes through it and produces a magnetic field. The movement of the armature contracts a return spring which then allows the armature to return to its original position when the current ceases. The solenoids can be linear or rotary, on/off or variable positioning and operated by d.c. or a.c. Such an arrangement can be used to

Figure 9.13 The basic forms of linear solenoids with (a) disk, (b) plunger, (c) conical plunger, (d) ball forms of armature. Not shown in the figures are the springs required to return the armature back to its original position when the current through the solenoid ceases.

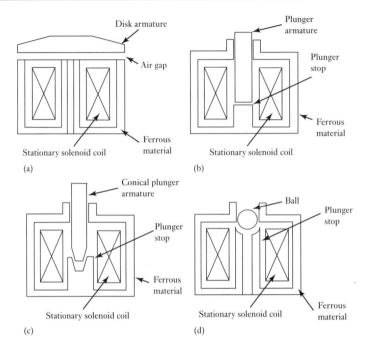

provide electrically operated actuators which are widely used for short stroke devices, typically up to 25 mm.

Figure 9.13 shows four examples of linear solenoids with different forms of armature. The form of the armature, the pole pieces and the central tube will depend on the use for which the actuator is designed. Disk armatures are useful where small distances of travel and fast action are required. Plunger armatures are widely used for applications requiring small distances of travel and fast action. Conical armatures are used for long-stroke applications, a typical application being for an automotive door lock mechanism. Ball armatures are used with fluid control applications, a typical application being an air bag deployment mechanism.

For a simple on/off device, there is no necessity for the design to give a linear characteristic. Where a proportional actuator is required, careful design is needed to give a movement of the armature proportional to the solenoid current. A simple example of the use of an on/off solenoid actuator is as a door lock with the lock either being actuated by the passage of a current through the solenoid or the reverse case when the passage of the current unlocks the door.

Solenoid valves are another example of such devices, being used to control fluid flow in hydraulic or pneumatic systems (see Figure 7.9). When a current passes through a coil, a soft iron plunger form of armature is pulled into the coil and, in doing so, can open or close ports to allow the flow of a fluid. The force exerted by the solenoid on the armature is a function of the current in the coil and the length of the armature within the coil. With on/off valves, i.e. those used for directional control, the current in the coil is controlled to be either on or off and the core is consequently in one of two positions. With proportional control valves, the current in the coil is controlled to give a plunger movement which is proportional to the size of the current.

Figure 9.14 A latching solenoid actuator.

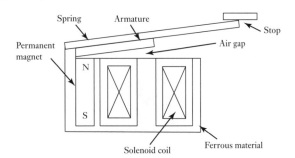

Solenoid actuators can be made to be latching, i.e. retain their actuated position when the solenoid current is switched off. Figure 9.14 illustrates this. A permanent magnet is added so that when there is no current through the solenoid it is not strong enough to pull the armature against its retaining spring into the closed position. However, when there is a current through the solenoid to give a magnetic field in the same direction as the permanent magnet then the armature is pulled into the closed position. When the current through the solenoid is switched off, the permanent magnet is strong enough to retain the armature in its closed position. To open it, the current through the solenoid has to be reversed to give a magnetic field in the opposite direction to that of the permanent magnet. Such a solenoid actuator can thus be used to switch on some device and leave it switched on until the reverse current signal is received.

9.5 Direct current motors

Electric motors are frequently used as the final control element in positional or speed control systems. Motors can be classified into two main categories: d.c. motors and a.c. motors, most motors used in modern control systems being d.c. motors. It is possible to divide d.c. motors into two main groups, those using brushes to make contact with a commutator ring assembly on the rotor to switch the current from one rotor winding to another and the brushless type. With the brush type of motor, the rotor has the coil winding and the stator can be either a permanent magnet or an electromagnet. With the brushless type, the arrangement is reversed in that the rotor is a permanent magnet and the stator has the coil winding.

9.5.1 Brush-type d.c. motor

A **brush-type d.c. motor** is essentially a coil of wire which is free to rotate, and so termed the rotor, in the field of a permanent magnet or an electromagnet, the magnet being termed the stator since it is stationary (Figure 9.15(a)). When a current is passed through the coil, the resulting forces acting on its sides at right angles to the field cause forces to act on those sides to give rotation. However, for the rotation to continue, when the coil passes through the vertical position the current direction through the coil has to be reversed and this is achieved by the use of brushes making contact with a split-ring commutator, the commutator rotating with the coil.

Figure 9.15   A d.c. motor:
(a) basics, (b) with two sets of
poles.

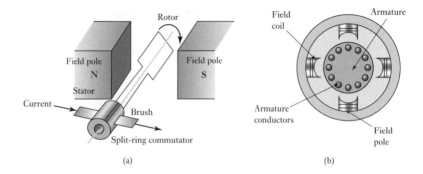

(a)   (b)

In the conventional d.c. motor, coils of wire are mounted in slots on a cylinder of magnetic material called the **armature**. The armature is mounted on bearings and is free to rotate. It is mounted in the magnetic field produced by **field poles**. These may be, for small motors, permanent magnets or electromagnets with their magnetism produced by a current through the **field coils**. Figure 9.15(b) shows the basic principle of a four-pole d.c. motor with the magnetic field produced by current-carrying coils. The ends of each armature coil are connected to adjacent segments of a segmented ring called the commutator with electrical contacts made to the segments through carbon contacts called brushes. As the armature rotates, the commutator reverses the current in each coil as it moves between the field poles. This is necessary if the forces acting on the coil are to remain acting in the same direction and if the rotation is to continue. The direction of rotation of the d.c. motor can be reversed by reversing either the armature current or the field current.

Consider a permanent magnet d.c. motor, the permanent magnet giving a constant value of flux density. For an armature conductor of length L and carrying a current i, the force resulting from a magnetic flux density B at right angles to the conductor is BiL (Figure 9.16(a)). The forces result in a torque T about the coil axis of Fb, with b being the breadth of the coil. Thus:

$$\text{Torque on a armature turn } T = BbLi = \Phi i$$

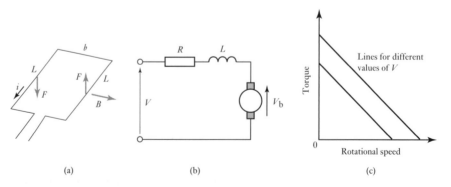

(a)   (b)   (c)

Figure 9.16   A d.c. motor: (a) forces on armature, (b) equivalent circuit, (c) torque–speed characteristic.

where Φ is the flux linked per armature turn. In practice there will be more than one armature turn and more than one set of poles, so we can write

$$\text{torque } T = k_t \Phi i$$

and k_t is a constant. The equation can be written as $T = K_t i$ where K_t is termed the torque constant for a motor. Since an armature coil is rotating in a magnetic field, electromagnetic induction will occur and a back e.m.f. will be induced. The back e.m.f. v_b is proportional to the rate at which the flux linked by the coil changes and hence, for a constant magnetic field, is proportional to the angular velocity ω of the rotation. Thus:

$$\text{back e.m.f. } v_b = k_v \Phi \omega$$

where k_v is a constant. The equation can be written as $v_b = K_v \omega$ where K_v is the back e.m.f. constant for a motor.

We can consider a d.c. motor to have the equivalent circuit shown in Figure 9.16(b), i.e. the armature coil being represented by a resistor R in series with an inductance L in series with a source of back e.m.f. If we neglect the inductance of the armature coil then the voltage providing the current i through the resistance is the applied voltage V minus the back e.m.f., i.e. $V - v_b$. Hence:

$$i = \frac{V - v_b}{R} = \frac{V - k_v \Phi \omega}{R} = \frac{V - K_v \omega}{R}$$

The torque T is thus:

$$T = k_t \Phi i = \frac{k_t \Phi}{R}(V - k_v \Phi \omega) = \frac{K_t}{R}(V - K_v \omega)$$

Graphs of the torque against the rotational speed ω are a series of straight lines for different voltage values (Figure 9.16(c)). The starting torque, i.e. the torque when $\omega = 0$, is, when putting this zero value in the derived equations, $K_t V/R$ and is thus proportional to the applied voltage, and the starting current is V/R. The torque decreases with increasing speed. If a permanent magnet motor developed a torque of 6 Nm with an armature current of 2 A, then, as $T = K_t i$, the torque developed with a current of 1 A would be 3 Nm.

The speed of a permanent magnet motor depends on the current through the armature coil and thus can be controlled by changing the armature current. The electrical power converted to mechanical power developed by a motor when operating under steady-state conditions is the product of the torque and the angular velocity. The power delivered to the motor in steady-state conditions is the sum of the power loss through the resistance of the armature coil and the mechanical power developed.

As an example, a small permanent magnet motor S6M41 by PMI Motors has K_t = 3.01 Ncm/A, K_V = 3.15 V per thousand rev/min, a terminal resistance of 1.207 Ω and an armature resistance of 0.940 Ω.

Figure 9.17   Direct current motors: (a) series, (b) shunt, (c) compound, (d) separately wound, (e) torque–speed characteristics.

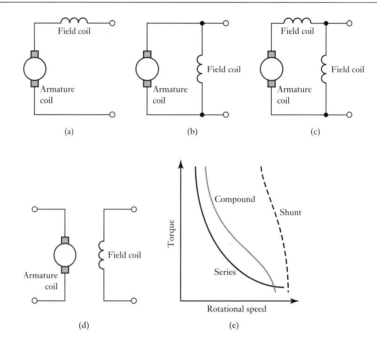

9.5.2   Brush-type d.c. motors with field coils

Direct current motors with field coils are classified as series, shunt, compound and separately excited according to how the field windings and armature windings are connected (Figure 9.17).

1   *Series wound motor* (Figure 9.17(a))

With the series-wound motor, the armature and field coils are in series and thus carry the same current. The flux Φ depends on the armature current i_a and so the torque acting on the armature is $k_t\Phi i_a = k i_a^2$. At start-up, when $\omega = 0$, $i_a = V/R$ and so the starting torque $= k(V/R)^2$. As such motors have a low resistance, they have a high starting torque and high no-load speed. As the speed is increased, so the torque decreases. Since Ri is small, $V = v_b + Ri \simeq v_b$ and so, since $v_b = k_v\Phi\omega$ and Φ is proportional to i, we have V proportional to $i\omega$. To a reasonable approximation V is constant and so the speed is inversely proportional to the current. The speed thus drops quite markedly when the load is increased. Reversing the polarity of the supply to the coils has no effect on the direction of rotation of the motor; it will continue rotating in the same direction since both the field and armature currents have been reversed. Such d.c. motors are used where large starting torques are required. With light loads there is a danger that a series-wound motor might run at too high a speed.

2   *Shunt-wound motor* (Figure 9.17(b))

With the shunt-wound motor, the armature and field coils are in parallel. It provides the lowest starting torque and a much lower no-load speed and has good speed regulation. The field coil is wound with many turns of fine wire and so has a much larger resistance than the armature coil. Thus, with a constant supply voltage, the field current is virtually constant. The torque at start-up is $k_t V/R$ and thus it provides a low starting torque and a low no-load speed. With V virtually constant, the motor gives almost constant speed regardless of load and such motors are very widely used because of this characteristic. To reverse the direction of rotation, either the armature or field supplied must be reversed.

3 *Compound motor* (Figure 9.17(c))
The compound motor has two field windings, one in series with the armature and one in parallel. The aim is to get the best features of the series- and shunt-wound motors, namely a high starting torque and good speed regulation.

4 *Separately excited motor* (Figure 9.17(d))
The separately excited motor has separate control of the armature and field currents and can be considered to be a special case of the shunt-wound motor.

The speed of such d.c. motors can be changed by changing either the armature current or the field current. Generally it is the armature current that is varied. This can be done by a series resistor. However, this method is very inefficient since the controller resistor consumes large amounts of power. An alternative is to control the armature voltage (see Section 9.5.3). D.C. motors develop a torque at standstill and so are self-starting. They can, however, require a starting resistance to limit the starting current as the starting current $i = (V - v_b)/R$. Since there is initially no back e.m.f. v_b to limit the current, the starting current can be very large.

The choice of motor will depend on its application. For example, with a robot manipulator, the robot wrist might use a series-wound motor because the speed decreases as the load increases. A shunt-wound motor would be used where a constant speed was required, regardless of the load.

9.5.3 Control of brush-type d.c. motors

The speed of a permanent magnet motor depends on the current through the armature coil. With a field coil motor the speed can be changed by varying either the armature current or the field current; generally it is the armature current that is varied. Thus speed control can be obtained by controlling the voltage applied to the armature. However, because fixed voltage supplies are often used, a variable voltage is obtained by an electronic circuit.

With an a.c. supply, the thyristor circuit of Figure 9.4(b) can be used to control the average voltage applied to the armature. However, we are often concerned with the control of d.c. motors by means of control signals emanating from microprocessors. In such cases the technique known as **pulse width modulation** (PWM) is generally used. This basically involves taking a constant d.c. supply voltage and chopping it so that the average value is varied (Figure 9.18).

Figure 9.19(a) shows how PWM can be obtained by means of a basic transistor circuit. The transistor is switched on or off by means of a signal applied to its base. The diode is to provide a path for current which arises when the transistor is off as a result of the motor acting as a generator. Such a circuit can only be used to drive the motor in one direction; a circuit (Figure 9.19(b)) involving four transistors, termed an H circuit, can be used

Figure 9.18 PWM:
(a) principles of PWM circuit,
(b) varying the armature voltage
by chopping the d.c. voltage.

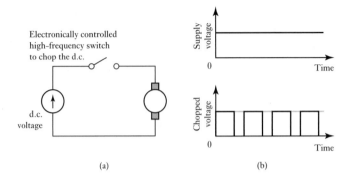

(a)　　　　　　　(b)

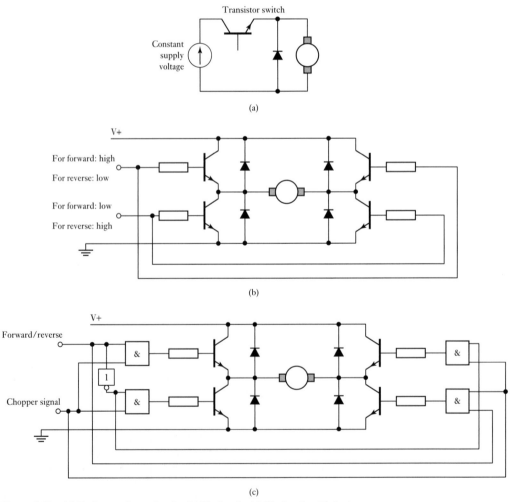

(a)

(b)

(c)

Figure 9.19 (a) Basic transistor circuit, (b) H-circuit, (c) H-circuit with logic gates.

to enable the motor to be operated in forward and reverse directions. This circuit can be modified by the use of logic gates so that one input controls the switching and one the direction of rotation (Figure 9.19(c)).

The above are examples of open-loop control; this assumes that conditions will remain constant, e.g. the supply voltage and the load driven by the motor. Closed-loop control systems use feedback to modify the motor speed if conditions change. Figure 9.20 shows some of the methods that might be employed.

Figure 9.20 Speed control with feedback.

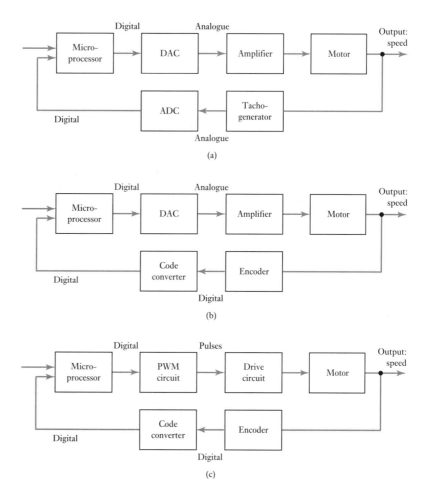

In Figure 9.20(a) the feedback signal is provided by a tachogenerator, this giving an analogue signal which has to be converted to a digital signal by an ADC for input to the microprocessor. The output from the microprocessor is converted to an analogue signal by an ADC and used to vary the voltage applied to the armature of the d.c. motor. In Figure 9.20(b) the feedback signal is provided by an encoder, this giving a digital signal which after code conversion can be directly inputted to the microprocessor. As in (a), the system shows an analogue voltage being varied to control the motor speed. In Figure 9.20(c) the system is completely digital and PWM is used to control the average voltage applied to the armature.

9.5.4   Brushless permanent magnet d.c. motors

A problem with d.c. motors is that they require a commutator and brushes in order periodically to reverse the current through each armature coil. The brushes make sliding contacts with the commutator and as a consequence sparks jump between the two and they suffer wear. Brushes thus have to be periodically changed and the commutator resurfaced. To avoid such problems **brushless motors** have been designed.

Essentially they consist of a sequence of stator coils and a permanent magnet rotor. A current-carrying conductor in a magnetic field experiences a force; likewise, as a consequence of Newton's third law of motion, the magnet will also experience an opposite and equal force. With the conventional d.c. motor, the magnet is fixed and the current-carrying conductors made to move. With the brushless permanent magnet d.c. motor, the reverse is the case: the current-carrying conductors are fixed and the magnet moves. The rotor is a ferrite or ceramic permanent magnet. Figure 9.21(a) shows the basic form of such a motor. The current to the stator coils is electronically switched by transistors in sequence round the coils, the switching being controlled by the position of the rotor so that there are always forces acting on the magnet causing it to rotate in the same direction. Hall sensors are generally used to sense the position of the rotor and initiate the switching by the transistors, the sensors being positioned around the stator.

Figure 9.21 (a) Brushless permanent magnet motor, (b) transistor switching.

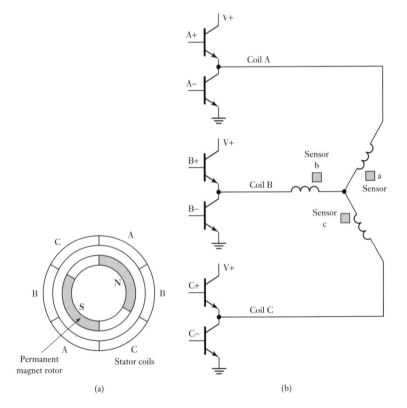

Figure 9.21(b) shows the transistor switching circuits that might be used with the motor shown in Figure 9.21(a). To switch the coils in sequence we need to supply signals to switch the transistors on in the right sequence. This is provided by the outputs from the three sensors operating through a decoder circuit to give the appropriate base currents. Thus when the rotor is in the vertical position, i.e. $0°$, there is an output from sensor c but none from a and b. This is used to switch on transistors A+ and B−. For the rotor in the $60°$ position there are signals from the sensors b and c and transistors A+ and C− are switched on. Table 9.1 shows the entire switching sequence. The entire circuit for controlling such a motor is available on a single integrated circuit.

Table 9.1 Switching sequence.

Rotor position	Sensor signals			Transistors on	
	a	b	c		
$0°$	0	0	1	A+	B−
$60°$	0	1	1	A+	C−
$120°$	0	1	0	B+	C−
$180°$	1	1	0	B+	A−
$240°$	1	0	0	C+	A−
$360°$	1	0	1	C+	B−

Brushless permanent magnet d.c. motors are becoming increasingly used in situations where high performance coupled with reliability and low maintenance are essential. Because of their lack of brushes, they are quiet and capable of high speeds.

9.6 Alternating current motors

It is possible to classify a.c. motors into two groups, single-phase and poly-phase, with each group being further subdivided into induction and synchronous motors. Single-phase motors tend to be used for low-power requirements while polyphase motors are used for higher powers. Induction motors tend to be cheaper than synchronous motors and are thus very widely used.

The **single-phase squirrel-cage induction motor** consists of a squirrel-cage rotor, this being copper or aluminium bars that fit into slots in end rings to form complete electric circuits (Figure 9.22(a)). There are no external electrical connections to the rotor. The basic motor consists of this rotor with a stator having a set of windings. When an alternating current passes through the stator windings an alternating magnetic field is produced. As a result of electromagnetic induction, e.m.f.s are induced in the conductors of the rotor and currents flow in the rotor. Initially, when the rotor is stationary, the forces on the current-carrying conductors of the rotor in the magnetic field of the stator are such as to result in no net torque. The motor is not self-starting. A number of methods are used to make the motor self-starting and give this initial impetus to start it; one is to use an auxiliary starting winding to give the rotor an initial push. The rotor rotates at a speed determined by the frequency of the alternating current applied to the stator. For a constant frequency supply to a two-pole single-phase motor

Figure 9.22 (a) Single-phase induction motor, (b) three-phase induction motor, (c) three-phase synchronous motor.

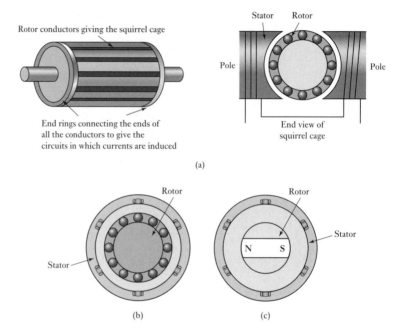

Rotor conductors giving the squirrel cage

End rings connecting the ends of all the conductors to give the circuits in which currents are induced

(a)

Stator   Rotor

Pole   Pole

End view of squirrel cage

Rotor

Stator

(b)

Rotor

Stator

N   S

(c)

the magnetic field will alternate at this frequency. This speed of rotation of the magnetic field is termed the **synchronous speed**. The rotor will never quite match this frequency of rotation, typically differing from it by about 1 to 3%. This difference is termed **slip**. Thus for a 50 Hz supply the speed of rotation of the rotor will be almost 50 revolutions per second.

The **three-phase induction motor** (Figure 9.22(b)) is similar to the single-phase induction motor but has a stator with three windings located 120° apart, each winding being connected to one of the three lines of the supply. Because the three phases reach their maximum currents at different times, the magnetic field can be considered to rotate round the stator poles, completing one rotation in one full cycle of the current. The rotation of the field is much smoother than with the single-phase motor. The three-phase motor has a great advantage over the single-phase motor of being self-starting. The direction of rotation is reversed by interchanging any two of the line connections, this changing the direction of rotation of the magnetic field.

Synchronous motors have stators similar to those described above for induction motors but a rotor which is a permanent magnet (Figure 9.22(c)). The magnetic field produced by the stator rotates and so the magnet rotates with it. With one pair of poles per phase of the supply, the magnetic field rotates through 360° in one cycle of the supply and so the frequency of rotation with this arrangement is the same as the frequency of the supply. Synchronous motors are used when a precise speed is required. They are not self-starting and some system has to be employed to start them.

Alternating current motors have the great advantage over d.c. motors of being cheaper, more rugged, reliable and maintenance free. However, speed control is generally more complex than with d.c. motors and as a consequence a speed-controlled d.c. drive generally works out cheaper than a speed-controlled a.c. drive, though the price difference is steadily dropping as a result of technological developments and the reduction in price of solid-state devices. Speed control of a.c. motors is based around the provision of a variable frequency supply, since the speed of such motors is determined by the frequency of the supply. The torque developed by an a.c. motor is constant when the ratio of the applied stator voltage to frequency is constant. Thus to maintain a constant torque at the different speeds when the frequency is varied, the voltage applied to the stator has also to be varied. With one method, the alternating current is first rectified to direct current by a **converter** and then **inverted** back to alternating current again but at a frequency that can be selected (Figure 9.23). Another method that is often used for operating slow-speed motors is the **cycloconverter**. This converts alternating current at one frequency directly to alternating current at another frequency without the intermediate d.c. conversion.

Figure 9.23 Variable speed a.c. motor.

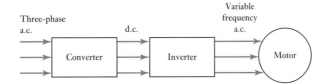

9.7 Stepper motors

The **stepper motor** is a device that produces rotation through equal angles, the so-called **steps**, for each digital pulse supplied to its input. Thus, for example, if with such a motor 1 pulse produces a rotation of 6° then 60 pulses will produce a rotation through 360°. There are a number of forms of stepper motor.

1 *Variable reluctance stepper*

Figure 9.24 shows the basic form of the variable reluctance stepper motor. With this form the rotor is made of soft steel and is cylindrical with four poles, i.e. fewer poles than on the stator. When an opposite pair of windings has current switched to them, a magnetic field is produced with lines of force which pass from the stator poles through the nearest set of poles on the rotor. Since lines of force can be considered to be

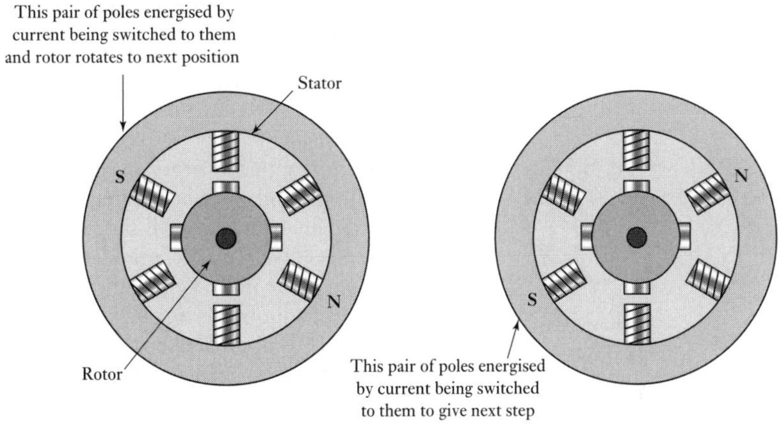

Figure 9.24 Variable reluctance stepper motor.

This pair of poles energised by current being switched to them and rotor rotates to next position

Stator

Rotor

This pair of poles energised by current being switched to them to give next step

Figure 9.25 Permanent magnet two-phase stepper motor with 90° steps. (a), (b), (c) and (d) show the positions of the magnet rotor as the coils are energised in different directions.

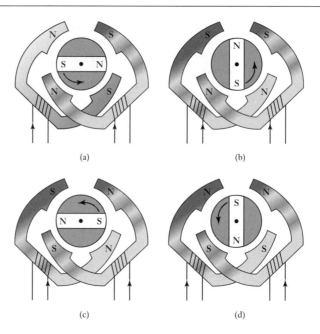

(a)

(b)

(c)

(d)

rather like elastic thread and always trying to shorten themselves, the rotor will move until the rotor and stator poles line up. This is termed the position of minimum reluctance. This form of stepper generally gives step angles of 7.5° or 15°.

2   *Permanent magnet stepper*

Figure 9.25 shows the basic form of the permanent magnet motor. The motor shown has a stator with four poles. Each pole is wound with a field winding, the coils on opposite pairs of poles being in series. Current is supplied from a d.c. source to the windings through switches. The rotor is a permanent magnet and thus when a pair of stator poles has a current switched to it, the rotor will move to line up with it. Thus for the currents giving the situation shown in the figure, the rotor moves to the 45° position. If the current is then switched so that the polarities are reversed, the rotor will move a further 45° in order to line up again. Thus by switching the currents through the coils, the rotor rotates in 45° steps. With this type of motor, step angles are commonly 1.8°, 7.5°, 15°, 30°, 34° or 90°.

3   *Hybrid stepper*

Hybrid stepper motors combine the features of both the variable reluctance and permanent magnet motors, having a permanent magnet encased in iron caps which are cut to have teeth (Figure 9.26). The rotor sets itself in the minimum reluctance position in response to a pair of stator coils being energised. Typical step angles are 0.9° and 1.8°. If a motor has n phases on the stator and m teeth on the rotor, the total number of steps per revolution is nm. Such stepper motors are extensively used in high-accuracy positioning applications, e.g. in computer hard disk drives.

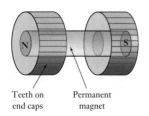

Teeth on   Permanent
end caps   magnet

Figure 9.26 Hybrid motor rotor.

9.7.1 Stepper motor specifications

The following are some of the terms commonly used in specifying stepper motors.

1 *Phase*

This term refers to the number of independent windings on the stator, e.g. a four-phase motor. The current required per phase and its resistance and inductance will be specified so that the controller switching output is specified. Two-phase motors, e.g. Figure 9.25, tend to be used in light-duty applications, three-phase motors tend to be variable reluctance steppers, e.g. Figure 9.24, and four-phase motors tend to be used for higher power applications.

2 *Step angle*

This is the angle through which the rotor rotates for one switching change for the stator coils.

3 *Holding torque*

This is the maximum torque that can be applied to a powered motor without moving it from its rest position and causing spindle rotation.

4 *Pull-in torque*

This is the maximum torque against which a motor will start, for a given pulse rate, and reach synchronism without losing a step.

5 *Pull-out torque*

This is the maximum torque that can be applied to a motor, running at a given stepping rate, without losing synchronism.

6 *Pull-in rate*

This is the maximum switching rate at which a loaded motor can start without losing a step.

7 *Pull-out rate*

This is the switching rate at which a loaded motor will remain in synchronism as the switching rate is reduced.

8 *Slew range*

This is the range of switching rates between pull-in and pull-out within which the motor runs in synchronism but cannot start up or reverse.

Figure 9.27 shows the general characteristics of a stepper motor.

9.7.2 Stepper motor control

Solid-state electronics is used to switch the d.c. supply between the pairs of stator windings. Two-phase motors, e.g. Figure 9.25, are termed **bipolar motors** when they have four connecting wires for signals to generate the switching sequence (Figure 9.28(a)). Such a motor can be driven by H-circuits (see Figure 9.19 and the associated discussion); Figure 9.28(b) shows the circuit and Table 9.2 shows the switching sequence required for the transistors to carry out the four steps, the sequence then being repeated for

Figure 9.27   Stepper motor
characteristics.

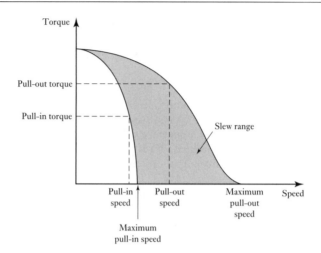

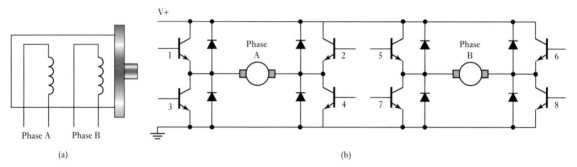

Figure 9.28   (a) Bipolar motor, (b) H-circuit.

Table 9.2   Switching sequence
for full-stepping bipolar stepper.

Step	Transistors			
	1 and 4	2 and 3	5 and 8	6 and 7
1	On	Off	On	Off
2	On	Off	Off	On
3	Off	On	Off	On
4	Off	On	On	Off

further steps. The sequence gives a clockwise rotation; for an anti-clockwise
rotation the sequence is reversed.

Half-steps, and hence finer resolution, are obtainable if, instead of the full-
stepping sequence needed to implement a pole reversal to get from one step
to the next, the coils are switched so that the rotor stops at a position halfway
to the next full step. Table 9.3 shows the sequence for half-stepping with the
bipolar stepper.

Two–phase motors are termed **unipolar** when they have six connecting
wires for the generation of the switching sequence (Figure 9.29). Each of
the coils has a centre-tap. With the centre-taps of the phase coils connected
together, such a form of stepper motor can be switched with just four

Table 9.3 Half-steps for bipolar stepper.

	Transistors			
Step	1 and 4	2 and 3	5 and 8	6 and 7
1	On	Off	On	Off
2	On	Off	Off	Off
3	On	Off	Off	On
4	Off	Off	Off	On
5	Off	On	Off	On
6	Off	On	Off	Off
7	Off	On	On	Off
8	Off	Off	On	Off

Table 9.4 Switching sequence for full-stepping unipolar stepper.

	Transistors			
Step	1	2	3	4
1	On	Off	On	Off
2	On	Off	Off	On
3	Off	On	Off	On
4	Off	On	On	Off

Table 9.5 Half-steps for unipolar stepper.

	Transistors			
Step	1	2	3	4
1	On	Off	On	Off
2	On	Off	Off	Off
3	On	Off	Off	On
4	Off	Off	Off	On
5	Off	On	Off	On
6	Off	On	Off	Off
7	Off	On	On	Off
8	Off	Off	On	Off

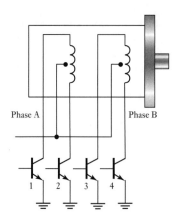

Phase A   Phase B

1   2   3   4

Figure 9.29 Unipolar motor.

transistors. Table 9.4 gives the switching sequence for the transistors in order to produce the steps for clockwise rotation, the sequence then being repeated for further steps. For anti-clockwise rotation the sequence is reversed. Table 9.5 shows the sequence when the unipolar is half-stepping.

Integrated circuits are available to provide the drive circuitry. Figure 9.30 shows the connections with the integrated circuit SAA 1027 for a four-phase stepper. The three inputs are controlled by applying high or low signals to them. When the set terminal is held high, the output from the integrated circuit changes state each time the trigger terminal goes from low to high. The sequence repeats itself at four-step intervals but can be reset to the zero condition at any time by applying a low signal to the trigger terminal. When the rotation input is held low there is clockwise rotation, when high it is anti-clockwise.

Figure 9.30 Integrated circuit
SAA 1027 for stepper motor.

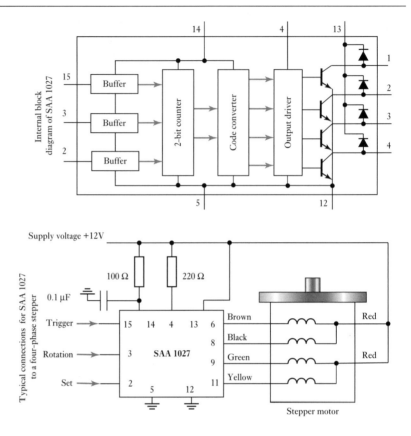

Some applications require very small step angles. While the step angle can
be made small by increasing the number of rotor teeth and/or the number
of phases, generally more than four phases and 50 to 100 teeth are not
used. Instead a technique known as **mini-stepping** is used. This involves
dividing each step into a number of equal size substeps. This is done by
using different currents to the coils so that the rotor moves to intermediate
positions between normal step positions. Thus, for example, a step of
1.8° might be subdivided into 10 equal steps.

Stepper motors can be used to give controlled rotational steps but also
can give continuous rotation with their rotational speed controlled by
controlling the rate at which pulses are applied to it to cause stepping.
This gives a very useful controlled variable speed motor which finds many
applications.

Because stepper coils have inductance and switched inductive loads can
generate large back e.m.f.s when switched, when steppers are connected to
microprocessor output ports it is necessary to include protection to avoid
damage to the microprocessor. This may take the form of resistors in the
lines to limit the current, though these must have values carefully chosen
both to provide the protection and not to limit the value of the current
needed to switch the transistors. Diodes across the coils prevent current
in the reverse direction and so give protection. An alternative is to use
optoisolators (see Section 3.3).

9.7.3 Selection of a stepper motor

The following should be considered in the selection of a stepper motor.

1 The operating torque requirements of the application. The rating torque must be high enough to accommodate the torque and slew range requirement. Also the torque-speed characteristics must be appropriate.
2 The step angle must be of high enough resolution to provide the required output motion increments.
3 Cost.

This will require looking at the data specifications for stepper motors. The following are some typical values taken from a manufacturer's data sheet for a unipolar stepper motor (Canon 42M048C1U-N):

d.c. operating voltage	5 V
Resistance per winding	9.1 Ω
Inductance per winding	8.1 mH
Holding torque	66.2 mNm/9.4 oz. in
Rotor moment of inertia	$12.5 \times 10^{-4} \text{g m}^2$
Detent torque	12.7 mNm/1.8 oz. in
Step angle	7.5°
Step angle tolerance	$\pm 0.5°$
Steps per revolution	48

The detent torque is the torque required to rotate the stepper motor when the motor windings are not energised.

Once a motor has been selected, a drive system will need to be found which is compatible with the motor. For example, for use with a unipolar motor the Cybernetics CY512 might be used if a maximum input voltage of 7 V and maximum current per phase of 80 mA is acceptable. The SAA1027 by Signetics is a widely used driver for small unipolar stepper motors with a maximum input voltage of 18 V and a maximum current per phase of 350 mA. For a two-phase bipolar or four-phase unipolar motor, the SCS-Thomson L297/L298 could be considered, it being a two-chip logic driver set. The L297 chip generates the motor phase sequences of four-phase TTL logic signals for two-phase and four-phase unipolar motors and the L298 is a bridge driver designed to accept such signals and drive inductive loads, in this case a stepper. A bipolar motor can be driven with winding currents up to 2 A.

When a pulse is supplied to a stepper motor we have essentially an input to an inductor–resistor circuit and the resulting torque is applied to the load, resulting in angular acceleration. As a consequence, the system will have a natural frequency; it will not go directly to the next step position but will generally have damped oscillations about it before settling down to the steady-state value (Figure 9.31). See Section 24.1.2 for a discussion of this and a derivation of the natural frequency and damping factor.

Figure 9.31   Oscillations about the steady-state angle.

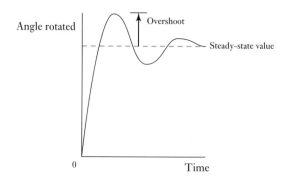

9.8   Motor selection

When selecting a motor for a particular application, factors that need to be taken into account include:

1   inertia matching;
2   torque requirements;
3   power requirements.

9.8.1   Inertia matching

The concept of impedance matching introduced in Section 3.8 for electrical impedances can be extended to mechanical systems, and an analogous situation to that described there for electrical circuits is that of a motor, a torque source, directly rotating a load (Figure 9.32(a)). The torque required to give a load with moment of inertia I_L and angular acceleration α is $I_L\alpha$. The torque required to accelerate the motor shaft is $T_M = I_M\alpha_M$ and that required to accelerate the load is $T_L = I_L\alpha_L$. The motor shaft will, in the absence of gearing, have the same angular acceleration and the same angular velocity. The power needed to accelerate the system as a whole is $T_M\omega + T_L\omega$, where ω is the angular velocity. Thus:

$$\text{power} = (I_M + I_L)\,\alpha\omega$$

This power is produced by the motor torque T_M and thus the power must equal $T_M\omega$. Hence:

$$T = (I_M + I_L)\alpha$$

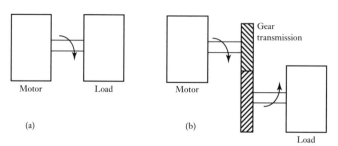

Figure 9.32   (a) Motor rotating load, (b) motor with gear transmission rotating load.

The torque to obtain a given angular acceleration will be minimised when $I_M = I_L$. Thus, for maximum power transfer, the moment of inertia of the load should be similar to that of the motor.

For the situation where the motor rotates the load through a gear transmission (Figure 9.32(b), the condition for maximum power transfer is that the moment of inertia of the motor equals the reflected moment of inertia of the load, this being $n^2 I_L$, where n is the gear ratio and I_L the moment of inertia of the load (see Section 17.2.2).

Thus for maximum power transfer, the moment of inertia of the motor should match that of the load or the reflected load when gears are used. This will mean that the torque to obtain a given acceleration will be minimised. This is particularly useful if the motor is being used for fast positioning. With a geared system, adjustment of the gear ratio can be used to enable matching to be obtained.

9.8.2 Torque requirements

Figure 9.33 shows the operating curves for a typical motor. For continuous running, the stall torque value should not be exceeded. This is the maximum torque value at which overheating will not occur. For intermittent use, greater torques are possible. As the angular speed is increased so the ability of the motor to deliver torque diminishes. Thus if higher speeds and torques are required than can be given by a particular motor, a more powerful motor needs to be selected.

Figure 9.33 Torque–speed graph

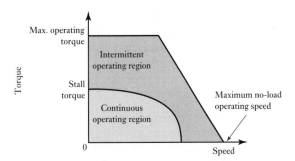

Suppose we require a motor to operate a drum-type hoist and lift a load (Figure 9.34). With a drum diameter of, say, 0.5 m and a maximum load m of 1000 kg, then the tension in the cable will be $mg = 1000 \times 9.81 = 9810$ N. The torque at the drum will be $9810 \times 0.25 = 24\,525$ Nm or about 2.5 kNm. If the hoist is operating at a constant speed v of 0.5 m/s then the drum angular velocity ω is $v/r = 0.5/0.25 = 2$ rad/s or $2/2\pi = 0.32$ revs/s. The motor is driving the shaft through a gear. We might decide that the gear ratio should be such that the maximum motor speed should be about 1500 rev/min or 25 rev/s. This means a gear ratio n of 25/0.32 or near enough 80:1. The load torque on the motor will be reduced by a factor of 80 from that on the drum and so will be $2500/80 = 31.25$ Nm. If we allow for some friction in the gear then the maximum torque on the motor which we should allow for is about 35 Nm.

Figure 9.34   A motor lifting a
load.

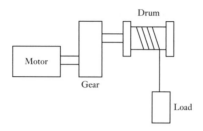

However, this is only the maximum torque when the load is being lifted at a constant speed. We need to add to this the torque needed to accelerate the load from rest to the speed of 0.5 m/s. If, say, we want to reach this speed from rest in 1 s then the accelerating torque needed is $I\alpha$, where I is the moment of inertia and α the angular acceleration. The effective moment of inertia of the load as seen by the motor via the gear is $(1/n^2) \times$ the moment of inertia of the load mr^2 and so is $(1/80)^2 \times 1000 \times 0.25^2 = 0.0098$ kg m^2 or about 0.01 kg m^2. The referred moment of inertia of the drum and gear might add 0.02 kg m^2. To find the total moment of inertia involved in the lifting we also have to add to this the moment of inertia of the motor. Manufacturers' data sheets might give a value of, say, 0.02 kg m^2 and so the total moment of inertia involved in the lifting might be $0.01 + 0.02 + 0.02 = 0.05$ kg m^2. The motor speed is required to rise from 0 to 25 rev/s in 1 s so the angular acceleration is $(25 \times 2\pi)/1 = 157$ rad/s^2 or about 160 rad/s^2. The accelerating torque required is thus $0.05 \times 160 = 8$ Nm. Hence the maximum torque we have to allow for is that required to lift the load at a constant velocity plus that need to accelerate it to this velocity from rest and so is $35 + 8 = 43$ Nm.

We can write the arguments involved in the above example algebraically as follows. The torque T_m required from a motor is that needed by the load T_L or T_L/n for a geared load with gear ratio n, and that needed to accelerate the motor $I_m\alpha_m$, where I_m is the moment of inertia of the motor and α_m its angular acceleration:

$$T_m = \frac{T_L}{n} + I_m\alpha_m$$

The angular acceleration of the load α_L is given by

$$\alpha_m = n\alpha_L$$

Because there will be torque T_f required to overcome the load friction, the torque used to accelerate the load will be $(T_L - T_f)$ and so

$$T_L - T_f = I_L\alpha_L$$

Thus we can write

$$T_m = \frac{1}{n}\left[T_f + \alpha_L(I_L + n^2 I_m)\right]$$

9.8.3   Power requirements

The motor needs to be able to run at the maximum required velocity without overheating. The total power P required is the sum of the power required to overcome friction and that needed to accelerate the load. As power is the product of torque and angular speed, then the power required to overcome the frictional torque T_f is $T_f\omega$ and that required to accelerate the load with angular acceleration α is $(I_L\alpha)\omega$, where I_L is the moment of inertia of the load. Thus:

$$P = T_f\omega + I_L\alpha\omega$$

Summary

Relays are electrically operated switches in which changing a current in one electric circuit switches a current on or off in another circuit.

A **diode** can be regarded as only passing a current in one direction, the other direction being very high resistance.

A **thyristor** can be regarded as a diode which has a gate controlling the conditions under which the diode can be switched on. A **triac** is similar to the thyristor and is equivalent to a pair of thyristors connected in reverse parallel on the same chip.

Bipolar transistors can be used as switches by switching the base current between zero and a value that drives the transistor into saturation. **MOSFETs** are similar and can also be used for switching.

The basic principle of a **d.c. motor** is of a loop of wire, the armature, which is free to rotate in the field of a magnet as a result of a current passing through the loop. The magnetic field may be provided by a permanent magnet or an electromagnet, i.e. a field coil. The speed of a permanent magnet motor depends on the current through the armature coil; with a field coil motor it depends on either the current through the armature coil or that through the field coil. Such d.c. motors require a commutator and brushes in order periodically to reverse the current through each armature coil. The **brushless permanent magnet d.c. motor** has a permanent magnet rotor and a sequence of stator coils through which the current is switched in sequence.

It is possible to classify a.c. motors into two groups, single-phase and polyphase, with each group being further subdivided into induction and synchronous motors. Single-phase motors tend to be used for low-power requirements while polyphase motors are used for higher powers. Induction motors tend to be cheaper than synchronous motors and are thus very widely used.

The **stepper motor** is a motor that produces rotation through equal angles, the so-called **steps**, for each digital pulse supplied to its input.

Motor selection has to take into account **inertia matching**, and the **torque** and **power requirements**.

Problems

9.1 Explain how the circuit shown in Figure 9.35 can be used to debounce a switch.

9.2 Explain how a thyristor can be used to control the level of a d.c. voltage by chopping the output from a constant voltage supply.

9.3 A d.c. motor is required to have (a) a high torque at low speeds for the movement of large loads, (b) a torque which is almost constant regardless of speed. Suggest suitable forms of motor.

Figure 9.35   Problem 9.1.

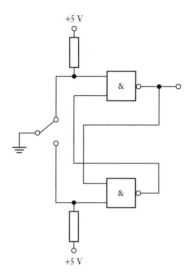

9.4   Suggest possible motors, d.c. or a.c., which can be considered for applications where (a) cheap, constant torque operation is required, (b) high controlled speeds are required, (c) low speeds are required, (d) maintenance requirements have to be minimised.

9.5   Explain the principle of the brushless d.c. permanent magnet motor.

9.6   Explain the principles of operation of the variable reluctance stepper motor.

9.7   If a stepper motor has a step angle of 7.5°, what digital input rate is required to produce a rotation of 10 rev/s?

9.8   What will be the step angle for a hybrid stepper motor with eight stator windings and ten rotor teeth?

9.9   A permanent magnet d.c. motor has an armature resistance of 0.5 Ω and when a voltage of 120 V is applied to the motor it reaches a steady-state speed of rotation of 20 rev/s and draws 40 A. What will be (a) the power input to the motor, (b) the power loss in the armature, (c) the torque generated at that speed?

9.10   If a d.c. motor produces a torque of 2.6 Nm when the armature current is 2 A, what will be the torque with a current of 0.5 A?

9.11   How many steps/pulses per second will a microprocessor need to output per second to a stepper motor if the motor is to give an output of 0.25 rev/s and has a step angle of 7.5°?

9.12   A stepper motor is used to rotate a pulley of diameter 240 mm and hence a belt which is moving a mass of 200 kg. If this mass is to be accelerated uniformly from rest to 100 mm/s in 2 s and there is a constant frictional force of 20 N, what will be the required pull-in torque for the motor?

Part IV
Microprocessor systems

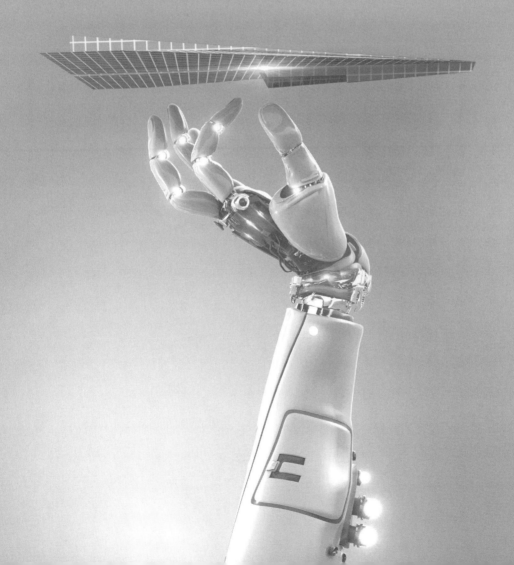

Chapter ten   Microprocessors and microcontrollers

Objectives

The objectives of this chapter are that, after studying it, the reader should be able to:
- Describe the basic structure of a microprocessor system.
- Describe the architecture of common microprocessors and how they can be incorporated in microprocessor systems.
- Describe the basic structure of microcontrollers and how their registers can be set up to carry out tasks.
- Explain how programs can be developed using flow charts and pseudocode.

10.1 Control

If we take a simple control problem, e.g. the sequencing of the red, amber, green lights at a traffic crossing, then it should be possible to solve it by an electronic control system involving combinational and sequential logic integrated circuits. However, with a more complex situation there might be many more variables to control in a more complex control sequence. The simplest solution now becomes not one of constructing a system based on hard-wired connections of combinational and sequential logic integrated circuits but of using a microprocessor and using software to make the 'interconnections'.

The microprocessor systems that we are concerned with in this book are for use as control systems and are termed **embedded microprocessors**. This is because such a microprocessor is dedicated to controlling a specific function and is self-starting, requiring no human intervention and completely self-contained with its own operating program. For the human, it is not apparent that the system is a microprocessor one. Thus, a modern washing machine contains a microprocessor but all that the operator has to do to operate it is to select the type of wash required by pressing the appropriate button or rotating a switch and then press the button to start.

This chapter is an overview of the structure of microprocessors and microcontrollers with the next two chapters discussing programming and Chapter 13 interfacing.

10.2 Microprocessor systems

Systems using microprocessors basically have three parts: a **central processing unit** (CPU) to recognise and carry out program instructions (this is the part which uses the microprocessor), **input and output interfaces** to handle communications between the microprocessor and the outside world (the term **port** is used for the interface), and **memory** to

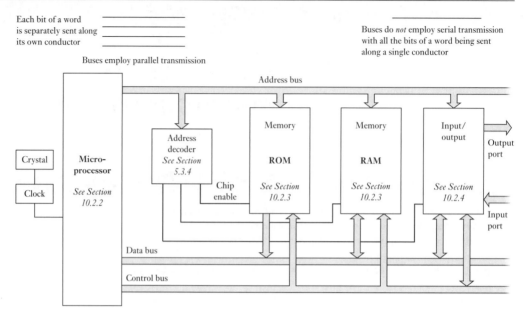

Figure 10.1   General form of a microprocessor system and its buses. All the components share the same data bus and address bus. This arrangement is known as the von Neumann architecture.

hold the program instructions and data. Figure 10.1 illustrates the general arrangement of a microprocessor system.

Microprocessors which have memory and various input/output arrangements all on the same chip are called **microcontrollers**.

10.2.1   Buses

Digital signals move from one section to another along paths called **buses**. A bus, in the physical sense, is just a number of parallel conductors along which electrical signals can be carried and are paths which can be shared by all the chips in the system. This is because if separate connections were used between the chips, there would be a very large number of connecting conductors. Using shared connection buses does mean that when one chip puts data on the bus, the other chips have to wait their turn until the data transfer is complete before one of them can put its data on the bus. Typically a bus has 16 or 32 parallel connections so that each can carry 1 bit of a data word simultaneously. This gives faster transmission than having a serial connection in which an entire word is sent in a sequence of bits along a single conductor.

There are three forms of bus in a microprocessor system.

1   *Data bus*
    The data associated with the processing function of the CPU is carried by the **data bus**. Thus, it is used to transport a word to or from the CPU and the memory or the input/output interfaces. Each wire in the bus carries a binary signal, i.e. a 0 or a 1. Thus with a four-wire bus we might have the

word 1010 being carried, each bit being carried by a separate wire in the bus, as:

Word	Bus wire
0 (least significant bit)	First data bus wire
1	Second data bus wire
0	Third data bus wire
1 (most significant bit)	Fourth data bus wire

The more wires the data bus has, the longer the word length that can be used. The range of values which a single item of data can have is restricted to that which can be represented by the word length. Thus with a word of length 4 bits the number of values is $2^4 = 16$. Thus if the data is to represent, say, a temperature, then the range of possible temperatures must be divided into 16 segments if we are to represent that range by a 4-bit word. The earliest microprocessors were 4-bit (word length) devices, and such 4-bit microprocessors are still widely used in such devices as toys, washing machines and domestic central heating controllers. They were followed by 8-bit microprocessors, e.g. the Motorola 6800, the Intel 8085A and the Zilog Z80. Now, 16-bit, 32-bit and 64-bit microprocessors are available; however, 8-bit microprocessors are still widely used for controllers.

2 *Address bus*

The **address bus** carries signals which indicate where data is to be found and so the selection of certain memory locations or input or output ports. Each storage location within a memory device has a unique identification, termed its address, so that the system is able to select a particular instruction or data item in the memory. Each input/output interface also has an address. When a particular address is selected by its address being placed on the address bus, only that location is open to the communications from the CPU. The CPU is thus able to communicate with just one location at a time. A computer with an 8-bit data bus has typically a 16-bit-wide address bus, i.e. 16 wires. This size of address bus enables 2^{16} locations to be addressed; 2^{16} is 65 536 locations and is usually written as 64K, where K is equal to 1024. The more memory that can be addressed, the greater the volume of data that can be stored and the larger and more sophisticated the programs that can be used.

3 *Control bus*

The signals relating to control actions are carried by the **control bus**. For example, it is necessary for the microprocessor to inform memory devices whether they are to read data from an input device or write data to an output device. The term READ is used for receiving a signal and WRITE for sending a signal. The control bus is also used to carry the system clock signals; these are to synchronise all the actions of the microprocessor system. The clock is a crystal-controlled oscillator and produces pulses at regular intervals.

10.2.2 The microprocessor

The microprocessor is generally referred to as the central processing unit (CPU). It is that part of the processor system which processes the data,

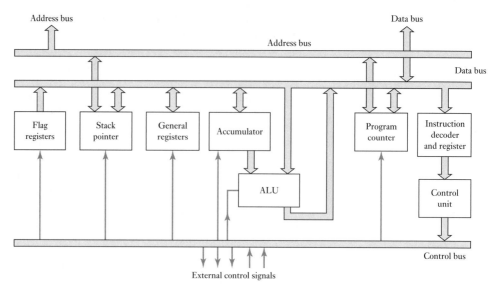

Figure 10.2 General internal architecture of a microprocessor.

fetching instructions from memory, decoding them and executing them. The internal structure – the term **architecture** is used – depends on the microprocessor concerned. Figure 10.2 indicates, in a simplified manner, the general architecture of a microprocessor.

The following are the functions of the constituent parts of a microprocessor.

1 *Arithmetic and logic unit* (ALU)
The arithmetic and logic unit is responsible for performing the data manipulation.

2 *Registers*
Internal data that the CPU is currently using is temporarily held in a group of **registers** while instructions are being executed. These are memory locations within the microprocessor and are used to store information involved in program execution. A microprocessor will contain a group of registers, each type of register having a different function.

3 *Control unit*
The **control unit** determines the timing and sequence of operations. It generates the timing signals used to fetch a program instruction from memory and execute it. The Motorola 6800 uses a clock with a maximum frequency of 1 MHz, i.e. a clock period of 1 μs, and instructions require between two and twelve clock cycles. Operations involving the microprocessor are reckoned in terms of the number of cycles they take.

There are a number of types of register, the number, size and types of registers varying from one microprocessor to another. The following are common types of registers.

1 *Accumulator register*
The accumulator register (A) is where data for an input to the arithmetic and logic unit is temporarily stored. In order for the CPU to be able to access, i.e. read, instructions or data in the memory, it has to supply the

address of the required memory word using the address bus. When this has been done, the required instructions or data can be read into the CPU using the data bus. Since only one memory location can be addressed at once, temporary storage has to be used when, for example, numbers are combined. For example, in the addition of two numbers, one of the numbers is fetched from one address and placed in the accumulator register while the CPU fetches the other number from the other memory address. Then the two numbers can be processed by the arithmetic and logic section of the CPU. The result is then transferred back into the accumulator register. The accumulator register is thus a temporary holding register for data to be operated on by the arithmetic and logic unit and also, after the operation, the register for holding the results. It is thus involved in all data transfers associated with the execution of arithmetic and logic operations.

2 *Status register, or condition code registeror flag register*
This contains information concerning the result of the latest process carried out in the arithmetic and logic unit. It contains individual bits with each bit having special significance. The bits are called **flags**. The status of the latest operation is indicated by each flag with each flag being set or reset to indicate a specific status. For example, they can be used to indicate whether the last operation resulted in a negative result, a zero result, a carry output occurs (e.g. the sum of two binary numbers such as 1010 and 1100 is (1)0110 which might be bigger than the microprocessor's word size and thus there is a carry of a 1), an overflow occurs or the program is to be allowed to be interrupted to allow an external event to occur. The following are common flags:

Flag	Set, i.e. 1	Reset, i.e. 0
Z	Result is zero	Result is not zero
N	Result is negative	Result is not negative
C	Carry is generated	Carry is not generated
V	Overflow occurs	Overflow does not occur
I	Interrupt is ignored	Interrupt is processed normally

As an illustration, consider the state of the Z, N, C and V flags for the operation of adding the hex numbers 02 and 06. The result is 08. Since it is not zero, then Z is 0. The result is positive, so N is 0. There is no carry, so C is 0. The unsigned result is within the range -128 to $+127$ and so there is no overflow and V is 0. Now consider the flags when the hex numbers added are F9 and 08. The result is (1)01. The result is not zero, so Z is 0. Since it is positive, then N is 0. The unsigned result has a carry and so C is 1. The unsigned result is within the range -128 to $+127$ and so V is 0.

3 *Program counter register* (PC) *or instruction pointer* (IP)
This is the register used to allow the CPU to keep track of its position in a program. This register contains the address of the memory location that contains the next program instruction. As each instruction is executed, the program counter register is updated so that it contains the address of the memory location where the next instruction to be executed is stored. The program counter is incremented each time so that the CPU executes instructions sequentially unless an instruction, such as JUMP or BRANCH, changes the program counter out of that sequence.

4   *Memory address register* (MAR)

This contains the address of data. Thus, for example, in the summing of two numbers the memory address register is loaded with the address of the first number. The data at the address is then moved to the accumulator. The memory address of the second number is then loaded into the memory address register. The data at this address is then added to the data in the accumulator. The result is then stored in a memory location addressed by the memory address register.

5   *Instruction register* (IR)

This stores an instruction. After fetching an instruction from the memory via the data bus, the CPU stores it in the instruction register. After each such fetch, the microprocessor increments the program counter by one with the result that the program counter points to the next instruction waiting to be fetched. The instruction can then be decoded and used to execute an operation. This sequence is known as the **fetch–execute cycle**.

6   *General-purpose registers*

These may serve as temporary storage for data or addresses and be used in operations involving transfers between other registers.

7   *Stack pointer register* (SP)

The contents of this register form an address which defines the top of the stack in RAM. The **stack** is a special area of the memory in which program counter values can be stored when a subroutine part of a program is being used.

The number and form of the registers depends on the microprocessor concerned. For example, the Motorola 6800 microprocessor (Figure 10.3) has two accumulator registers, a status register, an index register, a stack pointer register and a program counter register. The status register has flag bits to show negative, zero, carry, overflow, half-carry and interrupt. The Motorola 6802 is similar but includes a small amount of RAM and a built-in clock generator.

The Intel 8085A microprocessor is a development of the earlier 8080 processor; the 8080 required an external clock generator, whereas the 8085A has an in-built clock generator. Programs written for the 8080 can be run on the 8085A. The 8085A has six general-purpose registers B, C, D, E, H and L, a stack pointer, a program counter, a flag register and two temporary registers. The general-purpose registers may be used as six 8-bit registers or in pairs BC, DE and HL as three 16-bit registers. Figure 10.4 shows a block diagram representation of the architecture.

As will be apparent from Figure 10.3 and Figure 10.4, microprocessors have a range of timing and control inputs and outputs. These provide outputs when a microprocessor is carrying out certain operations and inputs to influence control operations. In addition there are inputs related to interrupt controls. These are designed to allow program operation to be interrupted as a result of some external event.

10.2.3   Memory

The memory unit in a microprocessor system stores binary data and takes the form of one or more integrated circuits. The data may be program instruction codes or numbers being operated on.

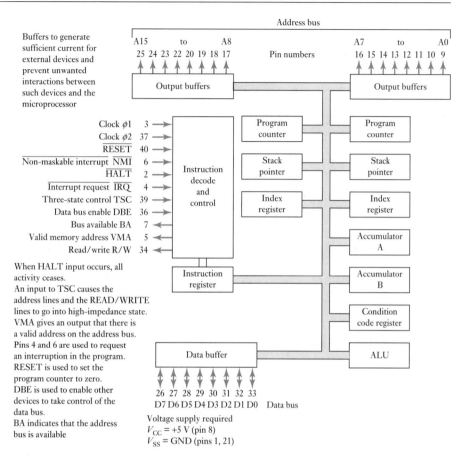

Figure 10.3 Motorola 6800 architecture.

The size of the memory is determined by the number of wires in the address bus. The memory elements in a unit consist essentially of large numbers of storage cells with each cell capable of storing either a 0 or a 1 bit. The storage cells are grouped in locations with each location capable of storing one word. In order to access the stored word, each location is identified by a unique address. Thus with a 4–bit address bus we can have 16 different addresses with each, perhaps, capable of storing 1 byte, i.e. a group of 8 bits (Figure 10.5).

The size of a memory unit is specified in terms of the number of storage locations available; 1K is $2^{10} = 1024$ locations and thus a 4K memory has 4096 locations.

There are a number of forms of memory unit.

1 *ROM*

For data that is stored permanently a memory device called a **read-only memory** (ROM) is used. ROMs are programmed with the required contents during the manufacture of the integrated circuit. No data can then be written into this memory while the memory chip is in the computer. The data can only be read and is used for fixed programs such as computer operating systems and programs for dedicated microprocessor applications. They do not lose their memory when power

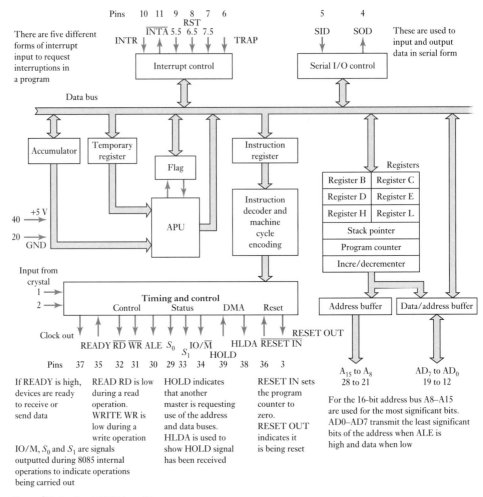

Figure 10.4　Intel 8085A architecture.

Address	Data contents
0000	
0001	
0010	
0011	
0100	
etc.	
1111	

Figure 10.5　Address bus size.

is removed. Figure 10.6(a) shows the pin connections of a typical ROM chip which is capable of storing $1K \times 8$ bits.

2 PROM

The term **programmable ROM** (PROM) is used for ROM chips that can be programmed by the user. Initially every memory cell has a fusible link which keeps its memory at 0. The 0 is permanently changed to 1 by sending a current through the fuse to open it permanently. Once the fusible link has been opened the data is permanently stored in the memory and cannot be further changed.

3 EPROM

The term **erasable and programmable ROM** (EPROM) is used for ROMs that can be programmed and their contents altered. A typical EPROM chip contains a series of small electronic circuits, cells, which can store charge. The program is stored by applying voltages to the integrated circuit connection pins and producing a pattern of charged

Figure 10.6 (a) ROM chip, (b) RAM chip.

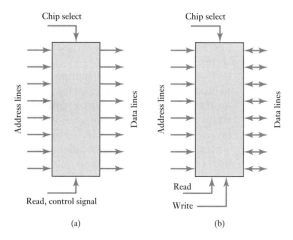

and uncharged cells. The pattern remains permanently in the chip until erased by shining ultraviolet light through a quartz window on the top of the device. This causes all the cells to become discharged. The chip can then be reprogrammed. The Intel 2716 EPROM has 11 address pins and a single chip enable pin which is active when taken low.

4 *EEPROM*
   Electrically erasable PROM (EEPROM) is similar to EPROM. Erasure is by applying a relatively high voltage rather than using ultraviolet light.

5 *RAM*
   Temporary data, i.e. data currently being operated on, is stored in a read/write memory referred to as a **random-access memory** (RAM). Such a memory can be read or written to. Figure 10.6(b) shows the typical pin connections for a 1K × 8-bit RAM chip. The Motorola 6810 RAM chip has seven address pins and six chip select pins of which four are active when low and two active when high and all must be made simultaneously active to enable the RAM.

When ROM is used for program storage, then the program is available and ready for use when the system is switched on. Programs stored in ROM are termed **firmware**. Some firmware must always be present. When RAM is used for program storage then such programs are referred to as **software**. When the system is switched on, software may be loaded into RAM from some other peripheral equipment such as a keyboard or hard disk or floppy disk or CD.

10.2.4 Input/output

The input/output operation is defined as the transfer of data between the microprocessor and the external world. The term **peripheral devices** is used for pieces of equipment that exchange data with a microprocessor system. Because the speeds and characteristics of peripheral devices can differ significantly from those of the microprocessor, they are connected via interface chips. A major function of an interface chip is thus to synchronise data transfers between the microprocessor and peripheral device. In input operations the input device places the data in the data register of the interface chip; this holds the data until it is read by the microprocessor. In output operations the microprocessor places the data in the register until it is read by the peripheral.

For the microprocessor to input valid data from an input device, it needs to be certain that the interface chip has correctly latched the input data. It can do this by **polling** or an **interrupt**. With polling, the interface chip uses a status bit set to 1 to indicate when it has valid data. The microprocessor keeps on checking the interface chip until it detects the status bit as 1. The problem with this method is that the microprocessor is having to wait for this status bit to show. With the interrupt method, the interface chip sends an interrupt signal to the microprocessor when it has valid data; the microprocessor then suspends execution of its main program and executes the routine associated with the interrupt in order to read the data.

10.2.5   Examples of systems

Figure 10.7 shows an example of a microprocessor system using the Intel 8085A microprocessor. It has an address latch 74LS373, a 3-line to 8-line address decoder 74LS138, two 1K $\times$ 4 RAM chips 2114, a 2K $\times$ 8 EPROM chip 2716 and input and output interface chips 74LS244 and 74LS374.

1 *Address latch*
   The output ALE (address latch enable) provides an output to the external hardware to indicate when the lines AD0–AD7 contain addresses and when they contain data. When ALE is made high, it activates the latch and the lines A0–A7 pass the lower part of the address to it where it becomes latched. Thus when ALE changes and goes back to being low, so that data can be outputted from the microprocessor, this part of the address remains latched in the 74LS373. The higher part of the address is sent through A8–A15 and is always valid, and the full address is given by the lower part from the latch and the upper part from the microprocessor address bus.

2 *Address decoder*
   The 74LS138 is a 3-line to 8-line decoder and provides an active low at one of its eight outputs; the output selected depends on the signals on its three input lines A, B and C. Before it can make such a selection it has to be enabled by enable 1 and enable 2 being low and enable 3 high.

3 *EPROM*
   Address bits A11, A12, A13 and A14 are used to select which device is to be addressed. This leaves the address bits on A0–A10 for addresses. Hence the EPROM can have $2^{11} = 2048$ addresses. This is the size of the Intel 2716 EPROM. The EPROM is selected whenever the microprocessor reads an address from 0000 to 07FF and outputs its 8-bit contents to the data bus via O0–O7. The output enable line OE is connected to the read output of the microprocessor to ensure that the EPROM is only written to.

4 *RAM*
   Two RAM chips are shown as being used, each being 1K $\times$ 4. Together they provide a memory for an 8-bit-wide signal. Both the chips use the

Figure 10.7 Intel 8085A system.

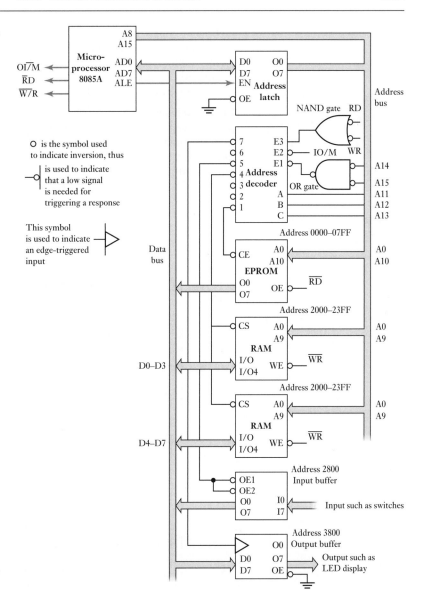

O is the symbol used
to indicate inversion, thus

is used to indicate
that a low signal
is needed for
triggering a response

This symbol
is used to indicate
an edge-triggered
input

same address bits A0–A9 for memory selection with one chip providing data D0–D3 and the other D4–D7. With 10 address bits we have $2^{10} = 1024$ different addresses 2000 to 23FF. The write enable WE input is used by the RAM to determine whether the RAM is being written to or read. If it is low then the selected RAM address is being written to and if high it is being read.

5 *Input buffer*

The input buffer 74LS244 is set to pass the binary value of the inputs over the data bus whenever OE1 and OE2 are low. It is accessed at any address from 2800 to 2FFF, thus we might use the address 2800. The buffer is to ensure that the inputs impose very little loading on the microprocessor.

6 *Output latch*

The 74LS374 is an output latch. It latches the microprocessor output so that the output devices have time to read it while the microprocessor can get on with other instructions in its program. The output latch is given a range of addresses from 3800 to 3FFF and thus might be addressed by using 3800.

Figure 10.8 shows an example of a system based on the use of a Motorola 6800 microprocessor and having just one RAM chip, one ROM chip and a programmable input/output. No address decoding is necessary with this system because of the small number of devices involved. For parallel inputs/outputs a peripheral interface adapter (PIA) (see Section 13.4) is used and for serial inputs/outputs an asynchronous interface adapter

Figure 10.8 M6800 system.

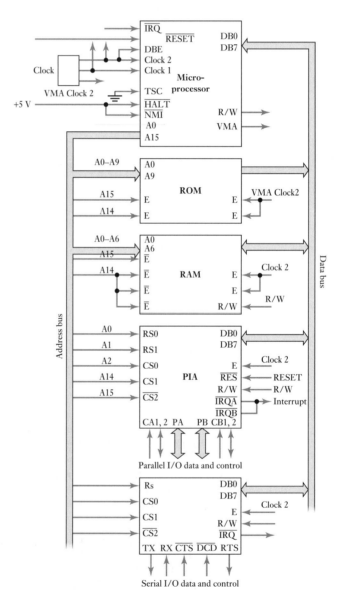

(ACIA) (see Section 13.5) is used. These can be programmed to deal with both inputs and outputs and give the required buffering.

1 *RAM*

Address lines A14 and A15 are connected to the enable inputs of the RAM chip. When both these lines are low then the RAM chip will be conversing with the microprocessor.

2 *ROM*

Address lines A14 and A15 are connected to the enable inputs of the ROM chip and when the signals on both these lines are high then the ROM chip is addressed.

3 *Inputs/outputs*

Address lines A14 and A15 are connected to the enable inputs of the PIA and ACIA. When the signal on line 15 is low and the signal on A14 high then the input/output interfaces are addressed. In order to indicate which of the devices is being enabled, address line A2 is taken high for the PIA and address line A3 is taken high for the ACIA.

10.3 Microcontrollers

For a microprocessor to give a system which can be used for control, additional chips are necessary, e.g. memory devices for program and data storage and input/output ports to allow it to communicate with the external world and receive signals from it. The microcontroller is the integration of a microprocessor with memory and input/output interfaces, and other peripherals such as timers, on a single chip. Figure 10.9 shows the general block diagram of a microcontroller.

The general microcontroller has pins for external connections of inputs and outputs, power, clock and control signals. The pins for the inputs and

Figure 10.9 Block diagram of a microcontroller.

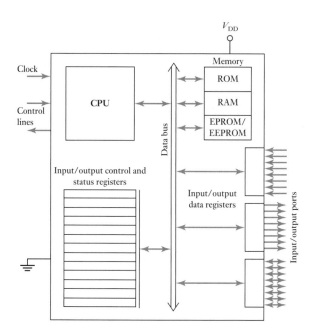

outputs are grouped into units called input/output ports. Usually such ports have eight lines in order to be able to transfer an 8-bit word of data. Two ports may be used for a 16-bit word, one to transmit the lower 8 bits and the other the upper 8 bits. The ports can be input only, output only or programmable to be either input or output.

The Motorola 68HC11, the Intel 8051 and the PIC16C6x/7x are examples of 8-bit microcontrollers in that the data path is 8 bits wide. The Motorola 68HC16 is an example of a 16-bit microcontroller and the Motorola 68300 a 32-bit microcontroller. Microcontrollers have limited amounts of ROM and RAM and are widely used for embedded control systems. A microprocessor system with separate memory and input/output chips is more suited to processing information in a computer system.

10.3.1 Motorola Freescale M68HC11

Motorola offers two basic 8-bit families of microcontrollers, the 68HC05 being the inexpensive core and the 68HC11 the higher performance core. The Motorola M68HC11 family (Figure 10.10), this being based on the Motorola 6800 microprocessor, is very widely used for control systems. Note that this microcontroller was introduced by Motorola and is now produced by Freescale Semiconductor.

There are a number of versions of this family, the differences being due to differences in the RAM, ROM, EPROM, EEPROM and configuration register features. For example, one version (68HC11A8) has 8K ROM,

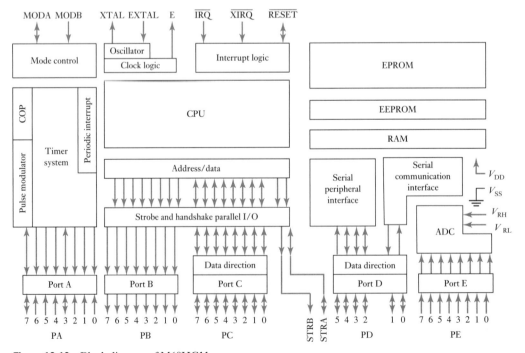

Figure 10.10 Block diagram of M68HC11.

512-byte EEPROM, 256-byte RAM, a 16-bit timer system, a synchronous serial peripheral interface, an asynchronous non-return-to-zero serial communication interface, an 8-channel, 8-bit analogue-to-digital (A/D) converter (ADC) for analogue inputs, and five ports A, B, C, D and E.

1 *Port A*
Port A has three input lines only, four output lines only and one line that can serve as either input or output. The port A data register is at address \$1000 (Figure 10.11) with a pulse accumulator control register (Figure 10.12) at address \$1026 being used to control the function of each bit in port A. This port also provides access to the internal timer of the microcontroller, the PAMOD, PEDGE, RTR1 and RTRO bits controlling the pulse accumulator and clock.

Figure 10.11 Port A register.

Port A data register \$1000

Bit	7	6	5	4	3	2	1	0

Figure 10.12 Pulse accumulator control register.

Bit	7	6	5	4	3	2	1	0
	DDRA7	PAEN6	PAMOD	PEDGE	0	0	RTR1	RTRO

DDRA7 Set as 0 for input and 1 for output

PAEN6 Set as 0 to disable the pulse accumulator and allow port A to be used for I/O, 1 to enable the pulse accumulator

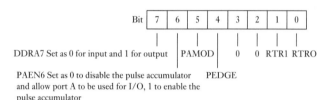

2 *Port B*
Port B is output only and has eight output lines (Figure 10.13). Input data cannot be put on port B pins. Its data register is at the address \$1004 and to output data it has to be written to this memory location.

Figure 10.13 Port B register.

Port B data register \$1004

Bit	7	6	5	4	3	2	1	0

3 *Port C*
Port C can be either input or output, data being written or read from its data register at address \$1003 (Figure 10.14). Its direction is controlled by the port data direction register at address \$1007. The 8 bits in this register correspond to the individual bits in port C and determine whether the lines are inputs or outputs; when the data direction register bit is set to 0 it is an input and when set to 1 it is an output. The lines STRA and STRB (when operating in single-chip mode) are associated with ports B and C and are used for handshake signals with those ports. Such lines control

Figure 10.14 Port C registers.

Port C data register \$1003

Bit	7	6	5	4	3	2	1	0

Port C data direction register \$1007

Bit	7	6	5	4	3	2	1	0

When a bit is set to 0 the corresponding bit in the port is an input, when set to 1 an output the timing of data transfers. The parallel input/output control register PIOC at address $1002 contains bits to control the handshaking mode and the polarity and active edges of the handshaking signals.

4 *Port D*

Port D contains just six lines; these can be either input or output and have a data register at address $1008 (Figure 10.15); the directions are controlled by a port data direction register at address $1009 with the corresponding bit being set to 0 for an input and 1 for an output. Port D also serves as the connection to the two serial subsystems of the microcontroller. The serial communication interface is an asynchronous system that provides serial communication that is compatible with modems and terminals. The serial peripheral interface is a high-speed synchronous system which is designed to communicate between the microcontroller and peripheral components that can access at such rates.

Figure 10.15 Port D registers.

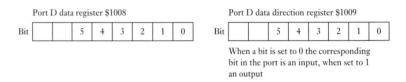

5 *Port E*

Port E is an 8-bit input-only port (Figure 10.16) which can be used as a general-purpose input port or for analogue inputs to the internal ADC. The two inputs V_{RH} and V_{RL} provide reference voltages to the ADC. The port E data register is at address $1002.

Figure 10.16 Port E register.

The 68HC11 has an internal ADC; port E bits 0, 1, 2, 3, 4, 5, 6 and 7 are the analogue input pins. Two lines V_{RH} and V_{LH} provide the reference voltages used by the ADC; the high reference voltage V_{RH} should not be lower than V_{DD}, i.e. 5 V, and the low reference voltage V_{LH} should not be lower than V_{SS}, i.e. 0 V. The ADC must be enabled before it can be used. This is done by setting the A/D power-up (ADPU) control bit in the OPTION register (Figure 10.17), this being bit 7. Bit 6 selects the clock source for the ADC. A delay of at least 100 μs is required after powering up to allow the system to stabilise.

Figure 10.17 OPTION register.

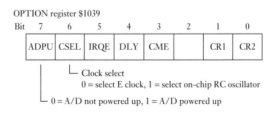

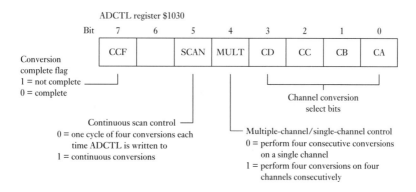

Figure 10.18 ADCTL register.

MULT = 0

CD	CC	CB	CA	Channel converted
0	0	0	0	PE0
0	0	0	1	PE1
0	0	1	0	PE2
0	0	1	1	PE3
0	1	0	0	PE4
0	1	0	1	PE5
0	1	1	0	PE6
0	1	1	1	PE7

MULT = 1

CD	CC	CB	CA	A/D result register			
				ADR1	ADR2	ADR3	ADR4
0	0	×	×	PE0	PE1	PE2	PE3
0	1	×	×	PE4	PE5	PE6	PE7

The A/D conversion is initiated by writing to the A/D control/ status register (ADCTL) after powering up and the stabilisation delay (Figure 10.18). This involves selecting the channels and operation modes. The conversion will then start one clock cycle later. For example, if single-channel mode is selected by setting MULT = 0 then four successive A/D conversions will occur of the channel selected by the CD–CA bits. The results of the conversion are placed in the A/D result registers ADR1–ADR4.

6 *Modes*

MODA and MODB are two pins that can be used to force the microcontroller into one of four modes at power-up, these modes being special bootstrap, special test, single chip and expanded:

MODB	MODA	Mode
0	1	Special bootstrap
0	1	Special test
1	0	Single chip
1	1	Expanded

In the single-chip mode the microcontroller is completely self-contained with the exception of an external clock source and a reset circuit. With such a mode the microcontroller may not have enough resources, e.g. memory, for some applications and so the expanded mode can be used so that the number of addresses can be increased. Ports B and C then provide address, data and control buses. Port B functions as the upper eight address pins and port C as the multiplexed data and low address pins. The bootstrap mode allows a manufacturer to load special programs in a special ROM for an M68HC11 customer. When the microcontroller is set in this mode, the special program is loaded. The special test mode is primarily used during Motorola's internal production testing.

After the mode has been selected, the MODA pin becomes a pin which can be used to determine if an instruction is starting to execute. The MODB pin has the other function of giving a means by which the internal RAM of the chip can be powered when the regular power is removed.

7 *Oscillator pins*
The oscillator system pins XTAL and EXTAL are the connections to access the internal oscillator. Figure 10.19 shows an external circuit that might be used. E is the bus clock and runs at one-quarter of the oscillator frequency and can be used to synchronise external events.

Figure 10.19   Oscillator output.

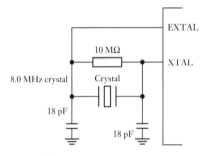

8 *Interrupt controller*
The interrupt controller is to enable the microcontroller to interrupt a program (see Section 13.3.3). An interrupt is an event that requires the CPU to stop normal program execution and perform some service related to the event. The two lines IRQ and XIRQ are for the external inputs of interrupt signals. RESET is for resetting the microcontroller and allowing an orderly system start-up. The state of the pin can be set either internally or externally. When a reset condition is detected, the pin signal is set low for four clock cycles. If after a further two cycles it is found to be still low, then an external reset is considered to have occurred. If a positive transition is detected on the power input V_{DD} a power-on reset occurs. This provides a 4064 cycle time delay. If the reset pin is low at the end of the power-on delay time, the microcontroller remains in the rest condition until it goes high.

9 *Timer*
The M68HC11 contains a timer system. This has a free-running counter, five-output compare function, the ability to capture the time when an external event occurs, a real-time periodic interrupt and a counter, called

the pulse accumulator, for external events. The free-running counter, called TCNT, is a 16-bit counter which starts counting at 0000 when the CPU is reset and continues thereafter continuously running and cannot be reset by the program. Its value can be read at any time. The source for the counter is the system bus clock and its output can be prescaled by setting the PR0 and PR1 bits as bits 0 and 1 in the TMSK2 register at address $1024 (Figure 10.20).

Figure 10.20   TMSK2 register.

Timer interrupt register 2 at address $1024

Bit	7	6	5	4	3	2	1	0

PR1 PR0

Prescale factors

			One count	
		Prescale	Bus frequency	
PR1	PR0	factor	2 MHz	1 MHz
0	0	1	0.5 ms	1 ms
0	1	4	2 ms	4 ms
1	0	8	4 ms	8 ms
1	1	16	8 ms	16 ms

The output compare functions allow times, i.e. timer counts, to be specified at which an output will occur when the preset count is reached. The input capture system captures the value of the counter when an input occurs and so the exact time at which the input occurs is captured. The pulse accumulator can be configured to operate as an event counter and count external clock pulses or as a gated time accumulator and store the number of pulses occurring in a particular time interval as a result of the counter being enabled and then, at some later time, disabled. The pulse accumulator control register PACTL (see Figure 10.12) at address $1026 is used to select the mode of operation. The bit PAEN is set to 0 to disable the pulse accumulator and to 1 to enable it, the bit PAMOD to 0 for the event counter mode and 1 for the gated time mode, and the bit PEDGE to 0 for the pulse accumulator to respond to a falling edge when in the event counter mode and 1 to respond to a rising edge in that mode. In gated mode, bit PEDGE set to 0 causes the count to be disabled when port A, bit 7 is 0 and to accumulate when 1, and when bit PEDGE is set to 1 in this mode, the count is disabled when port A, bit 7 is 1 and enabled when it is 0.

10 *COP*

Another timer function is COP, the computer operating properly function. This is a timer which times out and resets the system if an operation is not concluded in what has been deemed to be a reasonable time (see Section 16.2). This is often termed a **watchdog timer**.

11 *PWM*

Pulse width modulation (PWM) is used for the control of the speed of d.c. motors (see Sections 3.6 and 9.5.3) by using a square wave signal and, by varying the amount of time for which the signal is on, changing the average value of the signal. A square wave can be generated by a microcontroller by arranging for an output to come on every half-period.

However, some versions of M68HC11 have a PWM module and so, after the PWM module has been initialised and enabled, the PWM waveforms can be automatically outputted.

As will be apparent from the above, before a microcontroller can be used it is necessary to initialise it, i.e. set the bits in appropriate registers, so that it will perform as required.

10.3.2 Intel 8051

Another common family of microcontrollers is the Intel 8051, Figure 10.21 showing the pin connections and the architecture. The 8051 has four parallel input/output ports, ports 0, 1, 2 and 3. Ports 0, 2 and 3 also have alternative functions. The 8051AH version has 4K ROM bytes, 128-byte RAM, two timers and interrupt control for five interrupt sources.

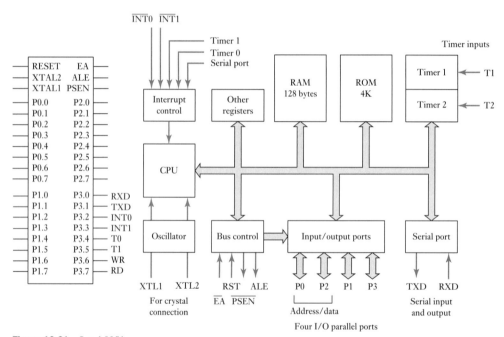

Figure 10.21 Intel 8051.

1 *Input/output ports*

   Port 0 is at address 80H, port 1 at address 90H, port 2 at address A0H and port 3 at address B0H (note the use of H, or h, after the address with Intel to indicate that it is in hex). When a port is to be used as an output port, the data is put into the corresponding special function register. When a port is to be used as an input port, the value FFH must first be written to it. All the ports are bit addressable. Thus we might, for example, use just bit 6 in port 0 to switch a motor on or off and perhaps bit 7 to switch a pump on or off.

Port 0 can be used as an input port or an output port. Alternatively it can be used as a multiplexed address and data bus to access external memory. Port 1 can be used as an input port or an output port. Port 2 can be used as an input port or an output port. Alternatively it can be used for the high address bus to access external memory. Port 3 can be used as an input port or an output port. Alternatively, it can be used as a special-purpose input/output port. The alternative functions of port 3 include interrupt and timer outputs, serial port input and output and control signals for interfacing with external memory. RXD is the serial input port, TXD is the serial output port, INT0 is the external interrupt 0, INT1 is the external interrupt 1, T0 is the timer/counter 0 external input, T1 is the timer/counter 1 external input, WR is the external memory write strobe and RD is the external memory read strobe. The term **strobe** describes a connection used to enable or disable a particular function. Port 0 can be used as either an input or output port. Alternatively it can be used to access external memory.

2 *ALE*

The ALE pin provides an output pulse for latching the low-order byte of the address during access to external memory. This allows 16-bit addresses to be used. Figure 10.22 illustrates this.

Figure 10.22 Use of ALE.

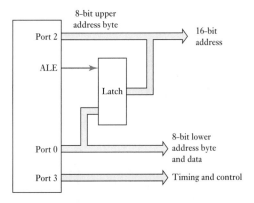

3 *PSEN*

The program store enable (PSEN) pin is the read signal pin for external program memory and is active when low. It is connected to the output enable pin of external ROM or EPROM.

4 *EA*

The external access (EA) pin is taken low for the microprocessor to access only external program code; when high it automatically accesses internal or external code depending on the address. Thus when the 8051 is first reset, the program counter starts at $0000 and points to the first program instruction in the internal code memory unless EA is tied low. Then the CPU issues a low on PSEN to enable the external code memory to be used. This pin is also used on a microcontroller with EPROM to receive the programming supply voltage for programming the EPROM.

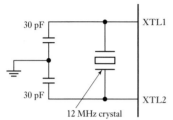

Figure 10.23   Crystal.

5   *XTAL1, XTAL2*

These are the connecting pins for a crystal or external oscillator. Figure 10.23 illustrates how they are used with a crystal. The most commonly used crystal frequency is 12 MHz.

6   *RESET*

A high signal on this pin for at least two machine cycles resets the microcontroller, i.e. puts in a condition to allow an orderly system start-up.

7   *Serial input/output*

Writing to the serial data buffer SBUF at address 99H loads data for transmission; reading SBUF accesses received data. The bit addressable serial port control register SCON at address 98H is used to control the various modes of operation.

8   *Timing*

The timer mode register TMOD at address 89H is used to set the operating mode for timer 0 and timer 1 (Figure 10.24). It is loaded at an entity and is not individually bit addressable. The timer control register TCON (Figure 10.25) contains status and control bits for timer 0 and timer 1. The upper 4 bits are used to turn the timers on and off or to signal a timer overflow. The lower 4 bits have nothing to do with timers but are used to detect and initiate external interrupts.

Figure 10.24   TMOD register.

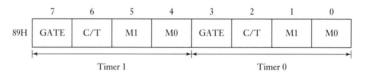

7	6	5	4	3	2	1	0
GATE	C/T	M1	M0	GATE	C/T	M1	M0

89H

Timer 1   Timer 0

Gate: 0 = timer runs whenever TR0/TR1 set
        1 = timer runs only when INT0/INT1 is high along with TR0/TR1

C/T: counter/timer select
        0 = input from system clock, 1 = input from TX0/TX1

M0 and M1 set the mode

M1	M0	Mode	
0	0	0	13-bit counter, lower 5 bits of TL0 and all 8 bits of TH0
0	1	1	16-bit counter
1	0	2	8-bit auto-reload timer/counter
1	1	3	TL0 is an 8-bit timer/counter controlled by timer 0 control bits. TH0 is an 8-bit timer controlled by timer 1 control bits. Timer 1 is off

Figure 10.25   TCON register.

7	6	5	4	3	2	1	0
TF1	TR1	TF0	TR0	IE1	IT1	IE0	IT0

88H

TF0, TF1   Timer overflow flag; set by hardware when time overflows and cleared by hardware when the processor calls the interrupt routine

TR0, TR1   Timer run control bits: 1 = timer on, 0 = timer off
IE0, IE1   Interrupt edge flag set by hardware when external interrupt edge or low level detected and cleared when interrupt processed

IT0, IT1   Interrupt type set by software: 1 = falling-edge-triggered interrupt, 0 = low-level-triggered interrupt

The source of the bits counted by each timer is set by the C/T bit; if the bit is low, the source is the system clock divided by 12, otherwise if high it is set to count an input from an external source. The timers can be started by setting TR0 or TR1 to 1 and stopping by making it 0. Another method of controlling a timer is by setting the GATE to 1 and so allowing a timer to be controlled by the INT0 or INT1 pin on the microcontroller going to 1. In this way an external device connected to one of these pins can control the counter on/off.

9 *Interrupts*
Interrupts force the program to call a subroutine located at a specified address in memory; they are enabled by writing to the interrupt enable register IE at address A8H (Figure 10.26).

Figure 10.26
IE register.

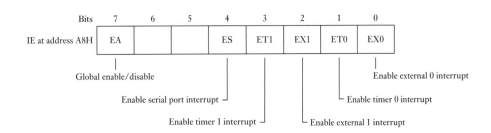

The term **special function registers** is used for the input/output control registers (Figure 10.27), like IE above, and these are located at addresses 80 to FF. Accumulator A (ACC) is the major register used for data operations; the B register is used for multiplication and division. P0, P1, P2 and P3 are the latch registers for ports 0, 1, 2 and 3.

8D	TH1	F0	B
8C	TH0	E0	ACC
8B	TL1	D0	PSW
8A	TL0	B8	IP
89	TMOD	B0	P3
88	TCON	A8	IE
87	PCON	A0	P2
83	DPH	99	SBUF
82	DPL	98	SCON
81	SP	90	P1
80	P0		

Figure 10.27 Registers.

10.3.3 Microchip™ microcontrollers

Another widely used family of 8-bit microcontrollers is that provided by Microchip™. The term peripheral interface controller (PIC) is used for its single-chip microcontrollers. These use a form of architecture termed **Harvard architecture**. With this architecture, instructions are fetched from program memory using buses that are distinct from the buses used for accessing variables (Figure 10.28). In the other microcontrollers so far discussed in this chapter, separate buses are not used and thus program data fetches have to wait for read/write and input/output operations to be completed before the next instruction can be received from memory. With Harvard architecture, instructions can be fetched every cycle without waiting, each instruction being executed during the cycle following its fetch. Harvard architecture enables faster execution speeds to be achieved for a given clock frequency. Figure 10.29 shows the pin connections for one version of the PIC16C74A and the 16F84 microcontroller, and Figure 10.30 the basic form of the architecture.

Figure 10.28 (a) Harvard architecture, (b) von Neumann architecture, (c) code execution.

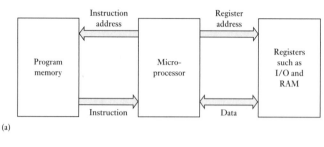

(a)

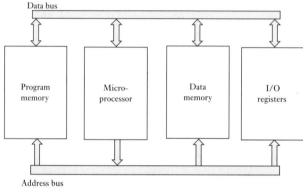

(b)

Instruction cycle	Operation		Instruction cycle	Operation
1	Fetch instruction 1		1	Fetch instruction 1
2	Fetch instruction 2	Execute instruction 1	2	Execute instruction 1
3	Fetch instruction 3	Execute instruction 2	3	Fetch instruction 2
4	Fetch instruction 4	Execute instruction 3	4	Execute instruction 2
Harvard architecture			Von Neumann architecture	

(c)

The basic features of the 16C74 microcontroller, other PIC microcontrollers having similar functions, are outlined below.

1 *Input/output ports*

Pins 2, 3, 4, 5, 6 and 7 are for the bidirectional input/output port A. As with the other bidirectional ports, signals are read from and written to via port registers. The direction of the signals is controlled by the TRIS direction registers; there is a TRIS register for each port. TRIS is set as 1 for read and 0 for write (Figure 10.31).

Pins 2, 3, 4 and 5 can also be used for analogue inputs, pin 6 for a clock input to timer 0; pin 7 can also be the slave select for the synchronous serial port (see later in this section).

Pins 33, 34, 35, 36, 37, 38, 39 and 40 are for the bidirectional input/output port B; the direction of the signals is controlled by a corresponding TRIS direction register. Pin 33 can also be the external interrupt pin. Pins 37, 38, 39 and 40 can also be the interrupt on change pins. Pin 39 can

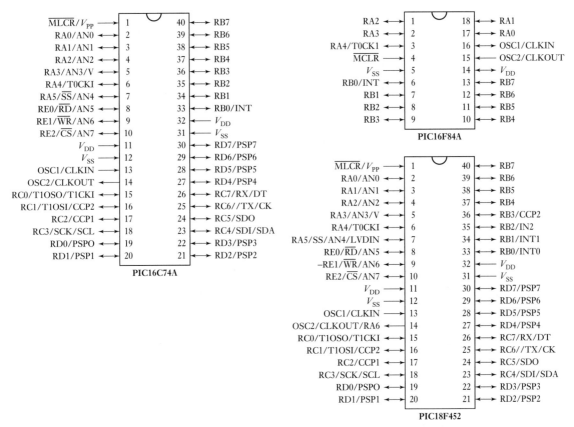

Figure 10.29 PIC pin diagrams.

also be the serial programming clock and pin 40 the serial programming data.

Pins 15, 16, 17, 18, 23, 24, 25 and 26 are for the bidirectional input/output port C; the direction of the signals is controlled by a corresponding TRIS direction register. Pin 15 can also be the timer 1 output or the timer 1 clock input. Pin 16 can also be the timer 1 oscillator input or Capture 2 input/Compare 2 output/PWM2 output.

Pins 19, 20, 21, 22, 27, 28, 29 and 30 are for the bidirectional input/output port D; the direction of the signals is controlled by a corresponding TRIS direction register.

Pins 8, 9 and 10 are for the bidirectional input/output port E; the direction of the signals is controlled by a corresponding TRIS direction register. Pin 8 can also be the read control for the parallel slave port or analogue input 5. The parallel slave port is a feature that facilitates the design of personal computer interface circuitry; when in use the pins of ports D and E are dedicated to this operation.

2 *Analogue inputs*
Pins 2, 3, 4, 5 and 7 of port A and pins 8, 9 and 10 of port E can also be used for analogue inputs, feeding through an internal ADC. Registers ADCON1 and TRISA for port A (TRISE for port E) must be initialised

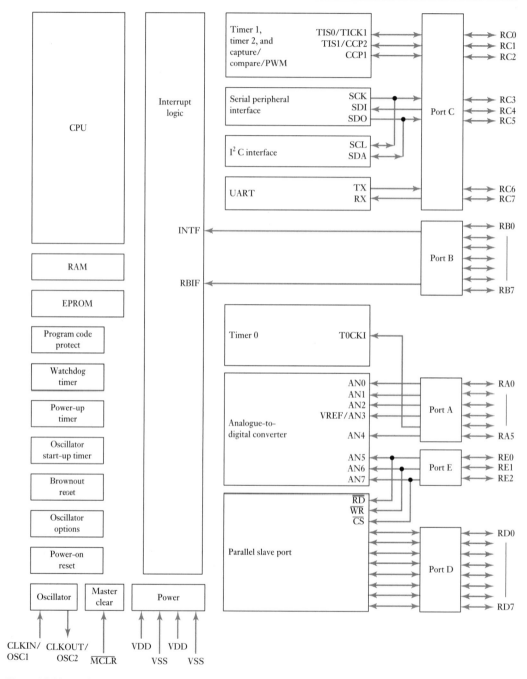

Figure 10.30 PIC 16C74/74A.

Figure 10.31 Port direction.

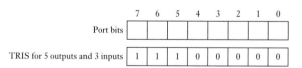

to select the reference voltage to be used for the conversion and select channels as inputs. Then ADCON0 has to be initialised using the settings indicated below:

ADCON0 bits			
5	4	3	For analogue input on
0	0	0	Port A, bit 0
0	0	1	Port A, bit 1
0	1	0	Port A, bit 2
0	1	1	Port A, bit 3
1	0	0	Port A, bit 5
1	0	1	Port E, bit 0
1	1	0	Port E, bit 1
1	1	1	Port E, bit 2

3 *Timers*

The microcontroller has three timers: timer 0, timer 1 and timer 2. Timer 0 is an 8-bit counter which can be written to or read from and can be used to count external signal transitions, generating an interrupt when the required number of events have occurred. The source of the count can be either the internal bus clock signal or an external digital signal. The choice of count source is made by the T0CS bit in the OPTION register (Figure 10.32).

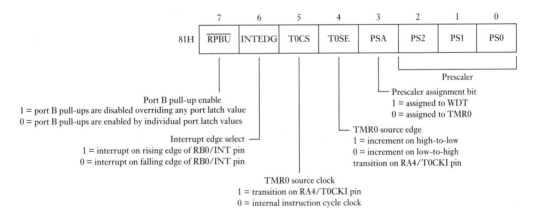

Figure 10.32 OPTION register.

If the prescaler is not selected then the count is incremented after every two cycles of the input source. A prescaler can be used so that signals are only passed to the counter after some other fixed number of clock cycles. The following shows the scaling rates possible. WDT gives the scaling factors selected when the watchdog timer is enabled. It is used to time out and reset the system if an operation is not concluded in a reasonable time; the default time is nominally 18 ms.

Prescalar bit values			TMR0 rate	WDT rate
PS2	PS1	PS0		
0	0	0	1 : 2	1 : 1
0	0	1	1 : 4	1 : 2
0	1	0	1 : 8	1 : 4
0	1	1	1 : 16	1 : 8
1	0	0	1 : 32	1 : 16
1	0	1	1 : 64	1 : 32
1	1	0	1 : 128	1 : 64
1	1	1	1 : 256	1 : 128

Timer 1 is the most versatile of the timers and can be used to monitor the time between signal transitions on an input pin or control the precise time of transitions on an output pin. When used with the capture or compare modes, it enables the microcontroller to control the timing of an output on pin 17.

Timer 2 can be used to control the period of a PWM output. PWM outputs are supplied at pins 16 and 17.

4 *Serial input/output*

The PIC microcontroller includes a synchronous serial port (SSP) module and a serial communications interface module (SCI). Pin 18 has the alternative functions of the synchronous serial clock input or output for SPI serial peripheral interface mode and I²C mode. The I²C bus provides a two-wire bidirectional interface that can be used with a range of other chips; it can also be used for connecting a master microcontroller to slave microcontrollers. UART, i.e. the universal asynchronous receiver transmitter, can be used to create a serial interface to a personal computer.

5 *Parallel slave port*

The parallel slave port uses ports D and E and enables the microcontroller to provide an interface with a PC.

6 *Crystal input*

Pin 13 is for the oscillator crystal input or external clock source input; pin 14 is for the oscillator crystal output. Figure 10.33(a) shows the arrangement that might be used for accurate frequency control,

Figure 10.33 Frequency control.

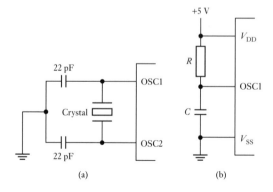

Figure 10.33 (b) that which might be used for a low-cost frequency control; for a frequency of 4 MHz we can have $R = 4.7\,k\Omega$ and $C = 33$ pF. The internal clock rate is the oscillator frequency divided by 4.

7 *Master clear and resets*
Pin 1 is the master clear MCLR, i.e. reset input, and is taken low to reset the device on demand and give an orderly start-up. When a V_{DD} rise is detected, a power-on-rest (POR) pulse is generated to provide a fixed time-out delay and keep the processor in the reset state. If the V_{DD} voltage goes below a specified voltage level for more than a certain amount of time, a brownout reset is activated. The watchdog timer is another way reset can occur. This is a timer which times out and rests the microcontroller if an operation is not concluded in what has been deemed a reasonable time.

The **special-purpose registers** (Figure 10.34) are used for input/output control, as illustrated above in relation to a few of these registers.

Figure 10.34 Special-purpose registers.

File address	Bank 0	Bank 1	File address
00h	INDF	INDF	80h
01h	TMR0	OPTION	81h
02h	PCL	PCL	82h
03h	STATUS	STATUS	83h
04h	FSR	FSR	84h
05h	PORTA	TRISA	85h
06h	PORTB	TRISB	86h
07h	PORTC	TRISC	87h
08h	PORTD	TRISD	88h
09h	PORTE	TRISE	89h
0Ah	PCLATH	PCLATH	8Ah
0Bh	INTCON	INTCON	8Bh
0Ch	PIR1	PIE1	8Ch
0Dh	PIR2	PIE2	8Dh
0Eh	TMR1L	PCON	8Eh
0Fh	TMR1H		8Fh
10h	T1CON		90h
11h	TMR2		91h
12h	T2CON	PR2	92h
13h	SSPBUF	SSPADD	93h
14h	SSPCON	SSPSTAT	94h
15h	CCPR1L		95h
16h	CCPR1H		96h
17h	CCP1CON		97h
18h	RCSTA	TXSTA	98h
19h	TXREG	SPBRG	99h
1Ah	RCREG		9Ah
1Bh	CCPR2L		9Bh
1Ch	CCPR2H		9Ch
1Dh	CCPR2CON		9Dh
1Eh	ADRES		9Eh
1Fh	ADCON0	ADCON1	9Fh
20h	General-purpose registers	General-purpose registers	A0h
7Fh			FFh

The registers for the PIC16C73/74 are arranged in two banks and before a particular register can be selected, the bank has to be chosen by setting a bit in the status register (Figure 10.35).

Figure 10.35 STATUS register.

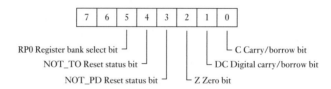

10.3.4 Atmel AVR microcontrollers and Arduino

Atmel AVR has a range of 8-bit microcontrollers and uses a modified Harvard architecture with the program and data stored in separate memory locations. **Arduino** is a small microcontroller board with complementary components which has been designed to facilitate the use of the microcontroller in control projects. The basic board, Arduino UNO Revision 3, uses an 8-bit Atmel microcontroller, Atmega328. This microcontroller has a memory system, input/output ports, timer/counter, pulse width modulation, ADC, an interrupt system and serial communication.

Arduino boards can be purchased pre-assembled, from a variety of retailers. The board has a universal serial bus (USB) plug to enable it to be directly connected to a computer and a number of connection sockets to enable it to be connected to external items such as motors, relays, etc. The board can be powered by connection to an external power supply, e.g. a 9 V battery, or through the USB connection from the computer. Pre-programmed into the on-board microcontroller chip is a boot loader that directly allows the uploading of programs into the microcontroller memory. The basic boards are supplemented by accessory boards, termed shield boards, e.g. a LCD (liquid crystal display) display board, a motor board and an Ethernet board, that can be plugged on top of the basic Arduino board and into board pin-headers. Multiple shields can be stacked.

The board is **open source**. This means that anyone is allowed to make Arduino-compatible boards. Starter kits are available, typically including such items as the Arduino board, a USB cable so that the board can be programmed from a computer, a breadboard to use for assembly of an external circuit with wires and commonly used components such as resistors, photoresistors potentiometers, capacitors, pushbuttons, temperature sensor, LCD alphanumeric display, light-emitting diodes (LEDs), DC motor, H-bridge motor driver, optocouplers, transistors and diodes.

Figure 10.36 shows the basic features of the Arduino UNO Revision 3 board. These are outlined below.

1 *Reset pin in the section power connectors*
  This resets the microcontroller so that it begins the program from the start. To perform a reset, this pin needs to be momentarily set low, i.e. connected to 0 V. This action can also be initiated by using the reset switch.

2 *Other pins in the power connectors*
  These provide different voltages, i.e. 3.5 V, 5 V, GND and 9 V.

Figure 10.36 The basic elements of an Arduino UNO Revision 3 board.

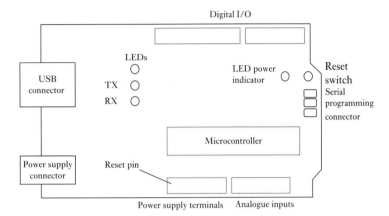

3 *Analogue inputs*
These are labelled A0 to A5 and can be used to detect voltage signals.
4 *Digital connections*
Labelled Digital 0 to 13, these can be used for either inputs or outputs. The first two of these connections are also labelled RX and TX for receive and transmit in communication.
5 *USB connector*
This is used to connect the board to a computer.
6 *Serial programming connector*
This offers a means of programming the Arduino without using the USB port.
7 *LEDs*
The board is equipped with three LEDs, one to indicate serial transmission (TX), one reception (RX) and one extra LED for use in projects.
8 *Power supply connector*
An external power supply can be connected to the board via a connector at the bottom left corner of the board. However, the board may be powered from the USB port during connection to a computer.
9 *Microcontroller*
The microcontroller ATmega328 comes pre-burned with a bootloader that allows new program code to be uploaded to it without the use of an external hardware programmer.

When purchased, an Arduino board usually has a sample program pre-installed. This program is to make an LED on the board flash. To start the program all you need to do is supply power, which can be done by plugging into the USB port of a computer. The LED should then flash, so verifying that the board is working. To install new software, termed sketches for Arduino boards, you need to install the Arduino software and load the USB drivers in your computer. Instructions for doing this and the software can be found at the Arduino website (www.arduino.cc). Once this is installed you can upload a program to the Arduino board.

10.3.5 Selecting a microcontroller

In selecting a microcontroller the following factors need to be considered.

1 *Number of input/output pins*
How many input/output pins are going to be needed for the task concerned?

2 *Interfaces required*
What interfaces are going to be required? For example, is PWM required? Many microcontrollers have PWM outputs, e.g. the PIC17C42 has two.

3 *Memory requirements*
What size memory is required for the task?

4 *The number of interrupts required*
How many events will need interrupts?

5 *Processing speed required*
The microprocessor takes time to execute instructions (see Section 11.2.2), this time being determined by the processor clock.

As an illustration of the variation of microcontrollers available, Table 10.1 shows details of members of the Intel 8051 family, Table 10.2 the PIC16Cxx family and Table 10.3 the M68HC11 family.

Table 10.1 Intel 8051 family members.

	ROM	EPROM	RAM	Timers	I/O ports	Interrupts
8031AH	0	0	128	2	4	5
8051AH	4K	0	128	2	4	5
8052AH	8K	0	256	3	4	6
8751H	0	4K	128	2	4	5

Table 10.2 PIC16C family members.

	I/O	EPROM	RAM	ADC channels	USART	CCP modules
PIC16C62A	22	2K	128	0	0	1
PIC16C63	22	4K	192	0	1	2
PIC16C64A	33	2K	128	0	0	1
PIC16C65A	33	4K	192	0	1	2
PIC16C72	22	2K	128	5	0	1
PIC16C73A	22	4K	192	5	1	2
PIC16C74A	33	4K	192	8	1	3

10.4 Applications

The following are two examples of how microcontrollers are used; more case studies are given in Chapter 24.

10.4.1 Temperature measurement system

As a brief indication of how a microcontroller might be used, Figure 10.37 shows the main elements of a temperature measurement system using an MC68HC11. The temperature sensor gives a voltage proportional to the

Table 10.3 M68HC11 family members.

	ROM	EEPROM	RAM	ADC	Timer	PWM	I/O	Serial	E-clock MHz
68HC11AO	0	0	256	8 ch., 8-bit	(1)	0	22	SCI, SPI	2
68HC11A1	0	512	256	8 ch., 8-bit	(1)	0	22	SCI, SPI	2
68HC11A7	8K	0	256	8 ch., 8-bit	(1)	0	38	SCI, SPI	3
68HC11A8	8K	512	256	8 ch., 8-bit	(1)	0	38	SCI, SPI	3
68HC11C0	0	512	256	4 ch., 4-bit	(2)	2 ch., 8-bit	36	SCI, SPI	2
68HC11D0	0	0	192	None	(2)	0	14	SCI, SPI	2

Timer: (1) is three-input capture, five-output compare, real-time interrupt, watchdog timer, pulse accumulator, (2) is three- or four-input capture, five- or four-output compare, real-time interrupt, watchdog timer, pulse accumulator. Serial: SCI is asynchronous serial communication interface, SPI is synchronous serial peripheral interface.

Figure 10.37 Temperature measurement system.

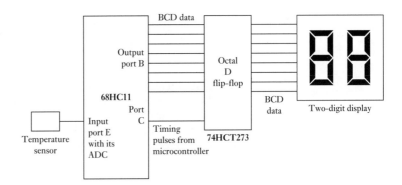

temperature (e.g. a thermotransistor such as LM35; see Section 2.9.4). The output from the temperature sensor is connected to an ADC input line of the microcontroller. The microcontroller is programmed to convert the temperature into a binary-coded decimal (BCD) output which can be used to switch on the elements of a two-digit, seven-element display. However, because the temperature may be fluctuating it is necessary to use a storage register which can hold data long enough for the display to be read. The storage register, 74HCT273, is an octal D-type flip-flop which is reset on the next positive-going edge of the clock input from the microcontroller.

10.4.2 Domestic washing machine

Figure 10.38 shows how a microcontroller might be used as the controller for a domestic washing machine. The microcontroller often used is the Motorola M68HC05B6; this is simpler and cheaper than the Motorola M68HC11 microcontroller discussed earlier in this chapter and is widely used for low-cost applications.

The inputs from the sensors for water temperature and motor speed are via the A/D input port. Port A provides the outputs for the various actuators used to control the machine and also the input for the water-level

Figure 10.38 Washing machine.

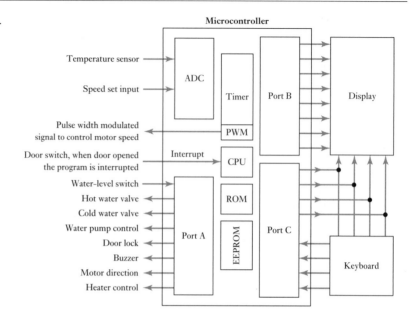

switch. Port B gives outputs to the display. Port C gives outputs to the display and also receives inputs from the keyboard used to input to the machine the various program selections. The PWM section of the timer provides a PWM signal to control the motor speed. The entire machine program is interrupted and stopped if the door of the washing machine is opened.

10.5   Programming

A commonly used method for the development of programs follows the following steps.

1 Define the problem, stating quite clearly what function the program is to perform, the inputs and outputs required, any constraints regarding speed of operation, accuracy, memory size, etc.
2 Define the algorithm to be used. An **algorithm** is the sequence of steps which define a method of solving the problem.
3 For systems with fewer than thousands of instructions a useful aid is to represent the algorithm by means of a **flow chart**. Figure 10.39(a) shows the standard symbols used in the preparation of flow charts. Each step of an algorithm is represented by one or more of these symbols and linked together by lines to represent the program flow. Figure 10.39(b) shows part of a flow chart where, following the program start, there is operation A, followed by a branch to either operation B or operation C depending on whether the decision to the query is yes or no. Another useful design tool is **pseudocode**. Pseudocode is a way of describing the steps in an algorithm in an informal way which can later be translated into a program (see Section 10.5.1).
4 Translate the flow chart/algorithm into instructions which the microprocessor can execute. This can be done by writing the instructions in some language, e.g. assembly language or perhaps C, and then

Figure 10.39 Flow chart:
(a) symbols, (b) example.

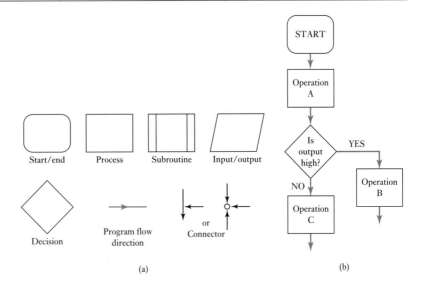

(a)

(b)

converting these, either manually or by means of an assembler computer program, into a code which is acceptable to the microprocessor, i.e. machine code.

5   Test and debug the program. Errors in programs are referred to as **bugs** and the process of tracking them down and eliminating them as **debugging**.

10.5.1   Pseudocode

Pseudocode is rather like drawing a flow chart and involves writing a program as a sequence of functions or operations with the decision element IF–THEN–ELSE and the repetition element WHILE–DO.

A sequence (Figure 10.40(a)) would be written as:

```
BEGIN A
   ...
END A
   ...
BEGIN B
   ...
END B
```

and a decision as:

```
IF X
THEN
   BEGIN A
   ...
   END A
ELSE
   BEGIN B
   ...
   END B
ENDIF X
```

Figure 10.40   (a) Sequence,
(b) IF–THEN–ELSE,
(c) WHILE–DO.

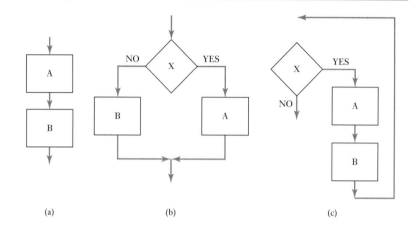

Figure 10.40(b) shows such a decision in a flow chart. A repetition is written as:

```
WHILE X
DO
  BEGIN A
  ...
  END A
  BEGIN B
  ...
  END B
ENDO WHILE X
```

Figure 10.40(c) shows the WHILE–DO as a flow chart. A program written in this way might then appear as:

```
BEGIN PROGRAM
  BEGIN A
    IF X
      BEGIN B
      END B
    ELSE
      BEGIN C
      END C
    ENDIF X
  END A
  BEGIN D
    IF Z
      BEGIN E
      END E
    ENDIF Z
  END D
```

Chapter 11 shows how programs can be written in assembly language and Chapter 12 in C.

Summary

Basically, systems involving **microprocessors** have three parts: a central processing unit (CPU), input and output interfaces and memory. Within a microprocessor, digital signals move along **buses**, these being parallel tracks for transmission of parallel rather than serial data.

Microcontrollers are the integration on a single chip of a microprocessor with memory, input/output interfaces and other peripherals such as timers.

An **algorithm** is the sequence of steps which define a method of solving a problem. **Flow charts** and **pseudocode** are two methods of describing such steps.

Problems

10.1 Explain, for a microprocessor, the roles of (a) accumulator, (b) status, (c) memory address, (d) program counter registers.

10.2 A microprocessor uses eight address lines for accessing memory. What is the maximum number of memory locations that can be addressed?

10.3 A memory chip has 8 data lines and 16 address lines. What will be its size?

10.4 How does a microcontroller differ from a microprocessor?

10.5 Draw a block diagram of a basic microcontroller and explain the function of each subsystem.

10.6 Which of the M68HC11 ports is used for (a) the ADC, (b) a bidirectional port, (c) serial input/output, (d) as just an 8-bit output-only port?

10.7 How many bytes of memory does the M68HC11A7 have for data memory?

10.8 For the Motorola M68HC11, port C is bidirectional. How is it configured to be (a) an input, (b) an output?

10.9 The Motorola M68HC11 can be operated in single-chip and in extended mode. Why these modes?

10.10 What is the purpose of the ALE pin connection with the Intel 8051?

10.11 What input is required to reset an Intel 8051 microcontroller?

10.12 Write pseudocode to represent the following:

(a) if A is yes then B, else C;

(b) while A is yes do B.

Chapter eleven Assembly language

Objectives

The objective of this chapter is that, after studying it, the reader should be able to use assembly language to write programs involving data transfers, arithmetic, logic, jumps, branches, subroutines, delays and look-up tables.

11.1 Languages

Software is the term used for the **instructions** that tell a microprocessor or microcontroller what to do. The collection of instructions that a microprocessor will recognise is its **instruction set**. The form of the instruction set depends on the microprocessor concerned. The series of instructions that is needed to carry out a particular task is called a **program**.

Microprocessors work in binary code. Instructions written in binary code are referred to as being in **machine code**. Writing a program in such a code is a skilled and very tedious process. It is prone to errors because the program is just a series of zeros and ones and the instructions are not easily comprehended by just looking at the pattern. An alternative is to use an easily comprehended form of shorthand code for the patterns of zeros and ones. For example, the operation of adding data to an accumulator might be represented by just ADDA. Such a shorthand code is referred to as a **mnemonic code**, a mnemonic code being a 'memory-aiding' code. The term **assembly language** is used for such a code. Writing a program using mnemonics is easier because they are an abbreviated version of the operation performed by the instruction. Also, because the instructions describe the program operations and can easily be comprehended, they are less likely to be used in error than the binary patterns of machine code programming. The assembler program has still, however, to be converted into machine code since this is all the microprocessor will recognise. This conversion can be done by hand using the manufacturer's data sheets which list the binary code for each mnemonic. However, computer programs are available to do the conversion, such programs being referred to as **assembler programs**.

High-level languages, e.g. BASIC, C, FORTRAN and PASCAL, are available which provide a type of programming language that is even closer to describing in easily comprehended language the types of operations required. Such languages have still, however, to be converted into machine code, by a computer program, for the microprocessor to be able to use. This chapter is an outline of how programs might be written in assembly language, Chapter 12 using C.

The following are commonly used instructions that may be given to a microprocessor, the entire list of such instructions being termed the instruction set. Appendix D gives instruction sets for three commonly encountered types of microcontroller. The instruction set differs from one microprocessor to another. The following are some of the commonly encountered instructions.

Data transfer/movement

1 *Load*

This instruction reads the contents of a specified memory location and copies it to a specified register location in the central processing unit (CPU) and is typically used with Motorola microprocessors, e.g. LDAA $0010:

Before instruction	After instruction
Data in location 0010	Data in location 0010
	Data from 0010 in accumulator A

2 *Store*

This instruction copies the current contents of a specified register into a specified memory location and is typically used with Motorola microprocessors, e.g. STA $0011:

Before instruction	After instruction
Data in accumulator A	Data in accumulator A
	Data copied to location 0011

3 *Move*

This instruction is used to move data into a register or copy data from one register to another and is used with peripheral interface controller (PIC) and Intel microprocessors, e.g. with PIC, MOV R5,A:

Before instruction	After instruction
Data in register A	Data in register A
	Data copied to register R5

4 *Clear*

This instruction resets all bits to zero, e.g. with Motorola, CLRA to clear accumulator A; with PIC, CLRF 06 to clear file register 06.

Arithmetic

5 *Add*

This instruction adds a number to the data in some register, e.g. with Intel, ADD A, #10h:

Before instruction	After instruction
Accumulator A with data	Accumulator A plus 10 hex

and with Motorola, ADDD $0020:

Before instruction	After instruction
Accumulator D with data	Accumulator D plus contents of location 0020

or the contents of a register to the data in a register, e.g. with Intel, ADD A, @R1:

Before instruction	After instruction
Accumulator A with data	Accumulator A plus contents of location R1

and with PIC, addwf 0C:

Before instruction	After instruction
Register 0C with data	Register 0C plus contents of location w

6 *Decrement*
This instruction subtracts 1 from the contents of a specified location. For example, we might have register 3 as the specified location and so, with Intel, DEC R3:

Before instruction	After instruction
Register R3 with data 0011	Register R3 with data 0010

7 *Increment*
This instruction adds 1 to the contents of a specified location, e.g. INCA with Motorola to increment the data in accumulator A by 1, incf 06 with PIC to increment the data in register 06 by 1.

8 *Compare*
This instruction indicates whether the contents of a register are greater than, less than or the same as the contents of a specified memory location. The result appears in the status register as a flag.

Logical

9 *AND*
This instruction carries out the logical AND operation with the contents of a specified memory location and the data in some register. Numbers are ANDed bit by bit, e.g. with Motorola, ANDA %1001:

Before instruction	After instruction
Accumulator A with data 0011 Memory location with data 1001	Accumulator A with data 0001

Only in the least significant bit in the above data do we have a 1 in both sets of data and the AND operation only gives a 1 in the least significant bit of the result. With PIC, ANDLW 01 adds the binary number 01 to the number in W and if the least significant bit is, say, 0 then the result is 0.

10 *OR*
This instruction carries out the logical OR operation with the contents of a specified memory location and the data in some register, bit by bit, e.g. with Intel, ORL A,#3Fh will OR the contents of register A with the hex number 3F.

11 *EXCLUSIVE-OR*
This instruction carries out the logical EXCLUSIVE-OR operation with the contents of a specified memory location and the data in some register,

bit by bit, e.g. with PIC, xorlw 81h (in binary 10000001):

Before instruction	After instruction
Register w with 10001110	Register w with 00001111

XORing with a 0 leaves a data bit unchanged whilst with a 1 the data bit is inverted.

12 *Logical shift (left or right)*
   Logical shift instructions involve moving the pattern of bits in the register one place to the left or right by moving a 0 into the end of the number. For example, for logical shift right a 0 is shifted into the most significant bit and the least significant bit is moved to the carry flag in the status register. With Motorola the instruction might be LSRA for shift to the right and LSLA for a shift to the left.

Before instruction	After instruction
Accumulator with 0011	Accumulator with 0001
	Status register indicates Carry 1

13 *Arithmetic shift (left or right)*
   Arithmetic shift instructions involve moving the pattern of bits in the register one place to the left or right but preserve the sign bit at the left end of the number, e.g. for an arithmetic shift right with the Motorola instruction ASRA:

Before instruction	After instruction
Accumulator with 1011	Accumulator with 1001
	Status register indicates Carry 1

14 *Rotate (left or right)*
   Rotate instructions involve moving the pattern of bits in the register one place to the left or right and the bit that spills out is written back into the other end, e.g. for a rotate right, Intel instruction RR A:

Before instruction	After instruction
Accumulator with 0011	Accumulator with 1001

Program control

15 *Jump or branch*
   This instruction changes the sequence in which the program steps are carried out. Normally the program counter causes the program to be carried out sequentially in strict numerical sequence. However, the jump instruction causes the program counter to jump to some other specified location in the program (Figure 11.1(a)). Unconditional jumps occur without the program testing for some condition to occur. Thus with Intel we can have LJMP

Figure 11.1 (a) Unconditional jump, (b) conditional jump.

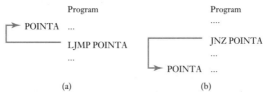

POINTA for the program to jump to the line in the program labelled POINTA, with Motorola the instruction would be JMP POINTA and with PIC it would be GOTO POINTA. Conditional jumps occur if some condition is realised (Figure 11.1(b)). With Intel we can have JNZ POINTA for the program to jump to the line in the program labelled POINTA if any bits in the accumulator are not zero, otherwise it continues with the next line. JZ POINTA is all the bits in the accumulator are zero. With PIC, a conditional jump can involve two lines of code: BTFC 05,1 to 'bit test and skip', i.e. test if bit 1 of file register 5, and if the result is 0 then it jumps the next program line, if 1 it executes it. The next line is GOTO POINTA. With Motorola branch is a conditional jump instruction for the program to determine which branch of a program will be followed if the specified conditions occur. For example, Motorola uses BEQ to branch if equal to zero, BGE for branch if greater than or equal to, BLE for branch if less than or equal to.

16 *Halt/stop*
This instruction stops all further microprocessor activity.

Numerical data may be binary, octal, hex or decimal. Generally in the absence of any indicator the assembler assumes that the number is decimal. With Motorola, a number is indicated by the prefix #; a binary number is preceded by % or followed by B; an octal number is preceded by @ or followed by O; a hex number is preceded by $ or followed by H; and a decimal number requires no indicating letter or symbol. With Intel, numerical values must be preceded by # to indicate a number and by B for binary, O or Q for octal, H or h for hex and D or nothing for decimal. With PIC microcontrollers the header file has R = DEC for decimal to be the default. Then for binary the number is enclosed in quotation marks and preceded by B and for hex by H.

11.2.1 Addressing

When a mnemonic, such as LDA, is used to specify an instruction it will be followed by additional information to specify the source and destination of the data required by the instruction. The data following the instruction is referred to as the **operand**.

　　There are several different methods that are used for specifying data locations, i.e. addressing, and hence the way in which the program causes the microprocessor to obtain its instructions or data. Different microprocessors have different addressing modes. The Motorola 68HC11 has the six addressing modes of immediate, direct, extended, indexed, inherent and relative; the Intel 8051 has the five modes of immediate, direct, register, indirect and indexed; the PIC microcontroller has the three modes of immediate, direct and indirect with the indirect mode allowing indexing. The following are commonly used addressing modes:

1 *Immediate addressing*
The data immediately following the mnemonic is the value to be operated on and is used for the loading of a predetermined value into a register or memory location. For example, with the Motorola code, LDA B #$25 means load the number 25 into accumulator B. The # signifies immediate mode and a number, the $ that the number is in hexadecimal notation. With the Intel code we might have MOV A,#25H to move the number 25 to the accumulator A. The # indicates a number and the H indicates a

hex number. With the PIC code we might have movlw H'25' to load the number 25 into the working register w, the H indicating it is a hex number.

2 *Direct, absolute, extended or zero-page addressing*
With this form of addressing, the data byte that follows the operation code directly gives an address that defines the location of the data to be used in the instruction. With Motorola the term **direct addressing** is used when the address given is only 8 bits wide; the term **extended addressing** is used when it is 16 bits wide. For example, with Motorola code, LDAA $25 means load the accumulator with the contents of memory location 0025; the 00 is assumed. With Intel code, for the same operation, we can have the direct address instruction MOV A,20H to copy the data at address 20 to the accumulator A. With the PIC code we might have movwf Reg1 to copy the contents of Reg1 into the working register, the address of Reg1 having been previously defined.

3 *Implied addressing, or inherent addressing*
With this mode of addressing, the address is implied in the instruction. For example, with Motorola code and Intel code, CLR A means clear accumulator A. With PIC code clrw means clear the working register.

4 *Register*
With this form of addressing, the operand is specified as the contents of one of the internal registers. For example, with Intel ADD R7,A to add the contents of the accumulator to register R7.

5 *Indirect*
This form of addressing means that that the data is to be found in a memory location whose address is given by the instruction. For example, with the PIC system the INDF and FSR registers are used. The address is first written to the FSR register and then this serves as an address pointer. A subsequent direct access of INDF with the instruction movf INDF,w will load the working register w using the contents of FSR as a pointer to the data location.

6 *Indexed addressing*
Indexed addressing means that the data is in a memory location whose address is held in an index register. The first byte of the instruction contains the operation code and the second byte contains the offset; the offset is added to the contents of the index register to determine the address of the operand. A Motorola instruction might thus appear as LDA A $FF,X; this means load accumulator A with data at the address given by adding the contents of the index register and FF. Another example is STA A $05,X; this means store the contents of accumulator A at the address given by the index register plus 05.

7 *Relative addressing*
This is used with branch instructions. The operation code is followed with a byte called the relative address. This indicates the displacement in address that has to be added to the program counter if the branch occurs. For example, Motorola code BEQ $F1 indicates that if the data is equal to zero then the next address in the program is F1 further on. The relative address of F1 is added to the address of the next instruction.

As an illustration, Table 11.1 shows some instructions with the modes of addressing used in Motorola systems.

Table 11.1   Examples of
addressing.

Address mode	Instruction	
Immediate	LDA A #$F0	Load accumulator A with data F0
Direct	LDA A $50	Load accumulator A with data at address 0050
Extended	LDA A$0F01	Load accumulator A with data at address 0F01
Indexed	LDA A $CF,X	Load accumulator with data at the address given by the index register plus CF
Inherent	CLR A	Clear accumulator A
Extended	CLR $2020	Clear address 2020, i.e. store all zeros at address 2020
Indexed	CLR $10,X	Clear the address given by the index register plus 10, i.e. store all zeros at that address

11.2.2   Data movement

The following is an example of the type of information that will be found in a manufacturer's (6800) instruction set sheet:

		Addressing modes					
		IMMED			DIRECT		
Operation	Mnemonic	OP	~	#	OP	~	#
Add	ADDA	8B	2	2	9B	3	2

~ is the number of microprocessor cycles required and # is the number of program bytes required.

This means that when using the immediate mode of addressing with this processor the Add operation is represented by the mnemonic ADDA. When the immediate form of addressing is used, the machine code for it is 8B and it will take two cycles to be fully expressed. The operation will require 2 bytes in the program. The **op-code** or **operation code** is the term used for the instruction that the microprocessor will act on and is expressed in hexa-decimal form. A byte is a group of eight binary digits recognised by the microprocessor as a word. Thus two words are required. With the direct mode of addressing, the machine code is 9B and takes three cycles and two program bytes.

To illustrate how information passes between memory and microprocessor, consider the following tasks. The addresses in RAM where a new program may be placed are just a matter of convenience. For the following examples, the addresses starting at 0010 have been used. For direct addressing to be used, the addresses must be on the zero page, i.e. addresses between 0000 and 00FF. The examples are based on the use of the instruction set for the M6800 microprocessor.

Task: Enter all zeros in accumulator A.

Memory address	Op-code	
0010	8F	CLR A

The next memory address that can be used is 0011 because CLR A only occupies one program byte. This is the inherent mode of addressing.

Task: Add to the contents of accumulator A the data 20.

Memory address	Op-code	
0010	8B 20	ADD A #$20

This uses the immediate form of addressing. The next memory address that can be used is 0012 because, in this form of addressing, ADD A occupies two program bytes.

Task: Load accumulator A with the data contained in memory address 00AF.

Memory address	Op-code	
0010	B6 00AF	LDA A $00AF

This uses the absolute form of addressing. The next memory address that can be used is 0013 because, in this form of addressing, LDA A occupies three program bytes.

Task: Rotate left the data contained in memory location 00AF.

Memory address	Op-code	
0010	79 00AF	ROL $00AF

This uses the absolute form of addressing. The next memory address that can be used is 0013 since ROL, in this mode, occupies three program bytes.

Task: Store the data contained in accumulator A into memory location 0021.

Memory address	Op-code	
0010	D7 21	STA A $21

This uses the direct mode of addressing. The next memory address that can be used is 0012 since STA A, in this mode, occupies two program bytes.

Task: Branch forward four places if the result of the previous instruction is zero.

Memory address	Op-code	
0010	27 04	BEQ $04

This uses the relative mode of addressing. The next memory address if the result is not zero is 0012 since BEQ, in this mode, occupies two program bytes. If the result is zero then the next address is 0012 + 4 = 0016.

11.3 Assembly language programs

An assembly language program should be considered as a series of instructions to an assembler which will then produce the machine code program. A program written in assembly language consists of a sequence of statements, one statement per line. A statement contains from one to four sections or **fields**, these being:

Label Op-code Operand Comment

A special symbol is used to indicate the beginning or end of a field, the symbols used depending on the microprocessor machine code assembler concerned. With the Motorola 6800 spaces are used. With the Intel 8080 there are a colon after a label, a space after the op-code, commas between entries in the address field and a semicolon before a comment. In general, a semicolon is used to separate comments from the operand.

The **label** is the name by which a particular entry in the memory is referred to. Labels can consist of letters, numbers and some other characters. With the Motorola 6800, labels are restricted to one to six characters, the first of which must be a letter, and cannot consist of a single letter A, B or X since these are reserved for reference to the accumulator or index register. With the Intel 8080, five characters are permitted with the first character a letter, @ or ?. The label must not use any of the names reserved for registers, instruction codes or pseudo-operations (see later in this section). All labels within a program must be unique. If there is no label then a space must be included in the label field. With the Motorola 6800, an asterisk (*) in the label field indicates that the entire statement is a comment, i.e. a comment included to make the purpose of the program clearer. As such, the comment will be ignored by the assembler during the assembly process for the machine code program.

The op-code specifies how data is to be manipulated and is specified by its mnemonic, e.g. LDA A. It is the only field that can never be empty. In addition, the op-code field may contain directives to the assembler. These are termed **pseudo-operations** since they appear in the op-code field but are not translated into instructions in machine code. They may define symbols, assign programs and data to certain areas of memory, generate fixed tables and data, indicate the end of the program, etc. Common assembly directives are:

Set program counter

ORG	This defines the starting memory address of the part of the program that follows. A program may have several origins.

Define symbols

EQU, SET, DEF	Equates/sets/defines a symbol for a numerical value, another symbol or an expression.

Reserve memory locations

RMB, RES	Reserves memory bytes/space.

Define constant in memory

FCB	Forms constant byte.
FCC	Forms constant character string.
FDB	Forms double byte constant.
BSW	Block storage of zeros.

The information included in the **operand** field depends on the mnemonic preceding it and the addressing mode used. It gives the address of the data to be operated on by the process specified by the op-code. It is thus often referred to as the **address field**. This field may be empty if the instructions given by the op-code do not need any data or address. Numerical data in this field may be hexadecimal, decimal, octal or binary. The assembler assumes that numbers are decimal unless otherwise specified. With the Motorola 6800 a hexadecimal number is preceded by $ or followed by H, an octal number preceded by @ or followed by O or Q, a binary number preceded by % or followed by B. With the Intel 8080, a hexadecimal number is followed by H, an octal number by O or Q, a binary number by B. Hexadecimal numbers must begin with a decimal digit, i.e. 0 to 9, to avoid confusion with names. With the Motorola 6800, the immediate mode of address can be indicated by preceding the operand by #, the indexed mode of address involving the operand being followed by X. No special symbols are used for direct or extended addressing modes. If the address is on the zero page, i.e. FF or less, then the assembler automatically assigns the direct mode. If the address is greater than FF, the assembler assigns the extended mode.

The comment field is optional and is there to allow the programmer to include any comments which may make the program more understandable by the reader. The comment field is ignored by the assembler during the production of the machine code program.

11.3.1 Examples of assembly language programs

The following examples illustrate how some simple programs can be developed.

Problem: the addition of the two 8-bit numbers located in different memory addresses and the storage of the result back into memory.

The algorithm is:

1 Start.
2 Load the accumulator with the first number. The accumulator is where the results of arithmetic operations are accumulated. It is the working register, i.e. like a notepad on which the calculations are carried out before the result is transferred elsewhere. Thus we have to copy data to the accumulator before we can carry out the arithmetic. With PIC the term working (w) register is used.
3 Add on the second number.
4 Store the sum in the designated memory location.
5 Stop.

Figure 11.2 shows these steps represented as a flow chart.

The following are the programs written for three different microcontrollers. In each, the first column gives the label, the second the op-code, the third the operand and the fourth the comment. Note that all comments are preceded by a semicolon.

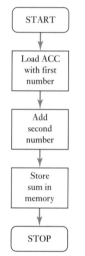

Figure 11.2 Flow chart for the addition of two numbers.

M68HC11 program

```
                 ; Addition of two numbers

NUM1    EQU      $00          ; location of number 1
NUM2    EQU      $01          ; location of number 2
SUM     EQU      $02          ; location for the sum

        ORG      $C000        ; address of start of user RAM
START   LDAA     $NUM1        ; load number 1 into acc. A
        ADDA     $NUM2        ; add number 2 to A
        STAA     SUM          ; save the sum to $02
        END
```

The first line in the program specifies the address of the first of the numbers to be added. The second line specifies the address of the number to be added to the first number. The third line specifies where the sum is to be put. The fourth line specifies the memory address at which the program should start. The use of the labels means that the operand involving that data does not have to specify the addresses but merely the labels.

The same program for an Intel 8051 would be as follows.

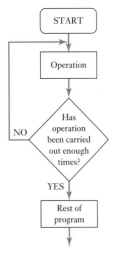

Figure 11.3 A loop.

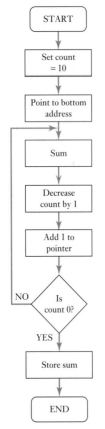

Figure 11.4 Flow chart for adding 10 numbers.

8051 program

```
                        ; Addition of two numbers

NUM1        EQU         20H         ; location of number 1
NUM2        EQU         21H         ; location of number 2
SUM         EQU         22H         ; location for the sum

            ORG         8000H       ; address of start of user RAM
START       MOV         A,NUM1      ; load number 1 into acc. A
            ADD         A,NUM2      ; add number 2 to A
            MOV         SUM,A       ; save the sum to address 22H
            END
```

The same program for a PIC microcontroller would be as follows.

PIC program

```
                        ; Addition of two numbers

Num1        equ         H'20'       ; location of number 1
Num2        equ         H'21'       ; location of number 2
Sum         equ         H'22'       ; location for the sum

            org         H'000'      ; address of start of user RAM
Start       movlw       Num1        ; load number 1 into w
            addlw       Num2        ; add number 2 to w
            movwf       Sum         ; save the sum H'22'
            End
```

In many programs there can be a requirement for a task to be carried out a number of times in succession. In such cases the program can be designed so that the operation passes through the same section a number of times. This is termed **looping**, a **loop** being a section of a program that is repeated a number of times. Figure 11.3 shows a flow diagram of a loop. With such a loop, a certain operation has to be performed a number of times before the program proceeds. Only when the number of such operations is completed can the program proceed. The following problem indicates such a looping.

Problem: the addition of numbers located at 10 different addresses (these might be the results of inputs from 10 different sensors which have each to be sampled).

The algorithm could be:

1 Start.
2 Set the count as 10.
3 Point to location of bottom address number.
4 Add bottom address number.
5 Decrease the count number by 1.
6 Add 1 to the address location pointer.
7 Is count 0? If not branch to 4. If yes proceed.
8 Store sum.
9 Stop.

Figure 11.4 shows the flow chart.

The program is:

```
COUNT      EQU      $0010
POINT      EQU      $0020
RESULT     EQU      $0050
           ORG      $0001
           LDA B    COUNT        ; Load count
           LDX      POINT        ; Initialise index register at start of
                                 ; numbers
SUM        ADD A    X            ; Add addend
           INX                   ; Add 1 to index register
           DEC B                 ; Subtract 1 from accumulator B
           BNE      SUM          ; Branch to sum
           STA A    RESULT       ; Store
           WAI                   ; Stop program
```

The count number of 10 is loaded into accumulator B. The index register gives the initial address of the data being added. The first summation step is to add the contents of the memory location addressed by the index register to the contents of the accumulator, initially assumed zero (a CLR A instruction could be used to clear it first). The instruction INX adds 1 to the index register so that the next address that will be addressed is 0021. DEC B subtracts 1 from the contents of accumulator B and so indicates that a count of 9 remains. BNE is then the instruction to branch to SUM if not equal to zero, i.e. if the Z flag has a 0 value. The program then loops and repeats the loop until ACC B is zero.

Problem: the determination of the biggest number in a list of numbers (it could be the determination of, say, the largest temperature value inputted from a number of temperature sensors).

The algorithm could be:

1 Clear answer address.
2 List starting address.
3 Load the number from the starting address.
4 Compare the number with the number in the answer address.
5 Store answer if bigger.
6 Otherwise save number.
7 Increase starting address by 1.
8 Branch to 3 if the address is not the last address.
9 Stop.

Figure 11.5 shows the flow chart. The program is:

```
FIRST      EQU      $0030
LAST       EQU      $0040
ANSW       EQU      $0041
           ORG      $0000
           CLR      ANSW         ; Clear answer
           LDX      FIRST        ; Load first address
NUM        LDA A    $30,X        ; Load number
           CMP A    ANSW         ; Compare with answer
           BLS      NEXT         ; Branch to NEXT if lower or same
           STA A    ANSW         ; Store answer
```

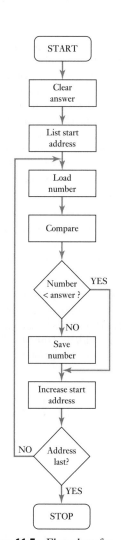

Figure 11.5 Flow chart for biggest number.

NEXT	INX		; Increment index register
	CPX	LAST	; Compare index register with LAST
	BNE	NUM	; Branch if not equal to zero
	WAI		; Stop program

The procedure is that first the answer address is cleared. The first address is then loaded and the number in that address put into accumulator A. LDA A $30,X means load accumulator A with data at address given by the index register plus 30. Compare the number with the answer, keeping the number if it is greater than the number already in the accumulator, otherwise branch to repeat the loop with the next number.

11.4 Subroutines

It is often the case that a block of programming, a subroutine, might be required a number of times in a program. For example, a block of programming might be needed to produce a time delay. It would be possible to duplicate the subroutine program a number of times in the main program. This, however, is an inefficient use of memory. Alternatively we could have a single copy of it in memory and branch or jump to it every time the subroutine was required. This, however, presents the problem of knowing, after completion of the subroutine, the point at which to resume in the main program. What is required is a mechanism for getting back to the main program and continuing with it from the point at which it was left to carry out the subroutine. To do this we need to store the contents of the program counter at the time of branching to the subroutine so that this value can be reloaded into the program counter when the subroutine is complete. The two instructions which are provided with most microprocessors to enable a subroutine to be implemented in this way are:

1 JSR (jump to routine), or CALL, which enables a subroutine to be called;
2 RTS (return from subroutine), or RET (return), which is used as the last instruction in a subroutine and returns it to the correct point in the calling program.

Subroutines may be called from many different points in a program. It is thus necessary to store the program counter contents in such a way that we have a last-in–first-out store (LIFO). Such a register is referred to as a **stack**. It is like a stack of plates in that the last plate is always added to the top of the pile of plates and the first plate that is removed from the stack is always the top plate and hence the last plate that was added to the stack. The stack may be a block of registers within a microprocessor or, more commonly, using a section of RAM. A special register within the microprocessor, called the stack pointer register, is then used to point to the next free address in the area of RAM being used for the stack.

In addition to the automatic use of the stack when subroutines are used, a programmer can write a program which involves the use of the stack for the temporary storage of data. The two instructions that are likely to be involved are:

1 PUSH, which causes data in specified registers to be saved to the next free location in the stack;
2 PULL, or POP, which causes data to be retrieved from the last used location in the stack and transferred to a specified register.

For example, prior to some subroutine, data in some registers may have to be saved and then, after the subroutine, the data restored. The program elements might thus be, with the Motorola 6800:

```
SAVE        PSH A   ; Save accumulator A to stack
            PSH B   ; Save accumulator B to stack
            TPA     ; Transfer status register to accumulator A
            PSH A   ; Save status register to stack
; Subroutine
RESTORE     PUL A   ; Restore condition code from stack to accumulator A
            TAP     ; Restore condition code from A to status register
            PUL B   ; Restore accumulator B from stack
            PUL A   ; Restore accumulator A from stack
```

11.4.1 Delay subroutine

Delay loops are often required when the microprocessor has an input from a device such as an analogue-to-digital converter. The requirement is often to signal to the converter to begin its conversion and then wait a fixed time before reading the data from the converter. This can be done by providing a loop which makes the microprocessor carry out a number of instructions before proceeding with the rest of the program. A simple delay program might be:

```
DELAY    LDA A    #$05      ; Load accumulator A with 05
LOOP     DEC A              ; Decrement accumulator A by 1
         BNE      LOOP      ; Branch if not equal to zero
         RTS                ; Return from subroutine
```

For each movement through the loop a number of machine cycles are involved. The delay program, when going through the loop five times, thus takes:

Instruction	Cycles	Total cycles
LDA A	2	2
DEC A	2	10
BNE	4	20
RTS	1	1

The total delay is thus 33 machine cycles. If each machine cycle takes, say, 1 μs then the total delay is 33 μs. For a longer delay a bigger number can be initially put into accumulator A.

An example of a time–delay subroutine for a PIC microcontroller is:

```
         movlw    Value     ; load count value required
         movwf    Count     ; loop counter
Delay    decfsz   Count     ; decrement counter
         goto     Delay     ; loop
```

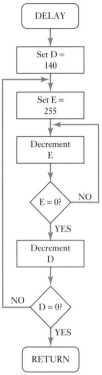

Figure 11.6 Nested loop delay.

The decfsz instruction takes one cycle and the goto instruction takes two cycles. This loop will be repeated (count − 1) times. In addition we have the movlw and movwf instructions, each taking one cycle, and when the count equals 1 we have decfsz which gives a further two cycles. Thus the total number of cycles is:

number of instruction cycles = 3(count − 1) + 4

Each instruction cycle takes four clock cycles and so the number of delay cycles introduced by this subroutine is:

number of clock cycles = 4[2(count − 1) + 4]

With a 4 MHz clock each clock cycle takes $1/(4 \times 10^6)$ s.

Often the delay obtained by using just the single loop described above is not enough. One way to obtain a longer delay is to use a nested loop. Figure 11.6 shows the flow chart for a nested loop delay. The inner loop is the same as the single-loop program described earlier. It will decrement register E 255 times before the looping is completed and the zero flag set. The outer loop causes the inner-loop routine to be repeatedly executed as register D is decremented down to zero. Thus with register D initially set with a loop count of, say, 140 then the time delay will be 140 × 2.298 = 321.71 ms.

The program is thus:

```
DELAY    MOV    D,8CH     ; set D to 8CH, i.e. 140
OLOOP    MOV    E,FFH     ; set E to FFH, i.e. 255
ILOOP    DEC    E         ; decrement E, i.e. inner-loop counter
         JNZ    ILOOP     ; repeat ILOOP 255 times
         DEC    D         ; decrement D, i.e. outer-loop counter
         JNZ    OLOOP     ; repeat OLOOP 140 times
```

The following are some examples of programs where time–delay subroutines are involved.

1 *Problem*: Switch an LED on and off repeatedly.
   With this problem a subroutine DELAY is used with loops to provide time delays; the microprocessor takes a finite amount of time to process the instructions in a loop and thus to complete the loop. The structure of the program is:

   1 If LED on
        Turn LED off
        While LED off, do subroutine TIME_DELAY
   2 ELSE
        Turn LED on
        Do subroutine TIME_DELAY

   Subroutine: TIME_DELAY
        Do an instruction, or instructions, or a loop, or a double loop depending on the length of time delay required.

Because of the length of time delay required, a double loop is likely to be used for the time delay. With Intel 8051 programming, it is possible to use the instruction DJNZ, decrement and jump if the result is not zero. It decrements the location indicated by the first operand and jumps to the second operand if the resulting value is not zero. The LED is connected to bit 0 of port 1 of the microcontroller. The program with Intel 8051 assembly instructions might thus be:

```
FLAG       EQU    0FH              ; flag set when LED is on
           ORG    8000H
START      JB     FLAG,LED_OFF     ; jump if LED_OFF bit set, i.e. LED is on
           SETB   FLAG             ; else set FLAG bit
           CLR    P1.0             ; turn LED on
           LCALL  DELAY            ; call up delay subroutine
           SJMP   START            ; jump to START
LED_OFF    CLR    FLAG             ; clear the LED on flag to indicate the
                                   ;   LED is off
           SETB   P1.0             ; turn LED off
           LCALL  DELAY            ; call up delay subroutine
           LJMP   START            ; jump to START
DELAY      MOV    R0,#0FFH         ; outer-loop delay value
ILOOP      MOV    R1,#0FFH         ; inner-loop delay value
OLOOP      DJNZ   R1,ILOOP         ; wait through inner loop
           DJNZ   R0,OLOOP         ; wait through outer loop
           RET                     ; return from subroutine
           END
```

2 *Problem*: Switch on in sequence eight LEDS.
   The rotate instruction can be used successively to turn on LEDs so that we have initially the bit pattern 0000 0001 which is then rotated to give 0000 0011, then 0000 0111, and so on. The following is a program in Motorola 68HC11 assembly language that can be used, the LEDs being connected to port B; a short delay is incorporated in the program:

```
COUNT    EQU    8               ; the count gives the number of loops
                                ; required, i.e. the number of
                                ; bits to be switched on
FIRST    EQU    %00000001       ; turn on 0 bit
PORTB    EQU    $1004           ; address of port B
         ORG    $C000
         LDAA   #FIRST          ; load initial value
         LDAB   #COUNT          ; load count
LOOP     STAA   PORTB           ; turn on bit 1 and so LED 1
         JSR    DELAY           ; jump to delay subroutine
         SEC                    ; set carry bit to rotate into least significant
                                ; bit to maintain bit as 1
         ROLA                   ; rotate left
         DECB                   ; decrement count
         BNE    LOOP            ; branch to loop eight times
DELAY    RTS                    ; simple short delay
         END
```

11.5 Look-up tables

Indexed addressing can be used to enable a program to look up values in a table. For example, in determining the squares of integers, a possible method is to look up the value corresponding to a particular integer in a table of squares, instead of doing the arithmetic to determine the square. Look-up tables are particularly useful when the relationship is non-linear and not described by a simple arithmetic equation, e.g. the engine management system described in Section 1.7.2 where ignition timing settings are a function of the angle of the crankshaft and the inlet manifold pressure. Here the microcontroller has to give a timing signal that depends on the input signals from the speed sensor and crankshaft sensors.

To illustrate how look-up tables are used, consider the problem of determining the squares of integers. We can place a table of squares of the integers 0, 1, 2, 3, 4, 5, 6, . . . in program memory and have the entries of the squares 0, 1, 4, 9, 16, 25, 36, . . . at successive addresses. If the number to be squared is 4 then this becomes the index for the index address of the data in the table, the first entry being index 0. The program adds the index to the base address of the table to find the address of the entry corresponding to the integer. Thus we have:

Index	0	1	2	3	4	5	6
Table entry	0	1	4	9	16	25	36

For example, with the Motorola 68HC11 microcontroller we might have the following look-up program to determine the squares:

```
REGBAS   EQU    $B600      ; base address for the table
         ORG    $E000
         LDAB   $20        ; load acc. B with the integer to be squared
         LDX    #REGBAS    ; point to table
         ABX               ; add contents of acc. B to index register X
         LDAA   $00,X      ; load acc. A with the indexed value
```

and we could have loaded the table into memory by using the pseudo-operation FDB:

```
ORG     $B600
FDB     $00,$01,$04,$09      ; giving values to the reserved memory
                            ; block
```

With the Intel 8051 microcontroller the instruction MOVC A,@A+ DPTR fetches data from the memory location pointed to by the sum of DPTR and the accumulator A and stores it in the accumulator. This instruction can be used to look up data in a table where the data pointer DPTR is initialised to the beginning of the table. As an illustration, suppose we want to use a table for the conversion of temperatures on the Celsius scale to the Fahrenheit scale. The program involves parameter passing of the temperature requiring conversion to a subroutine, so it might include the following instructions:

```
          MOV    A,#NUM          ; load the value to be converted
          CALL   LOOK_UP         ; call the LOOK_UP subroutine

LOOK_UP   MOV    DPTR,#TEMP      ; point to table
          MOVC   A,@A+DPTR       ; get the value from the table
          RET                    ; return from the subroutine
TMP       DB     32, 34, 36, 37, ; giving values to the table
                 39, 41, 43, 45
```

Another example of the use of a table is to sequence a number of outputs. This might be a sequence to operate traffic lights to give the sequence red, red plus amber, green, amber. The red light is illuminated when there is an output from RD0, the amber is illuminated from RD1, and the green is illuminated from RD2. The data table might then be:

	Red	Red + amber	Green	Amber
Index	0	1	2	3
	0000 0001	0000 0011	0000 0100	0000 0010

11.5.1 Delay with a stepper motor

In operating a stepper motor delays have to be used between each instruction to advance by a step to allow time for that step to occur before the next program instruction. A program to generate a continuous sequence of step pulses could thus have the algorithm:

1 Start.
2 State sequence of outputs needed to obtain the required step sequence.
3 Set to initial step position.
4 Advance a step.
5 Jump to delay routine to give time for the step to be completed.
6 Is this the last step in the step sequence for one complete rotation? If not, continue to next step; if yes, loop back to step 3.
7 Continue until infinity.

The following is a possible program for a stepper in the full–step configuration and controlled by the microcontroller M68HC11 using outputs from PB0, PB1, PB2 and PB3. A 'look-up' table is used for the output code sequence necessary from the outputs to drive the stepper in the step sequence. The following is the table used.

The code sequence required for full-step stepper operation is thus A, 9, 5, 6, A and so these values constitute the sequence that the pointer has to look up in the table. FCB is the op-code for 'form constant byte' and is used to initialise data bytes for the table.

Step	The outputs required from Port B				Code
	PB0	PB1	PB2	PB3	
1	1	0	1	0	A
2	1	0	0	1	9
3	0	1	0	1	5
4	0	1	1	0	6
1	1	0	1	0	4

```
BASE    EQU     $1000
PORTB   EQU     $4          ; Output port
TFLG1   EQU     $23         ; Timer interrupt flag register 1
TCNT    EQU     $0E         ; Timer counter register
TOC2    EQU     $18         ; Output compare 2 register
TEN_MS  EQU     20000       ; 10 ms on clock
```

```
              ORG     $0000
    STTBL     FCB     $A              ; This is the look-up table
              FCB     $9
              FCB     $5
              FCB     $6
    ENDTBL    FCB     $A              ; End of look-up table

              ORG     $C000
              LDX     #BASE
              LDAA    #$80
              STAA    TFLG1,X         ; Clear flag
    START     LDY     #STTBL
    BEG       LDAA    0,Y             ; Start with first position in table
              STAA    PORTB,X
              JSR     DELAY           ; Jump to delay
              INY                     ; Increment in table
              CPY     #ENTBL          ; Is it end of table?
              BNE     BEG             ; If not branch to BEG
              BRA     START           ; If yes, go to start again

    DELAY     LDD     TCNT,X
              ADDD    #TEN_MS         ; Add a 10 ms time delay
              STD     TOC2,X
    HERE      BRCLR   TFLG1, X, $80, HERE   ; Wait till time delay elapsed
              LDAA    #$80
              STAA    TFLG1,X         ; Clear flag
              RTS
```

Note that, in the above program, the label TEN_MS has the space underlined to indicate that both TEN and MS are part of the same label.

The delay in the above program is obtained by using the timer block in the microcontroller. A time delay of 10 ms is considered. For a microcontroller system with a 2 MHz clock a 10 ms delay is 20 000 clock cycles. Thus to obtain such a delay, the current value of the TCNT register is found and then 20 000 cycles added to it and the TOC2 register loaded with this value.

11.6   Embedded systems

Microprocessors and microcontrollers are often 'embedded' in systems so that control can be exercised. For example, a modern domestic washing machine has an embedded microcontroller which has been programmed with the different washing programs; all that the machine operator has to do is select the required washing program by means of a switch and the required program is implemented. The operator does not have to program the microcontroller. The term **embedded system** is used for a microprocessor-based system that is designed to control a function or range of functions and is not designed to be programmed by the system user. The programming has been done by the manufacturer and has been 'burnt' into the memory system and cannot be changed by the system user.

11.6.1 Embedding programs

In an embedded system the manufacturer makes a ROM containing the program. This is only economical if there is a need for a large number of these chips. Alternatively, for prototyping or low-volume applications, a program could be loaded into the EPROM/EEPROM of the application hardware. The following illustrates how the EPROM/EEPROM of microcontrollers can be programmed.

For example, to program the EPROM of the Intel 8051 microcontroller, the arrangement shown in Figure 11.7(a) is required. There must be a 4–6 MHz oscillator input. The procedure is outlined below.

1 The address of an EPROM location, to be programmed in the range 0000H to 0FFFH, is applied to port 1 and pins P2.0 and P2.1 of port 2; at the same time, the code byte to be programmed into that address is applied to port 0.
2 Pins P2.7, RST and ALE should be held high, pins P2.6 and PSEN low. For pins P2.4 and P2.5 it does not matter whether they are high or low.
3 Pin EA/V_{pp} is held at a logic high until just before ALE is to be pulsed, then it is raised to +21 V, ALE is pulsed low for 50 ms to program the code byte into the addressed location, and then EA is returned to a logic high.

Verification of the program, i.e. reading out of the program, is achieved by the arrangement shown in Figure 11.7(b).

1 The address of the program location to be read is applied to port 1 and pins P2.0 to P2.3 of port 2.
2 Pins EA/V_{pp}, RST and ALE should be held high, pins P2.7, P2.6 and PSEN low. For pins P2.4 and P2.5 it does not matter whether they are high or low.
3 The contents of the addressed location come out on port 0.

A security bit can be programmed to deny electrical access by any external means to the on-chip program memory. Once this bit has been programmed, it can only be cleared by the full erasure of the program memory. The same arrangement is used as for programming (Figure 11.7(a)) but P2.6 is held high.

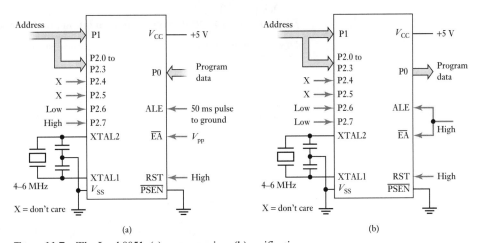

Figure 11.7 The Intel 8051: (a) programming, (b) verification.

Erasure is by exposure to ultraviolet light. Since sunlight and fluorescent lighting contain some ultraviolet light, prolonged exposure (about 1 week in sunlight or 3 years in room-level fluorescent lighting) should be avoided and the chip window should be shielded by an opaque label.

The Motorola 68HC11 microcontroller is available with an internal electrically erasable programmable read-only memory (EEPROM). The EEPROM is located at addresses $B600 to $B7FF. Like an EPROM, a byte is erased when all the bits are 1 and programming involves making particular bits 0. The EEPROM is enabled by setting the EEON bit in the CONFIG register (Figure 11.8) to 1 and disabled by setting it to 0. Programming is controlled by the EEPROM programming register (PPROG) (Figure 11.8).

Figure 11.8 CONFIG and PPROG.

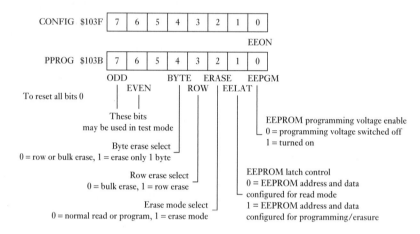

The procedure for programming is outlined below.

1 Write to the PPROG register to set the EELAT bit to 1 for programming.
2 Write data to the EEPROM address selected. This latches in the address and data to be programmed.
3 Write to the PPROG register to set the EEPGM bit to 1 to turn on the programming voltage.
4 Delay for 10 ms.
5 Write to the PPROG register to turn off, i.e. to 0, all the bits.

Here, in assembly language, is a programming subroutine for use with the MC68HC11:

```
EELAT     EQU     %00000010     ; EELAT bit
EEPGM     EQU     %00000001     ; EEPGM bit
PPROG     EQU     $1028         ; address of PPROG register

EEPROG
          PSHB
          LDAB    #EELAT
          STAB    PPROG         ; set EELAT = 1 and EEPGM = 0
          STAA    0,X           ; store data X to EEPROM address
          LDAB    #%00000011
          STAB    PPROG         ; set EELAT = 1 and EEPGM = 1
          JSR     DELAY_10      ; jump to delay 10 ms subroutine
```

```
                    CLR      PPROG      ; clear all the PPROG bits and return
                                          to the read mode
                    PULB
                    RTS
```

; Subroutine for approximately 10 ms delay
```
DELAY_10
                    PSHX
                    LDX      #2500      ; count for 20 000 cycles
DELAY               DEX
                    BNE      DELAY
                    PULX
                    RTS
```

The procedure for erasure is:

1   write to the PPROG register to select for erasure of a byte, row or the entire EEPROM;
2   write to an EEPROM address within the range to be erased;
3   write a 1 to the PPROG register to turn on the EEPGM bit and hence the erase voltage;
4   delay for 10 ms;
5   write zeros to the PPROG register to turn off all the bits.

With the built-in EEPROM with a PIC microcontroller, a program to write data into it is (Figure 11.9):

```
                bcf      STATUS, RP0     ; Change to Bank 0 for the data
                mov.f    Data, w         ; Load data to be written
                movwf    EEDATA
                movf     Addr, w         ; Load address of write data
                movwf    EEADR
                bsf      STATUS, RP0     ; Change to Bank 1
                bcf      INTCON, GIE     ; Disable interrupts
                bsf      EECON1, WREN    ; Enable for writing
                movlw    55h             ; Special sequence to enable
                                         ; writing
                movwf    EECON2
                movlw    0AAh
                movwf    EECON2
                bsf      EECON1, WR      ; Initiate write cycle
                bsf      INTCON, GIE     ; Re-enable interrupts
EE_EXIT         btfsc    EECON, WR       ; Check that the write is completed
                goto     EE_EXIT         ; If not, retry
                bsf      EECON, WREN     ; EEPROM write is complete
```

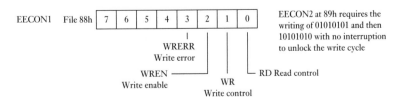

Figure 11.9 EECON registers.

EECON1   File 88h | 7 | 6 | 5 | 4 | 3 | 2 | 1 | 0 |

WRERR
Write error

WREN
Write enable

WR
Write control

RD Read control

EECON2 at 89h requires the writing of 01010101 and then 10101010 with no interruption to unlock the write cycle

Summary

The collection of instructions that a microprocessor will recognise is its **instruction set**. The series of instructions that is needed to carry out a particular task is called a **program**.

Microprocessors work in binary code. Instructions written in binary code are referred to as being in **machine code**. A shorthand code using simple, identifiable, terms rather than binary code is referred to as a **mnemonic code**, a mnemonic code being a 'memory-aiding' code. Such a code is termed **assembly language**. Assembly language programs consist of a sequence of statements, one per line, with each statement containing from one to four fields: label, op-code, operand and comment. The **label** is the name by which a particular entry in the memory is referred to. The **op-code** specifies how data is to be manipulated. The **operand** contains the address of the data to be operated on. The **comment** field is to allow the programmer to include comments which may make the program more understandable to the reader.

Problems

11.1   Using the following extract from a manufacturer's instruction set (6800), determine the machine codes required for the operation of adding with carry in (a) the immediate address mode, (b) the direct address mode.

| | | Addressing modes | | | | | |
| | | IMMED | | | DIRECT | | |
Operation	Mnemonic	OP	~	#	OP	~	#
Add with carry	ADC A	89	2	2	99	3	2

11.2   The clear operation with the Motorola 6800 processor instruction set has an entry only in the implied addressing mode column. What is the significance of this?

11.3   What are mnemonics for, say, the Motorola 6800 for (a) clear register A, (b) store accumulator A, (c) load accumulator A, (d) compare accumulators, (e) load index register?

11.4   Write a line of assembler program for (a) load the accumulator with 20 (hex), (b) decrement the accumulator A, (c) clear the address $0020, (d) ADD to accumulator A the number at address $0020.

11.5   Explain the operations specified by the following instructions: (a) STA B $35, (b) LDA A #$F2, (c) CLC, (d) INC A, (e) CMP A #$C5, (f) CLR $2000, (g) JMP 05,X.

11.6   Write programs in assembly language to:

(a) subtract a hexadecimal number in memory address 0050 from a hexadecimal number in memory location 0060 and store the result in location 0070;

(b) multiply two 8-bit numbers, located at addresses 0020 and 0021, and store the product, an 8-bit number, in location 0022;

(c) store the hexadecimal numbers 0 to 10 in memory locations starting at 0020;

(d) move the block of 32 numbers starting at address $2000 to a new start address of $3000.

11.7 Write, in assembly language, a subroutine that can be used to produce a time delay and which can be set to any value.

11.8 Write, in assembly language, a routine that can be used so that if the input from a sensor to address 2000 is high the program jumps to one routine starting at address 3000, and if low the program continues.

Chapter twelve C language

Objectives

The objectives of this chapter are that, after studying it, the reader should be able to:
- Comprehend the main features of C programs.
- Use C to write simple programs for microcontrollers.

12.1 Why C?

This chapter is intended to give an introduction to the C language and the writing of programs. C is a high-level language that is often used in place of assembly language (see Chapter 11) for the programming of microprocessors. It has the advantages when compared with the assembly language of being easier to use and that the same program can be used with different microprocessors; all that is necessary for this is that the appropriate compiler is used to translate the C program into the relevant machine language for the microprocessor concerned. Assembly language is different for the different microprocessors while C language is standardised, the standard being that of the American National Standards Institute (ANSI).

12.2 Program structure

Figure 12.1 gives an overview of the main elements of a C program. There is a pre-processor command that calls up a standard file, followed by the main function. Within this main function there are other functions which are called up as subroutines. Each function contains a number of statements.

Figure 12.1 Structure of a C program.

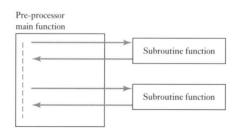

12.2.1 Key features

The following are key features of programs written in C language. Note that in C programs spaces and carriage returns are ignored by the compiler and purely used for the convenience of the programmer to make it easier to read the program.

1 *Keywords*

In C certain words are reserved as keywords with specific meanings. For example, *int* is used to indicate that integer values are concerned; *if* is used for when a program can change direction based on whether a decision is true or false. C requires that all keywords are in lower case letters. Such words should not be used for any other purpose in a C program. The following are the ANSI C standard keywords:

auto	double	int	struct
break	else	long	switch
case	enum	register	typedef
char	extern	return	union
const	float	short	unsigned
continue	for	signed	void
default	goto	sizeof	volatile
do	if	static	while

2 *Statements*

These are the entries which make up a program, every statement being terminated by a semicolon. Statements can be grouped together in blocks by putting them between braces, i.e. { }. Thus for a two-statement group we have:

```
{
    statement 1;
    statement 2;
}
```

3 *Functions*

The term **function** is used for a self-contained block of program code which performs a specific set of actions and has a name by which it can be referred (it is like a subroutine in an assembly language program). A function is written as a name followed by brackets, i.e. name( ). The brackets enclose arguments; a function's argument is a value that is passed to the function when the function is called. A function is executed by calling it up by its name in the program statement. For example, we might have the statement:

```
printf("Mechatronics");
```

This would mean that the word Mechatronics is passed to the function printf( ), a pre-written function which is called up by the pre-processor command, and, as a result, the word is displayed on screen. In order to indicate that characters form a sequence, e.g. those making up the word Mechatronics, they are enclosed within double quotes.

4 *Return*

A function may return a value to the calling routine. The **return type** appears in front of the function name, this specifying the type of value to be returned to the calling function when execution of the function is completed. For example, int main( ) is used for an integer return from the main function. The return type may be specified as void if the function does not return a value, e.g. void main(void). Often a header file will contain this return information and so it will not need to be specified for functions defined by the header file.

To return a value from a function back to the calling point, the keyword return is used, e.g. to return the result:

```
return result;
```

The return statement terminates the function.

5 *Standard library functions*

C packages are supplied with libraries containing a large number of predefined functions containing C code that have already been written and so saving you the effort of having to write them. These can be called up by naming them. In order to use the contents of any particular library, that library has to be specified in a header file. Examples of such library files are:

math.h for mathematical functions
stdio.h for input and output functions
time.h for date and time functions

For example, the function printf( ) is a function that can be called up from the stdio.h library and is the function for printing to the screen of the monitor. Another function is scanf( ) which can be used to read data from a keyboard.

6 *Pre-processor*

The **pre-processor** is a program that is identified by **pre-processor commands** so that it is executed prior to the compilation. All such commands are identified by having # at the beginning of the line. Thus we might have:

```
# include < >
```

to include the file named between the angle brackets < >. When this command is reached, the specified file will be inserted into the program. It is frequently used to add the contents of standard header files, these giving a number of declarations and definitions to enable standard library functions to be used. The entry would then be:

```
# include <stdio.h>
```

As an illustration, consider the simple program:

```
# include <stdio.h>
main( )
{
  printf("Mechatronics");
}
```

Before starting the main program the file stdio.h is added. Thus when the main program starts we are able to access the function printf( ) which results in the word Mechatronics being displayed on the screen.

Another type of pre-processor command is:

define pi 3.14

and this can be used to define values that will be inserted whenever a particular symbol is encountered in the program. Thus whenever pi is encountered the value 3.14 will be used.

define square(x) (x)*(x)

will replace the term square in the program by (x)*(x).

7 *Main function*
Every C program must have a function called main( ). This function is the one that exercises control when the program is executed and is the first function to be called up. Execution starts with its first statement. Other functions may be called up within statements, each one in turn being executed and control returned to the main function. The statement:

void main(void)

indicates that no result is to be returned to the main program and there is no argument. By convention a return value of 0 from main( ) is used to indicate normal program termination, i.e. the entry:

return 0;

8 *Comments*
/* and */ are used to enclose comments. Thus we might have an entry such as:

/* Main program follows */

Comments are ignored by the compiler and are just used to enable a programmer more easily to comprehend the program. Comments can span more than one line, e.g.

/* An example of a program used to
illustrate programming */

9 *Variables*
A **variable** is a named memory location that can hold various values. Variables that can hold a character are specified using the keyword *char*, such a variable being 8 bits long and generally used to store a single character. Signed integers, i.e. numbers with no fractional parts and which are signed to indicate positive or negative, are specified using the keyword *int*. The keyword *float* is used for floating-point numbers, these being numbers which have a fractional part. The keyword *double* is also

used for floating-point numbers but provides about twice the number of significant digits as *float*. To declare a variable the type is inserted before the variable name, e.g.:

    int counter;

This declares the variable 'counter' to be of the integer type. As another example we might have:

    float x, y;

This indicates that the variables x and y are both floating-point numbers.

10 *Assignments*
An assignment statement assigns the value of the expression to the right of the = sign to the variable on its left. For example, $a = 2$ assigns the value 2 to the variable a.

11 *Arithmetic operators*
The arithmetic operators used are: addition $+$, subtraction $-$, multiplication $*$, division $/$, modulus %, increment $+ +$ and decrement $- -$. Increment operators increase the value of a variable by 1, decrement operators decrease it by 1. The normal rules of arithmetic hold for the precedence of operations. For example, $2*4 + 6/2$ gives 11. An example of a program involving arithmetic operators is:

```
/*program to determine area of a circle*/

#include <stdio.h>  /*identifies the function library*/

int radius, area /*variables radius and area are integers*/

int main(void) /*starts main program, the int specifies
    that an integer value is returned, the void indicates
    that main( ) has no parameters*/
{
    printf("Enter radius:"); /*"Enter radius" on screen*/
    scanf("%d", &radius); /*Reads an integer from
    keyboard and assigns it to the variable radius*/
    area = 3.14 * radius * radius; /*Calculates area*/
    printf("\nArea = %d", area); /*On new line prints Area
    = and puts in numerical value of the area*/
    return 0; /*returns to the calling point*/
}
```

12 *Relational operators*
Relational operators are used to compare expressions, asking questions such as 'Is x equal to y?' or 'Is x greater than 10?' The relational operators are: is equal to $==$, is not equal to $!=$, less than $<$, less than or equal to $< =$, greater than $>$, greater than or equal to $> =$. Note that $==$ has to be used when asking if two variables are the same, $=$ is used for

assignment when you are stating that they are the same. For example, we might have the question 'Is x equal to 2?' and represent this by (a $= =$ 2).

13 *Logical operators*
The logical operators are:

Operator	Symbol
AND	&&
OR	||
NOT	!

Note that in C the outcome is equal to 1 if true and 0 if false.

14 *Bitwise operations*
The bitwise operators treat their operands as a series of individual bits rather than a numerical value, comparing corresponding bits in each operand, and only work with integer variables. The operators are:

Bitwise operation	Symbol
AND	&
OR	|
EXCLUSIVE-OR	^
NOT	~
Shift right	$\gg$
Shift left	$\ll$

Thus, for example, we might have the statement:

```
portA = portA | 0x0c;
```

The prefix 0x is used to indicate that the 0c is a hex value, being 0000 1100 in binary. The value ORed with port A is thus a binary number that forces bits 2 and 3 on, all the other bits remaining unchanged.

```
portA = portA ^ 1;
```

This statement causes all the bits except for bit 1 of port A to remain unchanged. If bit 0 is 1 in port A the XOR will force it to 0 and if it is 0 it will force it to 1.

15 *String*
A sequence of characters enclosed within double quotes, i.e. " ", is termed a string. As the term implies, the characters within the double quotes are treated as a linked entity. For example, we might have:

```
printf("Sum = %d", x)
```

The argument in ( ) specifies what is passed to the print function. There are two arguments, the two being separated by a comma. The first argument is the string between the double quotes and specifies how the output is to be presented, the %d specifying that the variable is to be displayed as a decimal integer. Other format specifiers are:

%c	character
%d	signed decimal integer
%e	scientific notation
%f	decimal floating point
%o	unsigned octal
%s	string of characters
%u	unsigned decimal integer
%x	unsigned hexadecimal
%%	prints a % sign

The other argument x specifies the value that is to be displayed.

As another example, the statement:

```
scanf("%d", &x);
```

reads a decimal integer from the keyboard and assigns it to the integer variable x. The & symbol in front of x is the 'address of' operator. When placed before the name of a variable, it returns the address of the variable. The command thus scans for data and stores the item using the address given.

16 *Escape sequences*

Escape sequences are characters that 'escape' from the standard interpretation of characters and are used to control the location of output on a display by moving the screen cursor or indicating special treatments. Thus we might have:

```
printf("\nSum = %d", d)
```

with the \n indicating that a new line is to be used when it is printed on the screen. Escape sequences commonly used are:

\a	sound a beep
\b	backspace
\n	new line
\t	horizontal tab
\\	backslash
\?	question mark
\'	single quotation

12.2.2 Example of a C program

An example of a simple program to illustrate the use of some of the above terms is:

```
/*A simple program in C*/
```

```
# include <stdio.h>
void main(void)
{
    int a, b, c, d; /*a, b, c and d are integers*/
    a = 4; /*a is assigned the value 4*/
    b = 3; /*b is assigned the value 3*/
    c = 5; /*c is assigned the value 5*/
    d = a * b * c; /*d is assigned the value of a * b * c*/
    printf("a * b * c = %d\n", d);
}
```

The statement int a, b, c, d; declares the variables a, b, c and d to be integer types. The statements a = 4, b = 3, c = 5 assign initial values to the variables, the = sign being used to indicate assignment. The statement d = a * b * c directs that a is to be multiplied by b and then by c and stored as d. The printf in the statement printf("a * b * c = %d\n", d) is the display on screen function. The argument contains %d and this indicates that it is to be converted to a decimal value for display. Thus it will print a * b * c = 60. The character \n at the end of the string is to indicate that a new line is to be inserted at that point.

12.3 Branches and loops

Statements to enable branching and looping in programs include *if*, *if/else*, *for*, *while* and *switch*.

1 *If*

The *if* statement allows branching (Figure 12.2(a)). For example, if an expression is true then the statement is executed, if not true it is not, and the program proceeds to the next statement. Thus we might have statements of the form:

```
if (condition 1 = = condition  2);
printf ("\nCondition is OK.");
```

Figure 12.2 (a) If, (b) if/else.

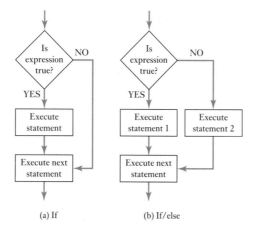

(a) If   (b) If/else

An example of a program involving if statements is:

```
#include  <studio.h>

int x, y;
main( )

{
    printf("\nInput an integer value for x: ");
    scanf("%d", &x);
    printf("\nInput an integer value for y: ");
    scanf(%d", &y);
    if( x = = y)
        printf("x is equal to y");
    if(x > y)
        printf("x is greater than y");
    if(x < y)
        printf("x is less than y");
        return 0;
}
```

The screen shows Input an integer value for x: and then a value is to be keyed in. The screen then shows Input a value for y: and then a value is to be keyed in. The if sequence then determines whether the keyed-in values are equal or which is greater than the other and displays the result on the screen.

2   *If/else*

The *if* statement can be combined with the *else* statement. This allows one statement to be executed if the result is yes and another if it is no (Figure 12.2(b)). Thus we might have:

```
#include <studio.h>

main( )
{
  int temp;
  if(temp > 50)
      printf("Warning");
  else
      printf("System OK");
}
```

3   *For*

The term *loop* is used for the execution of a sequence of statements until a particular condition reaches the required condition of being true, or false. Figure 12.3(a) illustrates this. One way of writing statements for a loop is to use the function *for*. The general form of the statement is:

```
for(initialising expression; test expression; increment  expression)
loop statement;
```

Thus we might have:

```
#include <studio.h>

int count
```

Figure 12.3 (a) For, (b) while.

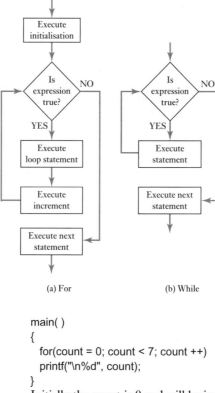

(a) For   (b) While

```
main( )
{
   for(count = 0; count < 7; count ++)
   printf("\n%d", count);
}
```

Initially the count is 0 and will be incremented by 1 and then looped to repeat the for statement as long as the count is less than 7. The result is that the screen shows 0 1 2 3 4 5 6 with each number being on a separate line.

4 *While*

The *while* statement allows for a loop to be continually repeated as long as the expression is true (Figure 12.3(b)). When the expression becomes false then the program continues with the statement following the loop. As an illustration we could have the following program where the while statement is used to count as long as the number is less than 7, displaying the results:

```
#include <studio.h>

int count;
int main( );
{
   count = 1;
   while(count < 7)
      {
         printf("\n%d", count);
         count ++;
      }
   return 0;
}
```

The display on the screen is 1 2 3 4 5 6 with each number on a separate line.

5 *Switch*

The *switch* statement allows for the selection between several alternatives, the test condition being in brackets. The possible choices are identified by case labels, these identifying the expected values of the test condition. For example, we might have the situation where if case 1 occurs we execute statement 1, if case 2 occurs we execute statement 2, etc. If the expression is not equal to any of the cases then the default statement is executed. After each case statement there is normally a break statement to transfer execution to the statement after the switch and stop the switch continuing down the list of cases. The sequence is thus (Figure 12.4):

```
switch(expression)
{
    case 1;
        statement 1;
    break
    case 2;
        statement 2;
        break;
    case 3;
```

Figure 12.4   Switch.

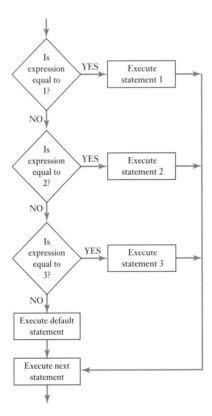

```
          statement 3;
          break;
      default;
          default statement;
  }
next statement
```

The following is an example of a program which recognises the numbers 1, 2 and 3 and will display whichever one is entered from the keyboard:

```
#include <stdio.h>

int main ( );
{
  int x;

  printf("Enter a number 0, 1, 2 or 3:  ");
  scanf("%d", &x);

  switch (x)
  {
    case 1:
        printf("One");
        break;
    case 2:
        printf("Two");
        break;
    case 3:
        printf("Three");
        break;
    default;
        printf("Not 1, 2 or 3");
  }
  return 0;
}
```

12.4   Arrays

Suppose we want to record the mid-day temperature for each day for a week and then later be able to find the temperature corresponding to any one particular day. This can be done using an array. An **array** is a collection of data storage locations with each having the same data type and referenced through the same name. To declare an array with the name Temperature to store values of type float we use the statement:

```
float Temperature[7];
```

The size of the array is indicated between square brackets [ ] immediately after the array name. In this case 7 has been used for the data for the seven days of the week. Individual elements in an array are referenced by an index value. The first element has the number 0, the second 1, and so on to the last element in an *n* sequence which will be *n* − 1. Figure 12.5(a) shows the form of a sequential array. To store values in the array we can write:

```
temperature[0] = 22.1;
temperature [1] = 20.4;
etc.
```

Figure 12.5   (a) A four-element sequential array, (b) a two-dimensional array.

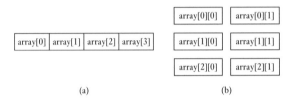

(a)   (b)

If you want to use scanf( ) to input a value into an array element, put & in front of the array name, e.g.:

```
scanf("%d", &temperature [3]);
```

The following is an example of a simple program to store and display the squares of the numbers 0, 1, 2, 3 and 4:

```
#Include <stdlo.h>

int main(void)
{
  int sqrs[5];
  int x;

  for(x = 1; x<5; x++)
    sqrs[x – 1] = x * x;
  for(x = 0; x < 4; x++)
    printf("%d", sqrs[x]);

  return 0;
}
```

Arrays can be given initial values when first declared, e.g.:

```
int array[7] = {10, 12, 15, 11, 10, 14, 12};
```

If you omit the array size, the compiler will create an array just large enough to hold the initialisation values:

```
int array[ ] = {10, 12, 15, 11, 10, 14, 12};
```

Multidimensional arrays can be used. For example, a table of data is a two-dimensional array (Figure 12.5(b)), where x represents the row and y the column, and is written as:

```
array[x][y];
```

12.5 Pointers

Each memory location has a unique address and this provides the means by which data stored at a location can be accessed. A **pointer** is a special kind of variable that can store the address of another variable. Thus if a variable called p contains the address of another variable called x, then p is said to **point** to x. Thus if x is at the address 100 in the memory then p would have the value 100. A pointer is a variable and, as with all other variables, has to be declared before it can be used. A pointer declaration takes the form:

```
type *name;
```

The * indicates that the name refers to a pointer. Often names used for pointers are written with the prefix p, i.e. in the form pname. Thus we might have:

```
int *pnumber;
```

To initialise a pointer and give it an address to point to, we can use &, which is the address-of operator, in a statement of the form:

```
pointer = &variable;
```

The following short program illustrates the above:

```
#include <stdio.h>

int main(void)
{
  int *p, x;
  x = 12;
  p = &x;   /*assigns p the address of x*/
  printf("%d", *p);   /*displays the value of x using pointer*/
  return 0;
{
```

The program thus displays the number 12 on the screen. Accessing the contents of a variable by using a pointer, as above, is called **indirect access**. The process of accessing the data in the variable pointed to by a pointer is termed **dereferencing** the pointer.

12.5.1 Pointer arithmetic

Pointer variables can have the arithmetic operators $+$, $-$, $++$ and $--$ applied to them. Incrementing or decrementing a pointer results in its pointing to the next or previous element in an array. Thus to increment a pointer to the next item in an array we can use:

```
pa++; /*using the increment by 1 operator*/
```

or:

```
pa = pa + 1; /*adding 1*/
```

12.5.2   Pointers and arrays

Pointers can be used to access individual elements in an array, the following program showing such access:

```
#include <stdio.h>

int main(void)
{
    int x[5] = (0, 2, 4, 6, 8);
    int *p;
    p = x; /*assigns to p the address of the start of x*/
    printf("%d %d", x[0], x[2]);

    return 0;
}
```

The statement printf("%d %d", x[0], x[2]); results in pointing to the address given by x and hence the values at the addresses [0] and [2] being displayed, i.e. 0 and 4, on separate lines.

12.6   Program development

In developing programs the aim is to end up with a set of machine language instructions which can be used to operate a microprocessor/microcontroller system. These instructions are called the **executable file**. In order to arrive at such a file, the following sequence of events occurs.

1   *Creating the source code*
   This is the writing of the sequence of statements in C that will constitute the program. Many compilers come with an editor and so the programmer can simply type in the source code from the keyboard. Otherwise a program such as Notepad with Microsoft Windows can be used. Using a word processor can present problems in that additional formatting information is included which can prevent compilation unless the file is saved without formatting information.

2   *Compiling the source code*
   Once the source code has been written, the programmer can direct the compiler to translate it into machine code. Before the compilation process starts, all the pre-processor commands are executed. The compiler can detect several different forms of error during the translation and generate messages indicating the errors. Sometimes a single error may result in the cascading sequence of errors all following from that single first error. Errors usually involve going back to the editor stage and re-editing the source code. The compiler then stores the resulting machine code in another file.

3   *Linking to create an executable file*
   Then the compiler is used to bring together, i.e. link, the generated code with library functions to give a single executable file. The program is then stored as the executable file.

12.6.1 Header files

Pre-processor commands are used at the beginning of a program to define the functions used in that program; this is so they can be referred to by simple labels. However, to save having to write long lists of standard functions for each program, a pre-processor instruction can be used to indicate that a file should be used which includes the relevant standard functions. All that is necessary is to indicate which file of standard functions should be used by the compiler; this file is a **header** since it comes at the head of the program. For example, <stdio.h> contains standard input and output functions such as gets (inputs, i.e. reads a line from a device), puts (outputs, i.e. writes a line to a device) and scanf (reads data); <math.h> contains mathematical functions such as cos, sin, tan, exp (exponential) and sqrt (square root).

Header files are also available to define the registers and ports of microcontrollers and save the programmer having to define each register and port by writing pre-processor lines for each. Thus for an Intel 8051 microcontroller we might have the header <reg.51.h>; this defines registers, e.g. the ports P0, P1, P2 and P3, and individual bits in bit-addressable registers, e.g. bits TF1, TR1, TF0, TR0, IE1, IT1, IE0 and IT0 in register TCON. Thus we can write instructions referring to port 0 inputs/outputs by purely using the label P0 or TF1 to refer to the TF1 bit in register TCON. Similarly, for a Motorola M68HC11E9 the header <hc11e9.h.> defines registers, e.g. PORTA, PORTB, PORTC and PORTD, and individual bits in bit-addressable registers, e.g. bits STAF, STAI, CWOM, HNDS, OIN, PLS, EGA and INVB in register PIOC. Thus, for example, we can write instructions referring to port A inputs/outputs by purely using the label PORTA. Libraries might also supply routines to help with the use of hardware peripheral devices such as keypads and liquid crystal displays.

The main program written for perhaps one specific microcontroller can, as a result of changing the header file, be easily adapted for running with another microcontroller. Libraries thus make C programs highly portable.

12.7 Examples of programs

The following are examples of programs written in C for microcontroller systems.

12.7.1 Switching a motor on and off

Consider the programming of a microcontroller, M68HC11, to start and stop a d.c. motor. Port C is used for the inputs and port B for the output to the motor, via a suitable driver (Figure 12.6). The start button is connected to PC0 to switch from a 1 to a 0 input when the motor is to be started. The stop button is connected to PC1 to switch from a 1 to a 0 input when the motor is to be stopped. The port C data direction register DDRC has to be set to 0 so that port C is set for inputs.

Figure 12.6   Motor control.

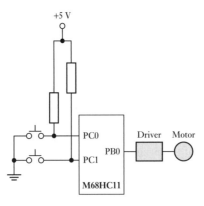

A program might be:

```
#include <hc11e9.h>   /*includes the header file*/

void main(void)
{
    PORTB.PB0 &=0; /*initially ensures motor off*/
    DDRC = 0; /*sets port C to be input*/
    while (1) /*repeats while this condition holds*/
    {
        if (PORTC.PC0 = =0)   /*is start button pressed?*/
            PORTB.PB0 |=1;  /*start output if pressed*/
        else if(PORTC.PC1 = =0)   /*is stop button pressed?*/
            PORTB.PB0 &=0;  /*stop output if pressed*/
    }
}
```

Note that | is the OR operator and sets a bit in the result to 0 only if the corresponding bits in both operands are 0, otherwise it sets to 1. It is used to turn on, or set, one or more bits in a value. Thus Port B.PB0 |=1 has 1 ORed with the value in PB0 and thus switches the motor on. It is a useful way of switching a number of bits in a port simultaneously. The & in PORTB.PB0 &=0 is used to AND the PB0 bit with 0 and so, since PB0 is already 1, assign to PORTB.PB0 the value 0.

12.7.2   Reading an ADC channel

Consider the task of programming a microcontroller (M68HC11) so that a single channel of its analogue-to-digital converter (ADC) can be read. The M68HC11 contains an eight-channel multiplexed, 8-bit, successive approximations ADC with inputs via port E (Figure 12.7). The ADC control/status register ADCTL contains the conversion complete flag CCF at bit 7 and other bits to control the multiplexer and the channel scanning. When CCF = 0 the conversion is not complete and when 1 it is complete.

Figure 12.7 ADC.

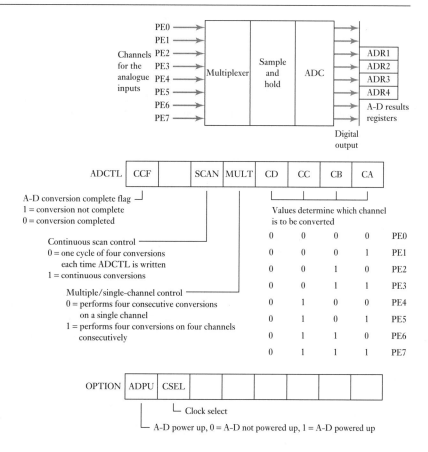

The analogue-to-digital conversion is initiated by writing a 1 to the analogue-to-digital power-up (ADPU) bit in the OPTION register. However, the ADC must have been turned on for at least 100 μs prior to reading a value.

To convert the analogue input to PE0, the first 4 bits in the ADCTL register, i.e. CA, CB, CC and CD, have to be all set to 0. When operating to convert just a single channel, bit 5 SCAN might be set to 0 and bit 4 MULT to 0. A simple program to read a particular channel might thus involve, after powering up the ADC, turning all the bits in the ADCTL register to zeros, putting in the channel number and then reading the input while CCF is 0.

The program might thus be as follows:

```
#include <hc11e9.h> /*the header file*/

void main(void)
{
    unsigned int k; /*this enters the channel number*/

    OPTION=0; /*this and following line turns the ADC on*/
    OPTION.ADPU=1;
    ADCTL &=~0x7;  /*clears the bits*/
```

```
ADCTL |=k; /*puts the channel number to be read*/
while (ADCTL.CCF==0);
return ADR1; /*returns converted value to address 1*/
}
```

Note that ~ is the complement operator and its action is to reverse every bit in its operand, changing all the zeros to ones and vice versa. Thus bit 7 is set. | is the OR operator and sets a bit in the result to 0 only if the corresponding bits in both operands are 0, otherwise it sets to 1. It is used to turn on, or set, one or more bits in a value. In this case with k = 1 it just sets CA to 1. A delay subroutine can be included to ensure that after power-up the value is not read too quickly.

12.8   Arduino programs

The term **sketch** is used for programs for use with the Arduino board. Arduino uses C language for its programs. The basic format of such programs consists of two functions, setup and loop. The setup function is executed at the beginning of a program and is used to configure pins, and to declare variables, constants, etc. The loop function is executed step-by-step and, when it reaches the end of the loop, it automatically returns to the first step of the loop function and goes on continuously repeating the loop until the program is stopped.

```
void setup( )
{
//set up code placed here
}

void loop ( )
{
// code steps provided here
}
```

Note that in Arduino programs the first { is sometimes placed on the line after the command and so the above program would appear as below:

```
void setup( ) {
//set up code placed here
}

void loop ( ) {
// code steps provided here
}
```

The setup function itself calls two built-in functions, pinMode and digitalWrite. The pinMode function sets a particular pin to be either an input or an output, since Arduino digital pins can function as either inputs or outputs. The digitalWrite function sets a pin HIGH or LOW. These two functions do not return a value so the program has to say they are void. As an illustration, consider a program to cause light-emitting diode (LED) on the board, which has been connected internally to pin 13, to blink on and off. Comments about programs are enclosed by /* and */ when on more than one line or just preceded by // when on a single line. Comments are not compiled into machine code for loading into the microcontroller.

```
//Turn on the internal LED for 0.5 s, then off for 0.5 s, repeatedly.

void setup ( )
{
  pinMode (13, OUTPUT);
}

void loop ( )
{
  digitalWrite (13, HIGH);
  delay (500);
  digital Write (13, LOW);
}
```

The delay function is to institute a delay of 0.500s between writing pin 13 HIGH and then writing LOW.

Suppose we want to turn an external LED on for 0.5s, then off for 0.5s, repeatedly. We need to specify which pin the LED is connected to and that the pin is to be regarded as an input. An important point when connecting an LED is that typically the voltage drop across an LED is limited to about 2V with a current of 20mA. As the board supplies 5V, there is a need to put a resistor in series with the LED so that the voltage drop across it will be 3V, leaving just a 2V drop for the LED. The minimum resistance needed is thus $V/I = 3/0.020 = 150\ \Omega$. Generally, a higher value will be used as the LED can still glow quite brightly with less current.

```
//Turn an external LED off for 0.5 s, then on for 0.5 s, repeatedly.

 #define ext_LED 12

void setup ( )
{
  pinMode (ext_LED, OUTPUT);
}

void loop ( )
{
  digitalWrite (ext_LED, LOW);
  delay (500);
  digitalWrite (ext_LED, HIGH);
  delay (500);
}
```

Now consider operating with an external and an internal LED.

```
/*Turn the internal LED on and the external LED off for 0.5 s, then the internal
LED off and the external LED on for 0.5 s, repeatedly.
*/
```

```
#define int_LED 13
#define ext_LED 12

void setup ( )
{
  pinMode (int_LED, OUTPUT);
  pinMode (ext_LED, OUTPUT);
}

void loop ( )
{
  digitalWrite (int_LED, HIGH);
  digitalWrite (ext_LED, LOW);
  delay (500);
  digitalWrite (int_LED, LOW);
  digitalWrite (ext_LED, HIGH);
  delay (500);
}
```

The above program thus has the internal and external LEDs repeatedly blinking alternately off and on. A possible enhancement of the above program is to have this taking place on and only if a switch is closed.

```
/*If a switch is closed turn the internal LED on and the external LED off for
0.5 s, then the internal LED off and the external LED on for 0.5 s, repeatedly.
*/
#define int_LED 13
#define ext_LED 12
#define ext_sw 11
Int switch_value;

void setup ( )
{
  pinMode (int_LED, OUTPUT);
  pinMode (ext_LED, OUTPUT);
  pinMode (ext_sw. INPUT);
}

void loop ( )
{
switch_value = digitalRead(ext_sw);
  if (switch_value ==LOW)
    {
    digitalWrite (int_LED, HIGH);
    digitalWrite (ext_LED, LOW);
    delay (500);
    digitalWrite (int_LED, LOW);
    digitalWrite (ext_LED, HIGH);
    delay (500);
    }
  else
    {
    digitalWrite(int_LED, LOW);
```

```
        digitalWrite(ext_LED, LOW)
      }
    }
```

The above gives just a simple introduction to writing programs for the Arduino. The procedures required are essentially just those outlined earlier in this chapter with C. Many already written programs, are freely available on the Arduino website.

To use a C language program with Arduino, a program called the Arduino Development Environment is first downloaded from the Arduino website into the host computer. This program enables C language code to be typed into the computer and it then compiles the program, checking that it conforms to the rules of C language and translating it into assembly language and then into machine code as this is the language that the Arduino board understands. When you first boot the Arduino board, it enters the bootloader, which is a chunk of code which has been downloaded into its memory at the factory and allows programs to be uploaded via its USB connector. If it then receives a command from the host computer to upload a program, the machine code program is loaded into the Arduino memory so that it becomes available for use by the Arduino microcontroller.

Basically the sequence of operations is:
1. download the Arduino Development Environment into the host computer from the Arduino website;
2. connect the Arduino board to the host computer via a USB cable;
3. start the Arduino Development Environment;
4. type the C program into the computer;
5. select the Upload button on the screen;
6. the program then runs on the Arduino board.

Summary

C is a high-level language which has the advantages when compared with the assembly language of being easier to use and that the same program can be used with different microprocessors; all that is necessary for this is that the appropriate compiler is used to translate the C program into the relevant machine language for the microprocessor concerned. Assembly language is different for the different microprocessors while C language is standardised.

C packages are supplied with libraries containing a large number of predefined functions containing C code that have already been written. In order to use the contents of any particular library, that library has to be specified in a header file. Every C program must have a function called main( ); this exercises control when the program is executed and is the first function to be called up. A program is made up of statements, every statement being terminated by a semicolon. Statements can be grouped together in blocks by putting them between braces, i.e. { }.

Problems

12.1   The following questions are all concerned with components of programs.

(a) State what is indicated by int in the statement:

```
int counter;
```

(b) State what the following statement indicates:

```
num = 10
```

(c) State what the result of the following statement will be:

```
printf("Name");
```

(d) State what the result of the following statement will be:

```
printf("Number %d", 12);
```

(e) State what the effect of the following is:

```
#include <stdio.h>
```

12.2   For the following program, what are the reasons for including the line (a) #include <stdio.h>, (b) the { and }, (c) the /d, and (d) what will appear on the screen when the program is executed?

```
#include <stdio.h>

main( )
{
    printf(/d"problem 3");
}
```

12.3   For the following program, what will be displayed on the screen?

```
#include <stdio.h>

int main(void);
{
    int num;
    num = 20;

printf("The number is %d", num);
return 0;
}
```

12.4   Write a program to compute the area of a rectangle given its length and width at screen prompts for the length and width and then display the answer preceded by the words 'The area is'.

12.5   Write a program that displays the numbers 1 to 15, each on a separate line.

12.6   Explain the reasons for the statements in the following program for the division of two numbers:

```c
#include <stdio.h>

int main(void);
{
    int num1, num2;

    printf("Enter first number:");
    scanf("%d", &num1);

    printf("Enter second number: ");
    scanf("%d", &num2);

    if(num2 = = 0)
        print f("Cannot divide by zero")
    else
        printf("Answer is: %d", num1/num2);
    return 0;
}
```

Chapter thirteen Input/output systems

Objectives

The objectives of this chapter are that, after studying it, the reader should be able to:
- Identify interface requirements and how they can be realised: buffers, handshaking, polling and serial interfacing.
- Explain how interrupts are used with microcontrollers.
- Explain the function of peripheral interface adapters and be able to program them for particular situations.
- Explain the function of asynchronous communication interface adapters.

13.1 Interfacing

When a microprocessor is used to control some system it has to accept input information, respond to it and produce output signals to implement the required control action. Thus there can be inputs from sensors to feed data in and outputs to such external devices as relays and motors. The term **peripheral** is used for a device, such as a sensor, keyboard, actuator, etc., which is connected to a microprocessor. It is, however, not normally possible to connect directly such peripheral devices to a microprocessor bus system due to a lack of compatibility in signal forms and levels. Because of such incompatibility, a circuit known as an interface is used between the peripheral items and the microprocessor. Figure 13.1 illustrates the arrangement. The interface is where this incompatibility is resolved.

Figure 13.1 The interfaces.

This chapter discusses the requirements of such interfaces and the very commonly used Motorola MC6820 Peripheral Interface Adapter and Motorola MC6850 Asynchronous Communications Interface Adapter.

13.2 Input/output addressing

There are two ways that microprocessors can select input/output devices. Some microprocessors, e.g. the Zilog Z80, have **isolated input/output**, and special input instructions such as IN are used to read from an input device and special output instructions such as OUT are used to output to an output device. For example, with the Z80 we might have:

IN A,(B2)

to read input device B2 and put the data in the accumulator A. An output instruction might be:

OUT (C), A

to write the data in accumulator A to port C.

More commonly, microprocessors do not have separate instructions for input and output but use the same instructions as they use for reading from or writing to memory. This is termed **memory-mapped input/output**. With this method, each input/output device has an address, just like a memory location. The Motorola 68HC11, Intel 8051 and PIC (peripheral interface controller) microcontrollers have no separate input/output instructions and use memory mapping. Thus, with memory mapping we might use:

LDAA $1003

to read the data input at address $1003 and:

STAA $1004

to write data to the output at address $1004.

Microprocessors use parallel ports to input or output bytes of data. Many peripherals often require several input/output ports. This can be because the data word of the peripheral is longer than that of the CPU. The CPU must then transfer the data in segments. For example, if we require a 16-bit output with an 8-bit CPU the procedure is:

1 the CPU prepares the eight most significant bits of the data;
2 the CPU sends the eight most significant bits of the data to the first port;
3 the CPU prepares the eight least significant bits of the data;
4 the CPU sends the eight least significant bits of the data to the second port;
5 thus, after some delay, all the 16 bits are available to the peripheral.

13.2.1 Input/output registers

The Motorola 68HC11 microcontroller has five ports A, B, C, D and E (see Section 10.3.1). Ports A, C and D are bidirectional and can be used for either input or output. Port B is output only and port E input only. Whether a bidirectional port is used for input or output depends on the setting of a bit in its control register. For example, port A at address $1000 is controlled by the pulse accumulator control register PACTL at address $1026. To set port A for input requires bit 7 to be 0; output requires bit 7 to be 1 (see Figure 10.12). Port C is bidirectional and the 8 bits in its register at address $1003 are controlled by the corresponding bits in its port data direction register at address $1007. When the corresponding data direction bit is set to 0 it is an input, when set to 1 it is an output. Port D is bidirectional and contains just six input/output lines at address $1008. It is controlled by a port direction register at address $1009. The direction of each line is controlled by the corresponding bit in the control register; it is set to 0 for an input and 1 for an output. Some of the ports can also be set to carry out other functions by setting other bits in their control registers.

For a fixed-direction port, e.g. port B in the Motorola 68HC11 is output only, the instructions needed to output some value, e.g. $FF, are simply those needed to load the data to that address. The instructions might be:

```
REGBAS    EQU      $1000      ; base address of I/O registers
PORTB     EQU      $04        ; offset of PORTB from REGBAS
          LDX      #REGBAS    ; load index register X
          LDAA     #$FF       ; load $FF into accumulator
          STAA     PORTB,X    ; store value at PORTB address
```

For the fixed-direction port E, which is input only, the instruction to read a byte from it might be:

```
REGBAS    EQU      $1000      ; base address of I/O registers
PORTE     EQU      $0A        ; offset of PORTE from REGBAS
          LDAA     PORTE,X    ; load value at PORTE into the
                             ; accumulator
```

For a bidirectional port such as C, before we can use it for an input we have to configure the port so that it acts as an input. This means setting all the bits to 0. Thus we might have:

```
REGBAS    EQU      $1000      ; base address of I/O registers
PORTC     EQU      $03        ; offset of PORTC from REGBAS
DDRC      EQU      $07        ; offset of data direction register from
                             ; REGBAS
          CLR      DDRC,X     ; set DDRS to all 0
```

For the Intel 8051 microcontroller (see Section 10.3.2) there are four parallel bidirectional input/output ports. When a port bit is to be used for output, the data is just put into the corresponding special function register bit; when it is used for input a 1 must be written to each bit concerned, thus FFH might be written for an entire port to be written to. Consider an example of Intel 8051 instructions to light an LED when a push-button is pressed. The push-button provides an input to P3.1 and an output to P3.0; the push-button pulls the input low when it is pressed:

```
          SETB     P3.1       ; make bit P3.1 a 1 and so an input
LOOP      MOV      C,P3.1     ; read the state of the push-button
                             ; and store it in the carry flag
          CPL      C          ; complement the carry flag
          MOV      P3.0, C    ; copy state of carry flag to output
          SJMP     LOOP       ; keep on repeating the sequence
```

With the PIC microcontroller the direction of the signals at its bidirectional ports is set by the TRIS direction registers (see Section 10.3.3). TRIS is set as 1 for read and 0 for write. The registers for the PIC16C73/74 are arranged in two banks and before a particular register can be selected the bank has to be chosen by setting bit 5 in the STATUS register. This register is in both banks and so we do not have to select the bank in order to use this register. The TRIS registers are in bank 1 and the PORT registers in bank 0. Thus to set port B as output we have first to select bank 1 and then set TRISB to 0. We can then select bank 0 and write the output to PORTB. The bank is selected by setting a bit in the STATUS register. The instructions to select port B as an output are thus:

```
Output    clrf    PORTB         ; clear all the bits in port B
          bsf     STATUS,RP0    ; use status register to select bank 1 by
                                ; setting RP0 to 1
          clrf    TRISB         ; clear bits so output
          bcf     STATUS,RP0    ; use status register to select bank 0
                                ; port B is now an output set to 0
```

13.3	Interface requirements

The following are some of the actions that are often required of an interface circuit.

1 *Electrical buffering / isolation*
This is needed when the peripheral operates at a different voltage or current to that on the microprocessor bus system or there are different ground references. The term **buffer** is used for a device that provides isolation and current or voltage amplification. For example, if the output of a microprocessor is connected to the base of a transistor, the base current required to switch the transistor is greater than that supplied by the microprocessor and so a buffer is used to step up the current. There also has often to be isolation between the microprocessor and the higher power system.

2 *Timing control*
Timing control is needed when the data transfer rates of the peripheral and the microprocessor are different, e.g. when interfacing a microprocessor to a slower peripheral. This can be achieved by using special lines between the microprocessor and the peripheral to control the timing of data transfers. Such lines are referred to as **handshake lines** and the process as **handshaking**.

3 *Code conversion*
This is needed when the codes used by the peripherals differ from those used by the microprocessor. For example, an LED display might require a decoder to convert the BCD output from a microprocessor into the code required to operate the seven display elements.

4 *Changing the number of lines*
Microprocessors operate on a fixed word length of 4 bits, 8 bits or 16 bits. This determines the number of lines in the microprocessor data bus. Peripheral equipment may have a different number of lines, perhaps requiring a longer word than that of the microprocessor.

5 *Serial-to-parallel, and vice versa, data transfer*
Within an 8-bit microprocessor, data is generally manipulated 8 bits at a time. To transfer 8 bits simultaneously to a peripheral thus requires eight data paths. Such a form of transfer is termed **parallel data transfer**. It is, however, not always possible to transfer data in this way. For example,

data transfer over the public telephone system can only involve one data path. The data has thus to be transferred sequentially 1 bit at a time. Such a form of transfer is termed **serial data transfer**. Serial data transfer is a slower method of data transfer than parallel data transfer. Thus, if serial data transfer is used, there will be a need to convert incoming serial data into parallel data for the microprocessor and vice versa for outputs from the microprocessor.

6 *Conversion from analogue to digital and vice versa*
The output from sensors is generally analogue and this requires conversion to digital signals for the microprocessor. The output from a microprocessor is digital and this might require conversion to an analogue signal in order to operate some actuator. Many microcontrollers have built-in analogue-to-digital converters, e.g. PIC 16C74/74A (see Figure 10.30) and Motorola M68HC11 (see Figure 10.10), so can handle analogue inputs. However, where required to give analogue outputs, the microcontroller output has generally to pass through an external digital-to-analogue converter (see Section 13.6.2 for an example).

13.3.1 Buffers

A **buffer** is a device that is connected between two parts of a system to prevent unwanted interference between the two parts. An important use of a buffer is in the microprocessor input port to isolate input data from the microprocessor data bus until the microprocessor requests it. The commonly used buffer is a **tristate buffer**. The tristate buffer is enabled by a control signal to provide logic 0 or 1 outputs, when not enabled it has a high impedance and so effectively disconnects circuits. Figure 13.2 shows the symbols for tristate buffers and the conditions under which each is enabled. Figure 13.2(a) and (b) shows the symbol for buffers that does not change the logic of the input and Figure 13.2(c) and (d) for ones that do.

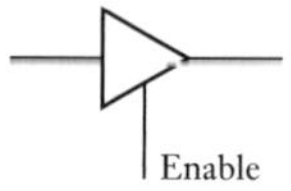

Figure 13.2 Buffers: (a) no logic change, enabled by 1, (b) no logic change, enabled by 0, (c) logic change, enabled by 1, (d) logic change, enabled by 0.

Enable	Input	Output
0	0	High impedance
0	1	High impedance
1	0	0
1	1	1

(a)

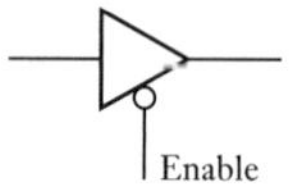

Enable	Input	Output
0	0	0
0	1	1
1	0	High impedance
1	1	High impedance

(b)

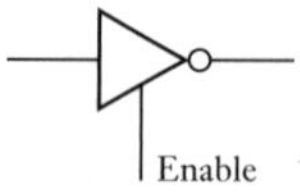

Enable	Input	Output
0	0	High impedance
0	1	High impedance
1	0	1
1	1	0

(c)

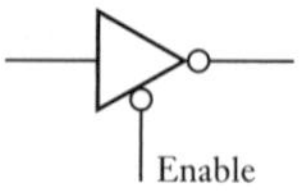

Enable	Input	Output
0	0	1
0	1	0
1	0	High impedance
1	1	High impedance

(d)

With PIC microcontrollers (see Section 10.3.3), the TRIS bit is connected to the enable input of a tristate buffer. If the bit is 0, the tristate buffer is enabled and simply passes its input value to its output, if it is 1 the tristate buffer is disabled and the output becomes high impedance (as in Figure 13.2(b)).

Such tristate buffers are used when a number of peripheral devices have to share the same data lines from the microprocessor, i.e. they are connected to the data bus, and thus there is a need for the microprocessor to be able to enable just one of the devices at a time with the others disabled. Figure 13.3 shows how such buffers can be used. Such buffers are available as integrated circuits, e.g. the 74125 with four non-inverting, active-low buffers and the 74126 with four non-inverting, active-high buffers.

Figure 13.3 Three-state buffer.

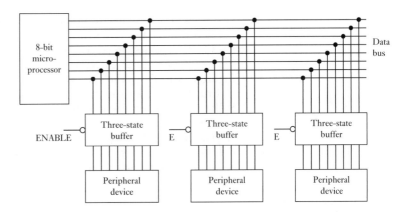

13.3.2 Handshaking

Unless two devices can send and receive data at identical rates, handshaking is necessary to exchange data. With handshaking the slower device controls the rate at which the data is transferred. For parallel data transfer **strobe-and-acknowledge** is the commonly used form of handshaking. The peripheral sends a DATA READY signal to the input/output section. The CPU then determines that the DATA READY signal is active. The CPU then reads the data from the input/output section and sends an INPUT ACKNOWLEDGED signal to the peripheral. This signal indicates that the transfer has been completed and thus the peripheral can send more data. For an output, the peripheral sends an OUTPUT REQUEST or PERIPHERAL READY signal to the input/output section. The CPU determines that the PERIPHERAL READY signal is active and sends the data to the peripheral. The next PERIPHERAL READY signal may be used to inform the CPU that the transfer has been completed.

With the microcontroller MC68HC11, the basic strobed input/output operates as follows. The handshaking control signals use pins STRA and STRB (Figure 13.4(a), also see Figure 10.10 for the full block model), port C is used for the strobed input and port B for the strobed output. When data is ready to be sent by the microcontroller a pulse is produced at STRA and sent to the peripheral device. When the microcontroller receives either a rising or falling edge to a signal on STRB, then the relevant output port of the microcontroller sends the data to the peripheral. When data is ready to

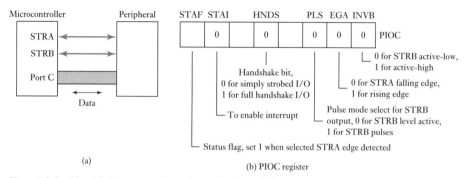

(a)

(b) PIOC register

Figure 13.4   Handshaking control: strobe–and–acknowledge.

be transmitted to the microcontroller, the peripheral sends a signal to STRA that it is ready and then a rising or falling edge to a signal on STRB is used to indicate readiness to receive. Before handshaking can occur, the parallel input/output register PIOC at address $1002 has to be first configured. Figure 13.4(b) shows the states required of the relevant bits in that register.

Full handshake input/output involves two signals being sent along STRB, the first being to indicate ready to receive data and the next one that the data has been read. This mode of operation requires that in PIOC the HNDS bit is set to 1 and if PLS is set to 0 the full handshake is said to be pulsed and if to 1 it is interlocked. With pulsed operation a pulse is sent as acknowledgement; with interlocked STRB there is a reset (Figure 13.5).

Figure 13.5   Full handshaking: (a) pulsed, (b) interlocked.

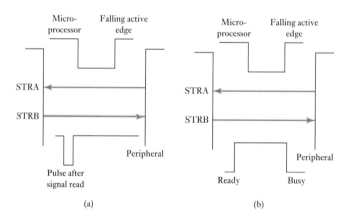

(a)

(b)

13.3.3   Polling and interrupts

Consider the situation where all input/output transfers of data are controlled by the program. When peripherals need attention they signal the microprocessor by changing the voltage level of an input line. The microprocessor can then respond by jumping to a program service routine for the device. On completion of the routine, a return to the main program occurs. Program control of inputs/outputs is thus a loop to read inputs and update outputs continuously, with jumps to service routines as required. This process of repeatedly checking each peripheral device to see if it is ready to send or accept a new byte of data is called **polling**.

An alternative to program control is **interrupt control**. An interrupt involves a peripheral device activating a separate interrupt request line. The reception of an interrupt results in the microprocessor suspending execution of its main program and jumping to the service routine for the peripheral. The interrupt must not lead to a loss of data and an interrupt handling routine has to be incorporated in the software so that the state of processor registers and the last address accessed in the main program are stored in dedicated locations in memory. After the interrupt service routine, the contents of the memory are restored and the microprocessor can continue executing the main program from where it was interrupted (Figure 13.6).

Figure 13.6 Interrupt control.

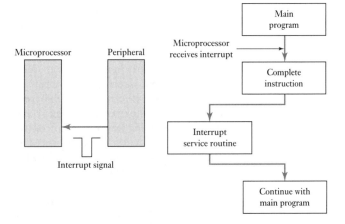

Thus, the following take place when an interrupt occurs.

1 The CPU waits until the end of the instruction it is currently executing before dealing with the interrupt.
2 All the CPU registers are pushed onto the stack and a bit set to stop further interrupts occurring during this interruption. The stack is a special area of memory in which program counter values can be stored when a subroutine is to be executed. The program counter gives the address of the next program instruction in a program and thus storing this value enables the program to be resumed at the place where it broke off to execute the interrupt.
3 The CPU then determines the address of the interrupt service routine to be executed. Some microprocessors have dedicated interrupt pins and the pin that is chosen determines which address is to be used. Other microprocessors have only one interrupt pin and the interrupting device must then supply data that tells the microprocessor where the interrupt service routine is located. Some microprocessors have both kinds of interrupt inputs. The starting address of an interrupt service routine is called an **interrupt vector**. The block of memory assigned to store the vectors is known as the **vector table**. Vector addresses are fixed by the chip manufacturer.
4 The CPU branches to the interrupt service routine.
5 After completion of this routine, the CPU registers are returned from the stack and the main program continues from the point it left off.

Unlike a subroutine call, which is located at a specific point in a program, an interrupt can be called from any point in the program. Note that the program does not control when an interrupt occurs; control lies with the interrupting event.

Input/output operations frequently use interrupts since often the hardware cannot wait. For example, a keyboard may generate an interrupt input signal when a key is pressed. The microprocessor then suspends the main program to handle the input from the keyboard; it processes the information and then returns to the main program to continue from where it left off. This ability to code a task as an interrupt service routine and tie it to an external signal simplifies many control tasks, enabling them to be handled without delay. For some interrupts it is possible to program the microprocessor to ignore the interrupt request signal unless an enable bit has been set. Such interrupts are termed **maskable**.

The Motorola 68HC1 has two external interrupt request inputs. XIRQ is a non-maskable interrupt and will always be executed on completion of the instruction currently being executed. When the XIRQ interrupt occurs, the CPU jumps to the interrupt service routine whose interrupt vector is held at address $FFF4/5 (the low and high bytes of the address). IRQ is a maskable interrupt. When the microcontroller receives a signal at the interrupt request pin IRQ by its going low, the microcontroller jumps to the interrupt service routine indicated by the interrupt vectors $FFF2/3. IRQ can be masked by the instruction set interrupt mask SEI and unmasked by the instruction clear interrupt mask CLI. At the end of an interrupt service routine the instruction RTI is used to return to the main program.

With the Intel 8051, interrupt sources are individually enabled or disabled through the bit-addressable register IE (interrupt enable) at address 0A8H (see Figure 10.26), a 0 disabling an interrupt and a 1 enabling it. In addition there is a global enable/disable bit in the IE register that is set to enable all external interrupts or cleared to disable all external interrupts. The TCON register (Figure 10.25) is used to determine the type of interrupt input signal that will initiate an interrupt.

Figure 13.7 INTCON.

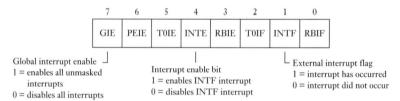

With the PIC microcontrollers, interrupts are controlled by the INTCON register (Figure 13.7). To use bit 0 of port B as an interrupt, it must be set as an input and the INTCON register must be initialised with a 1 in INTE and a 1 in GIE. If the interrupt is to occur on a rising edge then INTEDG (bit 6) in the OPTION register (see Figure 10.32) must be set to 1; if on a falling edge it must be set to 0. When an interrupt occurs, INTF is set. It can be cleared by the instruction bcf INTCON,INTF.

As an illustration of a program involving external interrupts, consider a simple on/off control program for a central heating system involving an Intel 8051 microcontroller (Figure 13.8). The central heating furnace is controlled by an output from P1.7 and two temperature sensors are used, one to determine when the temperature falls below, say, 20.5°C and the other when

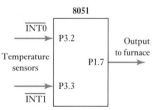

Figure 13.8 Central heating system.

it rises above 21.0°C. The sensor for the 21.0°C temperature is connected to interrupt INT0, port 3.2, and the sensor for the 20.5°C temperature is connected to interrupt INT1, port 3.3. By selecting the IT1 bit to be 1 in the TCON register, the external interrupts are edge triggered, i.e. activated when there is a change from 1 to 0. When the temperature rises to 21.0°C the external interrupt INT0 has an input which changes from 1 to 0 and the interrupt is activated to give the instruction CLR P1.7 for a 0 output to turn the furnace off. When the temperature falls to 20.5°C the external interrupt INT1 has an input which changes from 0 to 1 and the interrupt is activated to give the instruction SETB P1.7 for a 1 output to turn the furnace on. The MAIN program is just a set of instructions to configure and enable the interrupts, establish the initial condition of the furnace to be on if the temperature is less than 21.0° or off if above, and then to wait doing nothing until an interrupt occurs. With the program, a header file has been assumed:

```
        ORG    0
        LJMP   MAIN

        ORG    0003H        ; gives the entry address for ISR0
ISR0    CLR    P1.7         ; interrupt service routine to turn the
                            ; furnace off
        RETI                ; return from interrupt

        ORG    0013H        ; gives the entry address for ISR1
ISR1    SETB   P1.7         ; interrupt service routine to turn furnace off
        RETI                ; return from interrupt

        ORG    30H
MAIN    SETB   EX0          ; to enable external interrupt 0
        SETB   EX1          ; to enable external interrupt 1
        SETB   IT0          ; set to trigger when change from 1 to 0
        SETB   IT1          ; set to trigger when change from to 0
        SETB   P1.7         ; turn the furnace on
        JB     P3.2,HERE    ; if temperature greater than 21.0°C jump to
                            ; HERE and leave furnace on
        CLR    P1.7         ; turn the furnace off
HERE    SJMP   HERE         ; just doing nothing until an interrupt occurs
        END
```

Microcontrollers, in addition to the interrupt request, have a reset interrupt and a non-maskable interrupt. The reset interrupt is a special type of interrupt and when this occurs the system resets; thus when this is activated, all activity in the system stops, the starting address of the main program is loaded and the start-up routine is executed. The microcontroller M68HC11 has a computer operating properly (COP) watchdog timer. This is intended to detect software processing errors when the CPU is not executing certain sections of code within an allotted time. When this occurs the COP timer times out and a system reset is initiated.

The non-maskable interrupt cannot be masked and so there is no method of preventing the interrupt service routine being executed when it is connected to this line. An interrupt of this type is usually reserved for emergency routines such as those required when there is a power failure, e.g. switching to a back-up power supply.

13.3.4   Serial interfacing

With the parallel transmission of data, one line is used for each bit; serial systems, however, use a single line to transmit data in sequential bits. There are two basic types of serial data transfer: asynchronous and synchronous.

With **asynchronous transmission**, the receiver and the transmitter each use their own clock signals so it is not possible for a receiver to know when a word starts or stops. Thus it is necessary for each transmitted data word to carry its own start and stop bits so that it is possible for the receiver to tell where one word stops and another starts (Figure 13.9). With such a mode of transmission, the transmitter and receiver are typically remote (see Chapter 15 for details of standard interfaces). With **synchronous transmission**, the transmitter and receiver have a common clock signal and thus transmission and reception can be synchronised.

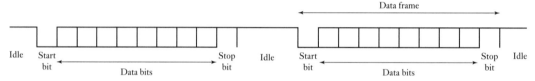

Figure 13.9   Asynchronous transmission.

The microcontroller MC68HC11 (see Figure 10.10) has a serial communications interface (SCI) which can be used for asynchronous transmission and thus can be used to communicate with remote peripheral devices. The SCI uses port D pin PD1 as a transmit line and port PD0 as a receive line. These lines can be enabled or disabled by the SCI control register. The microcontroller also has a serial peripheral interface (SPI) for synchronous transmission. This can be used for local serial communication, local meaning essentially inside the machine in which the chip is located.

13.4   Peripheral interface adapters

Interfaces can be specifically designed for particular inputs/outputs; however, programmable input/output interface devices are available which permit various different input and output options to be selected by means of software. Such devices are known as **peripheral interface adapters** (PIAs).

A commonly used PIA parallel interface is the Motorola MC 6821. It is part of the MC6800 family and thus can be directly attached to Motorola MC6800 and MC68HC11 buses. The device can be considered to be essentially just two parallel input/output ports, with their control logic, to link up with the host microprocessor. Figure 13.10 shows the basic structure of the MC6821 PIA and the pin connections.

The PIA contains two 8-bit parallel data ports, termed A and B. Each port has the following.

1 A *peripheral interface register*. An output port has to operate in a different way to an input port because the data must be held for the peripheral. Thus for output a register is used to store data temporarily. The register is said to be **latched**, i.e. connected, when a port is used for output and unlatched when used for input.
2 A *data direction register* that determines whether the input/output lines are inputs or outputs.

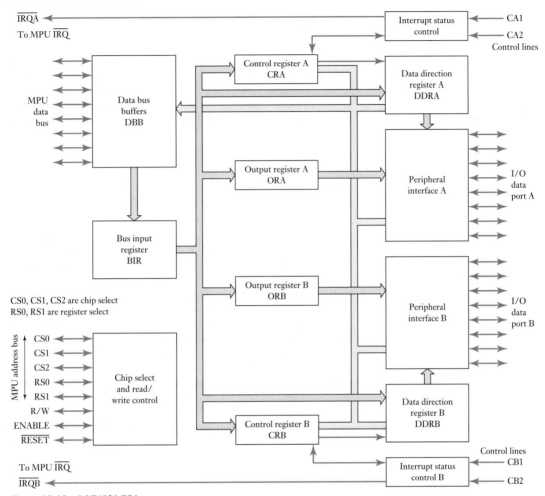

Figure 13.10 MC6821 PIA.

3 A *control register* that determines the active logical connections in the peripheral.
4 Two *control lines*, CA1 and CA2 or CB1 and CB2.

Two microprocessor address lines connect the PIA directly through the two register select lines RS0 and RS1. This gives the PIA four addresses for the six registers. When RS1 is low, side A is addressed and when it is high, side B. RS0 addresses registers on a particular side, i.e. A or B. When RS0 is high, the control register is addressed, when low the data register or the data direction register. For a particular side, the data register and the data direction register have the same address. Which of them is addressed is determined by bit 2 of the control register (see below).

Each of the bits in the A and B control registers is concerned with some features of the operation of the ports. Thus for the A control register we have the bits shown in Figure 13.11. A similar pattern is used for the B control register.

Figure 13.11 Control register.

B7	B6	B5	B4	B3	B2	B1	B0
IRQA1	IRQA2		CA2 control		DDRA access	CA1 control	

Bits 0 and 1

The first two bits control the way that CA1 or CB1 input control lines operate. Bit 0 determines whether the interrupt output is enabled. B0 = 0 disables the IRQA(B) microprocessor interrupt, B0 = 1 enables the interrupt. CA1 and CB1 are not set by the static level of the input but are edge triggered, i.e. set by a changing signal. Bit 1 determines whether bit 7 is set by a high-to-low transition (a trailing edge) or a low-to-high transition (a leading edge). B1 = 0 sets a high-to-low transition, B1 = 1 sets a low-to-high transition.

Bit 2

Bit 2 determines whether data direction registers or peripheral data registers are addressed. With B2 set to 0, data direction registers are addressed, with B2 set to 1, peripheral data registers are selected.

Bits 3, 4 and 5

These bits allow the PIA to perform a variety of functions. Bit 5 determines whether control line 2 is an input or an output. If bit 5 is set to 0, control line 2 is an input, if set to 1, it is an output. In input mode, both CA2 and CB2 operate in the same way. Bits 3 and 4 determine whether the interrupt output is active and which transitions set bit 6.

With B5 = 0, i.e. CA2(CB2) set as an input: B3 = 0 disables IRQA(B) microprocessor interrupt by CA2(CB2); B3 = 1 enables IRQA(B) microprocessor interrupt by CA2(CB2); B4 = 0 determines that the interrupt flag IRQA(B), bit B6, is set by a high-to-low transition on CA2(CB2); B4 = 1 determines that it is set by a low-to-high transition.

B5 = 1 sets CA2(CB2) as an output. In output mode CA2 and CB2 behave differently. For CA2: with B4 = 0 and B3 = 0, CA2 goes low on the first high-to-low ENABLE (E) transition following a microprocessor read of peripheral data register A and is returned high by the next CA1 transition; with B4 = 0 and B3 = 1, CA2 goes low on the first high-to-low ENABLE transition following a microprocessor read of the peripheral data register A and is returned to high by the next high-to-low ENABLE transition. For CB2: with B4 = 0 and B3 = 0, CB2 goes low on the first low-to-high ENABLE transition following a microprocessor write into peripheral data register B and is returned to high by the next CB1 transition; with B4 = 0 and B3 = 1, CB2 goes low on the first low-to-high ENABLE transition following a microprocessor write into peripheral data register B and is returned high by the next low-to-high ENABLE transition. With B4 = 1 and B3 = 0, CA2(CB2) goes low as the microprocessor writes B3 = 0 into the control register. With B4 = 0 and B3 = 1, CA2(CB2) goes high as the microprocessor writes B3 = 1 into the control register.

Bit 6

This is the CA2(CB2) interrupt flag, being set by transitions on CA2(CB2). With CA2(CB2) as an input (B5 = 0), it is cleared by a microprocessor read of the data register A(B). With CA2(CB2) as an output (B5 = 1), the flag is 0 and is not affected by CA2(CB2) transitions.

Bit 7

This is the CA1(CB1) interrupt flag, being cleared by a microprocessor read of data register A(B).

The process of selecting which options are to be used is termed **configuring** or **initialising** the PIA. The RESET connection is used to clear all the registers of the PIA. The PIA must then be initialised.

13.4.1 Initialising the PIA

Before the PIA can be used, a program has to be written and used so that the conditions are set for the desired peripheral data flow. The PIA program is placed at the beginning of the main program so that, thereafter, the microprocessor can read peripheral data. The initialisation program is thus only run once.

The initialisation program to set which port is to be input and which is to be output can have the following steps.

1 Clear bit 2 of each control register by a reset, so that data direction registers are addressed. Data direction register A is addressed as XXX0 and data direction register B as XXX2.
2 For A to be an input port, load all zeros into direction register A.
3 For B to be an output port, load all ones into direction register B.
4 Load 1 into bit 2 of both control registers. Data register A is now addressed as XXX0 and data register B as XXX2.

Thus an initialisation program in assembly language to make side A an input and side B an output could be, following a reset:

```
INIT    LDAA    #$00    ; Loads zeros
        STAA    $2000   ; Make side A input port
        LDAA    #$FF    ; Load ones
        STAA    $2000   ; Make side B output port
        LDAA    #$04    ; Load 1 into bit 2, all other bits 0
        STAA    $2000   ; Select port A data register
        STAA    $2002   ; Select port B data register
```

Peripheral data can now be read from input port A with the instruction LDAA 2000 and the microprocessor can write peripheral data to output port B with the instruction STAA 2002.

13.4.2 Connecting interrupt signals via a PIA

The Motorola MC6821 PIA (Figure 13.12) has two connections, IRQA and IRQB, through which interrupt signals can be sent to the microprocessor so that an interrupt request from CA1, CA2 or CB1, CB2 can drive the IRQ pin of the microprocessor to the active-low state. When the initialisation program for a PIA was considered in the previous section, only bit 2 of the control register was set as 1, the other bits being 0. These zeros disabled interrupt inputs. In order to use interrupts, the initialisation step which stores $04 into the control register must be modified. The form of the modification will depend on the type of change in the input which is required to initiate the interrupt.

Suppose, for example, we want CA1 to enable an interrupt when there is a high-to-low transition, with CA2 and CB1 not used and CB2 enabled and used for a set/reset output. The control register format to meet this specification is, for CA:

B0 is 1 to enable interrupt on CA1.
B1 is 0 so that the interrupt flag IRQA1 is set by a high-to-low transition on CA1.

Figure 13.12 Interfacing with a PIA.

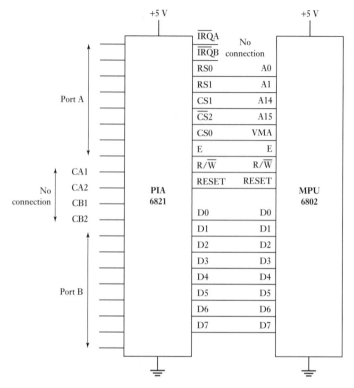

B2 is 1 to give access to the data register.
B3, B4, B5 are 0 because CA2 is disabled.
B6, B7 are read-only flags and thus a 0 or 1 may be used.

Hence the format for CA1 can be 00000101 which is 05 in hexadecimal notation. The control register format for CB2 is:

B0 is 0 to disable CB1.
B1 may be 0 or 1 since CB1 is disabled.
B2 is 1 to give access to the data register.
B3 is 0, B4 is 1 and B5 is 1, to select the set/reset.
B6, B7 are read-only flags and thus a 0 or 1 may be used.

Hence the format for CA1 can be 00110100 which is 34 in hexadecimal notation. The initialisation program might then read:

```
INIT    LDAA    #$00     ; Load zeros
        STAA    $2000    ; Make side A input port
        LDAA    #$FF     ; Load ones
        STAA    $2000    ; Make side B output port
        LDAA    #$05     ; Load the required control register format
        STAA    $2000    ; Select port A data register
        LDAA    #$34     ; Load the required control register format
        STAA    $2002    ; Select port B data register
```

13.4.3 An example of interfacing with a PIA

As an example of interfacing with a PIA, Figure 13.13 shows a circuit that can be used with a unipolar stepper motor (see Section 9.7.2). The inductive windings can generate a large back e.m.f. when switched so some way of isolating the windings from the PIA is required. Optoisolators, diodes or resistors might be used. Diodes give a cheap and simple interface, resistors not completely isolating the PIA.

Figure 13.13 Interfacing a stepper.

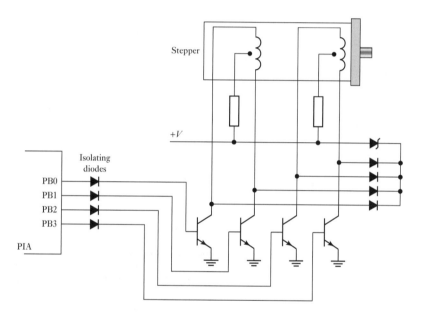

13.5 Serial communications interface

The **universal asynchronous receiver/transmitter** (UART) is the essential element of a serial communication system, the function being to change serial data to parallel for input and parallel data to serial for output. A common programmable form of a UART is the **asynchronous communications interface adapter** (ACIA) from Motorola MC6850; Figure 13.14 shows a block diagram of the constituent elements.

Data flow between the microprocessor and the ACIA is via eight bidirectional lines D0 to D7. The direction of the data flow is controlled by the microprocessor through the read/write input to the ACIA. The three chip select lines are used for addressing a particular ACIA. The register select line is used to select particular registers within the ACIA; if the register select line is high then the data transmit and data receive registers are selected, if low then the control and status registers are selected. The status register contains information on the status of serial data transfers as they occur and is used to read data carrier detect and clear-to-send lines. The control register is initially used to reset the ACIA and subsequently to define the serial data transfer rate and data format.

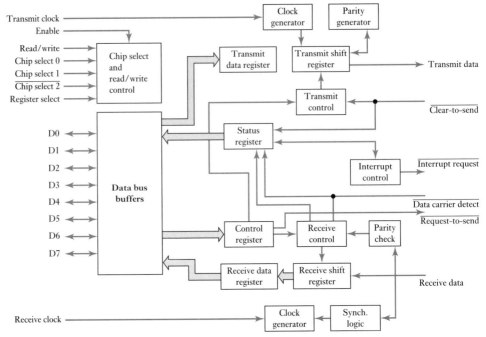

Figure 13.14   MC6850 ACIA.

The peripheral side of the ACIA includes two serial data lines and three control lines. Data is sent by the transmit data line and received by the receive data line. Control signals are provided by clear-to-send, data carrier detect and request-to-send. Figure 13.15 shows the bit formats of the control and Figure 13.16 the status registers.

Asynchronous serial data transfer is generally used for communications between two computers, with or without a modem, or a computer and a printer (see Chapter 15 for further discussion).

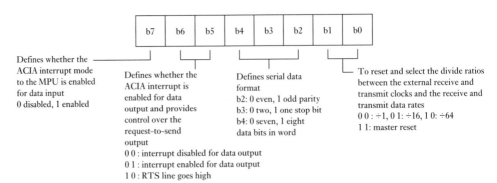

Figure 13.15   Control register.

Figure 13.16 Status register.

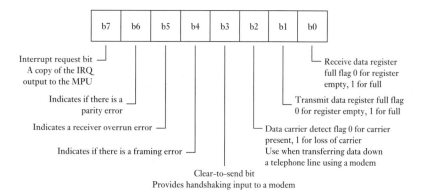

13.5.1 Serial interfaces of microcontrollers

Many microcontrollers have serial interfaces, i.e. built-in UARTs. For example, the Motorola M68HC11 has a serial peripheral interface (SPI), a synchronous interface, and a serial communication interface (SCI), an asynchronous interface (see Figure 10.10). The SPI requires the same clock signal to be used by the microcontroller and the externally connected device or devices (Figure 13.17(a)). The SPI allows several microcontrollers, with this facility, to be interconnected. The SCI is an asynchronous interface and so allows different clock signals to be used by the SCI system and the externally connected device (Figure 13.17(b)). General-purpose microprocessors do not have an SCI so a UART, e.g. Motorola MC6850, has then to be used to enable serial communication to take place. In some situations more than one SCI is required and thus the microcontroller M68HC11 requires supplementing with a UART.

Figure 13.17 (a) SPI, (b) SCI.

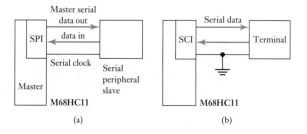

The SPI is initialised by bits in the SPI control register (SPCR) and the port D data direction control register (DDRD). The SPI status register contains status and error bits. The SCI is initialised by using the SCI control register 1, the SCI control register 2 and the baud rate control register. Status flags are in the SCI status register.

The Intel 8051 has a built-in serial interface with four modes of operation, these being selected by writing ones or zeros into SM0 and SM1 bits in the SCON (serial port control) register at address 98H (Figure 13.18) (Table 13.1).

In mode 0, serial data enters and leaves by RXD. Pin TXD outputs the shift clock and this is then used to synchronise the data transmission and

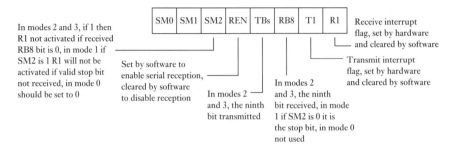

Figure 13.18   SCON register.

Table 13.1   Intel 8051 serial port modes.

SMO	SM1	Mode	Description	Baud rate
0	0	0	Shift register	Osc. freq./12
0	1	1	8-bit UART	Variable
1	0	2	9-bit UART	Osc. freq./12 or 64
1	1	3	9-bit UART	Variable

reception. Reception is initiated when REN is 1 and R1 is 0. Transmission is initiated when any data is written to SBUF, this being the serial port buffer at address 99H. In mode 1, 10 bits are transmitted on TXD or received on RXD; these are the start bit of 0, eight data bits and a stop bit of 1. Transmission is initiated by writing to SBUF and reception initiated by a 1 to 0 transition on RXD. In modes 2 and 3, 11 bits are transmitted on TXD or received on RXD.

PIC microcontrollers have an SPI (see Figure 10.30) which can be used for synchronous serial communications. When data is written to the SSBUF register it is shifted out of the SDO pin in synchronism with a clock signal on SCK and outputted through pin RC5 as a serial signal with the most significant bit appearing first and a clock signal through RC3. Input into the SSBUF register is via RC4. Many PIC microcontrollers also have a UART to create a serial interface for use with asynchronously transmitted serial data. When transmitting, each 8-bit byte is framed by a START bit and a STOP bit. When the START bit is transmitted, the RX line drops to a low and the receiver can then synchronise on this high-to-low transition. The receiver then reads the 8 bits of serial data.

13.6 Examples of interfacing

The following are examples of interfacing.

13.6.1   Interfacing a seven-segment display with a decoder

Consider where the output from a microcontroller is used to drive a seven-segment LED display unit (see Section 6.5). A single LED is an on/off indicator and thus the display number indicated will depend on which LEDs are on. Figure 13.19 shows how we can use a microcontroller to drive a

Figure 13.19 Driving a display.

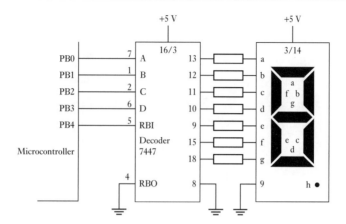

common anode display using a decoder driver, this being able to take in a BCD input and convert it to the appropriate code for the display.

For the 7447 decoder, pins 7, 1, 2 and 6 are the input pins of the decoder for the BCD input with pins 13, 12, 11, 10, 9, 15 and 14 being the outputs for the segments of the display. Pin 9 of the display is the decimal point. Table 13.2 shows the input and output signals for the decoder.

Table 13.2 The 7447 BCD decoder for a seven-segment display.

	Input pins				Output pins						
Display	6	2	1	7	13	12	11	10	9	15	14
0	L	L	L	L	ON	ON	ON	ON	ON	ON	OFF
1	L	L	L	H	OFF	ON	ON	OFF	OFF	OFF	OFF
2	L	L	H	L	ON	ON	OFF	ON	ON	OFF	ON
3	L	L	H	H	ON	ON	ON	ON	OFF	OFF	ON
4	L	H	L	L	OFF	ON	ON	OFF	OFF	ON	ON
5	L	H	H	L	ON	OFF	ON	ON	OFF	ON	ON
6	L	H	H	L	OFF	OFF	ON	ON	ON	ON	ON
7	L	H	H	H	ON	ON	ON	OFF	OFF	OFF	OFF
8	H	L	H	H	ON	ON	ON	OFF	OFF	OFF	OFF
9	H	L	H	L	ON	ON	ON	OFF	OFF	OFF	OFF

Blanking is when none of the segments are lit. This is used to prevent a leading 0 occurring when we have, say, three display units and want to display just 10 rather than 010 and so blank out the leading 0 and prevent it being illuminated. This is achieved by the ripple blanking input (RBI) being set low. When RBI is low and the BCD inputs A, B, C and D are low then the output is blanked. If the input is not zero the ripple blanking output (RBO) is high regardless of the RBI status. The RBO of the first digit in the display can be connected to the RBI of the second digit and the RBO of the second connected to the RBI of the third digit, thus allowing only the final 0 to be blanked (Figure 13.20).

Figure 13.20 Ripple blanking.

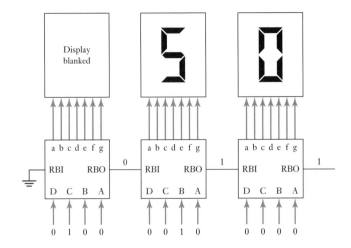

With displays having many display elements, rather than use a decoder for each element, multiplexing is used with a single decoder. Figure 13.21 shows the circuit for multiplexing a four-element common cathode type of display. The BCD data is outputted from port A and the decoder presents the decoder output to all the displays. Each display has its common cathode connected to ground through a transistor. The display cannot light up unless the transistor is switched on by an output from port B. Thus by switching between PB0, PB1, PB2 and PB3 the output from port A can be switched to the appropriate display. To maintain a constant display, a display is repeatedly turned on sufficiently often for the display to appear flicker-free.

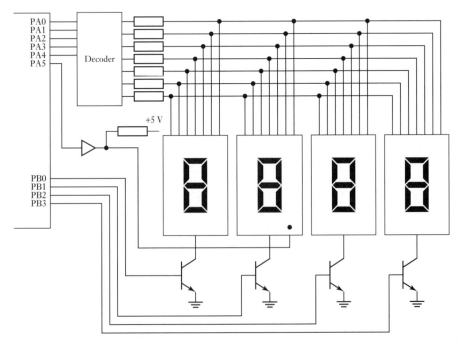

Figure 13.21 Multiplexing four displays.

Time division multiplexing can be used to enable more than one digit to be displayed at a time.

13.6.2 Interfacing analogue

Digital-to-analogue conversion is required when the output from a microprocessor or microcontroller is required to provide an analogue signal output. For example, the DAC Analog Devices AD557 can be used for this purpose. It produces an output voltage proportional to its digital input and input latches are available for microprocessor interfacing. If the latches are not required then pins 9 and 10 are connected to ground. Data is latched when there is a positive edge, i.e. a change from low to high, on the input to either pin 9 or pin 10. The data is then held until both these pins return to the low level. When this happens, the data is transferred from the latch to the DAC for conversion to an analogue voltage.

Figure 13.22 Waveform generation.

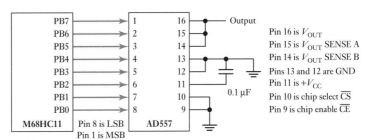

Figure 13.22 shows the AD557 with the latch not being used and connected to a Motorola M68HC11 so that, when the following program is run, it generates a sawtooth voltage output. Other voltage waveforms can readily be generated by changing the program:

```
BASE      EQU $1000        ; Base address of I/O registers
PORTB     EQU $04          ; Offset of PORTB from BASE

          ORG $C000
          LDX #BASE        ; Point X to register base
          CLR PORTB,X      ; Send 0 to the DAC
AGAIN     INC PORTB,X      ; Increment by 1
          BRA AGAIN        ; Repeat
          END
```

Summary

Interface requirements often mean electrical buffering/isolation, timing control, code conversion, changing the number of lines, serial to parallel and vice versa, conversion from analogue to digital and vice versa. Unless two devices can send and receive data at identical rates, **handshaking** is necessary.

Polling is the program control of inputs/outputs in which a loop is used continuously to read inputs and update outputs, with jumps to service routines as required, i.e. a process of repeatedly checking each peripheral device to see if it is ready to send or accept a new byte of data. An alternative to program control is **interrupt control**. An interrupt involves a peripheral device activating a separate interrupt request line. The reception of an interrupt results in the microprocessor suspending execution of its main program and jumping to the service routine for the peripheral. After the interrupt service routine, the contents of the memory are restored and the microprocessor can continue executing the main program from where it was interrupted.

There are two basic types of serial data transfer: asynchronous and synchronous. With **asynchronous transmission**, the receiver and the transmitter each use their own clock signals so it is not possible for a receiver to know when a word starts or stops. Thus it is necessary for each transmitted data word to carry its own start and stop bits so that it is possible for the receiver to tell where one word stops and another starts. With **synchronous transmission**, the transmitter and receiver have a common clock signal and thus transmission and reception can be synchronised.

Peripheral interface adapters (PIAs) are programmable input/output interface devices which permit various different input and output options to be selected by means of software.

The **universal asynchronous receiver/transmitter** (UART) is the essential element of a serial communication system, the function being to change serial data to parallel for input and parallel data to serial for output. A common programmable form of a UART is the **asynchronous communications interface adapter** (ACIA).

Problems

13.1   Describe the functions that can be required of an interface.

13.2   Explain the difference between a parallel and a serial interface.

13.3   Explain what is meant by a memory-mapped system for inputs/outputs.

13.4   What is the function of a peripheral interface adapter?

13.5   Describe the architecture of the Motorola MC6821 PIA.

13.6   Explain the function of an initialisation program for a PIA.

13.7   What are the advantages of using external interrupts rather than software polling as a means of communication with peripherals?

13.8   For a Motorola MC6821 PIA, what value should be stored in the control register if CA1 is to be disabled, CB1 is to be an enabled interrupt input and set by a low-to-high transition, CA2 is to be enabled and used as a set/reset output, and CB2 is to be enabled and go low on the first low-to-high E transition following a microprocessor write into peripheral data register B and return high by the next low-to-high E transition?

13.9   Write, in assembly language, a program to initialise the Motorola MC6821 PIA to achieve the specification given in problem 13.8.

13.10   Write, in assembly language, a program to initialise the Motorola MC6821 PIA to read 8 bits of data from port A.

Chapter fourteen

Programmable logic controllers

Objectives

The objectives of this chapter are that, after studying it, the reader should be able to:
- Describe the basic structure of PLCs and their operation.
- Develop ladder programs for a PLC involving logic functions, latching, internal relays and sequencing.
- Develop programs involving timers, counters, shift registers, master relays, jumps and data handling.

14.1 Programmable logic controller

A **programmable logic controller** (PLC) is a digital electronic device that uses a programmable memory to store instructions and to implement functions such as logic, sequencing, timing, counting and arithmetic in order to control machines and processes and has been specifically designed to make programming easy. The term logic is used because the programming is primarily concerned with implementing logic and switching operations. Input devices, e.g. switches, and output devices, e.g. motors, being controlled are connected to the PLC and then the controller monitors the inputs and outputs according to the program stored in the PLC by the operator and so controls the machine or process. Originally PLCs were designed as a replacement for hard-wired relay (e.g. Figure 9.2) and timer logic control systems. PLCs have the great advantage that it is possible to modify a control system without having to rewire the connections to the input and output devices, the only requirement being that an operator has to key in a different set of instructions. Also they are much faster than relay-operated systems. The result is a flexible system which can be used to control systems which vary quite widely in their nature and complexity. Such systems are widely used for the implementation of logic control functions because they are easy to use and program.

PLCs are similar to computers but have certain features which are specific to their use as controllers. These are:

1 they are rugged and designed to withstand vibrations, temperature, humidity and noise;
2 the interfacing for inputs and outputs is inside the controller;
3 they are easily programmed.

14.2 Basic PLC structure

Figure 14.1 shows the basic internal structure of a PLC. It consists essentially of a central processing unit (CPU), memory and input/output interfaces. The CPU controls and processes all the operations within the PLC. It is

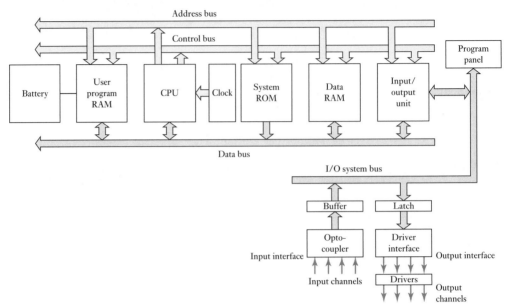

Figure 14.1   Architecture of a PLC.

supplied with a clock with a frequency of typically between 1 and 8 MHz. This frequency determines the operating speed of the PLC and provides the timing and synchronisation for all elements in the system. A bus system carries information and data to and from the CPU, memory and input/output units. There are several memory elements: a system ROM to give permanent storage for the operating system and fixed data, RAM for the user's program and temporary buffer stores for the input/output channels.

14.2.1   Input/output

The input and output units provide the interface between the system and the outside world and are where the processor receives information from external devices and communicates information to external devices. The input/output interfaces provide isolation and signal conditioning functions so that sensors and actuators can often be directly connected to them without the need for other circuitry. Inputs might be from limit switches which are activated when some event occurs, or other sensors such as temperature sensors, or flow sensors. The outputs might be to motor starter coils, solenoid valves, etc. Electrical isolation from the external world is usually by means of optoisolators (see Section 3.3).

Figure 14.2 shows the basic form of an input channel. The digital signal that is generally compatible with the microprocessor in the PLC is 5 V d.c. However, signal conditioning in the input channel, with isolation, enables a wide range of input signals to be supplied to it. Thus, with a larger PLC we might have possible input voltages of 5 V, 24 V, 110 V and 240 V. A small PLC is likely to have just one form of input, e.g. 24 V.

The output to the output unit will be digital with a level of 5 V. Outputs are specified as being of relay type, transistor type or triac type. With the relay type, the signal from the PLC output is used internally to operate a relay and so is able to switch currents of the order of a few amperes in an external

Figure 14.2 Input channel.

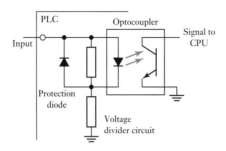

circuit. The relay isolates the PLC from the external circuit and can be used for both d.c. and a.c. switching. Relays are, however, relatively slow to operate. The transistor type of output uses a transistor to switch current through the external circuit. This gives a faster switching action. Optoisolators are used with transistor switches to provide isolation between the external circuit and the PLC. The transistor output is only for d.c. switching. Triac outputs can be used to control external loads which are connected to the a.c. power supply. Optoisolators are again used to provide isolation. Thus we can have outputs from the output channel which might be a 24 V, 100 mA switching signal, a d.c. voltage of 110 V, 1 A or perhaps 240 V, 1 A a.c., or 240 V, 2 A a.c., from a triac output channel. With a small PLC, all the outputs might be of one type, e.g. 240 V a.c., 1 A. With modular PLCs, however, a range of outputs can be accommodated by selection of the modules to be used.

The terms **sourcing** and **sinking** are used to describe the way in which d.c. devices are connected to a PLC. With sourcing, using the conventional current flow direction as from positive to negative, an input device receives current from the input module (Figure 14.3(a)). If the current flows from the output module to an output load then the output module is referred to as sourcing (Figure 14.3(b)). With sinking, an input device supplies current to the input module (Figure 14.3(c)). If the current flows to the output module from an output load then the output module is referred to as sinking (Figure 14.3(d)).

The input/output unit provides the interface between the system and the outside world, allowing for connections to be made through input/output channels to input devices such as sensors and output devices such as motors and solenoids. It is also through the input/output unit that programs are entered from a program panel. Every input/output point has a unique address which

Figure 14.3 (a), (b) Sourcing, (c), (d) sinking.

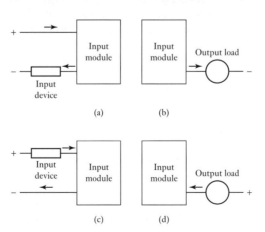

can be used by the CPU. It is like a row of houses along a road: number 10 might be the 'house' to be used for an input from a particular sensor while number '45' might be the 'house' to be used for the output to a particular motor.

14.2.2 Inputting programs

Programs are entered into the input/output unit from small hand-held programming devices, desktop consoles with a visual display unit (VDU), keyboard and screen display, or by means of a link to a personal computer (PC) which is loaded with an appropriate software package. Only when the program has been designed on the programming device and is ready is it transferred to the memory unit of the PLC.

The programs in RAM can be changed by the user. However, to prevent the loss of these programs when the power supply is switched off, a battery is likely to be used in the PLC to maintain the RAM contents for a period of time. After a program has been developed in RAM it may be loaded into an EPROM chip and so made permanent. Specifications for small PLCs often specify the program memory size in terms of the number of program steps that can be stored. A program step is an instruction for some event to occur. A program task might consist of a number of steps and could be, for example: examine the state of switch A, examine the state of switch B, if A and B are closed then energise solenoid P which then might result in the operation of some actuator. When this happens another task might then be started. Typically the number of steps that can be handled by a small PLC is of the order of 300 to 1000, which is generally adequate for most control situations.

14.2.3 Forms of PLCs

PLCs were first conceived in 1968. They are now widely used and extend from small self-contained units, i.e. single boxes, for use with perhaps 20 digital input/outputs to rack-mounted systems which can be used for large numbers of inputs/outputs, handle digital or analogue inputs/outputs, and also carry out proportional plus integral plus derivative (PID) control modes. The single-box type is commonly used for small programmable controllers and is supplied as an integral compact package complete with power supply, processor, memory and input/output units. Typically such a PLC might have 6, 8, 12 or 24 inputs and 4, 8 or 16 outputs and a memory which can store some 300 to 1000 instructions. For example, the MELSEC FX3U has models which can have 6, 8, 12 or 24 inputs and 4, 8 or 16 relay outputs and a memory which can store some 300 to 1000 instructions. Some systems are able to be extended to cope with more inputs and outputs by linking input/output boxes to them.

Systems with larger numbers of inputs and outputs are likely to be modular and designed to fit in racks. These consist of separate modules for power supply, processor, input/output, etc., and are mounted on rails within a metal cabinet. The rack type can be used for all sizes of programmable controllers and has the various functional units packaged in individual modules which can be plugged into sockets in a base rack. The mix of modules required for a particular purpose is decided by the user and the appropriate ones then plugged into the rack. So the number of input/output connections can be increased by just adding more input/output modules. For example, the SIMATIC S7-300/400 PLC is rack mounted with components for the power supply, the

CPU, input/output interface modules, signal modules which can be used to provide signal conditioning for inputs or outputs and communication modules which can be used to connect PLCs to each other or to other systems.

Another example of a modular system is that provided by the Allen-Bradley SLC-500 programmable logic controller system. This is a small, chassis-based, modular family of programmable controllers having multiple processor choices, numerous power supply options and extensive input/output capacity. The SLC 500 allows the creation of a system specifically designed for an application. PLC blocks are mounted in a rack, with interconnections between the blocks being via a backplane bus. The PLC power supply is the end box in a rack with the next box containing the microprocessor. The backplane bus has copper conductors and provides the means by which the blocks slotted into the rack receive electrical power and for exchanging data between the modules and the processor. The modules slide into the rack and engage connectors on the backplane. SLC 500 series PLC racks are available to take 4, 7, 10 or 13 modules. Modules are available providing 8, 16 or current sinking d.c. inputs, 8, 16 or 32 current sourcing d.c. outputs, 8, 16 or 32 current sourcing d.c. outputs, 4, 8 or 16 relay a.c./d.c. outputs, communication module to enable additional communications with other computers or PLCs. Software is available to allow for programming from a Windows environment.

14.3 Input/output processing

A PLC is continuously running through its program and updating it as a result of the input signals. Each such loop is termed a **cycle**. There are two methods that can be used for input/output processing: continuous updating and mass input/output copying.

14.3.1 Continuous updating

Continuous updating involves the CPU scanning the input channels as they occur in the program instructions. Each input point is examined individually and its effect on the program determined. There will be a built-in delay, typically about 3 ms, when each input is examined in order to ensure that only valid input signals are read by the microprocessor. This delay enables the microprocessor to avoid counting an input signal twice, or, more frequently, if there is contact bounce at a switch. A number of inputs may have to be scanned, each with a 3 ms delay, before the program has the instruction for a logic operation to be executed and an output to occur. The outputs are latched so that they retain their status until the next updating.

14.3.2 Mass input/output copying

Because, with continuous updating, there has to be a 3 ms delay on each input, the time taken to examine several hundred input/output points can become comparatively long. To allow a more rapid execution of a program, a specific area of RAM is used as a buffer store between the control logic and the input/output unit. Each input/output has an address in this memory. At the start of each program cycle the CPU scans all the inputs and copies their status into the input/output addresses in RAM. As the program is executed the stored, input data is read, as required, from RAM and the logic operations carried out. The resulting output signals are stored in the reserved input/output section of RAM. At the end of each program cycle all the outputs are transferred from RAM to the output channels. The outputs are latched so that they retain their status until the next updating. The sequence is:

1 scan all the inputs and copy into RAM;
2 fetch and decode and execute all program instructions in sequence, copying output instructions to RAM;
3 update all outputs;
4 repeat the sequence.

A PLC takes time to complete a cycle of scanning inputs and updating outputs according to the program instructions and so the inputs are not watched all the time but only examined periodically. A typical PLC cycle time is of the order of 10 to 50 ms and so the inputs and outputs are updated every 10 to 50 ms. This means that if a very brief input appears at the wrong moment in the cycle, it could be missed. Thus for a PLC with a cycle time of 40 ms, the maximum frequency of digital impulses that can be detected will be if one pulse occurs every 40 ms. The Mitsubishi compact PLC, MELSEC FX3U, has a quoted program cycle time of 0.065 μs per logical instruction and so the more complex the program, the longer the cycle time.

14.3.3 Input/output addresses

The PLC has to be able to identify each particular input and output and it does this by assigning addresses to each, rather like houses in a town have addresses to enable post to be delivered to the right families. With a small PLC the addresses are likely to be just a number preceded by a letter to indicate whether it is an input or output. For example, Mitsubishi and Toshiba have inputs identified as X400, X401, X402, etc., and outputs as Y430, Y431, etc. With larger PLCs having several racks of input and output channels and a number of modules in each rack, the racks and modules are numbered and so an input or output is identified by its rack number followed by the number of the module in that rack and then a number to show its terminal number in the module. For example, the Allen-Bradley PCL-5 has I:012/03 to indicate an input in rack 01 at module 2 and terminal 03.

14.4   Ladder programming

The form of programming commonly used with PLCs is **ladder programming**. This involves each program task being specified as though a rung of a ladder. Thus such a rung could specify that the state of switches A and B, the inputs, be examined and if A and B are both closed then a solenoid, the output, is energised. Figure 14.4 illustrates this idea by comparing it with an electric circuit.

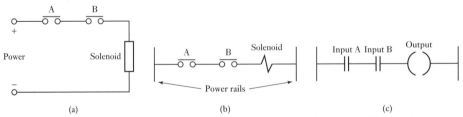

Figure 14.4   (a), (b) Alternative ways of drawing an electric circuit, (c) comparable rung in a ladder program.

The sequence followed by a PLC when carrying out a program can be summarised as follows.

1 Scan the inputs associated with one rung of the ladder program.
2 Solve the logic operation involving those inputs.
3 Set/reset the outputs for that rung.
4 Move on to the next rung and repeat operations 1, 2, 3.
5 Move on to the next rung and repeat operations 1, 2, 3.
6 Move on to the next rung and repeat operations 1, 2, 3.
7 And so on until the end of the program with each rung of the ladder program scanned in turn. The PLC then goes back to the beginning of the program and starts again.

PLC programming based on the use of **ladder diagrams** involves writing a program in a similar manner to drawing a switching circuit. The ladder diagram consists of two vertical lines representing the power rails. Circuits are connected as horizontal lines, i.e. the rungs of the ladder, between these two verticals. Figure 14.5 shows the basic standard symbols that are used and an example of rungs in a ladder diagram. In drawing the circuit line for a rung, inputs must always precede outputs and there must be at least one output on each line. Each rung must start with an input or a series of inputs and end with an output.

Figure 14.5 Ladder program.

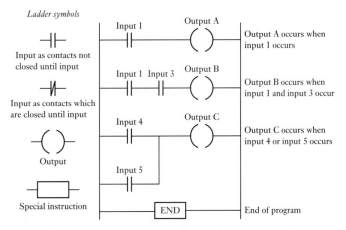

To illustrate the drawing of a ladder diagram, consider a situation where the output from the PLC is to energise a solenoid when a normally open start switch connected to the input is activated by being closed (Figure 14.6(a)). The program required is shown in Figure 14.6(b)). Starting with the input, we have the normally open symbol | |. This might have an input address X400. The line terminates with the output, the solenoid, with the symbol ( ). This might have the output address Y430. To indicate the end of the program, the end rung is marked. When the switch is closed the solenoid is activated. This might, for example, be a solenoid valve which opens to allow water to enter a vessel.

Figure 14.6 Switch controlling a solenoid.

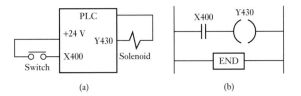

Another example might be an on/off temperature control (Figure 14.7(a)) in which the input goes from low to high when the temperature sensor reaches the set temperature. The output is then to go from on to off. The temperature sensor shown in the figure is a thermistor connected in a bridge arrangement with output to an operational amplifier connected as a comparator (see Section 3.2.7). The program (Figure 14.7(b)) shows the input as a normally closed pair of contacts, so giving the on signal and hence an output. When the contacts are opened to give the off signal then the output is switched off.

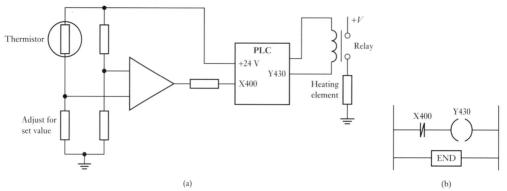

(a)   (b)

Figure 14.7   Temperature control system.

14.4.1 Logic functions

The logic functions can be obtained by combinations of switches (see Section 5.2) and the following shows how we can write ladder programs for such combinations (Figure 14.8).

1 AND
   Figure 14.8(a) shows a situation where a coil is not energised unless two, normally open, switches are both closed. Switch A and switch B have both to be closed, which thus gives an AND logic situation. The equivalent ladder diagram starts with | |, labelled Input 1, to represent switch A and in series with it | |, labelled Input 2, to represent switch B. The line then terminates with ( ) to represent the output.

2 OR
   Figure 14.8(b) shows a situation where a coil is not energised until either, normally open, switch A or B is closed. The situation is an OR logic gate. The equivalent ladder diagram starts with | |, labelled Input 1, to represent switch A and in parallel with it | |, labelled Input 2, to represent switch B. The line then terminates with ( ) to represent the output.

3 NOR
   Figure 14.8(c) shows how we can represent the ladder program line for a NOR gate. Since there has to be an output when neither A nor B have an input, and when there is an input to A or B the output ceases, the ladder program shows Input 1 in parallel with Input 2, with both being represented by normally closed contacts.

4 NAND
   Figure 14.8(d) shows a NAND gate. There is no output when both A and B have an input. Thus for the ladder program line to obtain an output we require no inputs to Input 1 and to Input 2.

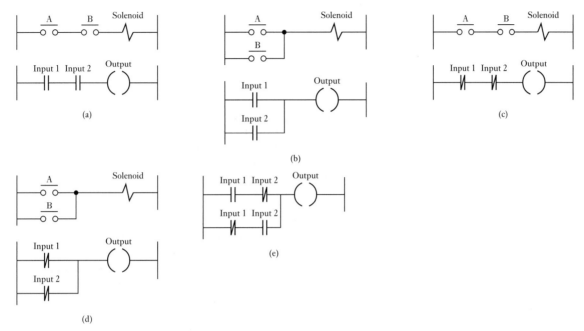

Figure 14.8 (a) AND, (b) OR, (c) NOR, (d) NAND, (e) XOR.

5 EXCLUSIVE-OR (XOR)

Figure 14.8(e) shows how we can draw the ladder program line for an XOR gate, there being no output when there is no input to Input 1 and Input 2 and when there is an input to both Input 1 and Input 2. Note that we have represented each input by two sets of contacts, one normally open and one normally closed.

Consider a situation where a normally open switch A must be activated and either of two other, normally open, switches B and C must be activated for a coil to be energised. We can represent this arrangement of switches as switch A in series with two parallel switches B and C (Figure 14.9(a)). For the coil to be energised we require A to be closed and either B or C to be closed. Switch A when considered with the parallel switches gives an AND logic situation. The two parallel switches give an OR logic situation. We thus have a combination of two gates. The truth table is:

Inputs			
A	B	C	Output
0	0	0	0
0	0	1	0
0	1	0	0
0	1	1	0
1	0	0	0
1	0	1	1
1	1	0	1
1	1	1	1

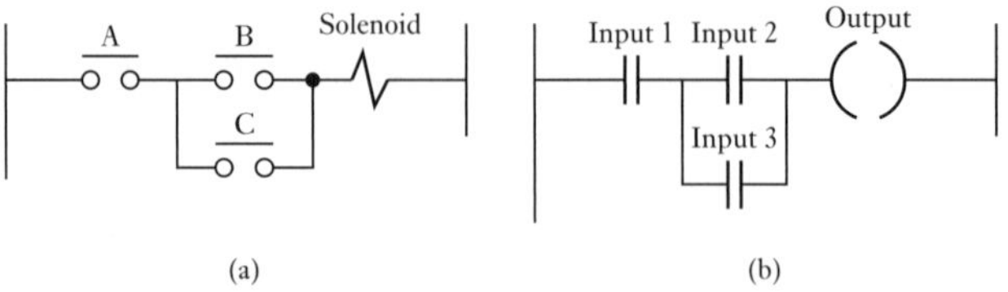

Figure 14.9 Switches controlling a solenoid.

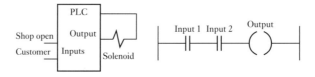

(a)　　　　　　　　　　　　　　　　(b)

For the ladder diagram, we start with | |, labelled Input 1, to represent switch A. This is in series with two | | in parallel, labelled Input 1 and Input 2, for switches B and C. The line then terminates with ( ) to represent the output, the coil. Figure 14.9(b) shows the line.

As a simple example of a program using logic gates, consider the requirement for there to be an output to the solenoid controlling the valve that will open a shop door when the shopkeeper has closed a switch to open the shop and a customer approaches the door and is detected by a sensor which then gives a high signal. The truth table for this system is thus:

Shop open switch	Customer approaching sensor	Solenoid output
Off	Off	Off
Off	On	Off
On	Off	Off
On	On	On

This truth table is that of an AND gate and thus the program for a PLC controlling the door is as shown in Figure 14.10.

Figure 14.10 Shop door system.

14.5　Instruction lists

Each horizontal rung on the ladder in a ladder program represents a line in the program and the entire ladder gives the complete program in 'ladder language'. The programmer can enter the program into the PLC using a keyboard with the graphic symbols for the ladder elements, or using a computer screen and a mouse to select symbols, and the program panel or computer then translates these symbols into machine language that can be stored in the PLC memory. There is an alternative way of entering a program and that is to translate the ladder program into an **instruction list** and then enter this into the programming panel or computer.

Instruction lists consist of a series of instructions with each instruction being on a separate line. An instruction consists of an operator followed by one or more operands, i.e. the subjects of the operator. In terms of ladder programs, each operator in a program may be regarded as a ladder element. Thus we might have for the equivalent of an input to a ladder program:

LD A (*Load input A*)

Table 14.1 Instruction code mnemonics.

IEC 1131-3	Mitsubishi	OMRON	Siemens	Operation	Ladder diagram
LD	LD	LD	A	Load operand into result register	Start a rung with open contacts
LDN	LDI	LD NOT	AN	Load negative operand into result register	Start a rung with closed contacts
AND	AND	AND	A	Boolean AND	A series element with open contacts
ANDN	ANI	AND NOT	AN	Boolean AND with negative operand	A series element with closed contacts
OR	OR	OR	O	Boolean OR	A parallel element with open contacts
ORN	ORI	OR NOT	ON	Boolean OR with negative operand	A parallel element with closed contacts
ST	OUT	OUT	=	Store result register into operand	An output from a rung

The operator is LD for loading, the operand A as the subject being loaded and the words preceded by and concluded by * in brackets are comments explaining what the operation is and are not part of the program operation instructions to the PLC, but to aid a reader in comprehending what the program is about.

The mnemonic codes used by different PLC manufacturers differ but an international standard (IEC 1131-3) has been proposed and is widely used. Table 14.1 shows common core mnemonics. In examples discussed in the rest of this chapter, where general descriptions are not used, the Mitsubishi mnemonics will be used. However, those used by other manufacturers do not differ widely from these and the principles involved in their use are the same.

14.5.1 Instruction lists and logic functions

The following shows how individual rungs on a ladder are entered using the Mitsubishi mnemonics where logic functions are involved (Figure 14.11).

14.5.2 Instruction lists and branching

The EXCLUSIVE-OR (XOR) gate shown in Figure 14.12 has two parallel arms with an AND situation in each arm. In such a situation Mitsubishi (Figure 14.12(a)) uses an ORB instruction to indicate 'OR together parallel branches'. The first instruction is for a normally open pair of contacts X400, the next instruction for a set of normally closed contacts X401, hence ANI X401. The third instruction describes a new line, its being recognised as a new line because it starts with LDI, all new lines starting with LD or LDI. Because the first line has not been ended by an output, the PLC recognises that a parallel line is involved for the second line and reads together the listed

Figure 14.11 (a) AND, (b) OR, (c) NOR, (d) NAND.

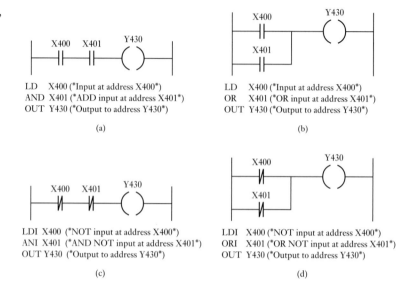

LD   X400 (*Input at address X400*)
AND X401 (*ADD input at address X401*)
OUT Y430 (*Output to address Y430*)

(a)

LD   X400 (*Input at address X400*)
OR   X401 (*OR input at address X401*)
OUT Y430 (*Output to address Y430*)

(b)

LDI X400 (*NOT input at address X400*)
ANI X401 (*AND NOT input at address X401*)
OUT Y430 (*Output to address Y430*)

(c)

LDI X400 (*NOT input at address X400*)
ORI X401 (*OR NOT input at address X401*)
OUT Y430 (*Output to address Y430*)

(d)

Figure 14.12 XOR.

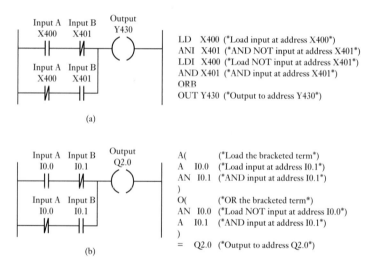

LD   X400 (*Load input at address X400*)
ANI X401 (*AND NOT input at address X401*)
LDI X400 (*Load NOT input at address X401*)
AND X401 (*AND input at address X401*)
ORB
OUT Y430 (*Output to address Y430*)

(a)

A((*Load the bracketed term*)
A   I0.0 (*Load input at address I0.1*)
AN   I0.1 (*AND input at address I0.1*)
)
O((*OR the bracketed term*)
AN   I0.0 (*Load NOT input at address I0.0*)
A   I0.1 (*AND input at address I0.1*)
)
=   Q2.0 (*Output to address Q2.0*)

(b)

elements until the ORB instruction is reached. ORB indicates to the PLC that it should OR the results of the first and second instructions with that of the new branch with the third and fourth instructions. The list concludes with the output OUT Y430. Figure 14.12(b) shows the Siemens version of XOR gate. Brackets are used to indicate that certain instructions are to be carried out as a block and are used in the same way as brackets in any mathematical equation. For example, $(1 + 2)/4$ means that the 1 and 2 must be added before dividing by 4. Thus with the Siemens instruction list the A(means that the load instruction A is only applied after the bracketed steps have been completed and) is reached. The IEC 1131-3 standard for such programming is to use brackets in the way used in the above Siemens example.

14.6 Latching and internal relays

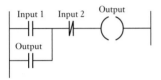

Figure 14.13 A latch circuit.

There are often situations where it is necessary to hold a coil energised, even when the input which energised it ceases. The term **latch circuit** is used for the circuit which carries out such an operation. It is a self-maintaining circuit in that, after being energised, it maintains that state until another input is received. It remembers its last state. An example of a latch circuit is shown in Figure 14.13. When Input 1 is energised and closes, there is an output. However, when there is an output, a set of contacts associated with the output is energised and closes. These contacts OR the Input 1 contacts. Thus, even if Input 1 contacts open, the circuit will still maintain the output energised. The only way to release the output is by operating the normally closed contact Input 2.

As an example of the use of a latching circuit, consider the requirement for a PLC to control a motor so that when the start signal button is momentarily pressed the motor starts and when the stop switch is used the motor switches off. Safety must be a priority in the design of a PLC system, so stop buttons must be hard-wired and not depend on the PLC software for implementation so that if there is a failure of the stop switch or PLC, the system is automatically safe. With a PLC system, a stop signal can be provided by a switch as shown in Figure 14.14(a). To start we momentarily close the press-button start switch and the motor internal control relay latches this closure and the output remains on. To stop we momentarily open the stop switch and this unlatches the start switch. However, if the stop switch cannot be operated then we cannot stop the system. Thus this system must *not* be used as it is unsafe, because if there is a fault and the switch cannot be operated, then no stop signal can be provided. What we require is a system that will still stop if a failure occurs in the stop switch. Figure 14.14(b) shows such a system. The program now has the stop switch as open contacts. However, because the hard-wired stop switch has normally closed contacts, then the program receives the signal to close the program contacts. Pressing the stop switch then opens the program contacts and stops the system.

Figure 14.14 Stop system: (a) unsafe, (b) safe.

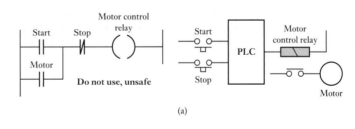

(a)

(b)

Figure 14.15 (a) An output controlled by two input arrangements, (b) starting of multiple outputs.

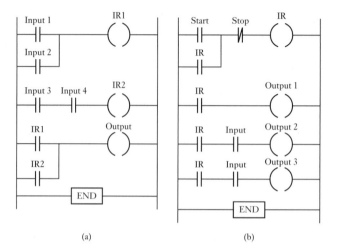

(a) (b)

14.6.1 Internal relays

The term **internal relay, auxiliary relay** or **marker** is used for what can be considered as an internal relay in the PLC. These behave like relays with their associated contacts, but in reality are not actual relays but simulations by the software of the PLC. Some have battery back-up so that they can be used in circuits to ensure a safe shut-down of plant in the event of a power failure. Internal relays can be very useful aids in the implementation of switching sequences.

Internal relays are often used when there are programs with multiple input conditions. Consider the situation where the excitation of an output depends on two different input arrangements. Figure 14.15(a) shows how we can draw a ladder diagram using internal relays. The first rung shows one input arrangement being used to control the coil of internal relay IR1. The second rung shows the other input arrangement controlling the coil of internal relay IR2. The contacts of the two relays are then put in an OR situation to control the output.

Another use of internal relays is for the starting of multiple outputs. Figure 14.15(b) shows such a ladder program. When the start contacts are closed, the internal relay is activated and latches the input. It also starts Output 1 and makes it possible for Outputs 2 and 3 to be activated.

Another example of the use of internal relays is resetting a latch. Figure 14.16 shows the ladder diagram. When the contacts of Input 1 are momentarily pressed, the output is energised. The contacts of the output are then closed and so latch the output, i.e. keep it on even when the contacts of the input are no longer closed. The output can be unlatched by the internal relay contacts opening. This will occur if Input 2 is closed and energises the coil of the internal relay.

An example of the use of a battery-backed internal relay is shown in Figure 14.17. When the contacts of Input 1 close, the coil of the battery-backed internal relay is energised. This closes the internal relay contacts and so even if contacts of the input open as a result of power failure, the internal relay contacts remain closed. This means that the output controlled by the internal relay remains energised, even when there is a power failure.

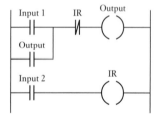

Figure 14.16 Resetting a latch.

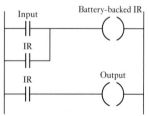

Figure 14.17 Use of a battery-backed internal relay.

14.7 Sequencing

There are often control situations where sequences of outputs are required, with the switch from one output to another being controlled by sensors. Consider the requirement for a ladder program for a pneumatic system (Figure 14.18) with double-solenoid valves controlling two double-acting cylinders A and B if limit switches a−, a+, b−, b+ are used to detect the limits of the piston rod movements in the cylinders and the cylinder activation sequence A+, B+, A−, B− is required. A possible program is shown in the figure. A start switch input has been included in the first rung. Thus cylinder extension for A, i.e. the solenoid A+ energised, only occurs when the start switch is closed and the b− switch is closed, this switch indicating that the B cylinder is retracted. When cylinder A is extended, the switch a+, which indicates the extension of A, is activated. This then leads to an output to solenoid B+ which results in B extending. This closes the switch indicating the extension of B, i.e. the b+ switch, and leads to the output to solenoid A− and the retraction of cylinder A. This retraction closes limit switch a− and so gives the output to solenoid B− which results in B retracting. This concludes the program cycle and leads to the first rung again, which awaits the closure of the start switch before being repeated.

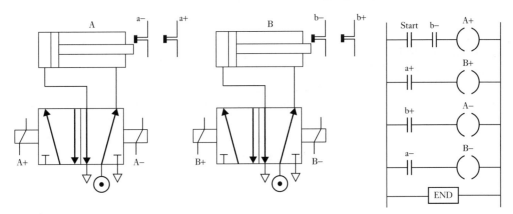

Figure 14.18 Cylinder sequencing.

As a further illustration, consider the problem of devising a ladder program to control a pneumatic system with double-solenoid-controlled valves and two cylinders A and B if limit switches a−, a+, b− and b+ are used to detect the limits of movement of the piston rod movements in the cylinders and the sequence required is for the piston rod in A to extend, followed by the piston rod in B extending, then the piston in B retracting and finally the cycle is completed by the piston in A retracting. An internal relay can be used to switch between groups of outputs to give the form of control for pneumatic cylinders, which is termed **cascade control** (see Section 7.5). Figure 14.19 shows a possible program. When the start switch is closed, the internal relay is activated. This energises solenoid A+ with the result that the piston in cylinder A extends. When extended it closes limit switch a+ and the piston in cylinder B extends. When this is extended it closes the limit switch b+. This activates the relay. As a result the B− solenoid is energised and the piston in B retracts. When this closes limit switch b−, solenoid A− is energised and the piston in cylinder A retracts.

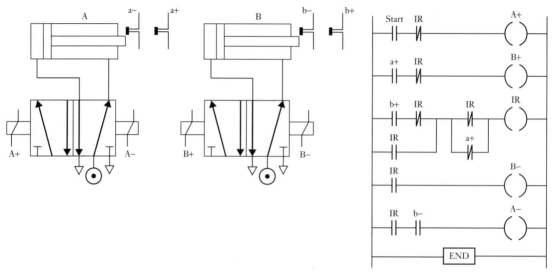

Figure 14.19   Cylinder sequencing.

14.8   Timers and counters

The previous sections in this chapter have been concerned with tasks requiring the series and parallel connections of input contacts. However, there are tasks which can involve time delays and event counting. These requirements can be met by the timers and counters which are supplied as a feature of PLCs. They can be controlled by logic instructions and represented on ladder diagrams.

14.8.1   Timers

A common approach used by PLC manufacturers is to consider timers to behave like relays with coils which when energised result in the closure or opening of contacts after some preset time. The timer is thus treated as an output for a rung with control being exercised over pairs of contacts elsewhere (Figure 14.20(a)). Others consider a timer as a delay block in a rung which delays signals in that rung reaching the output (Figure 14.20(b)).

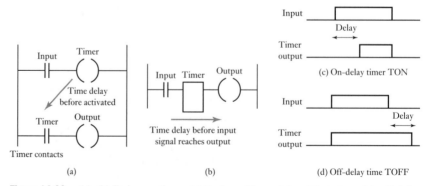

Figure 14.20   (a), (b) Delay-on timer, (c) timing with on-delay, (d) timing with off-delay.

Figure 14.21 Timed sequence.

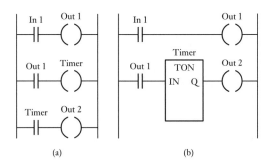

(a)   (b)

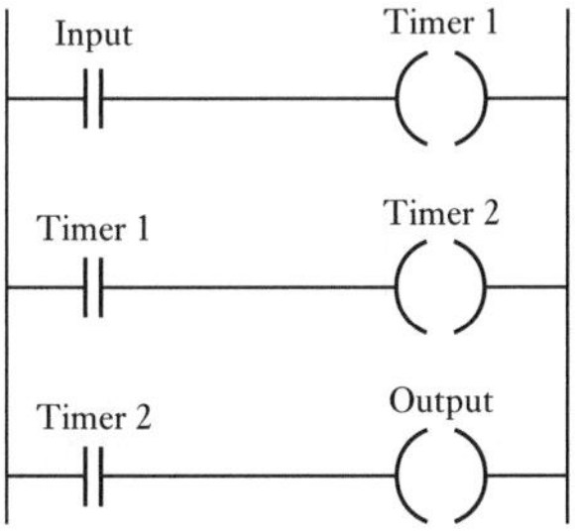

PLCs are generally provided with only a delay-on timer (TON), small PLCs possibly having only this type. Such a timer waits for a fixed delay period before turning on (Figure 14.20(c)), e.g. a period which can be set between 0.1 and 999 s in steps of 0.1 s. Other time-delay ranges and steps are possible.

As an illustration of the use of a timer for sequencing, consider the ladder diagram shown in Figure 14.21(a) or (b). When the input In 1 is on, the output Out 1 is switched on. The contacts associated with this output then start the timer. The contacts of the timer will close after the preset time delay. When this happens, output Out 2 is switched on.

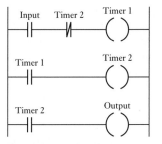

Figure 14.22 Cascaded timers.

Timers can be linked together, or **cascaded**, to give larger delay times than is possible with just one timer. Figure 14.22 shows such an arrangement. When the input contacts close, Timer 1 is started. After its time delay, its contacts close and Timer 2 is started. After its time delay, its contacts close and there is an output.

Figure 14.23 shows a program that can be used to cause an output to go on for 0.5 s, then off for 0.5 s, then on for 0.5 s, then off for 0.5 s, and so on. When the input contacts close, Timer 1 is started and comes on after 0.5 s, this being the time for which it was preset. After this time the Timer 1 contacts close and start Timer 2. It comes on after 0.5 s, its preset time, and opens its contacts. This results in Timer 1 being switched off. This results in its contact opening and switching off Timer 2. This then closes its contact and so starts the entire cycle again. The result is that timer contacts for Timer 1 are switched on for 0.5 s, then off for 0.5 s, on for 0.5 s, and so on. Thus the output is switched on for 0.5 s, then off for 0.5 s, on for 0.5 s, and so on.

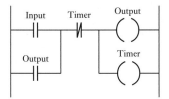

Figure 14.23 On/off cyclic timer.

Figure 14.24 shows how a delay-off timer, i.e. a timer which switches off an output after a time delay from being energised, can be devised. When the input contacts are momentarily closed the output is energised (Figure 14.20(d)) and the timer started. The output contacts latch the input and keep the output on. After the preset time of the timer, the timer comes on and breaks the latch circuit, so switching the output off.

Figure 14.24 Delay-off timer.

14.8.2 Counters

Counters are used when there is a need to count a specified number of contact operations, e.g. where items pass along a conveyor into boxes, and when the specified number of items has passed into a box, the next item is diverted into another box. Counter circuits are supplied as an internal

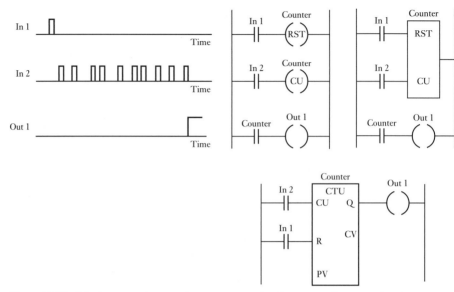

Figure 14.25   The inputs and output for a counter and various ways of representing the same program.

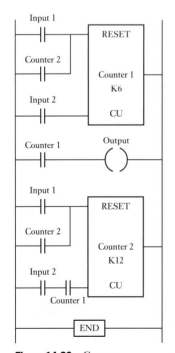

Figure 14.26   Counter.

feature of PLCs. In most cases the counter operates as a **down-counter**. This means that the counter counts down from the present value to zero, i.e. events are subtracted from the set value. When zero is reached the counter's contact changes state. An **up-counter** would count up to the preset value, i.e. events are added until the number reaches the set value. When the set value is reached the counter's contact changes state.

Different PLC manufacturers deal with counters in different ways. Some consider the counter to consist of two basic elements: one output coil to count input pulses and one to reset the counter, the associated contacts of the counter being used in other rungs, e.g. Mitsubishi and Allen-Bradley. Others treat the counter as an intermediate block in a rung from which signals emanate when the count is attained, e.g. Siemens. As an illustration, Figure 14.25 shows a basic counting circuit. When there is a pulse input to In 1, the counter is reset. When there is an input to In 2, the counter starts counting. If the counter is set for, say, 10 pulses, then when 10 pulse inputs have been received at In 2, the counter's contacts will close and there will be an output from Out 1. If at any time during the counting there is an input to In 1, the counter will be reset and start all over again and count for 10 pulses.

As an illustration of the use of a counter, consider the problem of the control for a machine which is required to direct 6 items along one path for packaging in a box, and then 12 items along another path for packaging in another box. Figure 14.26 shows the program that could be used. It involves two counters, one preset to count 6 and the other to count 12. Input 1 momentarily closes its contacts to start the counting cycle, resetting both counters. Input 2 contacts could be activated by a microswitch, which is activated every time an item passes up to the junction in the paths. Counter 1 counts 6 items and then closes its contact. This activates the output, which might be a solenoid used to activate a flap which closes one path and opens another. Counter 1 also has contacts which close and enables Counter 2 to start counting. When Counter 2 has counted 12 items it resets both the counters and opens the Counter 1 contacts, which then results in the output becoming deactivated and items no longer directed towards the box to contain 12 items.

14.9 Shift registers

A number of internal relays can be grouped together to form a register which can provide a storage area for a series sequence of individual bits. A 4-bit register would be formed by using four internal relays, an 8-bit using eight. The term **shift register** is used because the bits can be shifted along by 1 bit when there is a suitable input to the register. For example, with an 8-bit register we might initially have:

1	0	1	1	0	1	0	1

Then there is an input of a 0 shift pulse:

$0 \rightarrow$ | 0 | 1 | 0 | 1 | 1 | 0 | 1 | 0 | $\rightarrow 1$
|---|---|---|---|---|---|---|---|

with the result that all the bits shift along one place and the last bit overflows.

The grouping together of a number of auxiliary registers to form a shift register is done automatically by a PLC when the shift register function is selected at the control panel. With the Mitsubishi PLC, this is done by using the programming function SFT (shift) against the auxiliary relay number that is to be the first in the register array. This then causes the block of relays, starting from that initial number, to be reserved for the shift register. Thus, if we select M140 to be the first relay then the shift register will consist of M140, M141, M142, M143, M144, M145, M146 and M147.

Shift registers have three inputs: one to load data into the first element of the register (OUT), one as the shift command (SFT) and one for resetting (RST). With OUT, a logic level 0 or 1 is loaded into the first element of the shift register. With SFT, a pulse moves the contents of the register along 1 bit at a time, the final bit overflowing and being lost. With RST, a pulse of the closure of a contact resets the register contents to all zeros.

Figure 14.27 gives a ladder diagram involving a shift register when Mitsubishi notation is used; the principle is, however, the same with other manufacturers. M140 has been designated as the first relay of the register. When X400 is switched on, a logic 1 is loaded into the first element of the shift register, i.e. M140. We thus have the register with 10000000. The circuit shows that each element of the shift register has been connected as a contact in the circuit. Thus M140 contact closes and Y430 is switched on. When contact X401 is closed, then the bits in the register are shifted along the register by one place to give 11000000, a 1 being shifted into the register because X400 is still on. Contact M141 thus closes and Y430 is switched on. As each bit is shifted along, the outputs are energised in turn. Shift registers can thus be used to sequence events.

Figure 14.27   Shift register.

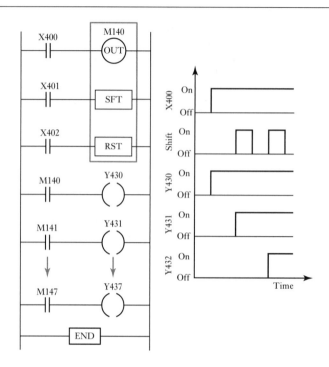

14.10   Master and jump controls

A whole block of outputs can be simultaneously turned off or on by using the same internal relay contacts in each output rung so that switching it on or off affects every one of the rungs. An alternative way of programming to achieve the same effect is to use a **master relay**. Figure 14.28 illustrates its use. We can think of it as controlling the power to a length of the vertical rails of the ladder. When there is an input to close Input 1 contacts, master relay MC1 is activated and then the block of program rungs controlled by that relay follows. The end of a master-relay-controlled section is indicated by the reset MCR. It is thus a branching program in that if there is Input 1 then branch to follow the MC1 controlled path; if not, follow the rest of the program and ignore the branch.

Figure 14.28   Master control relay.

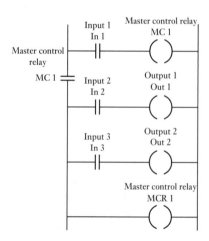

With a Mitsubishi PLC, an internal relay can be designated as a master control relay by programming it accordingly. Thus to program an internal relay M100 as a master control relay, the program instruction is:

MC M100

To indicate the end of the section controlled by a master control relay, the program instruction is:

MCR M100

14.10.1 Jumps

A function which is often provided with PLCs is the **conditional jump** function. Such a function enables programs to be designed so that if a certain condition exists then a section of the program is jumped. Figure 14.29 illustrates this on a flow diagram and with a section of ladder program. Following a section of program, A, the program rung is encountered with Input 1 and the conditional jump relay CJP. If Input 1 occurs then the program jumps to the rung with the end of jump relay coil EJP and so continues with that section of the program labelled as C, otherwise it continues with the program rungs labelled as program B.

Figure 14.29 Jump.

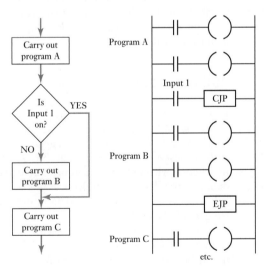

14.11   Data handling

With the exception of the shift register, the previous parts of this chapter have been concerned with the handling of individual bits of information, e.g. a switch being closed or not. There are, however, some control tasks where it is useful to deal with related groups of bits, e.g. a block of eight inputs, and so operate on them as a data word. Such a situation can arise when a sensor supplies an analogue signal which is converted to, say, an 8-bit word before becoming an input to a PLC.

The operations that may be carried out with a PLC on data words normally include:

1 moving data;
2 comparison of magnitudes of data, i.e. greater than, equal to, or less than;
3 arithmetic operations such as addition and subtraction;
4 conversions between binary-coded decimal (BCD), binary and octal.

As discussed earlier, individual bits have been stored in memory locations specified by unique addresses. For example, for the Mitsubishi PLC, input memory addresses have been preceded by an A, outputs by a Y, timers by a T, auxiliary relays by an M, etc. Data instructions also require memory addresses and the locations in the PLC memory allocated for data are termed **data registers**. Each data register can store a binary word of, usually, 8 or 16 bits and is given an address such as D0, D1, D2, etc. An 8-bit word means that a quantity is specified to a precision of 1 in 256, a 16-bit word a precision of 1 in 65536.

Each instruction has to specify the form of the operation, the source of the data used in terms of its data register and the destination data register of the data.

14.11.1 Data movement

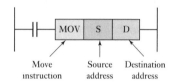

Figure 14.30 Move data.

For data movement the instruction will contain the move data instruction, the source address of the data and the destination address of the data. Thus the ladder rung could be of the form shown in Figure 14.30.

Such data transfers might be to move a constant into a data register, a time or count value to a data register, data from a data register to a timer or counter, data from a data register to an output, input data to a data register, etc.

14.11.2 Data comparison

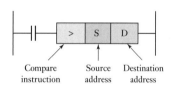

Figure 14.31 Compare data.

PLCs can generally make the data comparisons of *less than* (usually denoted by $<$ or LES), *equal to* ($=$ or EQU), *less than or equal to* ($\leq$ or $<=$ or LEQ), *greater than* ($>$ or GRT), *greater than or equal to* ($\geq$, $>=$ or GEQ) and *not equal to* ($\neq$ or $<>$ or NEQ). To compare data, the program instruction will contain the comparison instruction, the source address of the data and the destination address. Thus to compare the data in data register D1 to see if it is greater than data in data register D2, the ladder program rung would be of the form shown in Figure 14.31.

Such a comparison might be used when the signals from two sensors are to be compared by the PLC before action is taken. For example, an alarm might be required to be sounded if a sensor indicates a temperature above 80°C and remain sounding until the temperature falls below 70°C. Figure 14.32 shows the ladder program that could be used. The input temperature data is inputted to the source address and the destination address contains the set value. When the temperature rises to 80°C, or higher, the data value in the source address becomes $\geq$ the destination address value and there is an output to the alarm which latches the input. When the temperature falls to 70°C, or lower, the data value in the source address becomes $\leq$ the destination address value and there is an output to the relay which then opens its contacts and so switches the alarm off.

Figure 14.32 Temperature alarm.

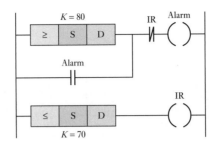

14.11.3 Arithmetic operations

Some PLCs can carry out just the arithmetic operations of addition and subtraction, others have even more arithmetic functions. The instruction to add or subtract generally states the instruction, the register containing the address of the value to be added or subtracted, the address of the value to which the addition or from which the subtraction is to be made and the register where the result is to be stored. Figure 14.33 shows the form used for the ladder symbol for addition with OMRON.

Addition or subtraction might be used to alter the value of some sensor input value, perhaps a correction or offset term, or alter the preset values of timers or counters.

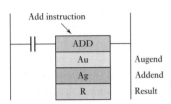

Figure 14.33 Add data.

14.11.4 Code conversions

All the internal operations in the CPU of a PLC are carried out using binary numbers. Thus, when the input is a signal which is decimal, conversion to BCD is used. Likewise, where a decimal output is required, conversion to decimal is required. Such conversions are provided with most PLCs. For example, with Mitsubishi, the ladder rung to convert BCD to binary is of the form shown in Figure 14.34. The data at the source address is in BCD and converted to binary and placed at the destination address.

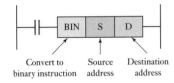

Figure 14.34 BCD to binary.

14.12 Analogue input/output

Many sensors generate analogue signals and many actuators require analogue signals. Thus, some PLCs may have an analogue-to-digital converter (ADC) module fitted to input channels and a digital-to-analogue converter (DAC) module fitted to output channels. An example of where such an item might be used is for the control of the speed of a motor so that its speed moves up to its steady value at a steady rate (Figure 14.35). The input is an on/off switch to start the operation. This opens the contacts for the data register and so it stores zero. Thus the output from the controller is zero and the analogue signal from the DAC is zero and hence motor speed is zero. The closing of the start contacts gives outputs to the DAC and the data register. Each time the program cycles through these rungs on the program, the data register is incremented by 1 and so the analogue signal is increased and hence the motor speed. Full speed is realised when the output from the data register is the word 11111111. The timer function of a PLC can be used to incorporate a delay between each of the output bit signals.

Figure 14.35 Ramping the speed of a motor.

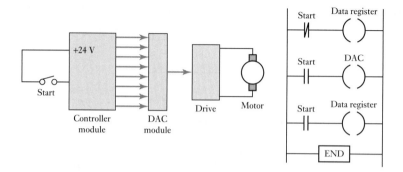

A PLC equipped with analogue input channels can be used to carry out a continuous control function, i.e. PID control (see Section 22.7). Thus, for example, to carry out proportional control on an analogue input the following set of operations can be used:

1 Convert the sensor output to a digital signal.
2 Compare the converted actual sensor output with the required sensor value, i.e. the set point, and obtain the difference. This difference is the error.
3 Multiply the error by the proportional constant K_P.
4 Move this result to the DAC output and use the result as the correction signal to the actuator.

An example of where such a control action might be used is with a temperature controller. Figure 14.36 shows a possibility. The input could be from a thermocouple, which after amplification is fed through an ADC into the PLC. The PLC is programmed to give an output proportional to the

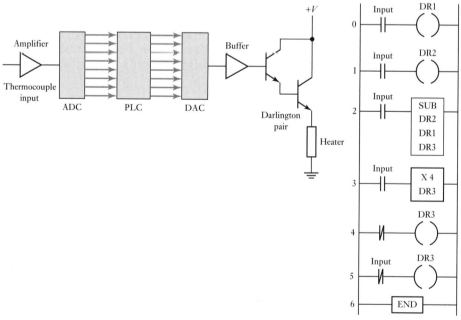

Figure 14.36 Proportional control of temperature.

error between the input from the sensor and the required temperature. The output word is then fed through a DAC to the actuator, a heater, in order to reduce the error.

With the ladder program shown, rung 0 reads the ADC and stores the temperature value in data register DR1. With rung 1, the data register DR2 is used to store the set point temperature. Rung 2 uses the subtract function to subtract the values held in data registers DR1 and DR2 and store the result in data register DR3, i.e. this data register holds the error value. With rung 3 a multiply function is used, in this case to multiply the value in data register DR3 by the proportional gain of 4. Rung 4 uses an internal relay which can be programmed to switch off DR3 if it takes a negative value. With rung 5 the data register DR3 is reset to zero when the input is switched off. Some PLCs have add-on modules which more easily enable PLC control to be used, without the need to write lists of instructions in the way outlined above.

Summary

A **programmable logic controller** (PLC) is a digital electronic device that uses a programmable memory to store instructions and to implement functions such as logic, sequencing, timing, counting and arithmetic in order to control machines and processes and has been specifically designed to make programming easy.

A PLC is continuously running through its program and updating it as a result of the input signals. Each such loop is termed a **cycle**. The form of programming commonly used with PLCs is **ladder programming**. This involves each program task being specified as though a rung of a ladder. There is an alternative way of entering a program and that is to translate the ladder program into an **instruction list**. Instruction lists consist of a series of instructions with each instruction being on a separate line. An instruction consists of an operator followed by one or more operands, i.e. the subjects of the operator.

A **latch circuit** is a circuit that, after being energised, maintains that state until another input is received. The term **internal relay, auxiliary relay** or **marker** is used for what can be considered as an internal relay in the PLC, these behaving like relays with their associated contacts. **Timers** can be considered to behave like relays with coils which when energised result in the closure or opening of contacts after some preset time or as a delay block in a rung which delays signals in that rung reaching the output. **Counters** are used to count a specified number of contact operations, being considered to be an output coil to count input pulses with a coil to reset the counter and the associated contacts of the counter being used in other rungs or as an intermediate block in a rung from which signals emanate when the count is attained. A **shift register** is a number of internal relays which have been grouped together to form a register for a series sequence of individual bits. A **master relay** enables a whole block of outputs to be simultaneously turned off or on. The **conditional jump** function enables a section of the program to be jumped if a certain condition exists. Operations that may be carried out with **data words** include moving data, comparison of magnitudes of data, arithmetic operations and conversions between binary-coded decimal (BCD), binary and octal.

Problems

14.1 What are the logic functions used for switches (a) in series, (b) in parallel?

14.2 Draw the ladder rungs to represent:

(a) two switches are normally open and both have to be closed for a motor to operate;

(b) either of two, normally open, switches have to be closed for a coil to be energised and operate an actuator;

(c) a motor is switched on by pressing a spring-return push-button start switch, and the motor remains on until another spring-return push-button stop switch is pressed.

14.3 Write the program instructions corresponding to the latch program shown in Figure 14.37.

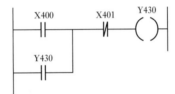

Figure 14.37   Problem 14.3.

14.4 Write the program instructions for the program in Figure 14.38 and state how the output varies with time.

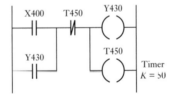

Figure 14.38   Problem 14.4.

14.5 Write the program instructions corresponding to the program in Figure 14.39 and state the results of inputs to the PLC.

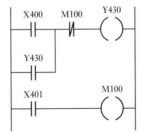

Figure 14.39   Problem 14.5.

14.6 Devise a timing circuit that will switch on an output for 1 s then off for 20 s, then on for 1 s, then off for 20 s, and so on.

14.7 Devise a timing circuit that will switch on an output for 10 s then switch it off.

14.8 Devise a circuit that can be used to start a motor and then after a delay of 100 s start a pump. When the motor is switched off there should be a delay of 10 s before the pump is switched off.

14.9 Devise a circuit that could be used with a domestic washing machine to switch on a pump to pump water for 100 s into the machine, then switch off and switch on a heater for 50 s to heat the water. The heater is then switched off and another pump is to empty the water from the machine for 100 s.

14.10 Devise a circuit that could be used with a conveyor belt which is used to move an item to a work station. The presence of the item at the work station is detected by means of breaking a contact activated by a beam of light to a photosensor. There the item stops for 100 s for an operation to be carried out before moving on and off the conveyor. The motor for the belt is started by a normally open start switch and stopped by a normally closed switch.

14.11 How would the timing pattern for the shift register in Figure 14.27 change if the data input X400 was of the form shown in Figure 14.40?

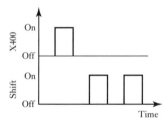

Figure 14.40 Problem 14.11.

14.12 Explain how a PLC can be used to handle an analogue input.

14.13 Devise a system, using a PLC, which can be used to control the movement of a piston in a cylinder so that when a switch is momentarily pressed, the piston moves in one direction and when a second switch is momentarily pressed, the piston moves in the other direction. Hint: you might consider using a 4/2 solenoid-controlled valve.

14.14 Devise a system, using a PLC, which can be used to control the movement of a piston in a cylinder using a 4/2 solenoid-operated pilot valve. The piston is to move in one direction when a proximity sensor at one end of the stroke closes contacts and in the other direction when a proximity sensor at the other end of the stroke indicates its arrival there.

Chapter fifteen Communication systems

Objectives

The objectives of this chapter are that, after studying it, the reader should be able to:
- Describe centralised, hierarchical and distributed control systems, network configurations and methods of transmitting data and protocols used.
- Describe the Open Systems Interconnection communication model.
- Describe commonly used communication interfaces: RS-232, IEEE 488, 20 mA current loop, I²C bus and CAN.

15.1 Digital communications

An **external bus** is a set of signal lines that interconnects microprocessors, microcontrollers, computers and programmable logic controllers (PLCs) and also connects them with peripheral equipment. Thus a computer needs to have a bus connecting it with a printer if its output is to be directed to the printer and printed. Multiprocessor systems are quite common. For example, in a car there are likely to be several microcontrollers with each controlling a different part of the system, e.g. engine management, braking and instrument panel, and communication between them is necessary. In automated plant not only is there a need for data to pass between programmable logic controllers, displays, sensors and actuators and allow for data and programs to be inputted by the operator, but there can also be data communications with other computers. There may, for example, be a need to link a PLC to a control system involving a number of PLCs and computers. Computer integrated manufacturing (CIM) is an example of a large network which can involve large numbers of machines linked together. This chapter is a consideration of how such data communications between computers can take place, whether it is just simply machine-to-machine or a large network involving large numbers of machines linked together, and the forms of standard communication interfaces.

15.2 Centralised, hierarchical and distributed control

Centralised computer control involves the use of one central computer to control an entire plant. This has the problem that failure of the computer results in the loss of control of the entire plant. This can be avoided by the use of dual computer systems. If one computer fails, the other one takes over. Such centralised systems were common in the 1960s and 1970s. The development of the microprocessor and the ever reducing costs of computers have led to multi-computer systems becoming more common and the development of hierarchical and distributed systems.

With the **hierarchical system,** there is a hierarchy of computers according to the tasks they carry out. The computers handling the more routine tasks are supervised by computers which have a greater decision-making role. For example, the computers which are used for direct digital control of systems are subservient to a computer which performs supervisory control of the entire system. The work is divided between the computers according to the function involved. There is specialisation of computers with some computers only receiving some information and others different information.

With the **distributed system,** each computer system carries out essentially similar tasks to all the other computer systems. In the event of a failure of one, or overloading of a particular computer, work can be transferred to other computers. The work is spread across all the computers and not allocated to specific computers according to the function involved. There is no specialisation of computers. Each computer thus needs access to all the information in the system.

In most modern systems there is generally a mixture of distributed and hierarchical systems. For example, the work of measurement and actuation may be distributed among a number of microcontrollers/computers which are linked together and provide the database for the plant. These may be overseen by a computer used for direct digital control or sequencing and this in turn may be supervised by one used for supervisory control of the plant as a whole. Typical levels in such a scheme are:

Level 1 Measurement and actuators
Level 2 Direct digital and sequence control
Level 3 Supervisory control
Level 4 Management control and design

Distributed/hierarchical systems have the advantage of allowing the task of measurement scanning and signal conditioning in control systems to be carried out by sharing it between a number of microprocessors. This can involve a large number of signals with a high frequency of scanning. If extra measurement loops are required, it is a simple matter to increase the capacity of the system by adding microprocessors. The units can be quite widely dispersed, being located near the source of the measurements. Failure of one unit does not result in failure of the entire system.

15.2.1 Parallel and serial data transmission

Data communication can be via parallel or serial transmission links.

1 *Parallel data transmission*
Within computers, data transmission is usually by **parallel data paths.** Parallel data buses transmit 8, 16 or 32 bits simultaneously, having a separate bus wire for each data bit and the control signals. Thus, if there are eight data bits to be transmitted, e.g. 11000111, then eight data wires are needed. The entire eight data bits are transmitted in the same time as it takes to transmit one data bit because each bit is on a parallel wire. Handshaking (see Section 13.3.2) lines are also needed, handshaking being used for each character transmitted with lines needed to indicate that data is available for transmission and that the receiving terminal is ready to receive. Parallel data transmission permits high data transfer rates but

is expensive because of the cabling and interface circuitry required. It is thus normally only used over short distances or where high transfer rates are essential.

2   *Serial data transmission*
This involves the transmission of data which, together with control signals, is sent bit by bit in sequence along a single line. Only two conductors are needed, to transmit data and to receive data. Since the bits of a word are transmitted sequentially and not simultaneously, the data transfer rate is considerably less than with parallel data transmission. However, it is cheaper since far fewer conductors are required. For example, with a car when a number of microcontrollers are used, the connections between them are by serial data transmission. Without the use of serial transmission the number of wires involved would be considerable. In general, serial data transmission is used for all but the shortest peripheral connections.

Consider the problem of sending a sequence of characters along a serial link. The receiver needs to know where one character starts and stops. Serial data transmission can be either asynchronous or synchronous. **Asynchronous transmission** implies that both the transmitter and receiver computers are not synchronised, each having its own independent clock signals. The time between transmitted characters is arbitrary. Each character transmitted along the link is thus preceded by a start bit to indicate to the receiver the start of a character, and followed by a stop bit to indicate its completion. This method has the disadvantage of requiring extra bits to be transmitted along with each character and thus reduces the efficiency of the line for data transmission. With **synchronous transmission** there is no need for start and stop bits since the transmitter and receiver have a common clock signal and thus characters automatically start and stop always at the same time in each cycle.

The **rate of data transmission** is measured in bits per second. If a group of n bits form a single symbol being transmitted and the symbol has a duration of T seconds then the data rate of transmission is n/T. The **baud** is the unit used. The baud rate is only the same as the number of bits per second transmitted if each character is represented by just one symbol. Thus a system which does not use start and stop pulses has a baud rate equal to the bit rate, but this will not be the case when there are such bits.

15.2.2   Serial data communication modes

Serial data transmission occurs in one of three modes.

1   *Simplex mode*
Transmission is only possible in one direction, from device A to device B, where device B is not capable of transmitting back to device A (Figure 15.1(a)). You can think of the connection between the devices as being like a one-way road. This method is usually only used for transmission to devices such as printers which never transmit information.

Figure 15.1   Communication modes.

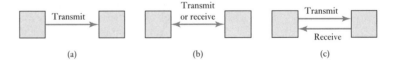

(a)   (b)   (c)

2 *Half-duplex mode*

Data is transmitted in one direction at a time but the direction can be changed (Figure 15.1(b)). Terminals at each end of the link can be switched from transmit to receive. Thus device A can transmit to device B and device B to device A but not at the same time. You can think of this as being like a two-lane road under repair with traffic from one lane being stopped by a traffic control to allow the traffic from the other lane through. Citizens Band (CB) radio is an example of half-duplex mode; a person can receive or talk but not do both simultaneously.

3 *Full-duplex mode*

Data may be transmitted simultaneously in both directions between devices A and B (Figure 15.1(c)). This is like a two-lane highway in which traffic can occur in both directions simultaneously. The telephone system is an example of full-duplex mode in that a person can talk and receive at the same time.

15.3 Networks

The term **network** is used for a system which allows two or more computers/microprocessors to be linked for the interchange of data. The logical form of the links is known as the network **topology**. The term **node** is used for a point in a network where one or more communication lines terminate or a unit is connected to the communication lines. The following are commonly used forms.

1 *Data bus*

This has a linear bus (Figure 15.2(a)) into which all the stations are plugged. This system is often used for multipoint terminal clusters. It is generally the preferred method for distances between nodes of more than 100 m.

2 *Star*

This has dedicated channels between each station and a central switching hub (Figure 15.2(b)) through which all communications must pass. This is the type of network used in the telephone systems (private branch exchanges (PBXs)) in many companies, all the lines passing through a central exchange. This system is also often used to connect remote and local terminals to a central mainframe computer. There is a major problem with this system in that if the central hub fails then the entire system fails.

3 *Hierarchy or tree*

This consists of a series of branches converging indirectly to a point at the head of the tree (Figure 15.2(c)). With this system there is only one transmission path between any two stations. This arrangement may be formed from a number of linked data bus systems. Like the bus method, it is often used for distances between nodes of more than 100 m.

Figure 15.2 Network topologies: (a) data bus, (b) star, (c) hierarchy, (d) ring, (e) mesh.

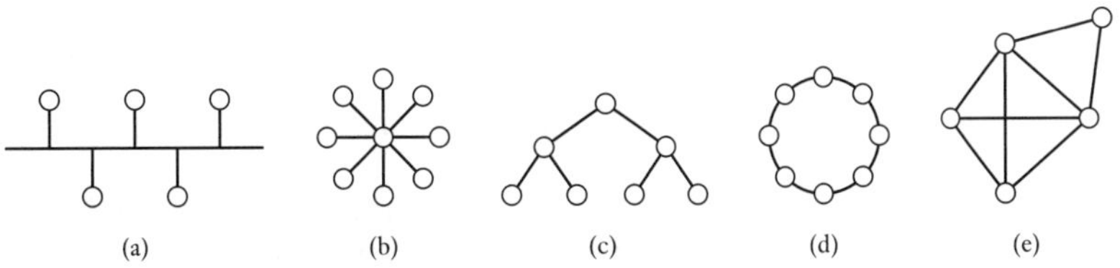

(a)   (b)   (c)   (d)   (e)

4 *Ring*

This is a very popular method for local area networks, involving each station being connected to a ring (Figure 15.2(d)). The distances between nodes are generally less than 100 m. Data put into the ring system continues to circulate round the ring until some system removes it. The data is available to all the stations.

5 *Mesh*

This method (Figure 15.2(e)) has no formal pattern to the connections between stations and there will be multiple data paths between them.

The term **local area network** (LAN) is used for a network over a local geographic area such as a building or a group of buildings on one site. The topology is commonly bus, star or ring. A **wide area network** is one that interconnects computers, terminals and local area networks over a national or international level. This chapter is primarily concerned with local area networks.

15.3.1 Network access control

Access control methods are necessary with a network to ensure that only one user of the network is able to transmit at any one time. The following are methods used.

With ring-based local area networks, two commonly used methods are:

1 *Token passing*

With this method a token, a special bit pattern, is circulated. When a station wishes to transmit it waits until it receives the token, then transmits the data with the token attached to its end. Another station wishing to transmit removes the token from the package of data and transmits its own data with the token attached to its end.

2 *Slot passing*

This method involves empty slots being circulated. When a station wishes to transmit data it deposits it in the first empty slot that comes along.

With bus or tree networks a method that is often used is:

3 *Carrier sense multiple access with collision detection (CSMA/CD)*

This method is generally identified with the **Ethernet LAN bus**. With the CSMA/CD method, stations have to listen for other transmissions before transmitting, with any station being able to gain control of the network and transmit, hence the term multiple access. If no activity is detected then transmission can occur. If there is activity then the system has to wait until it can detect no further activity. Despite this listening before transmission, it is still possible for two or more systems to start to transmit at the same time. If such a situation is detected, both stations cease submitting and wait a random time before attempting to retransmit.

15.3.2 Broadband and baseband

The term **broadband transmission** is used for a network in which information is modulated onto a radio frequency carrier which passes through the transmission medium such as a coaxial cable. Typically the topology of broadband local area networks is a bus with branches. Broadband transmission allows a number of modulated radio frequency carriers to be simultaneously transmitted and so offers a multichannel capability. The term **baseband transmission** is used when digital information is passed directly through the transmission medium. Baseband transmission networks can only support one information signal at a time. A LAN may be either baseband or broadband.

15.4 Protocols

Transmitted data will contain two types of information. One is the data which one computer wishes to send to another, the other is information termed **protocol data** and is used by the interface between a computer and the network to control the transfer of the data into the network or from the network into the computer. A protocol is a formal set of rules governing data format, timing, sequencing, access control and error control. The three elements of a protocol are:

1 *syntax*, which defines data format, coding and signal levels;
2 *semantics*, which deals with synchronisation, control and error handling;
3 *timing*, which deals with the sequencing of data and the choice of data rate.

When a sender communicates with a receiver then both must employ the same protocol, e.g. two microcontrollers with data to be serially transmitted between them. With simplex communication the data block can be just sent from sender to receiver. However, with half-duplex, each block of transmitted data, if valid, must be acknowledged (ACK) by the receiver before the next block of data can be sent (Figure 15.3(a)); if invalid a NAK, negative acknowledgement, signal is sent. Thus a continuous stream of data cannot be transmitted. The CRC bits, **cyclic redundancy check bits,** are a means of error detection and are transmitted immediately after a block of data. The data is transmitted as a binary number and at the transmitter the data is divided by a number and the remainder is used as the cyclic check code. At the receiver the incoming data, including the CRC, is divided by the same number and will give zero remainder if the signal is error-free. With full-duplex mode (Figure 15.3(b)), data can be continuously sent and received.

Figure 15.3 Protocols: (a) half-duplex, (b) full-duplex.

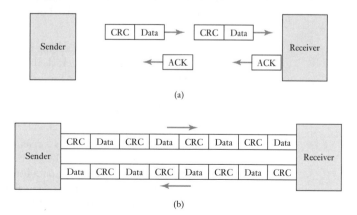

(a)

(b)

Figure 15.4 (a) Bisync,
(b) HDLC.

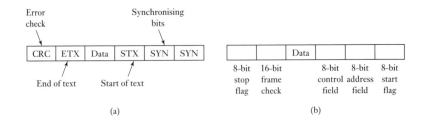

(a)

(b)

Within a package being sent, there is a need to include protocol information. For example, with asynchronous transmission there may be characters to indicate the start and end of data. With synchronous transmission and the **Bisync Protocol,** a data block is preceded by a synchronising sequence of bits, usually the ASCII character SYN (Figure 15.4(a)). The SYN characters are used by the receiver to achieve character synchronisation, preparing the receiver to receive data in 8-bit groupings. The Motorola MC6852 is a synchronous serial data adapter (SSDA) that is designed for use with 6800 microprocessors to provide a synchronous serial communications interface using the Bisync Protocol. It is similar to the asynchronous communications interface adapter described in Section 13.5. Another protocol is the **High-level Data Link Control** (HDLC) (Figure 15.4(b)). This is a full-duplex protocol with the beginning and end of a message being denoted by the bit pattern 01111110. Address and control fields follow the start flag. The address identifies the address of the destination station, the control field defines whether the frame is supervisory, information or unnumbered. Following the message is a 16-bit frame check sequence which is used to give a CRC. The Motorola 6854 is an example of a serial interface adapter using this HDLC Protocol.

15.5 Open Systems Interconnection communication model

Communication protocols have to exist on a number of levels. The International Organization for Standardization (ISO) has defined a seven-layer standard protocol system known as the **Open Systems Interconnection** (OSI) model. The model is a framework for developing a co-ordinated system of standards. The layers are as outlined below.

1 *Physical layer*
 This layer describes the means for bit transmission to and from physical components of the network. It deals with hardware issues, e.g. the types of cable and connectors to be used, synchronising data transfer and signal levels. Commonly used LAN systems defined at the physical layer are Ethernet and token ring.

2 *Data link layer*
 This layer defines the protocols for sending and receiving messages, error detection and correction and the proper sequencing of transmitted data. It is concerned with packaging data into packets and placing them on the cable and then taking them off the cable at the receiving end. Ethernet and token ring are also defined at this level.

3 *Network layer*

This deals with communication paths and the addressing, routing and control of messages on the network and thus making certain that the messages get to the right destinations. Commonly used network layer protocols are Internet Protocol (IP) and Novell's Internetwork Packet Exchange (IPX).

4 *Transport layer*

This provides for reliable end-to-end message transport. It is concerned with establishing and maintaining the connection between transmitter and receiver. Commonly used transport layer protocols are Internet Transmission Control Protocol (TCP) and Novell's Sequenced Packet Exchange (SPX).

5 *Session layer*

This layer is concerned with the establishment of dialogues between application processes which are connected together by the network. It is responsible for determining when to turn a communication between two stations on or off.

6 *Presentation layer*

This layer is concerned with allowing the encoded data transmitted to be presented in a suitable form for user manipulation.

7 *Application layer*

This layer provides the actual user information processing function and application-specific services. It provides such functions as file transfer or electronic mail which a station can use to communicate with other systems on the network.

15.5.1 Network standards

There are a number of network standards, based on the OSI layer model, that are commonly used. The following are examples.

In the United States, General Motors realised that the automation of its manufacturing activities posed a problem of equipment being supplied with a variety of non-standard protocols. GM thus developed a standard communication system for factory automation applications. The standard is referred to as the **Manufacturing Automation Protocol** (MAP) (Figure 15.5). The choice of protocols at the different layers reflects the requirement for the system to fit the manufacturing environment. Layers 1 and 2 are implemented in hardware electronics and levels 3 to 7 using software. For the physical layer, broadband transmission is used. The broadband method allows the system to be used for services in addition to those required for MAP communications. For the data link layer, the token system with a bus is used with logical link control (LLC) to implement such functions as error checking, etc. For the other layers, ISO standards are used. At layer 7, MAP includes manufacturing message services (MMS), an application relevant to factory floor communications which defines interactions between programmable logic controllers and numerically controlled machines or robots.

Figure 15.5 MAP.

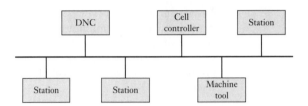

The **Technical and Office Protocol** (TOP) is a standard that was developed by Boeing Computer Services. It has much in common with MAP but can be implemented at a lower cost because it is a baseband system. It differs from MAP in layers 1 and 2, using either the token with a ring or the CSMA/CD method with a bus network. Also, at layer 7, it specifies application protocols that concern office requirements, rather than factory floor requirements. With the CSMA/CD method, stations have to listen for other transmissions before transmitting. TOP and MAP networks are compatible and a gateway device can be used to connect TOP and MAP networks. This device carries out the appropriate address conversions and protocol changes.

Systems Network Architecture (SNA) is a system developed by IBM as a design standard for IBM products. SNA is divided into seven layers; it, however, differs to some extent from the OSI layers (Figure 15.6). The data link control layer provides support for token ring for LANs. Five of the SNA layers are consolidated in two packages: the path control network for layers 2 and 3 and the network addressable units for layers 4, 5 and 6.

Figure 15.6 SNA.

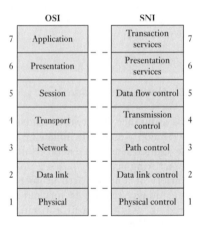

With PLC systems, it is quite common for the system used to be that marketed by the PLC manufacturer. For example, Allen-Bradley has the **Allen-Bradley data highway** which uses token passing to control message transmission; Mitsubishi has Melsec-Net and Texas Instruments has TIWAY. A commonly used system with PLC networks is the Ethernet. This is a single-bus system with CSMA/CD used to control access and is widely used with systems involving PLCs communicating with computers. The problem with using CSMA/CD is that, though this method works well when traffic is light, as network traffic increases the number of collisions and corresponding back-off of transmitters increases. Network throughput can thus slow down quite dramatically.

15.6 Serial communication interfaces

Serial interfacing can involve synchronous or asynchronous protocols. Commonly used asynchronous interfaces are RS-232, and later versions, the 20 mA current loop, I²C, CAN and USB.

15.6.1 RS-232

The most popular serial interface is **RS-232**; this was first defined by the American Electronic Industries Association (EIA) in 1962. The standard relates to data terminal equipment (DTE) and data circuit-terminating equipment (DCE). Data terminal equipment can send or receive data via the interface, e.g. a microcontroller. Data circuit-terminating equipment is devices which facilitate communication; a typical example is a modem. This forms an essential link between a microcomputer and a conventional analogue telephone line.

RS-232 signals can be grouped into three categories.

1 *Data*
RS-232 provides two independent serial data channels, termed primary and secondary. Both these channels are used for full-duplex operation.

2 *Handshake control*
Handshaking signals are used to control the flow of serial data over the communication path.

3 *Timing*
For synchronous operation it is necessary to pass clock signals between transmitters and receivers.

Table 15.1 gives the RS-232C connector pin numbers and signals for which each is used; not all the pins and signals are necessarily used in a particular set-up. The signal ground wire allows for a return path. The connector to a RS-232C serial port is via a 25-pin D-type connector; usually a male plug is used on cables and a female socket on the DCE or DTE.

For the simplest bidirectional link, only the two lines 2 and 3 for transmitted data and received data, with signal ground (7) for the return path of these signals, are needed (Figure 15.7(a)). Thus the minimum connection is via a three-wire cable. For a simple set-up involving a personal computer (PC) being linked with a visual display unit (VDU), pins 1, 2, 3, 4, 5, 6, 7 and 20 are involved (Figure 15.7(b)). The signals sent through pins 4, 5, 6 and 20 are used to check that the receiving end is ready to receive a signal; the transmitting end is ready to send and the data is ready to be sent.

RS-232 is limited concerning the distance over which it can be used as noise limits the transmission of high numbers of bits per second when the length of cable is more than about 15 m. The maximum data rate is about 20 kbits/s. Other standards such as RS-422 and RS-485 are similar to RS-232 and can be used for higher transmission rates and longer distances.

Table 15.1 RS-232 pin assignments.

Pin	Abbreviation	Direction: To	Signal/function
1	FG		Frame ground
2	TXD	DCE	Transmitted data
3	RXD	DTE	Received data
4	RTS	DCE	Request to send
5	CTS	DTE	Clear to send
6	DSR	DTE	DCE ready
7	SG		Signal ground/common return
8	DCD	DTE	Received line detector
12	SDCD	DTE	Secondary received line signal detector
13	SCTS	DTE	Secondary clear to send
14	STD	DCE	Secondary transmitted data
15	TC	DTE	Transmit signal timing
16	SRD	DTE	Secondary received data
17	RC	DTE	Received signal timing
18		DCE	Local loop-back
19	SRTS	DCE	Secondary request to send
20	DTR	DCE	Data terminal ready
21	SQ	DEC/DTE	Remote loop-back/signal quality detector
22	RI	DTE	Ring indicator
23		DEC/DTE	Data signal rate selector
24	TC	DCE	Transmit signal timing
25		DTE	Test mode

Figure 15.7 RS-232 connections: (a) minimum configuration, (b) PC connection.

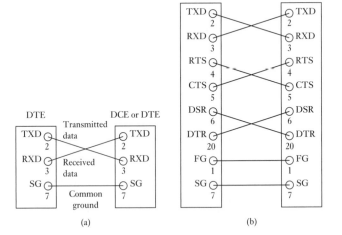

RS-422 uses a pair of lines for each signal and can operate up to about 1220 m or at higher transmission speeds up to 100 bits/s and in noisier environments; maximum speed and maximum distance cannot, however, be achieved simultaneously. RS-485 can be used up to about 1220 m with speeds of 100 kbits/s.

The serial communications interface of the Motorola microcontroller MC68HC11 is capable of full-duplex communications at a variety of baud rates. However, the input and output of this system use transistor–transistor logic (TTL) for which logic 0 is 0 V and logic 1 is +5 V. The RS-232C standards are +12 V for logic 0 and −12 V for logic 1. Thus conversion in signal levels is necessary. This can be achieved by using integrated circuit devices such as MC1488 for TTL to RS-232C conversion and MC1489 for RS-232C to TTL conversion (Figure 15.8).

Figure 15.8 Level conversion: (a) MC1488, (b) MC1489.

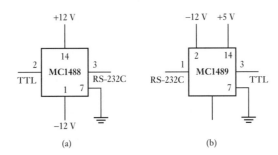

15.6.2 The 20 mA current loop

Another technique, based on RS-232 but not part of the standard, is the **20 mA current loop** (Figure 15.9). This uses a current signal rather than a voltage signal. A pair of separate wires is used for the transmission and the receiver loops with a current level of 20 mA used to indicate a logic 1 and 0 mA a logic 0. The serial data is encoded with a start bit, eight data bits and two stop bits. Such current signals enable a far greater distance, a few kilometres, between transmitter and receiver than with the standard RS-232 voltage connections.

Figure 15.9 The 20 mA current loop.

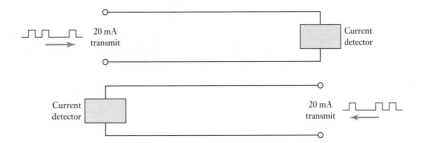

15.6.3 I²C bus

The **Inter-IC Communication bus**, referred to as the I²C bus, is a serial data bus designed by Philips for use for communications between integrated circuits or modules. The bus allows data and instructions to be exchanged between devices by means of just two wires. This results in a considerable simplification of circuits.

Figure 15.10   I²C bus.

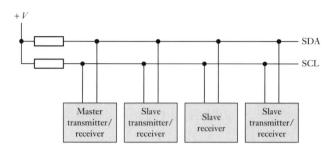

The two lines are a bidirectional data line (SDA) and a clock line (SCL). Both lines are connected to the positive power supply via resistors (Figure 15.10). The device generating the message is the transmitter and the device receiving the message the receiver. The device that controls the bus operation is the master and the devices which are controlled by the master are the slaves.

The following is the protocol used: a data transfer may only be initiated when the bus is not busy and during the data transfer, when the clock line is high, the data line must remain. Changes in the data line when the clock line is high are interpreted as control signals.

1   When both the data and clock lines are high the bus is not busy.
2   A change in the state of the data line from high to low while the clock is high is used to define the start of data transfer.
3   A change in the state of the data line from low to high while the clock is high defines the stop of data transfer.
4   Data is transferred between start and stop conditions.
5   After a start of data transfer, the data line is stable for the duration of the high periods of the clock signal, being able to change during the low periods of the clock signal.
6   There is one clock pulse per data bit transferred with no limit on the number of data bytes that can be transferred between the start and stop conditions; after each byte of data the receiver acknowledges with a ninth bit.
7   The acknowledge bit is a high level put on the bus by the transmitter, a low level by a receiver.

Figure 15.11 illustrates the above by showing the form of the clock signal and the outputs by a transmitter and a receiver.

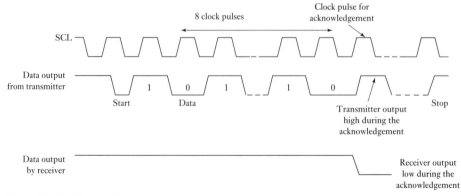

Figure 15.11   Bus conditions.

15.6.4 CAN bus

A modern automobile may have as many as seventy electronic control units (ECUs) for various subsystems, e.g. engine management systems, anti-lock brakes, traction control, active suspension, airbags, cruise control, windows, etc. This could involve a lot of wiring. However, an alternative approach is to use a common data bus with data transmitted along it and made available to all parts of the car. Bosch has thus developed a protocol known as **CAN** or **Controller Area Network**. The CAN bus is now also used as a fieldbus in other automation systems.

CAN is a multi-master serial bus standard for connecting ECUs. Each node in the system is able to both send and receive messages and requires the following.

1 A host processor to determine what received messages mean and which messages it wants to transmit. Sensors, actuators and control devices are not connected directly to the CAN bus but to a host processor and a CAN controller.
2 A CAN controller to store bits received serially from the bus until an entire message is available. After the CAN controller has triggered an interrupt call, the message can then be fetched by the host processor. The controller also stores messages ready for transmission serially onto the bus.
3 A transceiver, which is possibly integrated into the CAN controller, to adapt signal levels received from the bus to levels that the CAN controller expects and which has protective circuitry that protects the CAN controller. It also is used to convert the transmit-bit signal received from the CAN controller into a signal that is sent onto the bus.

Each message consists of an identification (ID) field, to identify the message-type or sender, and up to eight data bytes. Some means of arbitration is, however, needed if two or more nodes begin sending messages at the same time. A non-destructive arbitration method is used to determine which node may transmit, and it is the ID with 0s which is deemed dominant and allowed to win the conflict and transmit. Thus when a transmitting node puts a bit on the bus but detects that there is a more dominant one already on the bus, it disables its transmitter and waits until the end of the current transmission before attempting to start transmitting its own data. For example, suppose we have the 11-bit ID 11001100110 for message 1 and 10001101110 for message 2. By the time transmission has reached the fourth bit the arbitration indicates that message 1 is dominant and so message 2 ceases transmission.

The standard CAN data frame format for serial transmission consists of a message sandwiched between a start bit and a confirmation sent and end of frame bits. The message will have:

1 a 12-bit ID, the last bit being a remote transmission request bit;
2 a 6-bit control field consisting of an identifier extension bit, and a reserved bit, a 4-bit data length code to indicate the number of bytes of data;
3 the data field;
4 a 16-bit CRC field, i.e. a cyclic redundancy check for error detection.

15.6.5 USB

The *Universal Serial Bus* (USB) is designed to enable monitors, printers, modems and other input devices to be easily connected to PCs – the term plug-and-play is used. USB uses a star topology (see Section 15.3); thus only one device needs to be plugged into a PC with other devices then being able to be plugged into the resulting hub so resulting in a tiered star topology. We thus have a host hub at the PC into which other external hubs can be connected. Each port is a four-pin socket with two of the pins being for power and two for communications. The USB 1.0 and 2.0 provide a 5 V supply from which USB devices can draw power, although there is a limit of 500 mA current. USB devices requiring more power than is provided by a single port can use an external power supply.

The low-speed version USB 1.0 specification was introduced in 1996 and has a transfer rate of 12 Mbits/s and is limited to cable lengths of 3 m. The high-speed version USB 2.0 specification was introduced in April 2000 and has a data transfer rate of 480 Mbits/s and is limited to cable lengths of 5 m although, as up to five USB hubs can be used, a long chain of cables and hubs would enable distances up 30 m to be covered. A superspeed USB 3.0 specification was released by Intel and partners in August 2008 for a data transfer rate of 4.8 Gbits/s, and products using this specification are now becoming available. Data is transmitted in half-duplex mode for USB 1.0 and USB 2.0 with full-duplex being possible with USB 3.0 (see Section 15.2.2).

The root hub has complete control over all the USB ports. It initiates all communications with hubs and devices. No USB device can transfer any data onto the bus without a request from the host controller. In USB 2.0, the host controller polls the bus for traffic. For USB 3.0, connected devices are able to request service from the host. When a USB device is first connected to a USB host, an enumeration process is started by the host sending a reset signal to the USB device. After reset, the USB device's information is read by the host and the device is assigned a unique 7-bit address. If the device is supported by the host, the device driver needed for communicating with the device is loaded. The driver is used to supply the information about the device's needs, i.e. such things as speed, priority, function of the device and the size of packet needed for data transfer. When the application's software wants to send or receive some information from a device, it initiates a transfer via the device driver. The driver software then places the request in a memory location together with the requests that have been made by other device drivers. The host controller then takes all the requests and transfers it serially to the host hub ports. Since all the devices are in parallel on the USB bus, all of them hear the information. The host waits for a response. The relevant devices then respond with the appropriate information.

Packets sent out are in three basic types, namely handshaking, token and data, each having a different format and CRC (cyclic redundancy check, see Section 15.4). There are four types of token packet – start of frame, in and out packets to command a device to transmit or receive data, and set-up packet used for the initial set-up of a device.

15.6.6 Firewire

Firewire is a serial bus developed by Apple Computers, the specification being given by IEEE 1394. It offers plug-and-play capabilities and is used for applications such as disk drives, printers and cameras.

15.7 Parallel communication interfaces

For the parallel interface to a printer the Centronics parallel interface is commonly used. However, with instrumentation the most commonly used parallel interface in communications is the **General Purpose Instrument Bus** (GPIB), the IEEE 488 standard, originally developed by Hewlett Packard to link its computers and instruments and thus often referred to as the **Hewlett Packard Instrumentation Bus.** Each of the devices connected to the bus is termed a listener, talker or controller. Listeners are devices that accept data from the bus, talkers place data, on request, on the bus and controllers manage the flow of data on the bus by sending commands to talkers and listeners and carry out polls to see which devices are active (Figure 15.12(a)).

There is a total of 24 lines with the interface:

1 eight bidirectional lines to carry data and commands between the various devices connected to the bus;
2 five lines for control and status signals;
3 three lines for handshaking between devices;
4 eight lines are ground return lines.

Table 15.2 lists the functions of the lines and their pin numbers in a 25-way D-type connector. Up to 15 devices can be attached to the bus at any one time, each device having its own address.

The 8-bit parallel data bus can transmit data as one 8-bit byte at a time. Each time a byte is transferred the bus goes through a handshake cycle. Each device on the bus has its own address. Commands from the controller are signalled by taking the attention line (ATN) low. Commands are then directed to individual devices by placing addresses on the data lines; device addresses are sent via the data lines as a parallel 7-bit word with the lowest 5 bits providing the device address and the other 2 bits control information. If both these bits are 0 then the commands are sent to all addresses; if bit 6 is 1 and bit 7 is 0 the addressed device is switched to be a listener; if bit 6 is 0 and bit 7 is 1 then the device is switched to be a talker.

Handshaking uses the lines DAV, NRFD and NDAC, the three lines ensuring that the talker will only talk when it is being listened to by listeners

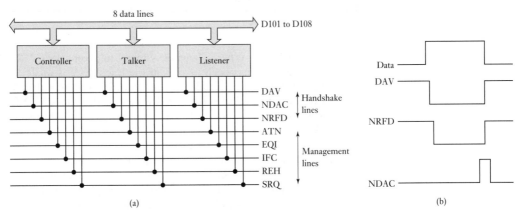

Figure 15.12 GPIB bus: (a) structure, (b) handshaking.

Table 15.2 IEEE 488 bus system.

Pin	Signal group	Abbreviation	Function
1	Data	D101	Data line 1
2	Data	D102	Data line 2
3	Data	D103	Data line 3
4	Data	D104	Data line 4
5	Management	EOI	End Or Identify. This is either used to signify the end of a message sequence from a talker device or used by the controller to ask a device to identify itself
6	Handshake	DAV	Data Valid. When the level is low on this line then the information on the data bus is valid and acceptable
7	Handshake	NRFD	Not Ready For Data. This line is used by listener devices taking it high to indicate that they are ready to accept data
8	Handshake	NDAC	Not Data Accepted. This line is used by listeners taking it high to indicate that data is being accepted
9	Management	IFC	Interface Clear. This is used by the controller to reset all the devices of the system to the start state
10	Management	SRQ	Service Request. This is used by devices to signal to the controller that they need attention
11	Management	ATN	Attention. This is used by the controller to signal that it is placing a command on the data lines
12		SHIELD	Shield
13	Data	D105	Data line 5
14	Data	D106	Data line 6
15	Data	D107	Data line 7
16	Data	D108	Data line 8
17	Management	REN	Remote Enable. This enables a device to indicate that it is to be selected for remote control rather than by its own control panel
18		GND	Ground/common (twisted pair with DAV)
19		GND	Ground/common (twisted pair with NRFD)
20		GND	Ground/common (twisted pair with NDAC)
21		GND	Ground/common (twisted pair with IFC)
22		GND	Ground/common (twisted pair with SRG)
23		GND	Ground/common (twisted pair with ATN)
24		GND	Signal ground

(Figure 15.12(b)). When a listener is ready to accept data, NRFD is made high. When data has been placed on the line, DAV is made low to notify devices that data is available. When a device accepts a data word it sets NDAC high to indicate that it has accepted the data and NRFD low to indicate that it is now not ready to accept data. When all the listeners have set NDAC high, then the talker cancels the data valid signal, DAV going high. This then results in NDAC being set low. The entire process can then be repeated for another word being put on the data bus.

The GPIB is a bus which is used to interface a wide range of instruments, e.g. digital multimeters and digital oscilloscopes, via plug-in boards (Figure 15.13) to computers with standard cables used to link the board with the instruments via interfaces.

Figure 15.13 GPIB hardware.

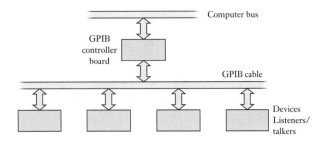

15.7.1 Other buses

Buses used to connect the central processing unit (CPU) to the input/output ports or other devices include the following.

1 The *XT computer bus* was introduced in 1983 for 8-bit data transfers with IBM PC/XT and compatible computers.

2 The *AT bus*, also referred to as the *industry standard architecture* (ISA) *bus*, was later introduced for use with 16-bit transfers with IBM PC and other compatible computers using 80286 and 80386 microprocessors. The AT bus is compatible with the XT bus so that plug-in XT boards can be used in AT bus slots.

3 The *extended industry standard architecture* (EISA) *bus* was developed to cope with 32-bit data transfers with IBM PC and other compatible computers using 80386 and 80486 microprocessors.

4 The *Micro Channel Architecture* (MCA) *bus* is a 16-bit or 32-bit data transfer bus designed for use with IBM Personal System/1 (PS/2) computers. Boards for use with this bus are not compatible with PC/XT/AT boards.

5 The *NuBus* is the 32-bit bus used in Apple's Macintosh II computers.

6 The *S-bus* is the 32-bit bus used in Sun Microsystem's SPARC stations.

7 The *TURBOchannel* is the 32-bit bus used in DECstation 5000 work stations.

8 The *VME bus* is the bus designed by Motorola for use with its 32-bit 68000-microprocessor-based system. Such a bus is now, however, widely used with other computer systems as the bus for use with instrumentation systems.

The above are called **backplane buses,** the term backplane being for the board (Figure 15.14) on which connectors are mounted and into which printed circuit boards containing a particular function, e.g. memory, can be plugged. The backplane provides the data, address and control bus signals to each board, so enabling systems to be easily expanded by the use of off-the-shelf boards. It is into such computer buses that data acquisition boards and boards used for interfacing instruments and other peripherals have to interface. Data acquisition and instrument boards are usually available in various configurations, depending on the computer with which they are to be used.

Figure 15.14   Backplane bus.

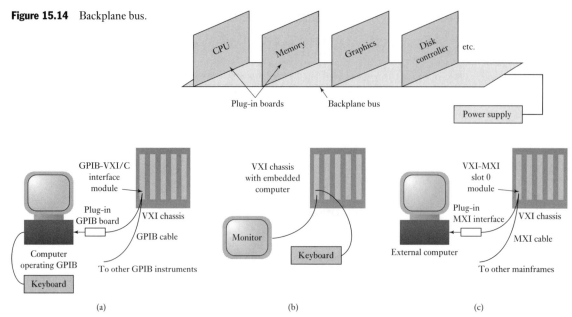

Figure 15.15   VXI options.

The VXIbus (VME Extensions for Instrumentation) is an extension of the specification of the VMEbus which has been designed for instrumentation applications such as automatic test equipment where higher speed communications are required than possible with the GPIB bus. It also provides better synchronisation and triggering and has been developed by a consortium of instrument manufacturers so that interoperability is possible between the products of different companies. The system involves VXI boards plugging into a mainframe. Figure 15.15 shows a number of possible system configurations that can be used. In Figure 15.15(a) a VXI mainframe is linked to an external controller, a computer, via a GPIB link. The controller talks across this link using a GPIB protocol to an interface board in the chassis which translates the GPIB Protocol into the VXI Protocol. This makes VXI instruments appear to the controller to be GPIB instruments and enables them to be programmed using GPIB methods. Figure 15.15(b) shows the complete computer embedded in the VXI chassis. This option offers the smallest possible physical size for the system and allows the computer to use directly the VXI backplane bus. Figure 15.15(c) uses a special high-speed system-bus-on-cable, the MXIbus, to link a computer and the VXI chassis, the MXI being 20 times faster than the GPIB.

15.8   Wireless protocols

IEEE 802.11 is a proposed standard for wireless LANs, specifying both the physical (PHY) and medium access control (MAC) layers of the network. The MAC layer specifies a carrier sense multiple access collision avoidance (CSMA/CA) protocol. With this, when a node has a packet ready for transmission, it first listens to ensure no other node is transmitting and if clear then transmits. Otherwise, it waits and then tries again. When a packet is transmitted, the transmitting node first sends out a ready-to-send (RTS) packet containing information about the length of the packet and then sends its packet. When the packet is successfully received, the receiving node transmits an acknowledgement (ACK) packet.

Bluetooth is a global standard for short-range radio transmission. When two Bluetooth-equipped devices are within 10 m of each other, a connection can be established. It is widely used for mobile phones and PCs.

Summary

An **external bus** is a set of signal lines that interconnects microprocessors, microcontrollers, computers and PLCs and also connects them with peripheral equipment.

Centralised computer control involves the use of one central computer to control an entire plant. With the **hierarchical system**, there is a hierarchical system of computers according to the tasks they carry out. With the **distributed system**, each computer system carries out essentially similar tasks to all the other computer systems.

Data communication can be via **parallel** or **serial transmission** links. Serial data transmission can be either asynchronous or synchronous transmission. **Asynchronous transmission** implies that both the transmitter and receiver computers are not synchronised, each having its own independent clock signals. Serial data transmission occurs in one of three modes: simplex, half-duplex and full-duplex.

The term **network** is used for a system which allows two or more computers/microprocessors to be linked for the interchange of data. Commonly used forms are data bus, star, hierarchy/tree, ring and mesh. Network access control is necessary to ensure that only one user is able to transmit at any one time; with ring-based networks, methods used are token passing and slot passing, while carrier sense multiple access with collision detection is used with bus or hierarchy networks. A **protocol** is a formal set of rules governing data format, timing, sequencing, access control and error control.

The International Organization for Standardization (ISO) has defined a seven-layer standard protocol system known as the **Open Systems Interconnection** (OSI) model.

Serial communication interfaces include RS-232 and its later versions, I^2C and CAN. **Parallel communication interfaces** include the General Purpose Instrument Bus (GPIB).

Problems

15.1 Explain the difference between centralised and distributed communication systems.

15.2 Explain the forms of bus/tree and ring networks.

15.3 A LAN is required with a distance between nodes of more than 100 m. Should the choice be bus or ring topology?

15.4   A multichannel LAN is required. Should the choice be broadband or baseband transmission?

15.5   What are MAP and TOP?

15.6   Explain what is meant by a communication protocol.

15.7   Briefly explain the two types of multiple-access control used with LANs.

15.8   A microcontroller M68HC11 is a 'listener' to be connected to a 'talker' via a GPIB bus. Indicate the connections to be made if full handshaking is to be used.

15.9   What problem has to be overcome before the serial data communications interface of the microcontroller M68HC11 can output data through an RS-232C interface?

15.10   What is a backplane bus?

Chapter sixteen Fault finding

Objectives

The objectives of this chapter are that, after studying it, the reader should be able to:
- Recognise the techniques used to identify faults in microprocessor-based systems, including both hardware and software.
- Explain the use of emulation and simulation.
- Explain how fault finding can be achieved with PLC systems.

16.1 Fault-detection techniques

This chapter is a brief consideration of the problems of fault detection with measurement, control and data communication systems. For details of the fault-finding checks required for specific systems or components, the manufacturer's manuals should be used.

There are a number of techniques that can be used to detect faults.

1 *Replication checks*

This involves duplicating or replicating an activity and comparing the results. In the absence of faults it is assumed that the results should be the same. It could mean, with transient errors, just repeating an operation twice and comparing the results or it could involve having duplicate systems and comparing the results given by the two. This can be an expensive option.

2 *Expected value checks*

Software errors are commonly detected by checking whether an expected value is obtained when a specific numerical input is used. If the expected value is not obtained then there is a fault.

3 *Timing checks*

This involves the use of timing checks that some function has been carried out within a specified time. These checks are commonly referred to as **watchdog timers**. For example, with a programmable logic controller (PLC), when an operation starts, a timer is also started and if the operation is not completed within the specified time a fault is assumed to have occurred. The watchdog timer trips, sets off an alarm and closes down part or the entire plant.

4 *Reversal checks*

Where there is a direct relationship between input and output values, the value of the output can be taken and the input which should have caused it computed. This can then be compared with the actual input.

5 *Parity and error coding checks*

This form of checking is commonly used for detecting memory and data transmission errors. Communication channels are frequently subject to interference which can affect data being transmitted. To detect whether data has been corrupted, a parity bit is added to the transmitted data word. The parity bit is chosen to make the resulting number of ones in the group either odd (odd parity) or even (even parity). If odd parity then the word can be checked after transmission to see if it is still odd. Other forms of checking involve codes added to transmitted data in order to detect corrupt bits.

6 *Diagnostic checks*

Diagnostic checks are used to test the behaviour of components in a system. Inputs are applied to a component and the outputs compared with those which should occur.

16.2 Watchdog timer

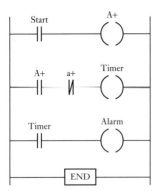

Figure 16.1 Watchdog timer.

A watchdog timer is basically a timer that the system must reset before it times out. If the timer is not reset in time then an error is assumed to have occurred.

As an illustration of such a timer, Figure 16.1 shows a simple ladder program that can be used to provide a PLC with a watchdog timer for an operation involving the movement of a piston in a cylinder. When the start switch is closed, the solenoid A+ is activated and starts the movement of the piston in the cylinder. It also starts a timer. When the piston is fully extended it opens the limit switch a+. This stops the timer. However, if a+ is not opened before the timer has timed out then the timer contacts close and an alarm is sounded. Thus the timer might be set for, say, 4 s on the assumption that the piston will be fully extended within that time. If, however, it sticks and fails to meet this deadline then the alarm sounds.

When a microprocessor executes instructions from its memory, a nearby electrical disturbance might momentarily upset the processor data bus and the wrong byte is accessed. Alternatively a software bug can result in the processor getting into problems when it returns from a subroutine. The consequence of such faults is that the system may crash with possibly dangerous consequences for actuators controlled by the microprocessor. To avoid this happening with critical systems, a watchdog timer is used to reset the microprocessor.

As an illustration of the provision in microprocessor-based systems of internal watchdog timers, consider the microcontroller MC68HC11 which includes an internal watchdog timer, called **computer operating properly (COP)**, to detect software processing errors. When the COP timer has been started it is necessary for the main program to reset the COP periodically before it times out. If the watchdog timer times out before being set to start timing all over again then a COP failure reset occurs. The COP timer can be reset to zero time by writing a $55 (0x55 in C language) to the COP reset register (COPRST) at address $103A (0x103A) followed later in the program by writing a $AA (0xAA) to clear the COP timer. If the program hangs in between the two instructions then COP times out and results in the COP failure reset routine being executed. The program, in assembly language, thus has the following lines:

```
LDAA    #$55      ; reset timer
STAA    $103A     ; writing $55 to COPRST
                  ; other program lines
```

```
LDAA    #$AA        ; clearing timer
STAA    $103A       ; writing $AA to COPRST
```

The COP operating period is set by setting CR1 and CR2 in the OPTION register, address $1039 (0x1039), to either 0 or 1. For example, with CR1 set to 0 and CR2 set to 0 a time out of 16.384 ms might be given, whereas with CR1 set to 1 and CR2 set to 0 a time out of 262.14 ms is given.

16.3 Parity and error coding checks

In order to try and detect when a data signal has been corrupted and has an error as a result of noise, error detection techniques such as **parity checks** are used.

In Section 4.5.2 a brief account was given of the parity method for error detection. With such a method an extra bit is added to a message to make the total number of ones an even number when even parity is used or an odd number when odd parity is used. For example, the character 1010000 would have a parity bit placed after the most significant bit of a 0 with an even-parity system, i.e. 01010000, or a 1 with odd parity, i.e. 11010000.

Such a method can detect the presence of a single error in the message but not the presence of two errors which result in no change in parity, e.g. with even parity and the above number a single error in, say, the third bit would be detected in 1101100 because the parity check bit would be wrong, but not if also there was an error in the first bit since 1101110 would have the correct parity bit.

If no error is detected the signal is acknowledged as being error-free by the return of the ACK character to the sending terminal; if an error is detected the signal NAK is used. This is called an **automatic repeat request** (ARQ). The NAK signal then results in the retransmission of the message.

The efficiency of error detection can be increased by the use of **block parity**. The message is divided into a number of blocks and each block has a block check character added at the end of the block. For example, with the following block, a check bit for even parity is placed at the end of each row and a further check bit at the foot of each column:

	Information bits				Check bit
First symbol	0	0	1	1	0
Second symbol	0	1	0	0	1
Third symbol	1	0	1	1	1
Fourth symbol	0	0	0	0	0
Block check bits	1	1	0	0	0

At the receiver the parity of each row and each column is checked and any single error is detected by the intersection of the row and column containing the error check bit.

Another form of error detection is the **cyclic redundancy check** (CRC). At the transmitting terminal the binary number representing the data being transmitted is divided by a predetermined number using modulo-2 arithmetic. The remainder from the division is the CRC character which is transmitted with the data. At the receiver the data plus the CRC character are divided by the same number. If no transmission errors have occurred there is no remainder.

A common CRC code is CRC-16, 16 bits being used for the check sequence. The 16 bits are considered to be the coefficients of a polynomial with the number of bits equal to the highest power of the polynomial. The data block is first multiplied by the highest power of the polynomial, i.e. x^{16}, and then divided by the CRC polynomial

$$x^{16} + x^{12} + x^5 + 1$$

using modulo-2 arithmetic, i.e. $x = 2$ in the polynomial. The CRC polynomial is thus 10001000000100001. The remainder of the division by this polynomial is the CRC.

As an illustration, suppose we have the data 10110111 or polynomial

$$x^7 + x^5 + x^4 + x^2 + x^1 + 1$$

and a CRC polynomial of

$$x^5 + x^4 + x^1 + 1$$

or 110011. The data polynomial is first multiplied by x^5 to give

$$x^{12} + x^{10} + x^9 + x^7 + x^6 + x^5$$

and so 1011011100000. Dividing this by the CRC polynomial gives

```
                    11010111
110011 |1011011100000
        110011
         110011
         110011
         100100
         110011
          101110
          110011
           111010
           110011
            01001
```

and so a remainder of 01001 which then becomes the CRC code transmitted with the data.

16.4 Common hardware faults

The following are some of the commonly encountered faults that can occur with specific types of components and systems.

16.4.1 Sensors

If there are faults in a measurement system then the sensor might be at fault. A simple test is to substitute the sensor with a new one and see what effect this has on the results given by the system. If the results change then it is likely that the original sensor was faulty; if the results do not change then the fault is elsewhere in the system. It is also possible to check that the voltage/current sources are supplying the correct voltages/currents, whether there is electrical continuity in connecting wires, that the sensor is correctly mounted and used under the conditions specified by the manufacturer's data sheet, etc.

16.4.2 Switches and relays

Dirt and particles of waste material between switch contacts are a common source of incorrect functioning of mechanical switches. A voltmeter used across a switch should indicate the applied voltage when the contacts are open and very nearly zero when they are closed. Mechanical switches used to detect the position of some item, e.g. the presence of a workpiece on a conveyor, can fail to give the correct responses if the alignment is incorrect or if the actuating lever is bent.

Inspection of a relay can disclose evidence of arcing or contact welding. The relay should then be replaced. If a relay fails to operate then a check can be made for the voltage across the coil. If the correct voltage is present then coil continuity can be checked with an ohmmeter. If there is no voltage across the coil then the fault is likely to be the switching transistor used with the relay.

16.4.3 Motors

Maintenance of both d.c. and a.c. motors involves correct lubrication. With d.c. motors the brushes wear and can require changing. Setting of new brushes needs to be in accordance with the manufacturer's specification. A single-phase capacitor start a.c. motor that is sluggish in starting probably needs a new starting capacitor. The three-phase induction motor has no brushes, commutator, slip rings or starting capacitor and, short of a severe overload, the only regular maintenance that is required is periodic lubrication.

16.4.4 Hydraulic and pneumatic systems

A common cause of faults with hydraulic and pneumatic systems is dirt. Small particles of dirt can damage seals, block orifices, cause valve spools to jam, etc. Thus filters should be regularly checked and cleaned, components should only be dismantled in clean conditions, and oil should be regularly checked and changed. With an electric circuit a common method of testing the circuit is to measure the voltages at a number of test points. Likewise, with a hydraulic and pneumatic system there needs to be points at which pressures can be measured. Damage to seals can result in hydraulic and pneumatic cylinders leaking, beyond that which is normal, and result in a drop in system pressure when the cylinder is actuated. This can be remedied by replacing the seals in the cylinders. The vanes in vane-type motors are subject to wear and can then fail to make a good seal with the motor housing, with the result of a loss of motor power. The vanes can be replaced. Leaks in hoses, pipes and fittings are common faults.

16.5 Microprocessor systems

Typical faults in microprocessor systems are:

1 *Chip failure*
Chips are fairly reliable but occasionally there can be failure.

2 *Passive component failure*
Microprocessor systems will usually include passive components such as resistors and capacitors. Failure of any of these can cause system malfunction.

3 *Open circuits*
Open circuits can result in a break in a signal path or in a power line. Typical reasons for such faults are unsoldered or faulty soldered joints, fracture of a printed circuit track, a faulty connection on a connector and breaks in cables.

4 *Short circuits*
Short circuits between points on a board which should not be connected often arise as a result of surplus solder bridging the gaps between neighbouring printed circuit tracks.

5 *Externally introduced interference*
Externally induced pulses will affect the operation of the system since they will be interpreted as valid digital signals. Such interference can originate from the mains supply having 'spikes' as a result of other equipment sharing the same mains circuit and being switched on or off. Filters in the mains supply to the system can be used to remove such 'spikes'.

6 *Software faults*
Despite extensive testing it is still quite feasible for software to contain bugs and under particular input or output conditions cause a malfunction.

16.5.1 Fault-finding techniques

Fault-finding techniques that are used with microprocessor-based systems include those listed below.

1 *Visual inspection*
Just carefully looking at a faulty system may reveal the source of a fault, e.g. an integrated circuit which is loose in its holder or surplus solder bridging tracks on a board.

2 *Multimeter*
This is of limited use with microprocessor systems but can be used to check for short- or open-circuit connections and the power supplies.

3 *Oscilloscope*
The oscilloscope is essentially limited to where repetitive signals occur and the most obvious such signal is the clock signal. Most of the other signals with a microprocessor system are not repetitive and depend on the program being executed.

4 *Logic probe*
The logic probe is a hand-held device (Figure 16.2(a)), shaped like a pen, which can be used to determine the logic level at any point in the

Figure 16.2 (a) Logic probe, (b) current tracer.

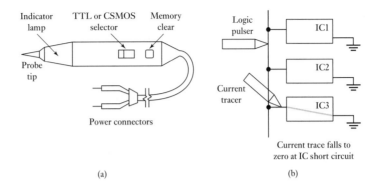

(a)

(b)

circuit to which it is connected. The selector switch is set for TTL or CSMOS operations and when the probe tip is touched to the point in question, the indicator lamp indicates whether it is below the logic level 0 threshold, above the logic 1 threshold or a pulsating signal. A pulse stretching circuit is often included with the probe in order to stretch the duration of a pulse to allow sufficient time for it to operate the indicator lamp and the effect to be noticed. A memory circuit can be used for detecting a single pulse, the memory clear button being pressed to turn off the indicator lamp and then any change in logic level is registered by the lamp.

5 *Logic pulser*
The logic pulser is a hand-held pulse generator, shaped like a pen, that is used to inject controlled pulses into circuits. The pulser probe tip is pressed against a node in the circuit and the button on the probe pressed to generate a pulse. It is often used with the logic probe to check the functions of logic gates.

6 *Current tracer*
The current tracer is similar to the logic probe but it senses pulsing current in a circuit rather than voltage levels. The tip of the current tracer is magnetically sensitive and is used to detect the changing magnetic field near a conductor carrying a pulsing current. The current tracer tip is moved along printed circuit tracks to trace out the low-impedance paths along which current is flowing (Figure 16.2(b)).

7 *Logic clip*
A logic clip is a device which clips to an integrated circuit and makes contact with each of the integrated circuit pins. The logic state of each pin is then shown by LED indicators, there being one for each pin.

8 *Logic comparator*
The logic comparator tests integrated circuits by comparing them to a good, reference, integrated circuit (Figure 16.3). Without removing the integrated circuit being tested from its circuit, each input pin is connected in parallel with the corresponding input pin on the reference integrated circuit; likewise each output pin is connected with the corresponding output pin on the reference integrated circuit. The two outputs are compared with an EXCLUSIVE-OR gate, which then gives an output when the two outputs differ. The pulse stretcher is used to extend the

Figure 16.3　Logic comparator.

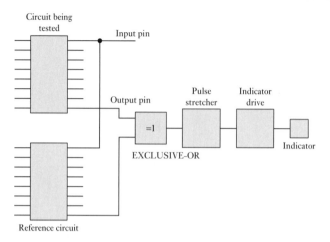

duration of the signal fed to the indicator so that very short-duration pulses will result in the indicator being on for a noticeable period.

9　*Signature analyser*

With analogue systems, fault finding usually involves tracing through the circuitry and examining the waveforms at various nodes, comparison of the waveforms with what would be expected enabling faults to be identified and located. With digital systems the procedure is more complex since trains of pulses at nodes all look very similar. To identify whether there is a fault, the sequence of pulses is converted into a more readily identifiable form, e.g. 258F, this being termed the **signature**. The signature obtained at a node can then be compared with that which should occur. When using the signature analyser with a circuit, it is often necessary for the circuit to have been designed so that data bus feedback paths can be broken easily for the test to stop faulty digital sequences being fed back during the testing. A short program, which is stored in ROM, is activated to stimulate nodes and enable signatures to be obtained. The microprocessor itself can be tested if the data bus is broken to isolate it from memory and it is then made to 'free-run' and give a 'no operation' (NO) instruction to each of its addresses in turn. The signatures for the microprocessor bus in this state can then be compared with those expected.

10　*Logic analyser*

The logic analyser is used to sample and store simultaneously in a 'first-in–first-out' (FIFO) memory the logic levels of bus and control signals in a unit under test. The point in the program at which the data capture starts or finishes is selected by the use of a 'trigger word'. The analyser compares its trigger word with the incoming data and only starts to store data when the word occurs in the program. Data capture then continues for a predetermined number of clock pulses and is then stopped. The stored data may then be displayed as a list of binary, octal, decimal or hexadecimal codes, or as a time display in which the waveforms are displayed as a function of time, or as a mnemonic display.

16.5.2 Systematic fault-location methods

The following are systematic fault-location methods.

1 *Input to output*
   A suitable input signal is injected into the first block of the system and then measurements are made in sequence, starting from the first block, at the output of each block in turn until the faulty block is found.

2 *Output to input*
   A suitable input signal is injected into the first block of the system and then measurements are made in sequence, starting from the last block, at the output of each block in turn until the faulty block is found.

3 *Half-split*
   A suitable input signal is injected into the first block of the system. The blocks constituting the system are split in half and each half tested to determine in which half the fault lies. The faulty half is then split into half and the procedure repeated.

16.5.3 Self-testing

Software can be used by a microprocessor-based system to institute a self-test program for correct functioning. Such programs are often initiated during the start-up sequence of a system when it is first switched on. For example, printers include microprocessors in their control circuits and generally the control program stored in ROM also includes test routines. Thus when first switched on, it goes through these test routines and is not ready to receive data until all tests indicate the system is fault-free.

A basic ROM test involves totalling all the data bytes stored in each location in ROM and comparing the sum against that already stored (the so-called **checksum test**). If there is a difference then the ROM is faulty; if they agree it is considered to be fault-free. A basic RAM test involves storing data patterns in which adjacent bits are at opposite logic levels, i.e. hex 55 and AA, into every memory location and then reading back each value stored to check that it corresponds to the data sent (the so-called **checker board test**).

16.6 Emulation and simulation

An emulator is a test board which can be used to test a microcontroller and its program. The board contains:

1 the microcontroller;
2 memory chips for the microcontroller to use as data and program memory;
3 an input/output port to enable connections to be made with the system under test;
4 a communications port to enable program code to be downloaded from a computer and the program operation to be monitored.

The program code can be written in a host computer and then downloaded through either a serial or a parallel link into the memory on the board. The microcontroller then operates as though this program were contained within its own internal memory. Figure 16.4 shows the general arrangement.

Figure 16.4   Using an emulator.

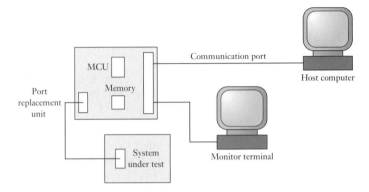

The input/output lines of the microcontroller are connected via an input/output port on the board to a plug-in device on the system under test so that it operates as though it had the microcontroller plugged into it. The board is already programmed with a monitor system which enables the operation of the program to be monitored and the contents of memory, registers, input/output ports to be checked and modified.

Figure 16.5 shows the basic elements of the evaluation board MC68HC11EVB that is provided by Motorola. This uses a monitor program called Bit User Fast Friendly Aid to Logical Operations (Buffalo). The 8K EPROM contains the Buffalo monitor. An MC6850 asynchronous communications interface adapter (ACIA) (see Section 13.5) is used for interfacing parallel and serial lines. A partial RS-232 interface is supplied with each of the two serial ports for connection to the host computer and the monitoring terminal.

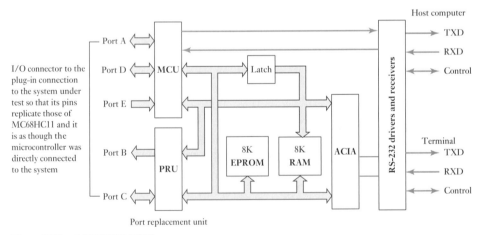

Figure 16.5   MC68HC11EVB.

16.6.1 Simulation

Instead of testing a program by running it with an actual microcontroller, one can test it by running it with a computer program that *simulates* the microcontroller. Such simulation can assist in the debugging of the program code. The display screen can be divided into a number of windows in which information is displayed such as the source code as it is executed, the central processing unit (CPU) registers and flags with their current states, the input/output ports, registers and timers, and the memory situation.

16.7 PLC systems

PLCs have a high reliability. They are electrically isolated by optoisolators or relays from potentially damaging voltages and currents at input/output ports; battery-backed RAM protects the application software from power failures or corruption; and the construction is so designed that the PLCs can operate reliably in industrial conditions for long periods of time. PLCs generally have several built-in fault procedures. Critical faults cause the CPU to stop, while other less critical faults cause the CPU to continue running but display a fault code on a display. The PLC manual will indicate the remedial action required when a fault code is displayed.

16.7.1 Program testing

The software checking program checks through a ladder program for incorrect device addresses, and provides a list on a screen, or as a printout, of all the input/output points used, counter and timer settings, etc., with any errors detected. Thus the procedure might involve:

1 opening and displaying the ladder program concerned;
2 selecting from the menu on the screen Ladder Test;
3 the screen might then display the message: Start from beginning of program (Y/N)?;
4 type Y and press Enter;
5 any error message is then displayed or the message of 'No errors found'.

For example, there might be a message for a particular output address that it is used as an output more than once in the program, a timer or counter is being used without a preset value, a counter is being used without a reset, there is no END instruction, etc. As a result of such a test, there may be a need to make changes to the program. Changes needed to rectify the program might be made by selecting exchange from the menu displayed on screen and following through the set of displayed screen messages.

16.7.2 Testing inputs and outputs

Most PLCs have the facility for testing inputs and outputs by what is termed forcing. Software is used to 'force' inputs and outputs on or off. To force inputs or outputs, a PLC has to be switched into the forcing or monitor mode by perhaps pressing a key marked FORCE or selecting the MONITOR mode on a screen display. As a result of forcing an input we can check that the consequential action of that input being on occurs. The installed program

Figure 16.6 Monitor mode symbols.

Open   Closed   Not energised   Energised

can thus be run and inputs and outputs simulated so that they, and all preset values, can be checked. However, care must be exercised with forcing in that an output might be forced that can result in a piece of hardware moving in an unexpected and dangerous manner.

As an illustration of the type of display obtained when forcing, Figure 16.6 shows how inputs might appear in the ladder program display when open and closed, and outputs when not energised and energised, and Figure 16.7(a) a selected part of a displayed ladder program and Figure 16.7(b) what happens when forcing occurs. Initially, Figure 16.7(a) shows rung 11, with inputs to X400, X401 and M100, but not X402, and with no output from Y430. For rung 12, the timer T450 contacts are closed, the display at the bottom of the screen indicating that there is no time left to run on T450. Because Y430 is not energised, the Y430 contacts are open and so there is no output from Y431. If we now force an input to X402 then the screen display changes to that shown in Figure 16.7(b) and Y430 is energised and consequently Y431 energised.

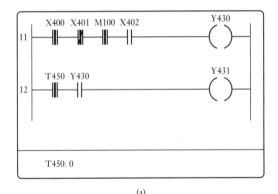
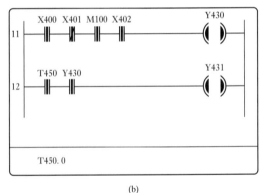

(a)   (b)

Figure 16.7 Forcing an input.

16.7.3 PLC as a monitor of systems

The PLC can also be used as a monitor of the system being controlled. It can be used to sound an alarm or light up a red light if inputs move outside prescribed limits, using the greater than, equal to, or less than functions, or its operations take longer than a prescribed time. See Figure 16.1 for an illustration of how a PLC ladder program might be used as a watchdog timer for some operation.

Often with PLC systems, status lamps are used to indicate the last output that has been set during a process and so, if the system stops, where the fault has occurred. The lamps are built into the program so that as each output occurs a lamp comes on and turns off the previous output status lamps. Figure 16.8 illustrates this.

Figure 16.8 Last output set diagnostic program.

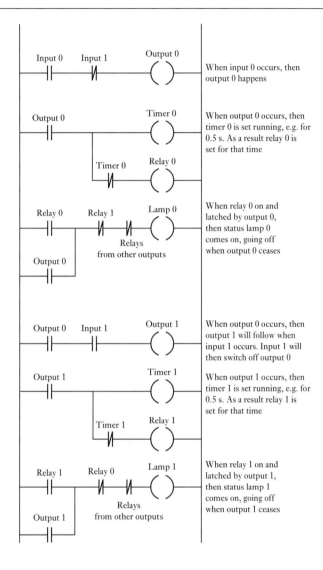

| | When input 0 occurs, then output 0 happens |
| Input 0 Input 1 Output 0 | |

When input 0 occurs, then
output 0 happens

When output 0 occurs, then
timer 0 is set running, e.g. for
0.5 s. As a result relay 0 is
set for that time

When relay 0 on and
latched by output 0,
then status lamp 0
comes on, going off
when output 0 ceases

When output 0 occurs, then
output 1 will follow when
input 1 occurs. Input 1 will
then switch off output 0

When output 1 occurs, then
timer 1 is set running, e.g. for
0.5 s. As a result relay 1 is
set for that time

When relay 1 on and
latched by output 1,
then status lamp 1
comes on, going off
when output 1 ceases

Summary

Techniques used to detect faults are replication checks, expected value checks, timing checks, i.e. watchdog timers, reversal checks, parity and error coding checks and diagnostic checks.

A **watchdog timer** is basically a timer that the system must reset before it times out. If the timer is not reset in time then an error is assumed to have occurred. **Parity checks** involve an extra bit being added to a message to make the total number of ones an even number when even parity is used or an odd number when odd parity is used. The efficiency of error detection can be increased by the use of **block parity** in which the message is divided into a number of blocks and each block has a check character added at the end of the block. The **cyclic redundancy check** (CRC) involves the binary number representing the data being transmitted being divided by a predetermined number using modulo-2 arithmetic. The remainder from the division is the CRC character which is transmitted with the data. At the receiver the data plus the CRC character are divided by the same number. If no transmission errors have occurred there is no remainder.

Software can be used by a microprocessor-based system to institute a self-test program for correct functioning. An **emulator** is a test board which can be used to test a microcontroller and its program. Instead of testing a program by running it with an actual microcontroller, one can test it by running it with a computer program that *simulates* the microcontroller.

PLCs generally have several built-in fault procedures. Critical faults cause the CPU to stop, while other less critical faults cause the CPU to continue running but display a fault code on a display. Most PLCs have the facility for testing inputs and outputs by what is termed **forcing**. Software is used to 'force' inputs and outputs on or off.

Problems

16.1   Explain what is meant by (a) replication checks, (b) expected value checks, (c) reversal checks, (d) parity checks.

16.2   Explain how a watchdog timer can be used with a PLC-controlled plant in order to indicate the presence of faults.

16.3   Explain the function of COP in the microcontroller MC68HC11.

16.4   The F2 series Mitsubishi PLC is specified as having:

Diagnosis: Programmable check (sum, syntax, circuit check), watchdog timer, battery voltage, power supply voltage

Explain the significance of the terms.

16.5   Explain how self-testing can be used by a microprocessor-based system to check its ROM and RAM.

Part V
System models

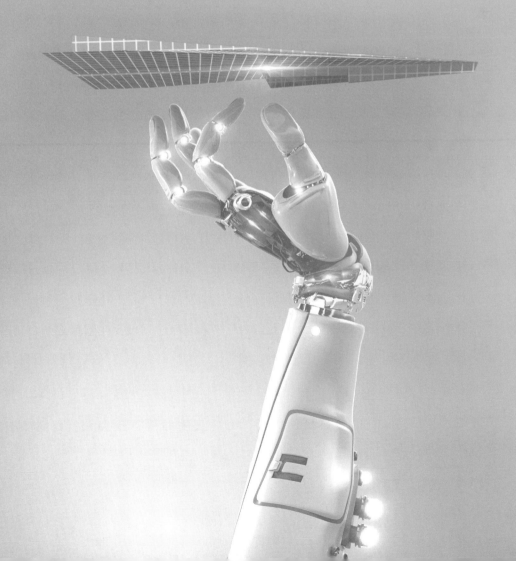

Chapter seventeen

Basic system models

Objectives

The objectives of this chapter are that, after studying it, the reader should be able to:
- Explain the importance of models in predicting the behaviour of systems.
- Devise models from basic building blocks for mechanical, electrical, fluid and thermal systems.
- Recognise analogies between mechanical, electrical, fluid and thermal systems.

17.1 Mathematical models

Consider the following situation. A microprocessor switches on a motor. How will the rotation of the motor shaft vary with time? The speed will not immediately assume the full-speed value but will only attain that speed after some time. Consider another situation. A hydraulic system is used to open a valve which allows water into a tank to restore the water level to that required. How will the water level vary with time? The water level will not immediately assume the required level but will only attain that level after some time.

In order to understand the behaviour of systems, **mathematical models** are needed. These are simplified representations of certain aspects of a real system. Such a model is created using equations to describe the relationship between the input and output of a system and can then be used to enable predictions to be made of the behaviour of a system under specific conditions, e.g. the outputs for a given set of inputs, or the outputs if a particular parameter is changed. In devising a mathematical model of a system it is necessary to make assumptions and simplifications and a balance has to be chosen between simplicity of the model and the need for it to represent the actual real-world behaviour. For example, we might form a mathematical model for a spring by assuming that the extension x is proportional to the applied force F, i.e. $F = kx$. This simplified model might not accurately predict the behaviour of a real spring where the extension might not be precisely proportional to the force and where we cannot apply this model regardless of the size of the force, since large forces will permanently deform the spring and might even break it and this is not predicted by the simple model.

The basis for any mathematical model is provided by the fundamental physical laws that govern the behaviour of the system. In this chapter a range of systems will be considered, including mechanical, electrical, thermal and fluid examples.

Like a child building houses, cars, cranes, etc., from a number of basic building blocks, systems can be made up from a range of building blocks.

Each building block is considered to have a single property or function. Thus, to take a simple example, an electric circuit system may be made up from building blocks which represent the behaviour of resistors, capacitors and inductors. The resistor building block is assumed to have purely the property of resistance, the capacitor purely that of capacitance and the inductor purely that of inductance. By combining these building blocks in different ways, a variety of electric circuit systems can be built up and the overall input/output relationships obtained for the system by combining in an appropriate way the relationships for the building blocks. Thus a mathematical model for the system can be obtained. A system built up in this way is called a **lumped parameter** system. This is because each parameter, i.e. property or function, is considered independently.

There are similarities in the behaviour of building blocks used in mechanical, electrical, thermal and fluid systems. This chapter is about the basic building blocks and their combination to produce mathematical models for physical, real, systems. Chapter 18 looks at more complex models. It needs to be emphasised that such models are only aids in system design. Real systems often exhibit non-linear characteristics and can depart from the ideal models developed in these chapters. This matter is touched on in Chapter 18.

17.2 Mechanical system building blocks

The models used to represent mechanical systems have the basic building blocks of springs, dashpots and masses. **Springs** represent the stiffness of a system, **dashpots** the forces opposing motion, i.e. frictional or damping effects, and **masses** the inertia or resistance to acceleration (Figure 17.1). The mechanical system does not have to be really made up of springs, dashpots and masses but have the properties of stiffness, damping and inertia. All these building blocks can be considered to have a force as an input and a displacement as an output.

The stiffness of a spring is described by the relationship between the forces F used to extend or compress a spring and the resulting extension or compression x (Figure 17.1(a)). In the case of a spring where the extension or compression is proportional to the applied forces, i.e. a linear spring,

$$F = kx$$

where k is a constant. The bigger the value of k, the greater the forces have to be to stretch or compress the spring and so the greater the stiffness. The

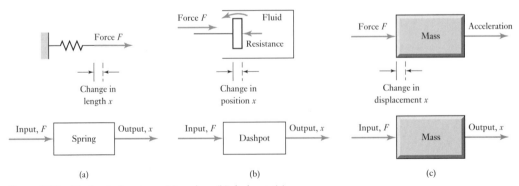

Figure 17.1   Mechanical systems: (a) spring, (b) dashpot, (c) mass.

object applying the force to stretch the spring is also acted on by a force, the force being that exerted by the stretched spring (Newton's third law). This force will be in the opposite direction and equal in size to the force used to stretch the spring, i.e. kx.

The dashpot building block represents the types of forces experienced when we endeavour to push an object through a fluid or move an object against frictional forces. The faster the object is pushed, the greater the opposing forces become. The dashpot which is used pictorially to represent these damping forces which slow down moving objects consists of a piston moving in a closed cylinder (Figure 17.1(b)). Movement of the piston requires the fluid on one side of the piston to flow through or past the piston. This flow produces a resistive force. In the ideal case, the damping or resistive force F is proportional to the velocity v of the piston. Thus

$$F = cv$$

where c is a constant. The larger the value of c, the greater the damping force at a particular velocity. Since velocity is the rate of change of displacement x of the piston, i.e. $v = dx/dt$, then

$$F = c\frac{dx}{dt}$$

Thus the relationship between the displacement x of the piston, i.e. the output, and the force as the input is a relationship depending on the rate of change of the output.

The mass building block (Figure 17.1(c)) exhibits the property that the bigger the mass, the greater the force required to give it a specific acceleration. The relationship between the force F and the acceleration a is (Newton's second law) $F = ma$, where the constant of proportionality between the force and the acceleration is the constant called the mass m. Acceleration is the rate of change of velocity, i.e. dv/dt, and velocity v is the rate of change of displacement x, i.e. $v = dx/dt$. Thus

$$F = ma = m\frac{dv}{dt} = m\frac{d(dx/dt)}{dt} = m\frac{d^2x}{dt^2}$$

Energy is needed to stretch the spring, accelerate the mass and move the piston in the dashpot. However, in the case of the spring and the mass we can get the energy back but with the dashpot we cannot. The spring when stretched stores energy, the energy being released when the spring springs back to its original length. The energy stored when there is an extension x is $\frac{1}{2}kx^2$. Since $F = kx$ this can be written as

$$E = \frac{1}{2}\frac{F^2}{k}$$

There is also energy stored in the mass when it is moving with a velocity v, the energy being referred to as kinetic energy, and released when it stops moving:

$$E = \frac{1}{2}mv^2$$

However, there is no energy stored in the dashpot. It does not return to its original position when there is no force input. The dashpot dissipates energy rather than storing it, the power P dissipated depending on the velocity v and being given by

$$P = cv^2$$

17.2.1   Rotational systems

The spring, dashpot and mass are the basic building blocks for mechanical systems where forces and straight line displacements are involved without any rotation. If there is rotation then the equivalent three building blocks are a **torsional spring**, a **rotary damper** and the **moment of inertia**, i.e. the inertia of a rotating mass. With such building blocks the inputs are torque and the outputs angle rotated. With a torsional spring the angle θ rotated is proportional to the torque T. Hence

$$T = k\theta$$

With the rotary damper a disc is rotated in a fluid and the resistive torque T is proportional to the angular velocity ω, and since angular velocity is the rate at which angle changes, i.e. $d\theta/dt$,

$$T = c\omega = c\frac{d\theta}{dt}$$

The moment of inertia building block has the property that the greater the moment of inertia I, the greater the torque needed to produce an angular acceleration α:

$$T = I\alpha$$

Thus, since angular acceleration is the rate of change of angular velocity, i.e. $d\omega/dt$, and angular velocity is the rate of change of angular displacement, then

$$T = I\frac{d\omega}{dt} = I\frac{d(d\theta/dt)}{dt} = I\frac{d^2\theta}{dt^2}$$

The torsional spring and the rotating mass store energy; the rotary damper just dissipates energy. The energy stored by a torsional spring when twisted through an angle θ is $\frac{1}{2}k\theta^2$ and since $T = k\theta$ this can be written as

$$E = \frac{1}{2}\frac{T^2}{k}$$

The energy stored by a mass rotating with an angular velocity ω is the kinetic energy E, where

$$E = \frac{1}{2}I\omega^2$$

The power P dissipated by the rotatory damper when rotating with an angular velocity ω is

$$P = c\omega^2$$

Table 17.1 summarises the equations defining the characteristics of the mechanical building blocks when there is, in the case of straight line

Table 17.1 Mechanical building blocks.

Building block	Describing equation	Energy stored or power dissipated
Translational		
Spring	$F = kx$	$E = \dfrac{1}{2}\dfrac{F^2}{k}$
Dashpot	$F = c\dfrac{dx}{dt} = cv$	$P = cv^2$
Mass	$F = m\dfrac{d^2x}{dt^2} = m\dfrac{dv}{dt}$	$E = \dfrac{1}{2}mv^2$
Rotational		
Spring	$T = k\theta$	$E = \dfrac{1}{2}\dfrac{T^2}{k}$
Rotational damper	$T = c\dfrac{d\theta}{dt} = c\omega$	$P = c\omega^2$
Moment of inertia	$T = I\dfrac{d^2\theta}{dt^2} = I\dfrac{d\omega}{dt}$	$E = \dfrac{1}{2}I\omega^2$

displacements (termed translational), a force input F and a displacement x output and, in the case of rotation, a torque T and angular displacement θ.

17.2.2 Building up a mechanical system

Many systems can be considered to be essentially a mass, a spring and dashpot combined in the way shown in Figure 17.2(a) and having an input of a force F and an output of displacement x (Figure 17.2(b)). To evaluate the relationship between the force and displacement for the system, the procedure to be adopted is to consider just one mass, and just the forces acting on that body. A diagram of the mass and just the forces acting on it is called a **free-body diagram** (Figure 17.2(c)).

When several forces act concurrently on a body, their single equivalent resultant can be found by vector addition. If the forces are all acting along the same line or parallel lines, this means that the resultant or net force acting on the block is the algebraic sum. Thus for the mass in Figure 17.2(c), if we consider just the forces acting on that block then the net force applied to the

Figure 17.2 (a) Spring–dashpot–mass, (b) system, (c) free-body diagram.

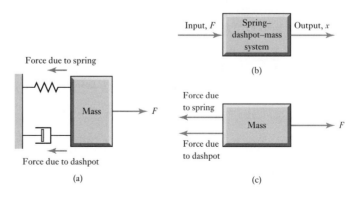

mass is the applied force F minus the force resulting from the stretching or compressing of the spring and minus the force from the damper. Thus

$$\text{net force applied to mass } m = F - kx - cv$$

where v is the velocity with which the piston in the dashpot, and hence the mass, is moving. This net force is the force applied to the mass to cause it to accelerate. Thus

$$\text{net force applied to mass } = ma$$

Hence

$$F - kx - c\frac{\mathrm{d}x}{\mathrm{d}t} = m\frac{\mathrm{d}^2x}{\mathrm{d}t^2}$$

or, when rearranged,

$$m\frac{\mathrm{d}^2x}{\mathrm{d}t^2} + c\frac{\mathrm{d}x}{\mathrm{d}t} + kx = F$$

This equation, called a **differential equation**, describes the relationship between the input of force F to the system and the output of displacement x. Because of the $\mathrm{d}^2x/\mathrm{d}t^2$ term, it is a second-order differential equation; a first-order differential equation would only have $\mathrm{d}x/\mathrm{d}t$.

There are many systems which can be built up from suitable combinations of the spring, dashpot and mass building blocks. Figure 17.3 illustrates some.

Figure 17.3(a) shows the model for a machine mounted on the ground and could be used as a basis for studying the effects of ground disturbances on the displacements of a machine bed. Figure 17.3(b) shows a model for the wheel and its suspension for a car or truck and can be used for the study of

Figure 17.3   Model for (a) a machine mounted on the ground, (b) the chassis of a car as a result of a wheel moving along a road, (c) the driver of a car as it is driven along a road.

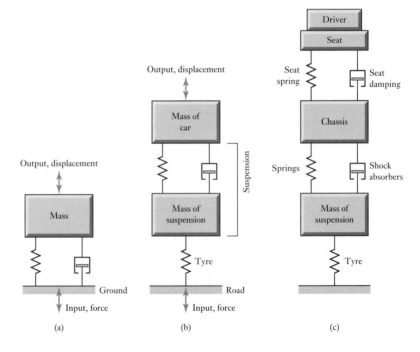

the behaviour that could be expected of the vehicle when driven over a rough road and hence as a basis for the design of the vehicle suspension. Figure 17.3(c) shows how this model can be used as part of a larger model to predict how the driver might feel when driven along a road. The procedure to be adopted for the analysis of such models is just the same as outlined above for the simple spring–dashpot–mass model. A free-body diagram is drawn for each mass in the system, such diagrams showing each mass independently and just the forces acting on it. Then for each mass the resultant of the forces acting on it is equated to the product of the mass and the acceleration of the mass.

To illustrate the above, consider the derivation of the differential equation describing the relationship between the input of the force F and the output of displacement x for the system shown in Figure 17.4.

The net force applied to the mass is F minus the resisting forces exerted by each of the springs. Since these are k_1x and k_2x, then

$$\text{net force} = F - k_1x - k_2x$$

Since the net force causes the mass to accelerate, then

$$\text{net force} = m\frac{\mathrm{d}^2x}{\mathrm{d}t^2}$$

Hence

$$m\frac{\mathrm{d}^2x}{\mathrm{d}t^2} + (k_1 + k_2)x = F$$

The procedure for obtaining the differential equation relating the inputs and outputs for a mechanical system consisting of a number of components can be summarised as:

1 isolate the various components in the system and draw free-body diagrams for each;
2 hence, with the forces identified for a component, write the modelling equation for it;
3 combine the equations for the various system components to obtain the system differential equation.

As an illustration, consider the derivation of the differential equation describing the motion of the mass m_1 in Figure 17.5(a) when a force F is applied. Consider the free-body diagrams (Figure 17.5(b)). For mass m_2 these are the force F and the force exerted by the upper spring. The force exerted by the upper spring is due to its being stretched by $(x_2 - x_3)$ and so is $k_2(x_3 - x_2)$. Thus the net force acting on the mass is

$$\text{net force} = F - k_2(x_3 - x_2)$$

This force will cause the mass to accelerate and so

$$F - k_2(x_3 - x_2) = m_2\frac{\mathrm{d}^2x_3}{\mathrm{d}t}$$

For the free-body diagram for mass m_1, the force exerted by the upper spring is $k_2(x_3 - x_2)$ and that by the lower spring is $k_1(x_1 - x_2)$. Thus the net force acting on the mass is

$$\text{net force} = k_1(x_2 - x_1) - k_2(x_3 - x_2)$$

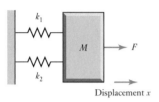

Figure 17.4 Example.

Figure 17.5   Mass–spring
system.

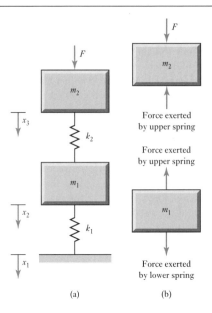

(a)   (b)

This force will cause the mass to accelerate and so

$$k_1(x_2 - x_1) - k_2(x_3 - x_2) = m_1\frac{\mathrm{d}^2x_2}{\mathrm{d}t}$$

We thus have two simultaneous second-order differential equations to describe the behaviours of the system.

Similar models can be constructed for rotating systems. To evaluate the relationship between the torque and angular displacement for the system the procedure to be adopted is to consider just one rotational mass block, and just the torques acting on that body. When several torques act on a body simultaneously, their single equivalent resultant can be found by addition in which the direction of the torques is taken into account. Thus a system involving a torque being used to rotate a mass on the end of a shaft (Figure 17.6(a)) can be considered to be represented by the rotational building blocks shown in Figure 17.6(b). This is a comparable situation with that analysed above (Figure 17.2) for linear displacements and yields a similar equation

$$I\frac{\mathrm{d}^2\theta}{\mathrm{d}t^2} + c\frac{\mathrm{d}\theta}{\mathrm{d}t} + k\theta = T$$

Figure 17.6   Rotating a mass
on the end of a shaft:
(a) physical situation,
(b) building block model.

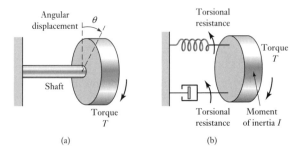

(a)   (b)

Figure 17.7 A two-gear train system.

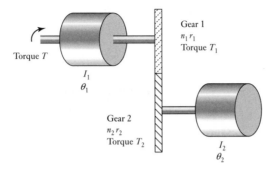

Gear 1
$n_1 r_1$
Torque T_1

Torque T

I_1
θ_1

Gear 2
$n_2 r_2$
Torque T_2

I_2
θ_2

Motors operating through gear trains to rotate loads are a feature of many control systems. Figure 17.7 shows a simple model of such a system. It consists of a mass of moment of inertia I_1 connected to gear 1 having n_1 teeth and a radius r_1 and a mass of moment of inertia I_2 connected to a gear 2 with n_2 teeth and a radius r_2. We will assume that the gears have negligible moments of inertia and also that rotational damping can be ignored.

If gear 1 is rotated through an angle θ_1 then gear 2 will rotate through an angle θ_2 where

$$r_1\theta_1 = r_2\theta_2$$

The ratio of the gear teeth numbers is equal to the ratio n of the gear radii:

$$\frac{r_1}{r_2} = \frac{n_1}{n_2} = n$$

If a torque T is applied to the system and torque T_1 is applied to gear 1 then the net torque is $T - T_1$ and so

$$T - T_1 = I_1\frac{d^2\theta_1}{dt^2}$$

If the torque T_2 occurs at gear 2 then

$$T_2 = I_2\frac{d^2\theta_2}{dt^2}$$

We will assume that the power transmitted by gear 1 is equal to that transmitted by gear 2 and so, as the power transmitted is the product of the torque and angular velocity, we have

$$T_1\frac{d\theta_1}{dt} = T_2\frac{d\theta_2}{dt}$$

Since $r_2\theta_1 = r_2\theta_2$ it follows that

$$r_1\frac{d\theta_1}{dt} = r_2\frac{d\theta_2}{dt^2}$$

and so

$$\frac{T_1}{T_2} = \frac{r_1}{r_2} = n$$

Thus we can write

$$T - T_1 = T - nT_2 = T - n\left(I_2\frac{d^2\theta_2}{dt^2}\right)$$

and so

$$T - n\left(I_2\frac{d^2\theta_2}{dt^2}\right) = I_1\frac{d^2\theta_1}{dt^2}$$

Since $\theta_2 = n\theta_1$, $d\theta_2/dt = nd\theta_1/dt$ and $d^2\theta_2/dt^2 = nd^2\theta_1/dt^2$ and so

$$T - n^2\left(I_2\frac{d^2\theta_1}{dt^2}\right) = I_1\frac{d^2\theta_1}{dt^2}$$

$$(I_1 + n^2I_2)\frac{d^2\theta_1}{dt^2} = T$$

Without the gear train we would have had simply

$$I_1\frac{d^2\theta_1}{dt^2} = T$$

Thus the moment of inertia of the load is reflected back to the other side of the gear train as an additional moment of inertia term n^2I_2.

17.3 Electrical system building blocks

The basic building blocks of electrical systems are inductors, capacitors and resistors (Figure 17.8).

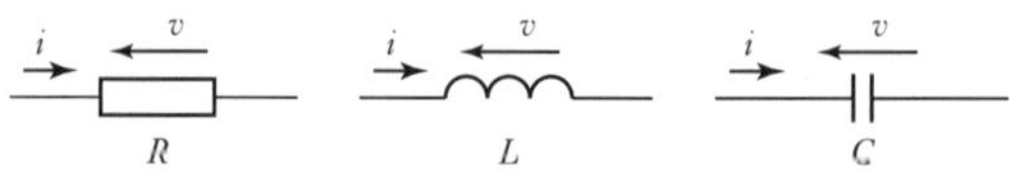

Figure 17.8   Electrical building blocks.

For an **inductor** the potential difference v across it at any instant depends on the rate of change of current (di/dt) through it:

$$v = L\frac{di}{dt}$$

where L is the inductance. The direction of the potential difference is in the opposite direction to the potential difference used to drive the current through the inductor, hence the term back e.m.f. The equation can be rearranged to give

$$i = \frac{1}{L}\int v\,dt$$

For a **capacitor**, the potential difference across it depends on the charge q on the capacitor plates at the instant concerned:

$$v = \frac{q}{C}$$

where C is the capacitance. Since the current i to or from the capacitor is the rate at which charge moves to or from the capacitor plates, i.e. $i = \mathrm{d}q/\mathrm{d}t$, then the total charge q on the plates is given by

$$q = \int i \, \mathrm{d}t$$

and so

$$v = \frac{1}{C} \int i \, \mathrm{d}t$$

Alternatively, since $v = q/C$ then

$$\frac{\mathrm{d}v}{\mathrm{d}t} = \frac{1}{C}\frac{\mathrm{d}q}{\mathrm{d}t} = \frac{1}{C}i$$

and so

$$i = C\frac{\mathrm{d}v}{\mathrm{d}t}$$

For a **resistor**, the potential difference v across it at any instant depends on the current i through it

$$v = Ri$$

where R is the resistance.

Both the inductor and capacitor store energy which can then be released at a later time. A resistor does not store energy but just dissipates it. The energy stored by an inductor when there is a current i is

$$E = \frac{1}{2}Li^2$$

The energy stored by a capacitor when there is a potential difference v across it is

$$E = \frac{1}{2}Cv^2$$

The power P dissipated by a resistor when there is a potential difference v across it is

$$P = iv = \frac{v^2}{R}$$

Table 17.2 summarises the equations defining the characteristics of the electrical building blocks when the input is current and the output is potential difference. Compare them with the equations given in Table 17.1 for the mechanical system building blocks.

Table 17.2 Electrical building blocks.

Building block	Describing equation	Energy stored or power dissipated
Inductor	$i = \dfrac{1}{L} \int v \, \mathrm{d}t$ $v = L\dfrac{\mathrm{d}i}{\mathrm{d}t}$	$E = \dfrac{1}{2}Li^2$
Capacitor	$i = C\dfrac{\mathrm{d}v}{\mathrm{d}t}$	$E = \dfrac{1}{2}Cv^2$
Resistor	$i = \dfrac{v}{R}$	$P = \dfrac{v^2}{R}$

17.3.1　Building up a model for an electrical system

The equations describing how the electrical building blocks can be combined are **Kirchhoff's laws,** and can be expressed as outlined below.

Law 1:　the total current flowing towards a junction is equal to the total current flowing from that junction, i.e. the algebraic sum of the currents at the junction is zero.

Law 2:　in a closed circuit or loop, the algebraic sum of the potential differences across each part of the circuit is equal to the applied e.m.f.

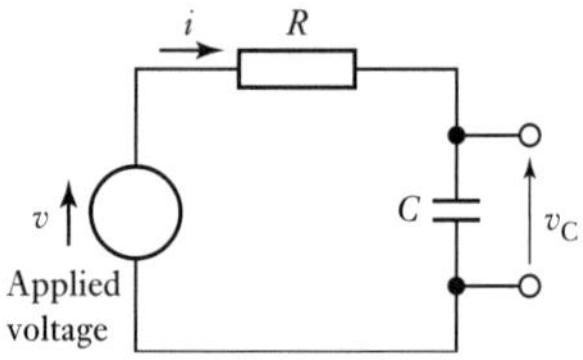

Figure 17.9

Resistor–capacitor system.

Now consider a simple electrical system consisting of a resistor and capacitor in series, as shown in Figure 17.9. Applying Kirchhoff's second law to the circuit loop gives

$$v = v_R + v_C$$

where v_R is the potential difference across the resistor and v_C that across the capacitor. Since this is just a single loop, the current i through all the circuit elements will be the same. If the output from the circuit is the potential difference across the capacitor, v_C, then since $v_R = iR$ and $i = C(dv_C/dt)$,

$$v = RC\frac{dv_C}{dt} + v_C$$

This gives the relationship between the output v_C and the input v and is a first-order differential equation.

　　Figure 17.10 shows a resistor–inductor–capacitor system. If Kirchhoff's second law is applied to this circuit loop,

$$v = v_R + v_L + v_C$$

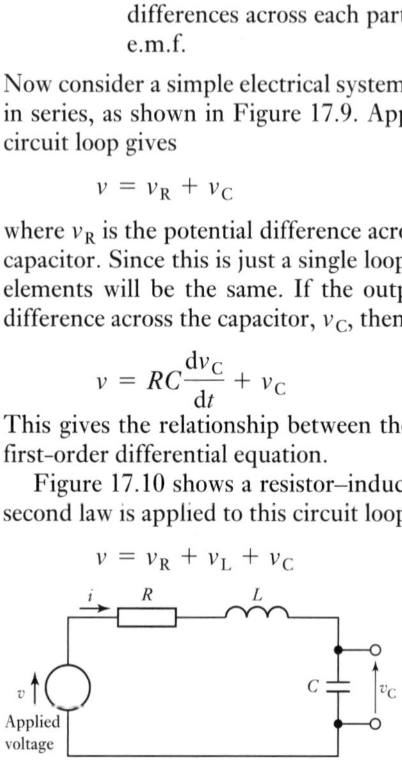

Figure 17.10

Resistor–inductor–capacitor system.

where v_R is the potential difference across the resistor, v_L that across the inductor and v_C that across the capacitor. Since there is just a single loop, the current i will be the same through all circuit elements. If the output from the circuit is the potential difference across the capacitor, v_C, then since $v_R = iR$ and $v_L = L(di/dt)$

$$v = iR + L\frac{di}{dt} + v_C$$

But $i = C(dv_C/dt)$ and so

$$\frac{di}{dt} = C\frac{d(dv_C/dt)}{dt} = C\frac{d^2v_C}{dt^2}$$

Hence

$$v = RC\frac{dv_C}{dt} + LC\frac{d^2v_C}{dt^2} + v_C$$

This is a second-order differential equation.

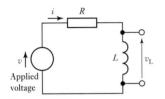

Figure 17.11
Resistor–inductor system.

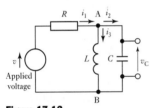

Figure 17.12
Resistor–capacitor–inductor
system.

As a further illustration, consider the relationship between the output, the potential difference across the inductor of v_L, and the input v for the circuit shown in Figure 17.11. Applying Kirchhoff's second law to the circuit loop gives

$$v = v_R + v_L$$

where v_R is the potential difference across the resistor R and v_L that across the inductor. Since $v_R = iR$,

$$v = iR + v_L$$

Since

$$i = \frac{1}{L} \int v_L dt$$

then the relationship between the input and output is

$$v = \frac{R}{L} \int v_L dt + v_L$$

As another example, consider the relationship between the output, the potential difference v_C across the capacitor and the input v for the circuit shown in Figure 17.12. Applying Kirchhoff's law 1 to node A gives

$$i_1 = i_2 + i_3$$

But

$$i_1 = \frac{v - v_A}{R}$$

$$i_2 = \frac{1}{L} \int v_A \, dt$$

$$i_3 = C \frac{dv_A}{dt}$$

Hence

$$\frac{v - v_A}{R} = \frac{1}{L} \int v_A \, dt + C \frac{dv_A}{dt}$$

But $v_C = v_A$. Hence, with some rearrangement,

$$v = RC \frac{dv_C}{dt} + v_C + \frac{R}{L} \int v_C \, dt$$

17.3.2 Electrical and mechanical analogies

The building blocks for electrical and mechanical systems have many similarities (Figure 17.13). For example, the electrical resistor does not store energy but dissipates it, with the current i through the resistor being given by $i = v/R$, where R is a constant, and the power P dissipated by $P = v^2/R$. The mechanical analogue of the resistor is the dashpot. It also does not store energy but dissipates it, with the force F being related to the velocity v by $F = cv$, where c is a constant, and the power P dissipated by $P = cv^2$. Both these sets of equations have similar forms. Comparing them, and taking

Figure 17.13 Analogous
systems.

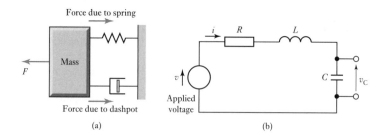

(a)   (b)

the current as being analogous to the force, then the potential difference is
analogous to the velocity and the dashpot constant c to the reciprocal of the
resistance, i.e. $(1/R)$. These analogies between current and force, potential
difference and velocity, hold for the other building blocks with the spring
being analogous to inductance and mass to capacitance.

The mechanical system in Figure 17.1(a) and the electrical system in
Figure 17.1(b) have input/output relationships described by similar
differential equations:

$$m\frac{d^2x}{dt^2} + c\frac{dx}{dt} + kx = F \quad \text{and} \quad RC\frac{dv_C}{dt} + LC\frac{d^2v_C}{dt^2} + v_C = v$$

The analogy between current and force is the one most often used.
However, another set of analogies can be drawn between potential difference
and force.

17.4 Fluid system building blocks

In fluid flow systems there are three basic building blocks which can be
considered to be the equivalent of electrical resistance, capacitance and
inductance. Fluid systems can be considered to fall into two categories:
hydraulic, where the fluid is a liquid and is deemed to be incompressible;
and pneumatic, where it is a gas which can be compressed and consequently
shows a density change.

Hydraulic resistance is the resistance to flow which occurs as a result
of a liquid flowing through valves or changes in a pipe diameter
(Figure 17.14(a)). The relationship between the volume rate of flow of
liquid q through the resistance element and the resulting pressure difference
$(p_1 - p_2)$ is

$$p_1 - p_2 = Rq$$

where R is a constant called the hydraulic resistance. The bigger the
resistance, the bigger the pressure difference for a given rate of flow. This
equation, like that for the electrical resistance and Ohm's law, assumes a
linear relationship. Such hydraulic linear resistances occur with orderly flow
through capillary tubes and porous plugs but non-linear resistances occur
with flow through sharp-edged orifices or if flow is turbulent.

Hydraulic capacitance is the term used to describe energy storage with
a liquid where it is stored in the form of potential energy. A height of liquid
in a container (Figure 17.14(b)), i.e. a so-called pressure head, is one form of
such a storage. For such a capacitance, the rate of change of volume V in the

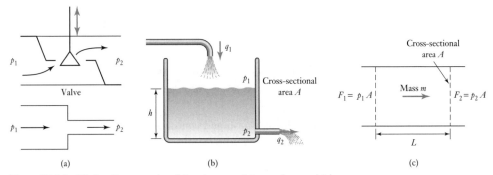

Figure 17.14 Hydraulic examples: (a) resistance, (b) capacitance, (c) inertance.

container, i.e. dV/dt, is equal to the difference between the volumetric rate at which liquid enters the container q_1 and the rate at which it leaves q_2,

$$q_1 - q_2 = \frac{dV}{dt}$$

But $V = Ah$, where A is the cross-sectional area of the container and h the height of liquid in it. Hence

$$q_1 - q_2 = \frac{d(Ah)}{dt} = A\frac{dh}{dt}$$

But the pressure difference between the input and output is p, where $p = h\rho g$ with ρ being the liquid density and g the acceleration due to gravity. Thus, if the liquid is assumed to be incompressible, i.e. its density does not change with pressure,

$$q_1 - q_2 = A\frac{d(p/\rho g)}{dt} = \frac{A}{\rho g}\frac{dp}{dt}$$

The hydraulic capacitance C is defined as being

$$C = \frac{A}{\rho g}$$

Thus

$$q_1 - q_2 = C\frac{dp}{dt}$$

Integration of this equation gives

$$p = \frac{1}{C}\int (q_1 - q_2)\,dt$$

Hydraulic inertance is the equivalent of inductance in electrical systems or a spring in mechanical systems. To accelerate a fluid and so increase its velocity, a force is required. Consider a block of liquid of mass m (Figure 17.14(c)). The net force acting on the liquid is

$$F_1 - F_2 = p_1 A - p_2 A = (p_1 - p_2)A$$

where $(p_1 - p_2)$ is the pressure difference and A the cross-sectional area. This net force causes the mass to accelerate with an acceleration a, and so

$$(p_1 - p_2)A = ma$$

But a is the rate of change of velocity dv/dt, hence

$$(p_1 - p_2)A = m\frac{dv}{dt}$$

But the mass of liquid concerned has a volume of AL, where L is the length of the block of liquid or the distance between the points in the liquid where the pressures p_1 and p_2 are measured. If the liquid has a density ρ then $m = AL\rho$ and so

$$(p_1 - p_2)A = AL\rho\frac{dv}{dt}$$

But the volume rate of flow $q = Av$, hence

$$(p_1 - p_2)A = L\rho\frac{dq}{dt}$$

$$p_1 - p_2 = I\frac{dq}{dt}$$

where the hydraulic inertance I is defined as

$$I = \frac{L\rho}{A}$$

With pneumatic systems the three basic building blocks are, as with hydraulic systems, resistance, capacitance and inertance. However, gases differ from liquids in being compressible, i.e. a change in pressure causes a change in volume and hence density. **Pneumatic resistance** R is defined in terms of the mass rate of flow dm/dt (note that this is often written as an m with a dot above it to indicate that the symbol refers to the mass rate of flow and not just the mass) and the pressure difference $(p_1 - p_2)$ as

$$p_1 - p_2 = R\frac{dm}{dt} = R\dot{m}$$

Pneumatic capacitance C is due to the compressibility of the gas, and is comparable with the way in which the compression of a spring stores energy. If there is a mass rate of flow dm_1/dt entering a container of volume V and a mass rate of flow of dm_2/dt leaving it, then the rate at which the mass in the container is changing is $(dm_1/dt - dm_2/dt)$. If the gas in the container has a density ρ then the rate of change of mass in the container is

$$\text{rate of change of mass in container} = \frac{d(\rho V)}{dt}$$

But, because a gas can be compressed, both ρ and V can vary with time. Hence

$$\text{rate of change of mass in container} = \rho\frac{dV}{dt} + V\frac{d\rho}{dt}$$

Since $(dV/dt) = (dV/dp)(dp/dt)$ and, for an ideal gas, $pV = mRT$ with consequently $p = (m/V)RT = \rho RT$ and $d\rho/dt = (1/RT)(dp/dt)$, then

$$\text{rate of change of mass in container} = \rho\frac{dV}{dp}\frac{dp}{dt} + \frac{V}{RT}\frac{dp}{dt}$$

where R is the gas constant and T the temperature, assumed to be constant, on the Kelvin scale. Thus

$$\frac{dm_1}{dt} - \frac{dm_2}{dt} = \left(\rho\frac{dV}{dp} + \frac{V}{RT}\right)\frac{dp}{dt}$$

The pneumatic capacitance due to the change in volume of the container C_1 is defined as

$$C_1 = \rho\frac{dV}{dp}$$

and the pneumatic capacitance due to the compressibility of the gas C_2 as

$$C_2 = \frac{V}{RT}$$

Hence

$$\frac{dm_1}{dt} - \frac{dm_2}{dt} = (C_1 + C_2)\frac{dp}{dt}$$

or

$$p_1 - p_2 = \frac{1}{C_1 + C_2}\int(\dot{m}_1 - \dot{m}_2)\,dt$$

Pneumatic inertance is due to the pressure drop necessary to accelerate a block of gas. According to Newton's second law, the net force is $ma = d(mv)/dt$. Since the force is provided by the pressure difference $(p_1 - p_2)$, then if A is the cross-sectional area of the block of gas being accelerated

$$(p_1 - p_2)A = \frac{d(mv)}{dt}$$

But m, the mass of the gas being accelerated, equals ρLA with ρ being the gas density and L the length of the block of gas being accelerated. And the volume rate of flow $q = Av$, where v is the velocity. Thus

$$mv = \rho LA\frac{q}{A} = \rho Lq$$

and so

$$(p_1 - p_2)A = L\frac{d(\rho q)}{dt}$$

But $\dot{m} = \rho q$ and so

$$p_1 - p_2 = \frac{L}{A}\frac{d\dot{m}}{dt}$$

$$p_1 - p_2 = I\frac{d\dot{m}}{dt}$$

with the pneumatic inertance I being $I = L/A$.

Table 17.3 shows the basic characteristics of the fluid building blocks, both hydraulic and pneumatic.

For hydraulics the volumetric rate of flow and for pneumatics the mass rate of flow are analogous to the electric current in an electrical system. For both hydraulics and pneumatics the pressure difference is analogous to the potential difference in electrical systems. Compare Table 17.3 with Table 17.2. Hydraulic and pneumatic inertance and capacitance are both energy storage elements; hydraulic and pneumatic resistance are both energy dissipaters.

17.4.1 Building up a model for a fluid system

Figure 17.15 shows a simple hydraulic system, a liquid entering and leaving a container. Such a system can be considered to consist of a capacitor, the liquid in the container, with a resistor, the valve.

Table 17.3 Hydraulic and pneumatic building blocks.

Building block	Describing equation	Energy stored or power dissipated
Hydraulic		
Inertance	$q = \dfrac{1}{L}\displaystyle\int (p_1 - p_2)\,dt$	$E = \dfrac{1}{2}Iq^2$
	$p = L\dfrac{dq}{dt}$	
Capacitance	$q = C\dfrac{d(p_1 - p_2)}{dt}$	$E = \dfrac{1}{2}C(p_1 - p_2)^2$
Resistance	$q = \dfrac{p_1 - p_2}{R}$	$P = \dfrac{1}{R}(p_1 - p_2)^2$
Pneumatic		
Inertance	$\dot{m} = \dfrac{1}{L}\displaystyle\int (p_1 - p_2)\,dt$	$E = \dfrac{1}{2}I\dot{m}^2$
Capacitance	$\dot{m} = C\dfrac{d(p_1 - p_2)}{dt}$	$E = \dfrac{1}{2}C(p_1 - p_2)^2$
Resistance	$\dot{m} = \dfrac{p_1 - p_2}{R}$	$P = \dfrac{1}{R}(p_1 - p_2)^2$

Figure 17.15 A fluid system.

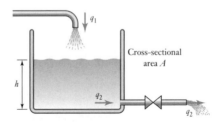

Cross-sectional area A

Inertance can be neglected since flow rates change only very slowly. For the capacitor we can write

$$q_1 - q_2 = C\frac{\mathrm{d}p}{\mathrm{d}t}$$

The rate at which liquid leaves the container q_2 equals the rate at which it leaves the valve. Thus for the resistor

$$p_1 - p_2 = Rq_2$$

The pressure difference $(p_1 - p_2)$ is the pressure due to the height of liquid in the container and is thus $h\rho g$. Thus $q_2 = h\rho g/R$ and so substituting for q_2 in the first equation gives

$$q_1 - \frac{h\rho g}{R} = C\frac{\mathrm{d}(h\rho g)}{\mathrm{d}t}$$

and, since $C = A/\rho g$,

$$q_1 = A\frac{\mathrm{d}h}{\mathrm{d}t} + \frac{\rho g h}{R}$$

This equation describes how the height of liquid in the container depends on the rate of input of liquid into the container.

A bellows is an example of a simple pneumatic system (Figure 17.16). Resistance is provided by a constriction which restricts the rate of flow of gas into the bellows and capacitance is provided by the bellows itself. Inertance can be neglected since the flow rate changes only slowly.

The mass flow rate into the bellows is given by

$$p_1 - p_2 = R\dot{m}$$

where p_1 is the pressure prior to the constriction and p_2 the pressure after the constriction, i.e. the pressure in the bellows. All the gas that flows into the bellows remains in the bellows, there being no exit from the bellows. The capacitance of the bellows is given by

$$\dot{m}_1 - \dot{m}_2 = (C_1 + C_2)\frac{\mathrm{d}p_2}{\mathrm{d}t}$$

The mass flow rate entering the bellows is given by the equation for the resistance and the mass leaving the bellows is zero. Thus

$$\frac{p_1 - p_2}{R} = (C_1 + C_2)\frac{\mathrm{d}p_2}{\mathrm{d}t}$$

Hence

$$p_1 = R(C_1 + C_2)\frac{\mathrm{d}p_2}{\mathrm{d}t} + p_2$$

This equation describes how the pressure in the bellows p_2 varies with time when there is an input of a pressure p_1.

The bellows expands or contracts as a result of pressure changes inside it. Bellows are just a form of spring and so we can write $F = kx$ for the relationship between the force F causing an expansion or contraction and the resulting displacement x, where k is the spring constant for the bellows. But

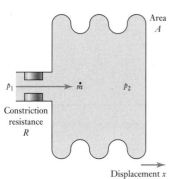

Area
A

p_1 $\dot{m}$ p_2

Constriction
resistance
R

Displacement x

Figure 17.16 A pneumatic system.

the force F depends on the pressure p_2, with $p_2 = F/A$ where A is the cross-sectional area of the bellows. Thus $p_2A = F = kx$. Hence substituting for p_2 in the above equation gives

$$p_1 = R(C_1 + C_2)\frac{k}{A}\frac{dx}{dt} + \frac{k}{A}x$$

This equation, a first-order differential equation, describes how the extension or contraction x of the bellows changes with time when there is an input of a pressure p_1. The pneumatic capacitance due to the change in volume of the container C_1 is $\rho dV/dp_2$ and since $V = Ax$, C_1 is $\rho A\, dx/dp_2$. But for the bellows $p_2A = kx$, thus

$$C_1 = \rho A\frac{dx}{d(kx/A)} = \frac{\rho A^2}{k}$$

C_2, the pneumatic capacitance due to the compressibility of the air, is $V/RT = Ax/RT$.

The following illustrates how, for the hydraulic system shown in Figure 17.17, relationships can be derived which describe how the heights of the liquids in the two containers will change with time. With this model inertance is neglected.

Container 1 is a capacitor and thus

$$q_1 - q_2 = C_1\frac{dp}{dt}$$

where $p = h_1\rho g$ and $C_1 = A_1/\rho g$ and so

$$q_1 - q_2 = A_1\frac{dh_1}{dt}$$

The rate at which liquid leaves the container q_2 equals the rate at which it leaves the valve R_1. Thus for the resistor,

$$p_1 - p_2 = R_1q_2$$

The pressures are $h_1\rho g$ and $h_2\rho g$. Thus

$$(h_1 - h_2)\rho g = R_1q_2$$

Using the value of q_2 given by this equation and substituting it into the earlier equation gives

$$q_1 - \frac{(h_1 - h_2)\rho g}{R_1} = A_1\frac{dh_1}{dt}$$

Figure 17.17 A fluid system.

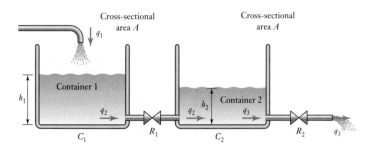

This equation describes how the height of the liquid in container 1 depends on the input rate of flow.

For container 2 a similar set of equations can be derived. Thus for the capacitor C_2,

$$q_2 - q_3 = C_2 \frac{dp}{dt}$$

where $p = h_2 \rho g$ and $C_2 = A_2/\rho g$ and so

$$q_2 - q_3 = A_2 \frac{dh_2}{dt}$$

The rate at which liquid leaves the container q_3 equals the rate at which it leaves the valve R_2. Thus for the resistor,

$$p_2 - 0 = R_2 q_3$$

This assumes that the liquid exits into the atmosphere. Thus, using the value of q_3 given by this equation and substituting it into the earlier equation gives

$$q_2 - \frac{h_2 \rho g}{R_2} = A_2 \frac{dh_2}{dt}$$

Substituting for q_2 in this equation using the value given by the equation derived for the first container gives

$$\frac{(h_1 - h_2)\rho g}{R_1} - \frac{h_2 \rho g}{R_2} = A_2 \frac{dh_2}{dt}$$

This equation describes how the height of liquid in container 2 changes.

17.5 Thermal system building blocks

There are only two basic building blocks for thermal systems: resistance and capacitance. There is a net flow of heat between two points if there is a temperature difference between them. The electrical equivalent of this is that there is only a net current i between two points if there is a potential difference v between them, the relationship between the current and potential difference being $i = v/R$, where R is the electrical resistance between the points. A similar relationship can be used to define **thermal resistance** R. If q is the rate of flow of heat and $(T_1 - T_2)$ the temperature difference, then

$$q = \frac{T_2 - T_1}{R}$$

The value of the resistance depends on the mode of heat transfer. In the case of conduction through a solid, for unidirectional conduction

$$q = Ak \frac{T_1 - T_2}{L}$$

where A is the cross-sectional area of the material through which the heat is being conducted and L the length of material between the points at which the temperatures are T_1 and T_2; k is the thermal conductivity. Hence, with this mode of heat transfer,

$$R = \frac{L}{Ak}$$

When the mode of heat transfer is convection, as with liquids and gases, then

$$q = Ah(T_2 - T_1)$$

where A is the surface area across which there is the temperature difference and h is the coefficient of heat transfer. Thus, with this mode of heat transfer,

$$R = \frac{1}{Ah}$$

Thermal capacitance is a measure of the store of internal energy in a system. Thus, if the rate of flow of heat into a system is q_1 and the rate of flow out is q_2, then

$$\text{rate of change of internal energy} = q_1 - q_2$$

An increase in internal energy means an increase in temperature. Since

$$\text{internal energy change} = mc \times \text{ change in temperature}$$

where m is the mass and c the specific heat capacity, then

$$\text{rate of change of internal energy } = mc \times \text{ rate of change of temperature}$$

Thus

$$q_1 - q_2 = mc\frac{\mathrm{d}T}{\mathrm{d}t}$$

where $\mathrm{d}T/\mathrm{d}t$ is the rate of change of temperature. This equation can be written as

$$q_1 - q_2 = C\frac{\mathrm{d}T}{\mathrm{d}t}$$

where C is the thermal capacitance and so $C = mc$. Table 17.4 gives a summary of the thermal building blocks.

Table 17.4 Thermal building blocks.

Building block	Describing equation	Energy stored
Capacitance	$q_1 - q_2 = C\dfrac{\mathrm{d}T}{\mathrm{d}t}$	$E = CT$
Resistance	$q = \dfrac{T_1 - T_2}{R}$	

17.5.1 Building up a model for a thermal system

Consider a thermometer at temperature T which has just been inserted into a liquid at temperature T_L (Figure 17.18).

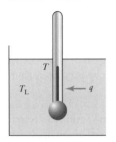

Figure 17.18 A thermal system.

If the thermal resistance to heat flow from the liquid to the thermometer is R, then

$$q = \frac{T_L - T}{R}$$

where q is the net rate of heat flow from liquid to thermometer. The thermal capacitance C of the thermometer is given by the equation

$$q_1 - q_2 = C\frac{\mathrm{d}T}{\mathrm{d}t}$$

Since there is only a net flow of heat from the liquid to the thermometer, $q_1 = q$ and $q_2 = 0$. Thus

$$q = C\frac{\mathrm{d}T}{\mathrm{d}t}$$

Substituting this value of q in the earlier equation gives

$$C\frac{\mathrm{d}T}{\mathrm{d}t} = \frac{T_L - T}{R}$$

Rearranging this equation gives

$$RC\frac{\mathrm{d}T}{\mathrm{d}t} + T = T_L$$

This equation, a first-order differential equation, describes how the temperature indicated by the thermometer T will vary with time when the thermometer is inserted into a hot liquid.

In the above thermal system the parameters have been considered to be lumped. This means, for example, that there has been assumed to be just one temperature for the thermometer and just one for the liquid, i.e. the temperatures are only functions of time and not position within a body.

To illustrate the above consider Figure 17.19 which shows a thermal system consisting of an electric fire in a room. The fire emits heat at the rate q_1 and the room loses heat at the rate q_2. Assuming that the air in the room is at a uniform temperature T and that there is no heat storage in the walls of the room, derive an equation describing how the room temperature will change with time.

If the air in the room has a thermal capacity C then

$$q_1 - q_2 = C\frac{\mathrm{d}T}{\mathrm{d}t}$$

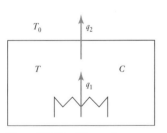

Figure 17.19 Thermal system.

If the temperature inside the room is T and that outside the room T_0 then

$$q_2 = \frac{T - T_0}{R}$$

where R is the resistivity of the walls. Substituting for q_2 gives

$$q_1 - \frac{T - T_0}{R} = C\frac{\mathrm{d}T}{\mathrm{d}t}$$

Hence

$$RC\frac{\mathrm{d}T}{\mathrm{d}t} + T = Rq_1 + T_0$$

Summary

A **mathematical model** of a system is a description of it in terms of equations relating inputs and outputs so that outputs can be predicted from inputs.

Mechanical systems can be considered to be made up from masses, springs and dashpots, or moments of inertia, springs and rotational dampers if rotational. Electrical systems can be considered to be made up from resistors, capacitors and inductors, hydraulic and pneumatic systems from resistance, capacitance and inertance, and thermal systems from resistance and capacitance.

There are many elements in mechanical, electrical, fluid and thermal systems which have similar behaviours. Thus, for example, mass in mechanical systems has similar properties to capacitance in electrical systems, capacitance in fluid systems and capacitance in thermal systems. Table 17.5 shows a comparison of the elements in each of these systems and their defining equations.

Table 17.5 System elements.

	Mechanical (translational)	Mechanical (rotational)	Electrical	Fluid (hydraulic)	Thermal
Element	Mass	Moment of inertia	Capacitor	Capacitor	Capacitor
Equation	$F = m\dfrac{d^2x}{dt^2}$	$T = I\dfrac{d^2\theta}{dt^2}$			
	$F = m\dfrac{dv}{dt}$	$T = I\dfrac{d\omega}{dt}$	$i = C\dfrac{dv}{dt}$	$q = C\dfrac{d(p_1 - p_2)}{dt}$	$q_1 - q_2 = C\dfrac{dT}{dt}$
Energy	$E = \dfrac{1}{2}mv^2$	$E = \dfrac{1}{2}I\omega^2$	$E = \dfrac{1}{2}Cv^2$	$E = \dfrac{1}{2}C(p_1 - p_2)^2$	$E = CT$
Element	Spring	Spring	Inductor	Inertance	None
Equation	$F = kx$	$T = k\theta$	$v = L\dfrac{di}{dt}$	$p = L\dfrac{dq}{dt}$	
Energy	$E = \dfrac{1}{2}\dfrac{F^2}{k}$	$E = \dfrac{1}{2}\dfrac{T^2}{k}$	$E = \dfrac{1}{2}Li^2$	$E = \dfrac{1}{2}Iq^2$	
Element	Dashpot	Rotational damper	Resistor	Resistance	Resistance
Equation	$F = c\dfrac{dx}{dt} = cv$	$T = c\dfrac{d\theta}{dt} = c\omega$	$i = \dfrac{v}{R}$	$q = \dfrac{p_1 - p_2}{R}$	$q = \dfrac{T_1 - T_2}{R}$
Power	$P = cv^2$	$P = c\omega^2$	$P = \dfrac{v^2}{R}$	$P = \dfrac{1}{R}(p_1 - p_2)^2$	

Problems

17.1 Derive an equation relating the input, force F, with the output, displacement x, for the systems described by Figure 17.20.

Figure 17.20 Problem 17.1.

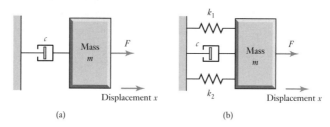

(a) (b)

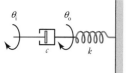

Figure 17.21
Problem 17.3.

17.2 Propose a model for the metal wheel of a railway carriage running on a metal track.

17.3 Derive an equation relating the input angular displacement θ_i with the output angular displacement θ_o for the rotational system shown in Figure 17.21.

17.4 Propose a model for a stepped shaft (i.e. a shaft where there is a step change in diameter) used to rotate a mass and derive an equation relating the input torque and the angular rotation. You may neglect damping.

17.5 Derive the relationship between the output, the potential difference across the resistor R of v_R, and the input v for the circuit shown in Figure 17.22 which has a resistor in series with a capacitor.

17.6 Derive the relationship between the output, the potential difference across the resistor R of v_R, and the input v for the series LCR circuit shown in Figure 17.23.

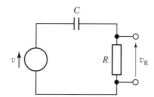 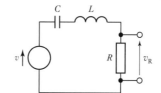 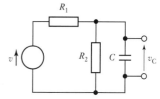

Figure 17.22 Problem 17.5. **Figure 17.23** Problem 17.6. **Figure 17.24** Problem 17.7.

17.7 Derive the relationship between the output, the potential difference across the capacitor C of v_C, and the input v for the circuit shown in Figure 17.24.

17.8 Derive the relationship between the height h_2 and time for the hydraulic system shown in Figure 17.25. Neglect inertance.

Figure 17.25
Problem 17.8.

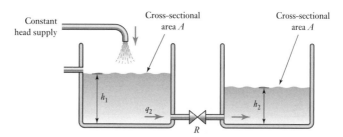

17.9   A hot object, capacitance C and temperature T, cools in a large room at temperature T_r. If the thermal system has a resistance R, derive an equation describing how the temperature of the hot object changes with time and give an electrical analogue of the system.

17.10   Figure 17.26 shows a thermal system involving two compartments, with one containing a heater. If the temperature of the compartment containing the heater is T_1, the temperature of the other compartment T_2 and the temperature surrounding the compartments T_3, develop equations describing how the temperatures T_1 and T_2 will vary with time. All the walls of the containers have the same resistance and negligible capacitance. The two containers have the same capacitance C.

Figure 17.26

Problem 17.10.

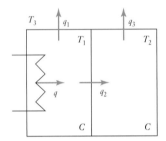

17.11   Derive the differential equation relating the pressure input p to a diaphragm actuator (as in Figure 7.23) to the displacement x of the stem.

17.12   Derive the differential equation for a motor driving a load through a gear system (Figure 17.27) which relates the angular displacement of the load with time.

Figure 17.27

Problem 17.12.

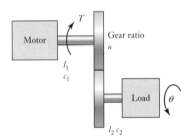

Chapter eighteen System models

Objectives

The objectives of this chapter are that, after studying it, the reader should be able to:
- Devise models for rotational–translational, electromechanical and hydraulic—mechanical systems.
- Linearise non-linear relationships in order to generate linear models.

18.1 Engineering systems

In Chapter 17 the basic building blocks of translational mechanical, rotational mechanical, electrical, fluid and thermal systems were separately considered. However, many systems encountered in engineering involve aspects of more than one of these disciplines. For example, an electric motor involves both electrical and mechanical elements. This chapter looks at how single-discipline building blocks can be combined to give models for such multidiscipline systems and also addresses the issue that often real components are not linear. For example, in considering a spring the simple model assumes that the force and extension are proportional, regardless of how large the force was. The mathematical model might thus be a simplification of a real spring. Non-linear models are, however, much more difficult to deal with and so engineers try to avoid them and a non-linear system might be approximated by a linear model.

18.2 Rotational–translational systems

There are many mechanisms which involve the conversion of rotational motion to translational motion or vice versa. For example, there are rack-and-pinion, shafts with lead screws, pulley and cable systems, etc.

To illustrate how such systems can be analysed, consider a rack-and-pinion system (Figure 18.1). The rotational motion of the pinion is transformed into translational motion of the rack. Consider first the pinion element. The net torque acting on it is $(T_{in} - T_{out})$. Thus, considering the moment of inertia element, and assuming negligible damping,

$$T_{in} - T_{out} = I \frac{d\omega}{dt}$$

where I is the moment of inertia of the pinion and ω its angular velocity. The rotation of the pinion will result in a translational velocity v of the rack. If the

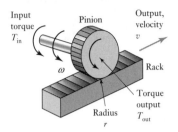

Figure 18.1 Rack-and-pinion.

pinion has a radius r, then $v = r\omega$. Hence we can write

$$T_{in} - T_{out} = \frac{I}{r}\frac{dv}{dt}$$

Now consider the rack element. There will be a force of T/r acting on it due to the movement of the pinion. If there is a frictional force of cv then the net force is

$$\frac{T_{out}}{r} - cv = m\frac{dv}{dt}$$

Eliminating T_{out} from the two equations gives

$$T_{in} - rcv = \left(\frac{I}{r} + mr\right)\frac{dv}{dt}$$

and so

$$\frac{dv}{dt} = \left(\frac{r}{1 + mr^2}\right)(T_{in} - rcv)$$

The result is a first-order differential equation describing how the output is related to the input.

18.3 Electro-mechanical systems

Electromechanical devices, such as potentiometers, motors and generators, transform electrical signals to rotational motion or vice versa. This section is a discussion of how we can derive models for such systems. A potentiometer has an input of a rotation and an output of a potential difference. An electric motor has an input of a potential difference and an output of rotation of a shaft. A generator has an input of rotation of a shaft and an output of a potential difference.

18.3.1 Potentiometer

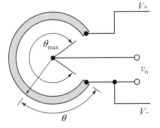

Figure 18.2 Rotary potentiometer.

The **rotary potentiometer** (Figure 18.2) is a potential divider and thus

$$\frac{v_0}{V} = \frac{\theta}{\theta_{max}}$$

where V is the potential difference across the full length of the potentiometer track and θ_{max} is the total angle swept out by the slider in being rotated from one end of the track to the other. The output is v_0 for the input θ.

18.3.2 Direct current motor

The d.c. motor is used to convert an electrical input signal into a mechanical output signal, a current through the armature coil of the motor resulting in a shaft being rotated and hence the load rotated (Figure 18.3).

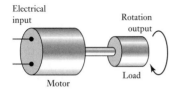

Figure 18.3 Motor driving a load.

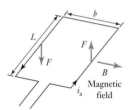

Figure 18.4 One wire of armature coil.

The motor basically consists of a coil, the armature coil, which is free to rotate. This coil is located in the magnetic field provided by a current through field coils or a permanent magnet. When a current i_a flows through the armature coil, then, because it is in a magnetic field, forces act on the coil and cause it to rotate (Figure 18.4). The force F acting on a wire carrying a current i_a and of length L in a magnetic field of flux density B at right angles to the wire is given by $F = Bi_aL$ and with N wires is $F = Nbi_aL$. The forces on the armature coil wires result in a torque T, where $T = Fb$, with b being the breadth of the coil. Thus

$$T = NBi_aLb$$

The resulting torque is thus proportional to (Bi_a), the other factors all being constants. Hence we can write

$$T = k_1Bi_a$$

Since the armature is a coil rotating in a magnetic field, a voltage will be induced in it as a consequence of electromagnetic induction. This voltage will be in such a direction as to oppose the change producing it and is called the back e.m.f. This back e.m.f. v_b is proportional to the rate or rotation of the armature and the flux linked by the coil, hence the flux density B. Thus

$$v_b = k_2B\omega$$

where ω is the shaft angular velocity and k_2 a constant.

Consider a d.c. motor which has the armature and field coils separately excited. With a so-called **armature-controlled motor** the field current i_f is held constant and the motor controlled by adjusting the armature voltage v_a. A constant field current means a constant magnetic flux density B for the armature coil. Thus

$$v_b = k_2B\omega = k_3\omega$$

where k_3 is a constant. The armature circuit can be considered to be a resistance R_a in series with an inductance L_a (Figure 18.5).

If v_a is the voltage applied to the armature circuit then, since there is a back e.m.f. of v_b, we have

$$v_a - v_b = L_a\frac{di_a}{dt} + R_ai_a$$

Figure 18.5 Direct current motor circuits.

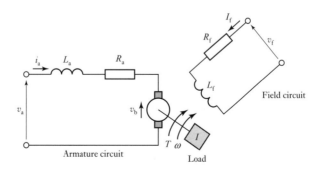

Figure 18.6 Direct current motors: (a) armature-controlled, (b) field-controlled.

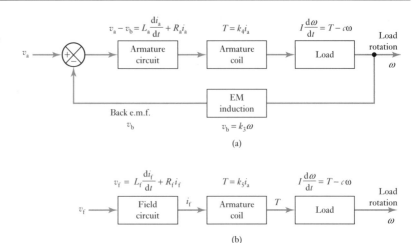

(a)

(b)

We can think of this equation in terms of the block diagram shown in Figure 18.6(a). The input to the motor part of the system is v_a and this is summed with the feedback signal of the back e.m.f. v_b to give an error signal which is the input to the armature circuit. The above equation thus describes the relationship between the input of the error signal to the armature coil and the output of the armature current i_a. Substituting for v_b,

$$v_a - k_3\omega = L_a\frac{di_a}{dt} + R_a i_a$$

The current i_a in the armature generates a torque T. Since, for the armature-controlled motor, B is constant we have

$$T = k_1 B i_a = k_4 i_a$$

where k_4 is a constant. This torque then becomes the input to the load system. The net torque acting on the load will be

$$\text{net torque} = T - \text{damping torque}$$

The damping torque is $c\omega$, where c is a constant. Hence, if any effects due to the torsional springiness of the shaft are neglected,

$$\text{net torque} = k_4 i_a - c\omega$$

This will cause an angular acceleration of $d\omega/dt$, hence

$$I\frac{d\omega}{dt} = k_4 i_a - c\omega$$

We thus have two equations that describe the conditions occurring for an armature-controlled motor, namely

$$v_a - k_3\omega = L_a\frac{di_a}{dt} + R_a i_a \text{ and } I\frac{d\omega}{dt} = k_4 i_a - c\omega$$

We can thus obtain the equation relating the output ω with the input v_a to the system by eliminating i_a. See the brief discussion of the Laplace transform in Chapter 20, or that in Appendix A, for details of how this might be done.

With a so-called **field-controlled motor** the armature current is held constant and the motor controlled by varying the field voltage. For the field circuit (Figure 18.5) there is essentially just inductance L_f in series with a resistance R_f. Thus for that circuit

$$v_f = R_f i_f + L_f \frac{di_f}{dt}$$

We can think of the field-controlled motor in terms of the block diagram shown in Figure 18.6(b). The input to the system is v_f. The field circuit converts this into a current i_f, the relationship between v_f and i_f being the above equation. This current leads to the production of a magnetic field and hence a torque acting on the armature coil, as given by $T = k_1 B i_a$. But the flux density B is proportional to the field current i_f and i_a is constant, hence

$$T = k_1 B i_a = k_5 i_f$$

where k_5 is a constant. This torque output is then converted by the load system into an angular velocity ω. As earlier, the net torque acting on the load will be

net torque $= T -$ damping torque

The damping torque is $c\omega$, where c is a constant. Hence, if any effects due to the torsional springiness of the shaft are neglected,

net torque $= k_5 i_f - c\omega$

This will cause an angular acceleration of $d\omega/dt$, hence

$$I \frac{d\omega}{dt} = k_5 i_f - c\omega$$

The conditions occurring for a field-controlled motor are thus described by the equations

$$v_f = R_f i_f + L_f \frac{di_f}{dt} \text{ and } I \frac{d\omega}{dt} = k_5 i_f - c\omega$$

We can thus obtain the equation relating the output ω with the input v_f to the system by eliminating i_f. See the brief discussion of the Laplace transform in Chapter 20, or that in Appendix A, for details of how this might be done.

18.4　Linearity

In combining blocks to create models of systems we are assuming that the relationship for each block is linear. The following is a brief discussion of linearity and how, because many real engineering items are non-linear, we need to make a linear approximation for a non-linear item.

The relationship between the force F and the extension x produced for an ideal spring is linear, being given by $F = kx$. This means that if force F_1 produces an extension x_1 and force F_2 produces an extension x_2, a force equal to $(F_1 + F_2)$ will produce an extension $(x_1 + x_2)$. This is called the **principle of superposition** and is a necessary condition for a system that can be termed a **linear system**. Another condition for a linear system is that if an input F_1 produces an extension x_1, then an input cF_1 will produce an output cx_1, where c is a constant multiplier.

A graph of the force F plotted against the extension x is a straight line passing through the origin when the relationship is linear (Figure 18.7(a)). Real springs, like any other real components, are not perfectly linear (Figure 18.7(b)). However, there is often a range of operation for which linearity can be assumed. Thus for the spring giving the graph in Figure 18.7(b), linearity can be assumed provided the spring is only used over the central part of its graph. For many system components, linearity can be assumed for operations within a range of values of the variable about some operating point.

Figure 18.7   Springs: (a) ideal, (b) real.

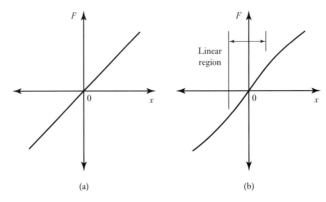

For some system components (Figure 18.8(a)) the relationship is non-linear. For such components the best that can be done to obtain a linear relationship is just to work with the straight line which is the slope of the graph at the operating point.

Figure 18.8   A non-linear relationship.

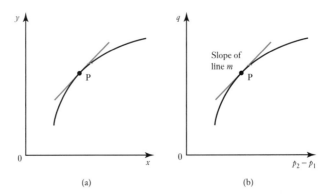

Thus for the relationship between y and x in Figure 18.8(a), at the operating point P where the slope has the value m

$$\Delta y = m \, \Delta x$$

where Δy and Δx are small changes in input and output signals at the operating point.

For example, the rate of flow of liquid q through an orifice is given by

$$q = c_d A \sqrt{\frac{2(p_1 - p_2)}{\rho}}$$

where c_d is a constant called the discharge coefficient, A the cross-sectional area of the orifice, ρ the fluid density and $(p_1 - p_2)$ the pressure difference. For a constant cross-sectional area and density the equation can be written as

$$q = C\sqrt{p_1 - p_2}$$

where C is a constant. This is a non-linear relationship between the rate of flow and the pressure difference. We can obtain a linear relationship by considering the straight line representing the slope of the rate of flow/pressure difference graph (Figure 18.8(b)) at the operating point. The slope m is $dq/d(p_1 - p_2)$ and has the value

$$m = \frac{dq}{d(p_1 - p_2)} = \frac{C}{2\sqrt{p_{o1} - p_{o2}}}$$

where $(p_{o1} - p_{o2})$ is the value at the operating point. For small changes about the operating point we will assume that we can replace the non-linear graph by the straight line of slope m and therefore can write $m = \Delta q/\Delta(p_1 - p_2)$ and hence

$$\Delta q = m \, \Delta(p_1 - p_2)$$

Hence, if we had $C = 2 \text{ m}^3/\text{s}$ per kPa, i.e. $q = 2(p_1 - p_2)$, then for an operating point of $(p_1 - p_2) = 4$ kPa with $m = 2/(2\sqrt{4}) = 0.5$, the linearised version of the equation would be

$$\Delta q = 0.5 \, \Delta(p_1 - p_2)$$

Linearised mathematical models are used because most of the techniques of control systems are based on there being linear relationships for the elements of such systems. Also, because most control systems are maintaining an output equal to some reference value, the variations from this value tend to be rather small and so the linearised model is perfectly appropriate.

18.5 Hydraulic–mechanical systems

Hydraulic–mechanical converters involve the transformation of hydraulic signals to translational or rotational motion, or vice versa. Thus, for example, the movement of a piston in a cylinder as a result of hydraulic pressure involves the transformation of a hydraulic pressure input to the system to a translational motion output.

Figure 18.9 shows a hydraulic system in which an input of displacement x_i is, after passing through the system, transformed into a displacement x_o of a load. The system consists of a spool valve and a cylinder. The input displacement x_i to the left results in the hydraulic fluid supply pressure p_s causing fluid to flow into the left-hand side of the cylinder. This pushes the piston in the cylinder to the right and expels the fluid in the right-hand side of the chamber through the exit port at the right-hand end of the spool valve.

The rate of flow of fluid to and from the chamber depends on the extent to which the input motion has uncovered the ports allowing the fluid to enter or leave the spool valve. When the input displacement x_i is to the right, the

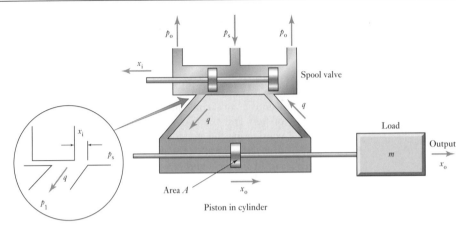

Figure 18.9 Hydraulic system and load.

spool valve allows fluid to move to the right-hand end of the cylinder and so results in a movement of the piston in the cylinder to the left.

The rate of flow of fluid q through an orifice, which is what the ports in the spool valve are, is a non-linear relationship depending on the pressure difference between the two sides of the orifice and its cross-sectional area A. However, a linearised version of the equation can be used (see the previous section for its derivation)

$$\Delta q = m_1 \Delta A + m_2 \Delta \text{ (pressure difference)}$$

where m_1 and m_2 are constants at the operating point. For the fluid entering the chamber, the pressure difference is $(p_s - p_1)$ and for the exit $(p_2 - p_o)$. If the operating point about which the equation is linearised is taken to be the point at which the spool valve is central and the ports connecting it to the cylinder are both closed, then for this condition q is zero, and so $\Delta q = q$, A is proportional to x_s if x_s is measured from this central position, and the change in pressure on the inlet side of the piston is $-\Delta p_1$ relative to p_s and on the exit side Δp_2 relative to p_o. Thus, for the inlet port the equation can be written as

$$q = m_1 x_i + m_2(-\Delta p_1)$$

and for the exit port

$$q = m_1 x_i + m_2 \Delta p_2$$

Adding the two equations gives

$$2q = 2m_1 x_i - m_2(\Delta p_1 - \Delta p_2)$$
$$q = m_1 x_i - m_3(\Delta p_1 - \Delta p_2)$$

where $m_3 = m_2/2$.

For the cylinder, the change in the volume of fluid entering the left-hand side of the chamber, or leaving the right-hand side, when the piston moves a distance x_o is Ax_o, where A is the cross-sectional area of the piston. Thus the rate at which the volume is changing is $A(dx_o/dt)$. The rate at which fluid is entering the left-hand side of the cylinder is q. However, since there is some leakage flow of fluid from one side of the piston to the other,

$$q = A\frac{dx_o}{dt} + q_L$$

where q_L is the rate of leakage. Substituting for q gives

$$m_1x_i - m_3(\Delta p_1 - \Delta p_2) = A\frac{dx_o}{dt} + q_L$$

The rate of leakage flow q_L is a flow through an orifice, the gap between the piston and the cylinder. This is of constant cross-section and has a pressure difference $(\Delta p_1 - \Delta p_2)$. Hence, using the linearised equation for such a flow,

$$q_L = m_4(\Delta p_1 - \Delta p_2)$$

Thus, using this equation to substitute for q_L,

$$m_1x_i - m_3(\Delta p_1 - \Delta p_2) = A\frac{dx_o}{dt} + m_4(\Delta p_1 - \Delta p_2)$$

$$m_1x_i - (m_3 + m_4)(\Delta p_1 - \Delta p_2) = A\frac{dx_o}{dt}$$

The pressure difference across the piston results in a force being exerted on the load, the force exerted being $(\Delta p_1 - \Delta p_2)A$. There is, however, some damping of motion, i.e. friction, of the mass. This is proportional to the velocity of the mass, i.e. (dx_o/dt). Hence the net force acting on the load is

$$\text{net force} = (\Delta p_1 - \Delta p_2)A - c\frac{dx_o}{dt}$$

This net force causes the mass to accelerate, the acceleration being (d^2x_o/dt^2). Hence

$$m\frac{d^2x_o}{dt^2} = (\Delta p_1 - \Delta p_2)A - c\frac{dx_o}{dt}$$

Rearranging this equation gives

$$\Delta p_1 - \Delta p_2 = \frac{m}{A}\frac{d^2x_o}{dt^2} + \frac{c}{A}\frac{dx_o}{dt}$$

Using this equation to substitute for the pressure difference in the earlier equation,

$$m_1x_i - (m_3 + m_4)\left(\frac{m}{A}\frac{d^2x_o}{dt^2} + \frac{c}{A}\frac{dx_o}{dt}\right) = A\frac{dx_o}{dt}$$

Rearranging gives

$$\frac{(m_3 + m_4)m}{A}\frac{d^2x_o}{dt^2} + \left(A + \frac{c(m_3 + m_4)}{A}\right)\frac{dx_o}{dt} = m_1x_i$$

and rearranging this equation leads to

$$\frac{(m_3 + m_4)m}{A^2 + c(m_3 + m_4)} \frac{d^2x_o}{dt^2} + \frac{dx_o}{dt} = \frac{Am_1}{A^2 + c(m_3 + m_4)} x_i$$

This equation can be simplified by introducing two constants k and τ, the latter constant being called the time constant (see Chapter 12). Hence

$$\tau \frac{d^2x_o}{dt^2} + \frac{dx_o}{dt} = kx_i$$

Thus the relationship between input and output is described by a second-order differential equation.

Summary

Many systems encountered in engineering involve aspects of more than one discipline and these can be considered by examining how the system can be built up from single-discipline building blocks.

A system is said to be linear when its basic equations, whether algebraic or differential, are such that the magnitude of the output produced is directly proportional to the input. For an algebraic equation, this means that the graph of output plotted against input is a straight line passing through the origin. So doubling the input doubles the output. For a linear system we can obtain the output of the system to a number of inputs by adding the outputs of the system to each individual input considered separately. This is called the **principle of superposition**.

Problems

18.1   Derive a differential equation relating the input voltage to a d.c. servo motor and the output angular velocity, assuming that the motor is armature controlled and the equivalent circuit for the motor has an armature with just resistance, its inductance being neglected.

18.2   Derive differential equations for a d.c. generator. The generator may be assumed to have a constant magnetic field. The armature circuit has the armature coil, having both resistance and inductance, in series with the load. Assume that the load has both resistance and inductance.

18.3   Derive differential equations for a permanent magnet d.c. motor.

Chapter nineteen

Dynamic responses of systems

Objectives

The objectives of this chapter are that, after studying it, the reader should be able to:
- Model dynamic systems by means of differential equations.
- Determine the outputs of first-order systems to inputs and determine time constants.
- Determine the outputs of second-order systems to inputs and identify the under-damped, critically damped and over-damped conditions.
- Describe the characteristics of second-order system responses in terms of rise time, overshoot, subsidence ratio, decrement and settling time.

19.1 Modelling dynamic systems

The most important function of a model devised for measurement or control systems is to be able to predict what the output will be for a particular input. We are not just concerned with a static situation, i.e. that after some time when the steady state has been reached an output of x corresponds to an input of y. We have to consider how the output will change with time when there is a change of input or when the input changes with time. For example, how will the temperature of a temperature-controlled system change with time when the thermostat is set to a new temperature? For a control system, how will the output of the system change with time when the set value is set to a new value or perhaps increased at a steady rate?

Chapters 17 and 18 were concerned with models of systems when the inputs varied with time, with the results being expressed in terms of differential equations. This chapter is about how we can use such models to make predictions about how outputs will change with time when the input changes with time.

19.1.1 Differential equations

To describe the relationship between the input to a system and its output we must describe the relationship between inputs and outputs which are both possible functions of time. We thus need a form of equation which will indicate how the system output will vary with time when the input is varying with time. This can be done by the use of a differential equation. Such an equation includes derivatives with respect to time and so gives information about how the response of a system varies with time. A derivative dx/dt describes the rate at which x varies with time; the derivative d^2x/dt^2 states how dx/dt varies with time. Differential equations can be classed as first-order, second-order, third-order, etc., according to the highest order of the derivative in the equation. For a first-order equation the highest order will be dx/dt, with a second-order d^2x/dt^2, with a third-order d^3x/dt^3, with nth-order d^nx/dt^n.

This chapter is about the types of responses we can expect from first-order and second-order systems and the solution of such differential equations in order that the response of the system to different types of input can be obtained. This chapter uses the 'try a solution' approach in order to find a solution; the Laplace transformation method is introduced in Chapter 20.

19.2 Terminology

In this section we look at some of the terms that are used in describing the dynamic responses of systems.

19.2.1 Natural and forced responses

The term **natural response** is used for a system when there is no input to the system forcing the variable to change but it is just changing naturally. As an illustration, consider the first-order system of water being allowed naturally to flow out of a tank (Figure 19.1(a)).

Figure 19.1 Water flowing out of a tank: (a) naturally with no input, (b) with forcing input.

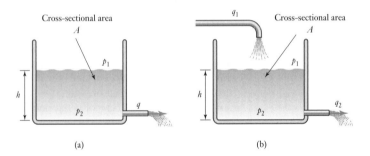

(a)   (b)

For such a system we have

$$p_1 - p_2 = Rq$$

where R is the hydraulic resistance. But $p_1 - p_2 = h\rho g$, where ρ is the density of the water, and q is the rate at which water leaves the tank and so is $-dV/dt$, with V being the volume of water in the tank and so being Ah. Thus $q = -d(Ah)/dt = -Adh/dt$ and so the above equation can be written as

$$h\rho g = -RA\frac{dh}{dt}$$

This is the natural response in that there is no input to the system forcing the variable h to change; it is just naturally changing with time. We can draw attention to this by writing the differential equation with all the output terms, i.e. h, on the same side of the equals sign and the input term of zero on the right, i.e.

$$RA\frac{dh}{dt} + (\rho g)h = 0$$

In Section 17.4.1 the differential equation was derived for a water tank from which water was flowing but also into which there was a flow of water (Figure 19.1(b)). This equation has a forcing input function of q_1 and can be written as

$$RA\frac{dh}{dt} + (\rho g)h = q_1$$

As another example, consider a thermometer being placed in a hot liquid at some temperature T_L. The rate at which the reading of the thermometer T changes with time was derived in Section 17.5.1 as being given by the differential equation

$$RC\frac{dT}{dt} + T = T_L$$

Such a differential equation has a forcing input of T_L.

19.2.2 Transient and steady-state responses

The total response of a control system, or element of a system, can be considered to be made up of two aspects, the steady-state response and the transient response. The **transient response** is that part of a system response which occurs as a result of a change in input and which dies away after a short interval of time. The **steady-state response** is the response that remains after all transient responses have died down.

To give a simple illustration of this, consider a vertically suspended spring (Figure 19.2) and what happens when a weight is suddenly suspended from it. The deflection of the spring abruptly increases and then may well oscillate until after some time it settles down to a steady value. The steady value is the steady-state response of the spring system; the oscillation that occurs prior to this steady state is the transient response.

Figure 19.2 Transient and steady-state responses of a spring system.

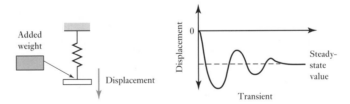

19.2.3 Forms of inputs

The input to the above spring system, the weight, is a quantity which varies with time. Up to some particular time there is no added weight, i.e. no input, then after that time there is an input which remains constant for the rest of the time. This type of input is known as a **step input** and is of the form shown in Figure 19.3(a).

The input signal to systems can take other forms, e.g. impulse, ramp and sinusoidal signals. An **impulse** is a very short-duration input (Figure 19.3(b)); a **ramp** is a steadily increasing input (Figure 19.3(c)) and can be described by

Figure 19.3 Inputs: (a) step at time 0, (b) impulse at some time, (c) ramp starting at time 0.

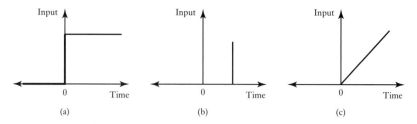

(a)   (b)   (c)

an equation of the form $y = kt$, where k is a constant; and a **sinusoidal** input can be described by an equation of the form $y = k \sin \omega t$, with ω being the so-called angular frequency and equal to $2\pi f$ where f is the frequency.

Both the input and the output are functions of time. One way of indicating this is to write them in the form $f(t)$, where f is the function and (t) indicates that its value depends on time t. Thus for the weight W input to the spring system we could write $W(t)$ and for the deflection d output $d(t)$. The term $y(t)$ is commonly used for an input and $x(t)$ for an output.

19.3 First-order systems

Consider a first-order system with $y(t)$ as the input to the system and $x(t)$ the output and which has a forcing input $b_0 y$ and can be described by a differential equation of the form

$$a_1 \frac{dx}{dt} + a_0 x = b_0 y$$

where a_1, a_0 and b_0 are constants.

19.3.1 Natural response

The input $y(t)$ can take many forms. Consider first the situation when the input is zero. Because there is no input to the system we have no signal forcing the system to respond in any way other than its natural response with no input. The differential equation is then

$$a_1 \frac{dx}{dt} + a_0 x = 0$$

We can solve this equation by using the technique called **separation of variables**. The equation can be written with all the x variables on one side and all the t variables on the other:

$$\frac{dx}{x} = -\frac{a_0}{a_1} dt$$

Integrating this between the initial value of $x = 1$ at $t = 0$, i.e. a unit step input, and x at t gives

$$\ln x = -\frac{a_0}{a_1} t$$

and so we have

$$x = e^{-a_0 t/a_1}$$

We could, however, have recognised that the differential equation would have a solution of the form $x = Ae^{st}$, where A and s are constants. We then have $dx/dt = sAe^{st}$ and so when these values are substituted in the differential equation we obtain

$$a_1 s A e^{st} + a_0 A e^{st} = 0$$

and so $a_1 s + a_0 = 0$ and $s = -a_0/a_1$. Thus the solution is

$$x = A e^{-a_0 t/a_1}$$

This is termed the natural response since there is no forcing function. We can determine the value of the constant A given some initial (boundary) condition. Thus if $x = 1$ when $t = 0$ then $A = 1$. Figure 19.4 shows the natural response, i.e. an exponential decay:

$$x = e^{-a_0 t/a_1}$$

Figure 19.4 Natural response of a first-order system.

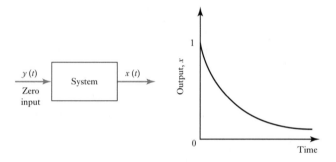

19.3.2 Response with a forcing input

Now consider the differential equation when there is a **forcing function**, i.e.

$$a_1 \frac{dx}{dt} + a_0 x = b_0 y$$

Consider the solution to this equation to be made up of two parts, i.e. $x = u + v$. One part represents the transient part of the solution and the other the steady-state part. Substituting this into the differential equation gives

$$a_1 \frac{d(u + v)}{dt} + a_0(u + v) = b_0 y$$

Rearranging this gives

$$\left(a_1 \frac{du}{dt} + a_0 u \right) + \left(a_1 \frac{dv}{dt} + a_0 v \right) = b_0 y$$

If we let

$$a_1 \frac{dv}{dt} + a_0 v = b_0 y$$

then we must have

$$a_1 \frac{du}{dt} + a_0 u = 0$$

and so two differential equations, one of which contains a forcing function and one which is just the natural response equation. This last equation is just the natural equation which we solved earlier in this section and so has a solution of the form

$$u = Ae^{-a_0t/a_1}$$

The other differential equation contains the forcing function y. For this differential equation the form of solution we try depends on the form of the input signal y. For a step input when y is constant for all times greater than 0, i.e. $y = k$, we can try a solution $v = A$, where A is a constant. If we have an input signal of the form $y = a + bt + ct^2 + \ldots$, where a, b and c are constants which can be zero, then we can try a solution which is of the form $v = A + Bt + Ct^2 + \ldots$. For a sinusoidal signal we can try a solution of the form $v = A \cos \omega t + B \sin \omega t$.

To illustrate this, assume there is a step input at a time of $t = 0$ with the size of the step being k (Figure 19.5(a)). Then we try a solution of the form $v = A$. Differentiating a constant gives zero; thus when this solution is substituted into the differential equation we obtain $a_0 A = b_0 k$ and so $v = (b_0/a_0)k$.

The full solution will be given by $x = u + v$ and so will be

$$x = Ae^{-a_0t/a_1} + \frac{b_0}{a_0} k$$

We can determine the value of the constant A given some initial (boundary) conditions. Thus if the output $x = 0$ when $t = 0$ then

$$0 = A + \frac{b_0}{a_0} k$$

Thus $A = -(b_0/a_0)k$. The solution then becomes

$$x = \frac{b_0}{a_0} k(1 - e^{-a_0t/a_1})$$

When $t \to \infty$ the exponential term tends to zero. The exponential term thus gives that part of the response which is the transient solution. The steady-state response is the value of x when $t \to \infty$ and so is $(b_0/a_0)k$. Thus the equation can be written as

$$x = \text{steady-state value} \times (1 - e^{-a_0t/a_1})$$

Figure 19.5(b) shows a graph of how the output x varies with time for the step input.

19.3.3 Examples of first-order systems

As a further illustration of the above, consider the following examples of first-order systems.

An electrical transducer system consists of a resistance in series with a capacitor and, when subject to a step input of size V, gives an output of a potential difference across the capacitor v, which is given by the differential equation

$$RC\frac{dv}{dt} + v = V$$

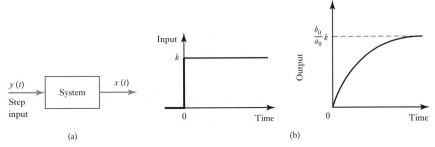

Figure 19.5 (a) Step input, (b) resulting output.

Comparing the differential equation with the equation solved earlier: $a_1 = RC$, $a_0 = 1$ and $b_0 = 1$. Then the solution is of the form

$$v = V(1 - e^{-t/RC})$$

Now consider an electric circuit consisting of a 1 MΩ resistance in series with a 2 μF capacitance. At a time $t = 0$ the circuit is subject to a ramp voltage of $4t$ V, i.e. the voltage increases at the rate of 4 V every 1 s. The differential equation will be of a similar form to that given in the previous example but with the step voltage V of that example replaced by the ramp voltage $4t$, i.e.

$$RC\frac{dv}{dt} + v = 4t$$

Thus, using the values given above,

$$2\frac{dv}{dt} + v = 4t$$

Taking $v = v_n + v_f$, i.e. the sum of the natural and forced responses, we have for the natural response

$$2\frac{dv_n}{dt} + v_n = 0$$

and for the forced response

$$2\frac{dv_f}{dt} + v_f = 4t$$

For the natural response differential equation we can try a solution of the form $v_n = Ae^{st}$. Hence, using this value

$$2Ase^{st} + Ae^{st} = 0$$

Thus $s = -\frac{1}{2}$ and so $v_n = Ae^{-t/2}$. For the forced response differential equation, since the right-hand side of the equation is $4t$ we can try a solution of the form $v_f = A + Bt$. Using this value gives $2B + A + Bt = 4t$. Thus we must have $B = 4$ and $A = -2B = -8$. Hence the solution is $v_f = -8 + 4t$. Thus the full solution is

$$v = v_n + v_f = Ae^{-t/2} - 8 + 4t$$

Since $v = 0$ when $t = 0$ we must have $A = 8$. Hence

$$v = 8e^{-t/2} - 8 + 4t$$

As a further example, consider a motor when the relationship between the output angular velocity ω and the input voltage v for the motor is given by

$$\frac{IR}{k_1 k_2} \frac{d\omega}{dt} + \omega = \frac{1}{k_1} v$$

Comparing the differential equation with the equation solved earlier, then $a_1 = IR/k_1 k_2$, $a_0 = 1$ and $b_0 = 1/k_1$. The steady-state value for a step input of size 1 V is thus $(b_0/a_0) = 1/k_1$.

19.3.4 The time constant

For a first-order system subject to a step input of size k we have an output y which varies with time t according to

$$x = \frac{b_0}{a_0} k \left(1 - e^{-a_0 t/a_1} \right)$$

or

$$x = \text{steady-state value} \times \left(1 - e^{-a_0 t/a_1} \right)$$

When the time $t = (a_1/a_0)$ then the exponential term has the value $e^{-1} = 0.37$ and

$$x = \text{steady-state value} \times (1 - 0.37)$$

In this time the output has risen to 0.63 of its steady-state value. This time is called the **time constant** τ:

$$\tau = \frac{a_1}{a_0}$$

In a time of $2(a_1/a_0) = 2\tau$, the exponential term becomes $e^{-2} = 0.14$ and so

$$x = \text{steady-state value} \times (1 - 0.14)$$

In this time the output has risen to 0.86 of its steady-state value. In a similar way, values can be calculated for the output after 3τ, 4τ, 5τ, etc. Table 19.1 shows the results of such calculations and Figure 19.6 the graph of how the output varies with time for a unit step input.

Table 19.1 Response of a first-order system to a step input.

Time t	Fraction of steady-state output
0	0
1τ	0.63
2τ	0.86
3τ	0.95
4τ	0.98
5τ	0.99
∞	1

Figure 19.6 Response of a first-order system to a step input.

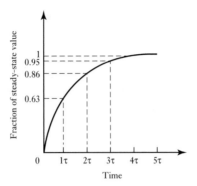

In terms of the time constant τ, we can write the equation describing the response of a first-order system as

$$x = \text{steady-state value} \times \left(1 - e^{-t/\tau}\right)$$

The time constant τ is (a_1/a_0), thus we can write our general form of the first-order differential equation

$$a_1 \frac{\mathrm{d}x}{\mathrm{d}t} + a_0 x = b_0 y$$

as

$$\tau \frac{\mathrm{d}x}{\mathrm{d}t} + x = \frac{b_0}{a_0} y$$

But b_0/a_0 is the factor by which the input y is multiplied to give the steady-state value. We can term it the **steady-state gain** since it is the factor stating by how much bigger the output is than the input under steady-state conditions. Thus if we denote this by G_{SS} then the differential equation can be written in the form

$$\tau \frac{\mathrm{d}x}{\mathrm{d}t} + x = G_{SS} y$$

To illustrate this, consider Figure 19.7 which shows how the output v_o of a first-order system varies with time when subject to a step input of 5 V. The

Figure 19.7 Example.

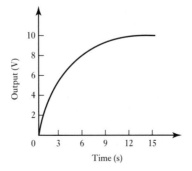

time constant is the time taken for a first-order system output to change from 0 to 0.63 of its final steady-state value. In this case this time is about 3 s. We can check this value, and that the system is first order, by finding the value at 2, i.e. 6 s. With a first-order system it should be 0.86 of the steady-state value. In this case it is. The steady-state output is 10 V. Thus the steady-state gain G_{SS} is (steady-state output/input) $= 10/5 = 2$. The differential equation for a first-order system can be written as

$$\tau \frac{\mathrm{d}x}{\mathrm{d}t} + x = G_{SS}y$$

Thus, for this system, we have

$$3 \frac{\mathrm{d}v_o}{\mathrm{d}t} + v_o = 2v_i$$

19.4   Second-order systems

Many second-order systems can be considered to be analogous to essentially just a stretched spring with a mass and some means of providing damping. Figure 19.8 shows the basis of such a system.

Figure 19.8   Spring–dashpot–mass system.

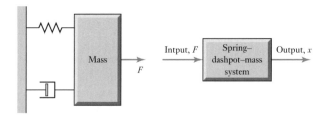

Such a system was analysed in Section 17.2.2. The equation describing the relationship between the input of force F and the output of a displacement x is

$$m \frac{\mathrm{d}^2x}{\mathrm{d}t^2} + c \frac{\mathrm{d}x}{\mathrm{d}t} + kx = F$$

where m is the mass, c the damping constant and k the spring constant.

The way in which the resulting displacement x will vary with time will depend on the amount of damping in the system. Thus if the force was applied as a step input and there was no damping at all then the mass would freely oscillate on the spring and the oscillations would continue indefinitely. No damping means $c = 0$ and so the $\mathrm{d}x/\mathrm{d}t$ term is zero. However, damping will cause the oscillations to die away until a steady displacement of the mass is obtained. If the damping is high enough there will be no oscillations and the displacement of the mass will just slowly increase with time and gradually the mass will move towards its steady displacement position. Figure 19.9 shows the general way that the displacements, for a step input, vary with time with different degrees of damping.

Figure 19.9 Effect of damping with a second-order system.

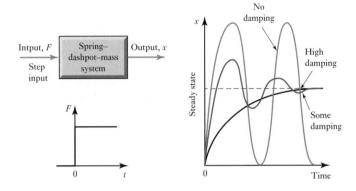

19.4.1 Natural response

Consider a mass on the end of a spring. In the absence of any damping and left to oscillate freely without being forced, the output of the second-order system is a continuous oscillation (simple harmonic motion). Thus, suppose we describe this oscillation by the equation

$$x = A \sin \omega_n t$$

where x is the displacement at a time t, A the amplitude of the oscillation and ω_n the angular frequency of the free undamped oscillations. Differentiating this gives

$$\frac{dx}{dt} = \omega_n A \cos \omega_n t$$

Differentiating a second time gives

$$\frac{d^2x}{dt^2} = -\omega_n^2 A \sin \omega_n t = -\omega_n^2 x$$

This can be reorganised to give the differential equation

$$\frac{d^2x}{dt^2} + \omega_n^2 x = 0$$

But for a mass m on a spring of stiffness k we have a restoring force of kx and thus

$$m\frac{d^2x}{dt^2} = -kx$$

This can be written as

$$\frac{d^2x}{dt^2} + \frac{k}{m}x = 0$$

Thus, comparing the two differential equations, we must have

$$\omega_n^2 = \frac{k}{m}$$

and $x = A \sin \omega_n t$ is the solution to the differential equation.

Now consider when we have damping. The motion of the mass is then described by

$$m\frac{d^2x}{dt^2} + c\frac{dx}{dt} + kx = 0$$

To solve this equation we can try a solution of the form $x_n = Ae^{st}$. This gives $dx_n/dt = Ase^{st}$ and $d^2x_n/dt^2 = As^2e^{st}$. Thus, substituting these values in the differential equation gives

$$mAs^2e^{st} + cAse^{st} + kAe^{st} = 0$$

$$ms^2 + cs + k = 0$$

Thus $x_n = Ae^{st}$ can only be a solution provided the above equation equals zero. This equation is called the **auxiliary equation**. The roots of the equation can be obtained by factoring or using the formula for the roots of a quadratic equation. Thus

$$s = \frac{-c \pm \sqrt{c^2 - 4mk}}{2m} = -\frac{c}{2m} \pm \sqrt{\left(\frac{c}{2m}\right)^2 - \frac{k}{m}}$$

$$= -\frac{c}{2m} \pm \sqrt{\frac{k}{m}\left(\frac{c^2}{4mk}\right) - \frac{k}{m}}$$

But $\omega_n^2 = k/m$ and so, if we let $\zeta^2 = c^2/4mk$, we can write the above equation as

$$s = -\zeta\omega_n \pm \omega_n\sqrt{\zeta^2 - 1}$$

ζ is termed the **damping factor**.

The value of s obtained from the above equation depends very much on the value of the square root term. Thus when ζ^2 is greater than 1 the square root term gives a square root of a positive number, and when ζ^2 is less than 1 we have the square root of a negative number. The damping factor determines whether the square root term is a positive or negative number and so the form of the output from the system.

1　*Over-damped*

　With $\zeta > 1$ there are two different real roots s_1 and s_2:

$$s_1 = -\zeta\omega_n + \omega_n\sqrt{\zeta^2 - 1}$$

$$s_2 = -\zeta\omega_n - \omega_n\sqrt{\zeta^2 - 1}$$

　and so the general solution for x_n is

$$x_n = Ae^{s_1 t} + Be^{s_2 t}$$

　For such conditions the system is said to be **over-damped**.

2　*Critically damped*
　When $\zeta = 1$ there are two equal roots with $s_1 = s_2 = -\omega_n$. For this condition, which is called **critically damped**,

$$x_n = (At + B)e^{-\omega_n t}$$

　It may seem that the solution for this case should be $x_n = Ae^{st}$, but two constants are required and so the solution is of this form.

3 *Under-damped*

With $\zeta < 1$ there are two complex roots since the roots both involve the square root of (-1):

$$s = -\zeta\omega_n \pm \omega_n\sqrt{\zeta^2 - 1} = -\zeta\omega_n \pm \omega_n\sqrt{-1}\sqrt{1 - \zeta^2}$$

and so writing j for $\sqrt{-1}$,

$$s = -\zeta\omega_n \pm j\omega_n\sqrt{1 - \zeta^2}$$

If we let

$$\omega = \omega_n\sqrt{1 - \zeta^2}$$

then we can write $s = -\zeta\omega_d \pm j\omega$ and so the two roots are

$$s_1 = -\zeta\omega_d + j\omega \text{ and } s_2 = -\zeta\omega_d - j\omega$$

The term ω is the angular frequency of the motion when it is in the damped condition specified by ζ. The solution under these conditions is thus

$$x_n = Ae^{(\zeta\omega_n + j\omega)t} + Be^{(-\zeta\omega_n - j\omega)t} = e^{-\zeta\omega_n t}(Ae^{j\omega t} + Be^{-j\omega t})$$

But $e^{j\omega t} = \cos\omega t + j\sin\omega t$ and $e^{-j\omega t} = \cos\omega t - j\sin\omega t$. Hence

$$x_n = e^{-\zeta\omega_n t}(A\cos\omega t + jA\sin\omega t + B\cos\omega t - jB\sin\omega t)$$

$$= e^{-\zeta\omega_n t}[(A + B)\cos\omega t + j(A - B)\sin\omega t)]$$

If we substitute constants P and Q for $(A + B)$ and $j(A - B)$, then

$$x_n = e^{-\zeta\omega_n t}(P\cos\omega t + Q\sin\omega t)$$

For such conditions the system is said to be **under-damped**.

19.4.2 Response with a forcing input

When we have a forcing input F the differential equation becomes

$$m\frac{d^2x}{dt^2} + c\frac{dx}{dt} + kx = F$$

We can solve this second-order differential equation by the same method used earlier for the first-order differential equation and consider the solution to be made up of two elements, a transient (natural) response and a forced response, i.e. $x = x_n + x_f$. Substituting for x in the above equation then gives

$$m\frac{d^2(x_n + x_f)}{dt^2} + c\frac{d(x_n + x_f)}{dt} + k(x_n + x_f) = F$$

If we let

$$m\frac{d^2x_n}{dt^2} + c\frac{dx_n}{dt} + kx_n = 0$$

then we must have

$$m \frac{d^2 x_f}{dt^2} + c \frac{dx_f}{dt} + kx_f = F$$

The previous section gave the solutions for the natural part of the solution. To solve the forcing equation,

$$m \frac{d^2 x_f}{dt^2} + c \frac{dx_f}{dt} + kx_f = F$$

we need to consider a particular form of input signal and then try a solution. Thus for a step input of size F at time $t = 0$ we can try a solution $x_f = A$, where A is a constant (see Section 19.3.2 on first-order differential equations for a discussion of the choice of solutions). Then $dx_f/dt = 0$ and $d^2 x_f/dt^2 = 0$. Thus, when these are substituted in the differential equation, $0 + 0 + kA = F$ and so $A = F/k$ and $x_f = F/k$. The complete solution, the sum of natural and forced solutions, is thus for the over-damped system

$$x = Ae^{s_1 t} + Be^{s_2 t} + \frac{F}{k}$$

for the critically damped system

$$x = (At + B)e^{-\omega_n t} + \frac{F}{k}$$

and for the under-damped system

$$x = e^{-\zeta \omega_n t}(P \cos \omega t + Q \sin \omega t) + \frac{F}{k}$$

When $t \to \infty$ the above three equations all lead to the solution $x = F/k$. This is the **steady-state condition.**

Thus a second-order differential equation in the form

$$a_2 \frac{d^2 x}{dt^2} + a_1 \frac{dx}{dt} + a_0 x = b_0 y$$

has a natural frequency given by

$$\omega_n^2 = \frac{a_0}{a_2}$$

and a damping factor given by

$$\zeta^2 = \frac{a_1^2}{4 a_2 a_0}$$

19.4.3 Examples of second-order systems

The following examples illustrate the points made above.

Consider a series RLC circuit (Figure 19.10) with $R = 100 \, \Omega$, $L = 2.0 \, H$ and $C = 20 \, \mu F$. When there is a step input V, the current i in the circuit is given by (see the text associated with Figure 17.9)

$$\frac{d^2 i}{dt^2} + \frac{R}{L} \frac{di}{dt} + \frac{1}{LC} i = \frac{V}{LC}$$

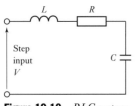

Figure 19.10 RLC system.

If we compare the equation with the general second-order differential equation of

$$a_2 \frac{d^2 x}{dt^2} + a_1 \frac{dx}{dt} + a_0 x = b_0 y$$

then the natural angular frequency is given by

$$\omega_n^2 = \frac{1}{LC} = \frac{1}{2.0 \times 20 \times 10^{-6}}$$

and so $\omega_n = 158$ Hz. Comparison with the general second-order equation also gives

$$\zeta^2 = \frac{(R/L)^2}{4 \times (1/LC)} = \frac{R^2 C}{4L} = \frac{100^2 \times 20 \times 10^{-6}}{4 \times 2.0}$$

Thus $\zeta = 0.16$. Since ζ is less than 1 the system is under-damped. The damped oscillation frequency ω is

$$\omega = \omega_n \sqrt{1 - \zeta^2} = 158\sqrt{1 - 0.16^2} = 156 \text{ Hz}$$

Because the system is under-damped the solution will be of the same form as

$$x = e^{-\zeta \omega_n t}(P \cos \omega t + Q \sin \omega t) + \frac{F}{k}$$

and so

$$i = e^{-0.16 \times 158 t}(P \cos 156t + Q \sin 156t) + V$$

Since $i = 0$ when $t = 0$, then $0 = 1(P + 0) + V$. Thus $P = -V$. Since $di/dt = 0$ when $t = 0$, then differentiating the above equation and equating it to zero gives

$$\frac{di}{dt} = e^{-\zeta \omega_n t}(\omega P \sin \omega t - \omega Q \cos \omega t) - \zeta \omega_n e^{-\zeta \omega_n t}(P \cos \omega t + Q \cos \omega t)$$

Thus $0 = 1(0 - \omega Q) - \zeta \omega_n (P + 0)$ and so

$$Q = \frac{\zeta \omega_n P}{\omega} = -\frac{\zeta \omega_n V}{\omega} = -\frac{0.16 \times 158 V}{156} \approx -0.16 \text{ V}$$

Thus the solution of the differential equation is

$$i = V - V e^{-25.3t}(\cos 156t + 0.16 \sin 156t)$$

Now consider the system shown in Figure 19.11. The input, a torque T, is applied to a disc with a moment of inertia I about the axis of the shaft. The shaft is free to rotate at the disc end but is fixed at its far end. The shaft rotation is opposed by the torsional stiffness of the shaft, an opposing torque of $k\theta_o$ occurring for an input rotation of θ_o. k is a constant. Frictional forces damp the rotation of the shaft and provide an opposing torque of $c \, d\theta_o/dt$, where c is a constant. Suppose we need to determine the condition for this system to be critically damped.

We first need to obtain the differential equation for the system. The net torque is

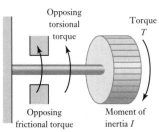

Figure 19.11　Torsional system.

$$\text{net torque} = T - c \frac{d\theta_o}{dt} - k\theta_o$$

The net torque is $I\,\mathrm{d}^2\theta_o/\mathrm{d}t^2$, hence

$$I\frac{\mathrm{d}^2\theta_o}{\mathrm{d}t^2} = T - c\frac{\mathrm{d}\theta_o}{\mathrm{d}t} - k\theta_o$$

$$I\frac{\mathrm{d}^2\theta_o}{\mathrm{d}t^2} + c\frac{\mathrm{d}\theta_o}{\mathrm{d}t} + k\theta_o = T$$

The condition for critical damping is given when the damping ratio ζ equals 1. Comparing the above differential equation with the general form of the second-order differential equation, then

$$\zeta^2 = \frac{a_1^2}{4a_2 a_0} = \frac{c^2}{4Ik}$$

Thus for critical damping we must have $c = \sqrt{(Ik)}$.

19.5 Performance measures for second-order systems

Figure 19.12 shows the typical form of the response of an under-damped second-order system to a step input. Certain terms are used to specify such a performance.

The **rise time** t_r is the time taken for the response x to rise from 0 to the steady-state value x_{SS} and is a measure of how fast a system responds to the input. This is the time for the oscillating response to complete a quarter of a cycle, i.e. $\frac{1}{2}\pi$. Thus

$$\omega t_r = \frac{1}{2}\pi$$

The rise time is sometimes specified as the time taken for the response to rise from some specified percentage of the steady-state value, e.g. 10%, to another specified percentage, e.g. 90%.

Figure 19.12   Step response of an under-damped system.

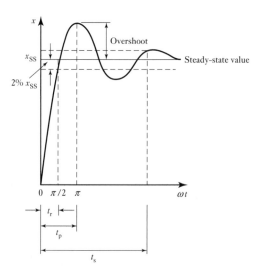

The **peak time** t_p is the time taken for the response to rise from 0 to the first peak value. This is the time for the oscillating response to complete one half cycle, i.e. π. Thus

$$\omega t_p = \pi$$

The **overshoot** is the maximum amount by which the response overshoots the steady-state value. It is thus the amplitude of the first peak. The overshoot is often written as a percentage of the steady-state value. For the under-damped oscillations of a system we can write

$$x = e^{-\zeta \omega_n t}(P \cos \omega t + Q \sin \omega t) + \text{steady-state value}$$

Since $x = 0$ when $t = 0$, then $0 = 1(P + 0) + x_{SS}$ and so $P = -x_{SS}$. The overshoot occurs at $\omega t = \pi$ and thus

$$x = e^{-\zeta \omega_n \pi / \omega}(P + 0) + x_{SS}$$

The overshoot is the difference between the output at that time and the steady-state value. Hence

$$\text{overshoot} = x_{SS} e^{-\zeta \omega_n \pi / \omega}$$

Since $\omega = \omega_n \sqrt{(1 - \zeta^2)}$ then we can write

$$\text{overshoot} = x_{SS} \exp\left(\frac{-\zeta \omega_n \pi}{\omega_n \sqrt{1 - \zeta^2}}\right) = x_{SS} \exp\left(\frac{-\zeta \pi}{\sqrt{1 - \zeta^2}}\right)$$

Expressed as a percentage of x_{SS},

$$\text{percentage overshoot} = \exp\left(\frac{-\zeta \pi}{\sqrt{1 - \zeta^2}}\right) \times 100\%$$

Table 19.2 gives values of the percentage overshoot for particular damping ratios.

Table 19.2 Percentage peak overshoot.

Damping ratio	Percentage overshoot
0.2	52.7
0.4	25.4
0.6	9.5
0.8	1.5

An indication of how fast oscillations decay is provided by the **subsidence ratio** or **decrement**. This is the amplitude of the second overshoot divided by that of the first overshoot. The first overshoot occurs when we have $\omega t = \pi$, the second overshoot when $\omega t = 3\pi$. Thus,

$$\text{first overshoot} = x_{SS} \exp\left(\frac{-\zeta \pi}{\sqrt{1 - \zeta^2}}\right)$$

$$\text{second overshoot} = x_{SS} \exp\left(\frac{-3\zeta \pi}{\sqrt{1 - \zeta^2}}\right)$$

and so

$$\text{subsidence ratio} = \frac{\text{second overshoot}}{\text{first overshoot}} = \exp\left(\frac{-2\zeta\pi}{\sqrt{1 - \zeta^2}}\right)$$

The **settling time** t_s is used as a measure of the time taken for the oscillations to die away. It is the time taken for the response to fall within and remain within some specified percentage, e.g. 2%, of the steady-state value (see Figure 19.12). This means that the amplitude of the oscillation should be less than 2% of x_{SS}. We have

$$x = e^{-\zeta\omega_n t}(P \cos \omega t + Q \sin \omega t) + \text{steady-state value}$$

and, as derived earlier, $P = -x_{SS}$. The amplitude of the oscillation is $(x - x_{SS})$ when x is a maximum value. The maximum values occur when ωt is some multiple of π and thus we have $\cos \omega t = 1$ and $\sin \omega t = 0$. For the 2% settling time, the settling time t_s is when the maximum amplitude is 2% of x_{SS}, i.e. $0.02 x_{SS}$. Thus

$$0.02 x_{SS} = e^{-\zeta\omega_n t_s}(x_{SS} \times 1 + 0)$$

Taking logarithms gives $\ln 0.02 = -\zeta\omega_n t_s$ and since $\ln 0.02 = -3.9$ or approximately -4, then

$$t_s = \frac{4}{\zeta\omega_n}$$

The above is the value of the settling time if the specified percentage is 2%. If the percentage is 5% the equation becomes

$$t_s = \frac{3}{\zeta\omega_n}$$

Since the time taken to complete one cycle, i.e. the periodic time, is $1/f$, where f is the frequency, and since $\omega = 2\pi f$, then the time to complete one cycle is $2\pi/f$. In a settling time of t_s the number of oscillations that occur is

$$\text{number of oscillations} = \frac{\text{settling time}}{\text{periodic time}}$$

and thus for a settling time defined for 2% of the steady-state value,

$$\text{number of oscillations} = \frac{4/\zeta\omega_n}{2\pi/\omega}$$

Since $\omega = \omega_n\sqrt{(1 - \zeta^2)}$, then

$$\text{number of oscillations} = \frac{2\omega_n\sqrt{1 - \zeta^2}}{\pi\zeta\omega_n} = \frac{2}{\pi}\sqrt{\frac{1}{\zeta^2} - 1}$$

To illustrate the above, consider a second-order system which has a natural angular frequency of 2.0 Hz and a damped frequency of 1.8 Hz. Since $\omega = \omega_n\sqrt{(1 - \zeta^2)}$, then the damping factor is given by

$$1.8 = 2.0\sqrt{1 - \zeta^2}$$

and $\zeta = 0.44$. Since $\omega t_r = \frac{1}{2}\pi$, then the 100% rise time is given by

$$t_r = \frac{\pi}{2 \times 1.8} = 0.87 \text{ s}$$

The percentage overshoot is given by

$$\text{percentage overshoot} = \exp\left(\frac{-\zeta\pi}{\sqrt{1-\zeta^2}}\right) \times 100\%$$

$$= \exp\left(\frac{-0.44\pi}{\sqrt{1-0.44^2}}\right) \times 100\%$$

The percentage overshoot is thus 21%. The 2% settling time is given by

$$t_s = \frac{4}{\zeta\omega_n} = \frac{4}{0.44 \times 2.0} = 4.5 \text{ s}$$

The number of oscillations occurring within the 2% settling time is given by

$$\text{number of oscillations} = \frac{2}{\pi}\sqrt{\frac{1}{\zeta^2}-1} = \frac{2}{\pi}\sqrt{\frac{1}{0.44^2}-1} = 1.3$$

19.6 System identification

In Chapters 17 and 18 models were devised for systems by considering them to be made up of simple elements. An alternative way of developing a model for a real system is to use tests to determine its response to some input, e.g. a step input, and then find the model that fits the response. This process of determining a mathematical model is known as **system identification**. Thus if we obtain a response to a step input of the form shown in Figure 19.5 then we might assume that it is a first-order system and determine the time constant from the response curve. For example, suppose the response takes a time of 1.5 s to reach 0.63 of its final height and the final height of the signal is five times the size of the step input. Table 19.1 indicates a time constant of 1.5 s and so the differential equation describing the model is

$$1.5\frac{dx}{dt} + x = 5y$$

An under-damped second-order system will give a response to a step input of the form shown in Figure 19.12. The damping ratio can be determined from measurements of the first and second overshoots with the ratio of these overshoots, i.e. the subsidence ratio, giving the damping ratio. The natural frequency can be determined from the time between successive overshoots. We can then use these values to determine the constants in the second-order differential equation.

Summary

The **natural response** of a system is when there is no input to the system forcing the variable to change but it is just changing naturally. The **forced response** of a system is when there is an input to the system forcing it to change.

A first-order system with no forcing input has a differential equation of the form

$$a_1\frac{dx}{dt} + a_0 x = 0$$

and this has the solution $x = e^{-a_0 t / a_1}$.

When there is a **forcing function** the differential equation is of the form

$$a_1 \frac{dx}{dt} + a_0 x = b_0 y$$

and the solution is $x =$ steady-state value $\times \left(1 - e^{-a_0 t / a_1}\right)$.

The **time constant** τ is the time the output takes to rise to 0.63 of its steady-state value and is (a_1 / a_0).

A second-order system with no forcing input has a differential equation of the form

$$m \frac{d^2 x}{dt^2} + c \frac{dx}{dt} + kx = 0$$

The natural angular frequency is given by $\omega_n^2 = k/m$ and the damping constant by $\zeta^2 = c^2 / 4mk$. The system is **over-damped** when we have $\zeta > 1$ and the general solution for x_n is

$$x_n = A e^{s_1 t} + B e^{s_2 t} \text{ with } s = -\zeta \omega_n \pm \omega_n \sqrt{\zeta^2 - 1}$$

When $\zeta = 1$ the system is **critically damped** and

$$x_n = (At + B) e^{-\omega_n t}$$

and with $\zeta < 1$ the system is **under-damped** and

$$x_n = e^{-\zeta \omega_n t} (P \cos \omega t + Q \sin \omega t)$$

When we have a forcing input F the second-order differential equation becomes

$$m \frac{d^2 x}{dt^2} + c \frac{dx}{dt} + kx = F$$

and for the over-damped system

$$x = A e^{s_1 t} + B e^{s_2 t} + \frac{F}{k}$$

for the critically damped system

$$x = (At + B) e^{-\omega_n t} + \frac{F}{k}$$

and for the under-damped system

$$x = e^{-\zeta \omega_n t} (P \cos \omega t + Q \sin \omega t) + \frac{F}{k}$$

The **rise time** t_r is the time taken for the response x to rise from 0 to the steady-state value x_{SS} and is a measure of how fast a system responds to the input and is given by $\omega t_r = \frac{1}{2}\pi$. The **peak time** t_p is the time taken for the response to rise from 0 to the first peak value and is given by $\omega t_p = \pi$. The **overshoot** is the maximum amount by which the response overshoots the steady-state value and is

$$\text{overshoot} = x_{SS} \exp\left(\frac{-\zeta \pi}{\sqrt{1 - \zeta^2}}\right)$$

The **subsidence ratio** or **decrement** is the amplitude of the second overshoot divided by that of the first overshoot and is

$$\text{subsidence ratio} = \exp\!\left(\frac{-2\zeta\pi}{\sqrt{1-\zeta^2}}\right)$$

The **settling time** t_s is the time taken for the response to fall within and remain within some specified percentage, e.g. 2%, of the steady-state value, this being given by

$$t_s = \frac{4}{\zeta\omega_n}$$

Problems

19.1 A first-order system has a time constant of 4 s and a steady-state transfer function of 6. What is the form of the differential equation for this system?

19.2 A mercury-in-glass thermometer has a time constant of 10 s. If it is suddenly taken from being at 20°C and plunged into hot water at 80°C, what will be the temperature indicated by the thermometer after (a) 10 s, (b) 20 s?

19.3 A circuit consists of a resistor R in series with an inductor L. When subject to a step input voltage V at time $t = 0$ the differential equation for the system is

$$\frac{di}{dt} + \frac{R}{L}i = \frac{V}{L}$$

What is (a) the solution for this differential equation, (b) the time constant, (c) the steady-state current i?

19.4 Describe the form of the output variation with time for a step input to a second-order system with a damping factor of (a) 0, (b) 0.5, (c) 1.0, (d) 1.5.

19.5 An RLC circuit has a current i which varies with time t when subject to a step input of V and is described by

$$\frac{d^2i}{dt^2} + 10\frac{di}{dt} + 16i = 16V$$

What is (a) the undamped frequency, (b) the damping ratio, (c) the solution to the equation if $i = 0$ when $t = 0$ and $di/dt = 0$ when $t = 0$?

19.6 A system has an output x which varies with time t when subject to a step input of y and is described by

$$\frac{d^2x}{dt^2} + 10\frac{dx}{dt} + 25x = 50y$$

What is (a) the undamped frequency, (b) the damping ratio, (c) the solution to the equation if $x = 0$ when $t = 0$ and $dx/dt = -2$ when $t = 0$ and there is a step input of size 3 units?

19.7 An accelerometer (an instrument for measuring acceleration) has an undamped angular frequency of 100 Hz and a damping factor of 0.6. What will be (a) the maximum percentage overshoot and (b) the rise time when there is a sudden change in acceleration?

19.8   What will be (a) the undamped angular frequency, (b) the damping factor, (c) the damped angular frequency, (d) the rise time, (e) the percentage maximum overshoot and (f) the 0.2% settling time for a system which gave the following differential equation for a step input y?

$$\frac{d^2x}{dt^2} + 5\frac{dx}{dt} + 16x = 16y$$

19.9   When a voltage of 10 V is suddenly applied to a moving-coil voltmeter it is observed that the pointer of the instrument rises to 11 V before eventually settling down to read 10 V. What is (a) the damping factor and (b) the number of oscillations the pointer will make before it is within 0.2% of its steady-state value?

19.10   A second-order system is described by the differential equation

$$\frac{d^2x}{dt^2} + c\frac{dx}{dt} + 4x = F$$

What value of damping constant c will be needed if the percentage overshoot is to be less than 9.5%?

19.11   Observation of the oscillations of a damped system when responding to an input indicates that the maximum displacement during the second cycle is 75% of the first displacement. What is the damping factor of the system?

19.12   A second-order system is found to have a time of 1.6 s between the first overshoot and the second overshoot. What is the natural frequency of the system?

Chapter twenty

System transfer functions

Objectives

The objectives of this chapter are that, after studying it, the reader should be able to:

- Define the transfer function and determine it from differential equations for first- and second-order systems.
- Determine the transfer functions for systems with feedback loops.
- Determine, using Laplace transforms, the responses of first- and second-order systems to simple inputs.
- Determine the effect of pole location on the responses of systems.

20.1 The transfer function

For an amplifier system it is customary to talk of the **gain** of the amplifier. This states how much bigger the output signal will be when compared with the input signal. It enables the output to be determined for specific inputs. Thus, for example, an amplifier with a voltage gain of 10 will give, for an input voltage of 2 mV, an output of 20 mV; or if the input is 1 V an output of 10 V. The gain states the mathematical relationship between the output and the input for the block. We can indicate when a signal is in the time domain, i.e. is a function of time, by writing it as $f(t)$. Thus, for an input of $y(t)$ and an output of $x(t)$ (Figure 20.1(a)),

$$\text{gain} = \frac{\text{output}}{\text{input}} = \frac{x(t)}{y(t)}$$

However, for many systems the relationship between the output and the input is in the form of a differential equation and so a statement of the function as just a simple number like the gain of 10 is not possible. We cannot just divide the output by the input because the relationship is a differential equation and not a simple algebraic equation. We can, however, transform a differential equation into an algebraic equation by using what is termed the **Laplace transform**. Differential equations describe how systems behave with time and are transformed by means of the Laplace transform into simple algebraic equations, not involving time, where we can carry out normal algebraic manipulations of the quantities. We talk of behaviour in the **time domain** being transformed to the *s*-**domain**. When in the *s*-domain a function is written, since it is a function of *s*, as $F(s)$. It is usual to use a capital letter F for the Laplace transform and a lower case letter f for the time-varying function $f(t)$.

We then define the relationship between output and input in terms of a **transfer function**, this stating the relationship between the Laplace transform of the output and the Laplace transform of the input. Suppose the input to a linear system has a Laplace transform of $Y(s)$ and the Laplace transform of the output is $X(s)$. The transfer function $G(s)$ of the system is then defined as

$$\text{transfer function} = \frac{\text{Laplace transform of output}}{\text{Laplace transform of input}}$$

$$G(s) = \frac{X(s)}{Y(s)}$$

with all the initial conditions being zero, i.e. we assume zero output when zero input and zero rate of change of output with time when zero rate of change of input with time. Thus the output transform is $X(s) = G(s)Y(s)$, i.e. the product of the input transform and the transfer function. If we represent a system in the s-domain by a block diagram (Figure 20.1(b)) then $G(s)$ is the function in the box which takes an input of $Y(s)$ and converts it to an output of $X(s)$.

Figure 20.1   Block diagrams: (a) in time domain, (b) in s-domain.

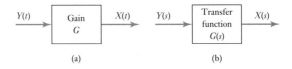

(a)   (b)

20.1.1   Laplace transforms

To obtain the Laplace transform of a differential equation, which includes quantities that are functions of time, we can use tables coupled with a few basic rules (Appendix A includes such a table and gives details of the rules). Figure 20.2 shows basic transforms for common forms of inputs.

Figure 20.2   Laplace transforms for common inputs.

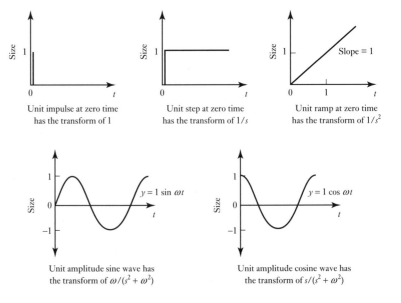

The following are some of the basic rules involved in working with Laplace transforms.

1 If a function of time is multiplied by a constant then the Laplace transform is multiplied by the same constant, i.e.

$af(t)$ has the transform of $aF(s)$

For example, the Laplace transform of a step input of 6 V to an electrical system is just six times the transform for a unit step and thus $6s$.

2 If an equation includes the sum of, say, two separate quantities which are functions of time, then the transform of the equation is the sum of the two separate Laplace transforms, i.e.

$f(t) + g(t)$ has the transform $F(s) + G(s)$

3 The Laplace transform of a first derivative of a function is

$$\text{transform of } \left\{ \frac{d}{dt} f(t) \right\} = sF(s) - f(0)$$

where $f(0)$ is the initial value of $f(t)$ when $t = 0$. However, when we are dealing with a transfer function we have all initial conditions zero.

4 The Laplace transform for the second derivative of a function is

$$\text{transform of } \left\{ \frac{d^2}{dt^2} f(t) \right\} = s^2 F(s) - sf(0) - \frac{d}{dt} f(0)$$

where $df(0)/dt$ is the initial value of the first derivative of $f(t)$ when we have $t = 0$. However, when we are dealing with a transfer function we have all initial conditions zero.

5 The Laplace transform of an integral of a function is

$$\text{transform of } \left\{ \int_0^t f(t) \, dt \right\} = \frac{1}{s} F(s)$$

Thus, in obtaining the transforms of differential or integral equations when all the initial conditions are zero, we:

replace a function of time $f(t)$ by $F(s)$,
replace a first derivative $df(t)/dt$ by $sF(s)$,
replace a second derivative $d^2f(t)/dt^2$ by $s^2F(s)$,
replace an integral $\int f(t) \, dt$ by $F(s)/s$.

When algebraic manipulations have occurred in the s-domain, then the outcome can be transformed back to the time domain by using the table of transforms in the inverse manner, i.e. finding the time domain function which fits the s-domain result. Often the transform has to be rearranged to be put into a form given in the table. The following are some useful such inversions; for more inversions see Appendix A:

Laplace transform	*Function of time*
1 $\dfrac{1}{s + a}$	e^{-at}
2 $\dfrac{a}{s(s + a)}$	$(1 - e^{-at})$

$$3 \quad \frac{b - a}{(s + a)(s + b)} \qquad\qquad e^{-at} - e^{-bt}$$

$$4 \quad \frac{s}{(s + a)^2} \qquad\qquad (1 - at)e^{-at}$$

$$5 \quad \frac{a}{s^2(s + a)} \qquad\qquad t - \frac{1 - e^{-at}}{a}$$

The following sections illustrate the application of the above to first-order and second-order systems.

20.2 First-order systems

Consider a system where the relationship between the input and the output is in the form of a first-order differential equation. The differential equation of a first-order system is of the form

$$a_1 \frac{dx}{dt} + a_0 x = b_0 y$$

where a_1, a_0, b_0 are constants, y is the input and x the output, both being functions of time. The Laplace transform of this, with all initial conditions zero, is

$$a_1 s X(s) + a_0 X(s) = b_0 Y(s)$$

and so we can write the transfer function $G(s)$ as

$$G(s) = \frac{X(s)}{Y(s)} = \frac{b_0}{a_1 s + a_0}$$

This can be rearranged to give

$$G(s) = \frac{b_0/a_0}{(a_1/a_0)s + 1} = \frac{G}{\tau s + 1}$$

where G is the gain of the system when there are steady-state conditions, i.e. there is no dx/dt term; (a_1/a_0) is the time constant τ of the system (see Section 19.3.4).

20.2.1 First-order system with step input

When a first-order system is subject to a unit step input, then $Y(s) = 1/s$ and the output transform $X(s)$ is

$$X(s) = G(s)Y(s) = \frac{G}{s(\tau s + 1)} = G\frac{(1/\tau)}{s(s + 1/\tau)}$$

Hence, since we have the transform in the form $a/s(s + a)$, using the inverse transformation listed as item 2 in the previous section gives

$$x = G(1 - e^{-t/\tau})$$

20.2.2 Examples of first-order systems

The following examples illustrate the above points in the consideration of the transfer function of a first-order system and its behaviour when subject to a step input.

1 Consider a circuit which has a resistance R in series with a capacitance C. The input to the circuit is v and the output is the potential difference v_C across the capacitor. The differential equation relating the input and output is

$$v = RC\frac{dv_C}{dt} + v_C$$

Determine the transfer function.

Taking the Laplace transform, with all initial conditions zero, then

$$V(s) = RCsV_C(s) + V_C(s)$$

Hence the transfer function is

$$G(s) = \frac{V_C(s)}{V(s)} = \frac{1}{RCs + 1}$$

2 Consider a thermocouple which has a transfer function linking its voltage output V and temperature input of

$$G(s) = \frac{30 \times 10^{-6}}{10s + 1} \text{V/°C}$$

Determine the response of the system when subject to a step input of size 100°C and hence the time taken to reach 95% of the steady-state value.

Since the transform of the output is equal to the product of the transfer function and the transform of the input, then

$$V(s) = G(s) \times \text{input}(s)$$

A step input of size 100°C, i.e. the temperature of the thermocouple is abruptly increased by 100°C, is $100/s$. Thus

$$V(s) = \frac{30 \times 10^{-6}}{10s + 1} \times \frac{100}{s} = \frac{30 \times 10^{-4}}{10s(s + 0.1)}$$

$$= 30 \times 10^{-4}\frac{0.1}{s(s + 0.1)}$$

The fraction element is of the form $a/s(s + a)$ and so the inverse transform is

$$V = 30 \times 10^{-4}(1 \times e^{-0.1t}) \text{ V}$$

The final value, i.e. the steady-state value, is when $t \to \infty$ and so is when the exponential term is zero. The final value is therefore 30×10^{-4} V. Thus the time taken to reach, say, 95% of this is given by

$$0.95 \times 30 \times 10^{-4} = 30 \times 10^{-4}(1 \times e^{-0.1t})$$

Thus $0.05 = e^{-0.1t}$ and $\ln 0.05 = -0.1t$. The time is thus 30 s.

3 Consider a ramp input to the above thermocouple system of $5t$ °C/s, i.e. the temperature is raised by 5°C every second. Determine how the voltage of the thermocouple varies with time and hence the voltage after 12 s.
   The transform of the ramp signal is $5/s^2$. Thus

$$V(s) = \frac{30 \times 10^{-6}}{10s + 1} \times \frac{5}{s^2} = 150 \times 10^{-6} \frac{0.1}{s^2(s + 0.1)}$$

The inverse transform can be obtained using item 5 in the list given in the previous section. Thus

$$V = 150 \times 10^{-6} \left(t - \frac{1 - e^{-0.1t}}{0.1} \right)$$

After a time of 12 s we would have $V = 7.5 \times 10^{-4}$ V.

4 Consider an impulse input of size 100°C, i.e. the thermocouple is subject to a momentary temperature increase of 100°C. Determine how the voltage of the thermocouple varies with time and hence the voltage after 2 s.
   The impulse has a transform of 100. Hence

$$V(s) = \frac{30 \times 10^{-6}}{10s + 1} \times 100 = 3 \times 10^{-4} \frac{1}{s + 0.1}$$

Hence $V = 3 \times 10^{-4} e^{-0.1t}$ V. After 2 s, the thermocouple voltage $V = 1.8 \times 10^{-4}$ V.

20.3 Second-order systems

For a second-order system, the relationship between the input y and the output x is described by a differential equation of the form

$$a_2 \frac{d^2x}{dt^2} + a_1 \frac{dx}{dt} + a_0 x = b_0 y$$

where a_2, a_1, a_0 and b_0 are constants. The Laplace transform of this equation, with all initial conditions zero, is

$$a_2 s^2 X(s) + a_1 s X(s) + a_0 X(s) = b_0 Y(s)$$

Hence

$$G(s) = \frac{X(s)}{Y(s)} = \frac{b_0}{a_2 s^2 + a_1 s + a_0}$$

An alternative way of writing the differential equation for a second-order system is

$$\frac{d^2x}{dt^2} + 2\zeta \omega_n \frac{dx}{dt} + \omega_n^2 x = b_0 \omega_n^2 y$$

where ω_n is the natural angular frequency with which the system oscillates and ζ is the damping ratio. The Laplace transform of this equation gives

$$G(s) = \frac{X(s)}{Y(s)} = \frac{b_0 \omega_n^2}{s^2 + 2\zeta \omega_n s + \omega_n^2}$$

The above are the general forms taken by the transfer function for a second-order system.

20.3.1 Second-order system with step input

When a second-order system is subject to a unit step input, i.e. $Y(s) = 1/s$, then the output transform is

$$X(s) = G(s)\,Y(s) = \frac{b_0\omega_n^2}{s(s^2 + 2\zeta\omega_n s + \omega_n)}$$

This can be rearranged as

$$X(s) = \frac{b_0\omega_n^2}{s(s + p_1)(s + p_2)}$$

where p_1 and p_2 are the roots of the equation

$$s^2 + 2\zeta\omega_n s + \omega_n^2 = 0$$

Hence, using the equation for the roots of a quadratic equation,

$$p = \frac{-2\zeta\omega_n \pm \sqrt{4\zeta^2\omega_n^2 - 4\omega_n^2}}{2}$$

and so the two roots p_1 and p_2 are

$$p_1 = -\zeta\omega_n + \omega_n\sqrt{\zeta^2 - 1} \qquad p_2 = -\zeta\omega_n - \omega_n\sqrt{\zeta^2 - 1}$$

With $\zeta > 1$ the square root term is real and the system is over-damped. To find the inverse transform we can either use partial fractions (see Appendix A) to break the expression down into a number of simple fractions or use item 14 in the table of transforms in Appendix A; the result in either case is

$$x = \frac{b_0\omega_n^2}{p_1 p_2}\left[1 - \frac{p_2}{p_2 - p_1}\,e^{-p_2 t} + \frac{p_1}{p_2 - p_1}\,e^{-p_1 t}\right]$$

With $\zeta = 1$ the square root term is zero and so $p_1 = p_2 = -\omega_n$. The system is critically damped. The equation then becomes

$$X(s) = \frac{b_0\omega_n^2}{s(s + \omega_n)^2}$$

This equation can be expanded by means of partial fractions (see Appendix A) to give

$$Y(s) = b_0\left[\frac{1}{s} - \frac{1}{s + \omega_n} - \frac{\omega_n}{(s + \omega_n)^2}\right]$$

Hence

$$x = b_0[1 - e^{-\omega_n t} - \omega_n t\,e^{-\omega_n t}]$$

With $\zeta < 1$, then

$$x = b_0\left[1 - \frac{e^{-\zeta\omega_n t}}{\sqrt{1 - \zeta^2}}\sin(\omega_n\sqrt{(1 - \zeta^2)}t + \phi)\right]$$

where $\cos\phi = \zeta$. This is an under-damped oscillation.

20.3.2 Examples of second-order systems

The following examples illustrate the above.

1 What will be the state of damping of a system having the following transfer function and subject to a unit step input?

$$G(s) = \frac{1}{s^2 + 8s + 16}$$

For a unit step input $Y(s) = 1/s$ and so the output transform is

$$X(s) = G(s)Y(s) = \frac{1}{s(s^2 + 8s + 16)} = \frac{1}{s(s+4)(s+4)}$$

The roots of $s^2 + 8s + 16$ are thus $p_1 = p_2 = -4$. Both the roots are real and the same and so the system is critically damped.

2 A robot arm having the following transfer function is subject to a unit ramp input. What will be the output?

$$G(s) = \frac{K}{(s+3)^2}$$

The output transform $X(s)$ is

$$X(s) = G(s)Y(s) = \frac{K}{(s+3)^2} \times \frac{1}{s^2}$$

Using partial fractions (see Appendix A) this becomes

$$X(s) = \frac{K}{9s^2} - \frac{2K}{9(s+3)} + \frac{K}{9(s+3)^2}$$

Hence the inverse transform is

$$x = \frac{1}{9}Kt - \frac{2}{9}Ke^{-3t} + \frac{1}{9}Kte^{-3t}$$

20.4 Systems in series

If a system consists of a number of subsystems in series, as in Figure 20.3, then the transfer function $G(s)$ of the system is given by

$$G(s) = \frac{X(s)}{Y(s)} = \frac{X_1(s)}{Y(s)} \times \frac{X_2(s)}{X_1(s)} \times \frac{X(s)}{X_2(s)}$$

$$= G_1(s) \times G_2(s) \times G_3(s)$$

The transfer function of the system as a whole is the product of the transfer functions of the series elements.

Figure 20.3 Systems in series.

20.4.1 Examples of systems in series

The following examples illustrate this. It has been assumed that when subsystems are linked together no interaction occurs between the blocks which would result in changes in their transfer functions, e.g. with electric circuits there can be problems when subsystem circuits interact and load each other.

1 What will be the transfer function for a system consisting of three elements in series, the transfer functions of the elements being 10, $2/s$ and $4/(s + 3)$?

   Using the equation developed above,

$$G(s) = 10 \times \frac{2}{s} \times \frac{4}{s + 3} = \frac{80}{s(s + 3)}$$

2 A field-controlled d.c. motor consists of three subsystems in series: the field circuit, the armature coil and the load. Figure 20.4 illustrates the arrangement and the transfer functions of the subsystems. Determine the overall transfer function.

Figure 20.4 Field-controlled d.c. motor.

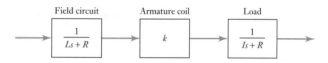

The overall transfer function is the product of the transfer functions of the series elements. Thus

$$G(s) = \frac{1}{Ls + R} \times k \times \frac{1}{Is + c} = \frac{k}{(Ls + R)(Is + c)}$$

20.5 Systems with feedback loops

Figure 20.5 shows a simple system having negative feedback. With **negative feedback** the system input and the feedback signals are subtracted at the summing point. The term **forward path** is used for the path having the transfer function $G(s)$ in the figure and **feedback path** for the one having $H(s)$. The entire system is referred to as a **closed-loop system**.

For the negative feedback system, the input to the subsystem having the forward-path transfer function $G(s)$ is $Y(s)$ minus the feedback signal. The feedback loop has a transfer function of $H(s)$ and has as its input $X(s)$, thus the feedback signal is $H(s)X(s)$. Thus the $G(s)$ element has an input of $Y(s) - H(s)X(s)$ and an output of $X(s)$ and so

$$G(s) = \frac{X(s)}{Y(s) - H(s)X(s)}$$

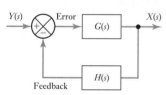

Figure 20.5 Negative feedback system.

This can be rearranged to give

$$\frac{X(s)}{Y(s)} = \frac{G(s)}{1 + G(s)H(s)}$$

Hence the overall transfer function for the negative feedback system $T(s)$ is

$$T(s) = \frac{X(s)}{Y(s)} = \frac{G(s)}{1 + G(s)H(s)}$$

20.5.1 Examples of systems with negative feedback

The following examples illustrate the above.

1 What will be the overall transfer function for a closed-loop system having a forward-path transfer function of $2/(s + 1)$ and a negative feedback-path transfer function of $5s$?

Using the equation developed above,

$$T(s) = \frac{G(s)}{1 + G(s)H(s)} = \frac{2/(s + 1)}{1 + [2/(s + 1)]5s} = \frac{2}{11s + 1}$$

2 Consider an armature-controlled d.c. motor (Figure 20.6). This has a forward path consisting of three elements: the armature circuit with a transfer function $1/(Ls + R)$, the armature coil with a transfer function k and the load with a transfer function $1/(Is + c)$. There is a negative feedback path with a transfer function K. Determine the overall transfer function for the system.

Figure 20.6 Armature-controlled d.c. motor.

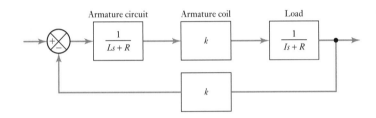

The forward-path transfer function for the series elements is the product of the transfer functions of the series elements, i.e.

$$G(s) = \frac{1}{Ls + R} \times k \times \frac{1}{Is + c} = \frac{k}{(Ls + R)(Is + c)}$$

The feedback path has a transfer function of K. Thus the overall transfer function is

$$T(s) = \frac{G(s)}{1 + G(s)H(s)} = \frac{\dfrac{k}{(Ls + R)(Is + c)}}{1 + \dfrac{kK}{(Ls + R)(Is + c)}}$$

$$= \frac{k}{(Ls + R)(Is + c) + kK}$$

20.6 Effect of pole location on transient response

We can define a system as being **stable** if, when it is given an input, it has transients which die away with time and leave the system in its steady-state condition. A system is said to be **unstable** if the transients do not die away with time but increase in size and so the steady-state condition is never attained.

Consider an input of a unit impulse to a first-order system with a transfer function of $G(s) = 1/(s + 1)$. The system output $X(s)$ is

$$X(s) = \frac{1}{s + 1} \times 1$$

and thus $x = e^{-t}$. As the time t increases so the output dies away eventually to become zero. Now consider the unit impulse input to a system with the transfer function $G(s) = 1/(s - 1)$. The output is

$$X(s) = \frac{1}{s - 1} \times 1$$

and so $x = e^{t}$. As t increases, so the output increases with time. Thus a momentary impulse to the system results in an ever increasing output; this system is unstable.

For a transfer function, the values of s which make the transfer function infinite are termed its **poles**; they are the roots of the characteristic equation. Thus for $G(s) = 1/(s + 1)$, there is a pole of $s = -1$. For $G(s) = 1/(s - 1)$, there is a pole of $s = +1$. Thus, for a first-order system the system is stable if the pole is negative, and **unstable** if the pole is positive (Figure 20.7).

Figure 20.7 First-order systems: (a) negative pole, (b) positive pole.

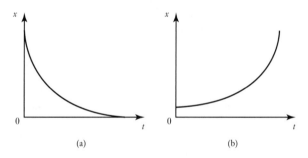

(a)   (b)

For a second-order system with transfer function

$$G(s) = \frac{b_0\omega_n^2}{s^2 + 2\zeta\omega_n s + \omega_n^2}$$

when subject to a unit impulse input,

$$X(s) = \frac{b_0\omega_n^2}{(s + p_1)(s + p_2)}$$

where p_1 and p_2 are the roots of the equation

$$s^2 + 2\zeta\omega_n s + \omega_n = 0$$

Using the equation for the roots of a quadratic equation,

$$p = \frac{-2\zeta\omega_n \pm \sqrt{4\zeta^2\omega_n^2 - 4\omega_n^2}}{2} = -\zeta\omega_n \pm \omega_n\sqrt{\zeta^2 - 1}$$

Depending on the value of the damping factor, the term under the square root sign can be real or imaginary. When there is an imaginary term the output involves an oscillation. For example, suppose we have a second-order system with transfer function

$$G(s) = \frac{1}{[s - (-2 + j1)][s - (-2 - j1)]}$$

i.e. $p = -2 \pm j1$. When subject to a unit impulse input the output is $e^{-2t} \sin t$. The amplitude of the oscillation, i.e. e^{-2t}, dies away as the time increases and so the effect of the impulse is a gradually decaying oscillation (Figure 20.8(a)). The system is stable.

Figure 20.8 Second-order systems.

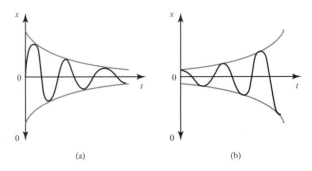

(a)   (b)

Suppose, however, we have a system with transfer function

$$G(s) = \frac{1}{[s - (2 + j1)][s - (2 - j1)]}$$

i.e. $p = +2 \pm j1$. When subject to a unit impulse input, the output is $e^{2t} \sin t$. The amplitude of the oscillation, i.e. e^{2t}, increases as the time increases (Figure 20.8(b)). The system is unstable.

In general, when an impulse is applied to a system, the output is in the form of the sum of a number of exponential terms. If just one of these terms is of exponential growth then the output continues to grow and the system is unstable. When there are pairs of poles involving plus or minus imaginary terms then the output is an oscillation.

A system is stable if the real part of all its poles is negative.
A system is unstable if the real part of any of its poles is positive.

20.6.1 The s-plane

We can plot the positions of the poles of a system on a graph with the x-axis being the real parts and the y-axis the imaginary parts. Such a graph is termed the s-**plane**. The location of the poles on the plane determines the stability of a system. Figure 20.9 shows such a plane and how the location of roots affects the response of a system.

20.6.2 Compensation

The output from a system might be unstable or perhaps the response is too slow or there is too much overshoot. Systems can have their responses to

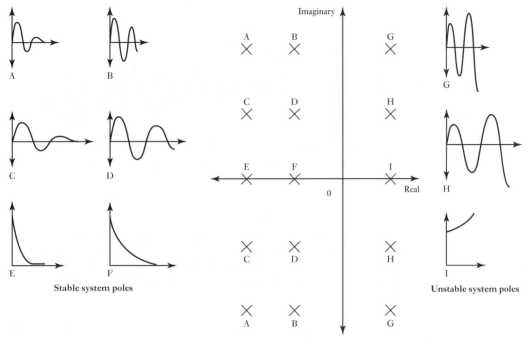

Figure 20.9 The s-plane.

inputs altered by including **compensators**. A compensator is a block which is incorporated in the system so that it alters the overall transfer function of the system in such a way as to obtain the required characteristics.

As an illustration of the use of a compensator, consider a position control system which has a negative feedback path with a transfer function of 1 and two subsystems in its forward path: a compensator with a transfer function of K and a motor/drive system with a transfer function of $1/s(s + 1)$. What value of K is necessary for the system to be critically damped? The forward path has a transfer function of $K/s(s + 1)$ and the feedback path a transfer function of 1. Thus the overall transfer function of the system is

$$T(s) = \frac{G(s)}{1 + G(s)H(s)} = \frac{\dfrac{K}{s(s + 1)}}{1 + \dfrac{K}{s(s + 1)}} = \frac{K}{s(s + 1) + K}$$

The denominator is thus $s^2 + s + K$. This will have the roots

$$s = \frac{-1 \pm \sqrt{1 - 4K}}{2}$$

To be critically damped we must have $1 - 4K = 0$ and hence the compensator must have the proportional gain of $K = \frac{1}{4}$.

Summary

The **transfer function** $G(s)$ of a system is (Laplace transform of the output)/(Laplace transform of the input). To obtain the transforms of differential or integral equations when all the initial conditions are zero we: replace a function of time $f(t)$ by $F(s)$, replace a first derivative $df(t)/dt$ by $sF(s)$, replace a second derivative $d^2f(t)/dt^2$ by $s^2F(s)$, replace an integral $\int f(t)\, dt$ by $F(s)/s$.

A **first-order system** has a transfer function of the form $G/(\tau s + 1)$, where τ is the time constant. A **second-order system** has a transfer function of the form

$$G(s) = \frac{b_0 \omega_n^2}{s^2 + 2\zeta \omega_n s + \omega_n^2}$$

where ζ is the damping factor and ω_n the natural angular frequency.

The values of s which make the transfer function infinite are termed its **poles**; they are the roots of the characteristic equation. A system is stable if the real part of all its poles is negative and unstable if the real part of any of its poles is positive.

Problems

20.1   What are the transfer functions for systems giving the following input/output relationships?

(a) A hydraulic system has an input q and an output h where

$$q = A\frac{dh}{dt} + \frac{\rho g h}{R}$$

(b) A spring-dashpot-mass system with an input F and an output x, where

$$m\frac{d^2x}{dt^2} + c\frac{dx}{dt} + kx = F$$

(c) An RLC circuit with an input v and output v_C, where

$$v = RC\frac{dv_C}{dt} + LC\frac{d^2v_C}{dt^2} + v_C$$

20.2   What are the time constants of the systems giving the transfer functions (a) $G(s) = 5/(3s + 1)$, (b) $G(s) = 2/(2s + 3)$?

20.3   Determine how the outputs of the following systems vary with time when subject to a unit step input at time $t = 0$: (a) $G(s) = 2/(s + 2)$ (b) $G(s) = 10/(s + 5)$.

20.4   What is the state of the damping for the systems having the following transfer functions?

(a) $G(s) = \dfrac{5}{s^2 - 6s + 16}$,   (b) $G(s) = \dfrac{10}{s^2 + s + 100}$,

(c) $G(s) = \dfrac{2s + 1}{s^2 + 2s + 1}$,   (d) $G(s) = \dfrac{3s + 20}{s^2 + 2s + 20}$

20.5   What is the output of a system with the transfer function $s/(s + 3)^2$ and subject to a unit step input at time $t = 0$?

20.6   What is the output of a system having the transfer function $G = 2/[(s + 3) \times (s + 4)]$ and subject to a unit impulse?

20.7   What are the overall transfer functions of the following negative feedback systems?

	Forward path	*Feedback path*
(a)	$G(s) = \dfrac{4}{s(s + 1)}$	$H(s) = \dfrac{1}{s}$
(b)	$G(s) = \dfrac{2}{s + 1}$	$H(s) = \dfrac{1}{s + 2}$
(c)	$G(s) = \dfrac{4}{(s + 2)(s + 3)}$	$H(s) = 5$

(d) two series elements $G_1(s) = 2/(s + 2)$
     and $G_2(s) = 1/s$ $H(s) = 10$

20.8   What is the overall transfer function for a closed-loop system having a forward-path transfer function of $5/(s + 3)$ and a negative feedback-path transfer function of 10?

20.9   A closed-loop system has a forward path having two series elements with transfer functions 5 and $1/(s + 1)$. If the feedback path has a transfer function $2/s$, what is the overall transfer function of the system?

20.10   A closed-loop system has a forward path having two series elements with transfer functions of 2 and $1/(s + 1)$. If the feedback path has a transfer function of s, what is the overall transfer function of the system?

20.11   A system has a transfer function of $1/[(s + 1)(s + 2)]$. What are its poles?

20.12   Which of the following systems are stable or unstable?

(a) $G(s) = 1/[(s + 5)(s + 2)]$,
(b) $G(s) = 1/[(s - 5)(s + 2)]$,
(c) $G(s) = 1/[(s - 5)(s - 5)]$,
(d) $G(s) = 1/(s^2 + s + 1)$,
(e) $G(s) = 1/(s^2 - 2s + 3)$.

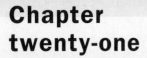

Chapter twenty-one

Frequency response

21.1 Sinusoidal input

In the previous two chapters, the response of systems to step, impulse and ramp inputs has been considered. This chapter extends this to when there is a sinusoidal input. While for many control systems a sinusoidal input might not be encountered normally, it is a useful testing input since the way the system responds to such an input is a very useful source of information to aid the design and analysis of systems. It is also useful because many other signals can be considered as the sum of a number of sinusoidal signals. In 1822 Jean Baptiste Fourier proposed that any periodic waveform, e.g. a square waveform, can be made up of a combination of sinusoidal waveforms and by considering the behaviour of a system to each individual sinusoidal waveform it is possible to determine the response to the more complex waveform.

21.1.1 Response of a system to a sinusoidal input

Consider a first-order system which is described by the differential equation

$$a_1 \frac{dx}{dt} + a_0 x = b_0 y$$

where y is the input and x the output. Suppose we have the unit amplitude sinusoidal input of $y = \sin \omega t$. What will the output be? Well we must end up with the sinusoid $b_0 \sin \omega t$ when we add $a_1 \, dx/dt$ and $a_0 x$. But sinusoids have the property that when differentiated, the result is also a sinusoid and with the same frequency (a cosine is a sinusoid, being just $\sin(\omega t + 90°)$). This applies no matter how many times we carry out the differentiation. Thus we should expect that the steady-state response x will also be sinusoidal and with the same frequency. The output may, however, differ in amplitude and phase from the input.

21.2   Phasors

In discussing sinusoidal signals it is convenient to use **phasors**. Consider a sinusoid described by the equation $v = V\sin(\omega t + \phi)$, where V is the amplitude, ω the angular frequency and ϕ the phase angle. The phasor can be represented by a line of length $|V|$ making an initial angle of ϕ with the phase reference axis (Figure 21.1). The $|\ |$ lines are used to indicate that we are only concerned with the magnitude or size of the quantity when specifying its length. To completely specify a phasor quantity requires a magnitude and angle to be stated. The convention generally adopted is to write a phasor in bold, non–italic, print, e.g. **V**. When such a symbol is seen it implies a quantity having both a magnitude and an angle.

Figure 21.1   Representing a sinusoidal signal by a phasor.

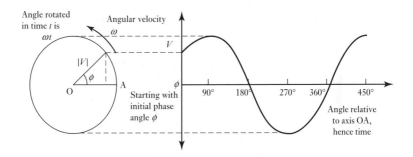

Such a phasor can be described by means of complex number notation. A complex quantity can be represented by $(x + jy)$, where x is the real part and y the imaginary part of the complex number. On a graph with the imaginary component as the y-axis and the real part as the x-axis, x and y are Cartesian co-ordinates of the point representing the complex number (Figure 21.2(a)).

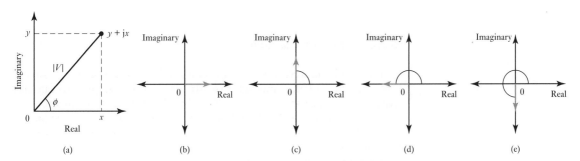

Figure 21.2   (a) Complex representation of a phasor, (b) 0°, (c) 90°, (d) 270°, (e) 360°.

If we take the line joining this point to the graph origin to represent a phasor, then we have the phase angle ϕ of the phasor represented by

$$\tan\phi = \frac{y}{x}$$

and its length by the use of Pythagoras theorem as

$$\text{length of phasor } |V| = \sqrt{x^2 + y^2}$$

Since $x = |V| \cos \phi$ and $y = |V| \sin \phi$, then we can write

$$\mathbf{V} = x + \mathrm{j}y = |V| \cos \theta + \mathrm{j}|V| \sin \theta = |V|(\cos \theta + \mathrm{j} \sin \theta)$$

Thus a specification of the real and imaginary parts of a complex number enables a phasor to be specified.

Consider a phasor of length 1 and phase angle $0°$ (Figure 21.2(b)). It will have a complex representation of $1 + \mathrm{j}0$. Now consider the same length phasor but with a phase angle of $90°$ (Figure 21.2(c)). It will have a complex representation of $0 + \mathrm{j}1$. Thus rotation of a phasor anti-clockwise by $90°$ corresponds to multiplication of the phasor by j. If we now rotate this phasor by a further $90°$ (Figure 21.2(d)), then following the same multiplication rule we have the original phasor multiplied by j^2. But the phasor is just the original phasor in the opposite direction, i.e. just multiplied by -1. Hence $\mathrm{j}^2 = -1$ and so $\mathrm{j} = \sqrt{(-1)}$. Rotation of the original phasor through a total of $270°$, i.e. $3 \times 90°$, is equivalent to multiplying the original phasor by $\mathrm{j}^3 = \mathrm{j}(\mathrm{j}^2) = -\mathrm{j}$.

To illustrate the above, consider a voltage v which varies sinusoidally with time according to the equation

$$v = 10 \sin(\omega t + 30°) \text{ V}$$

When represented by a phasor, what are (a) its length, (b) its angle relative to the reference axis, (c) its real and imaginary parts when represented by a complex number?

(a) The phasor will have a length scaled to represent the amplitude of the sinusoid and so is 10 V.
(b) The angle of the phasor relative to the reference axis is equal to the phase angle and so is $30°$.
(c) The real part is given by the equation $x = 10 \cos 30° = 8.7$ V and the imaginary part by $y = 10 \sin 30° = 5.0$ V. Thus the phasor is specified by $8.7 + \mathrm{j}5.0$ V.

21.2.1   Phasor equations

Consider a phasor as representing the unit amplitude sinusoid of $x = \sin \omega t$. Differentiation of the sinusoid gives $\mathrm{d}x/\mathrm{d}t = \omega \cos \omega t$. But we can also write this as $\mathrm{d}x/\mathrm{d}t = \omega \sin(\omega t + 90°)$. In other words, differentiation just results in a phasor with a length increased by a factor of ω and which is rotated round by $90°$ from the original phasor. Thus, in complex notation, we have multiplied the original phasor by $\mathrm{j}\omega$, since multiplication by j is equivalent to a rotation through $90°$.

Thus the differential equation

$$a_1 \frac{\mathrm{d}x}{\mathrm{d}t} + a_0 x = b_0 y$$

can be written, in complex notation, as a **phasor equation**

$$\mathrm{j}\omega a_1 \mathbf{X} + a_0 \mathbf{X} = b_0 \mathbf{Y}$$

where the bold, non-italic, letters indicate that the data refers to phasors. We can say that the differential equation, which was an equation in the time domain, has been transformed into an equation in the **frequency domain**. The frequency domain equation can be rewritten as

$$(j\omega a_1 + a_0)X = b_0 Y$$

$$\frac{X}{Y} = \frac{b_0}{j\omega a_1 + a_0}$$

But, in Section 20.2, when the same differential equation was written in the s-domain, we had

$$G(s) = \frac{X(s)}{Y(s)} = \frac{b_0}{a_1 s + a_0}$$

If we replace s by $j\omega$ we have the same equation. It turns out that we can always do this to convert from the s-domain to the frequency domain. This thus leads to a definition of a **frequency-response function** or **frequency transfer function** $G(j\omega)$, for the steady state, as

$$G(j\omega) = \frac{\text{output phasor}}{\text{input phasor}}$$

To illustrate the above, consider the determination of the frequency-response function for a system having a transfer function of

$$G(s) = \frac{1}{s+1}$$

The frequency-response function is obtained by replacing s by $j\omega$. Thus

$$G(j\omega) = \frac{1}{j\omega + 1}$$

21.3 Frequency response

The procedure for determining the frequency response of a system is thus:

1 replace s in the transfer function by $j\omega$ to give the frequency-response function;
2 the amplitude ratio between the output and the input is then the magnitude of the complex frequency-response function, i.e. $\sqrt{(x^2 + y^2)}$;
3 the phase angle between the output and the input is given by $\tan \phi = y/x$ or the ratio of the imaginary and real parts of the complex number representing the frequency-response function.

21.3.1 Frequency response for a first-order system

A first-order system has a transfer function which can be written as

$$G(s) = \frac{1}{1 + \tau s}$$

where τ is the time constant of the system (see Section 20.2). The frequency-response function $G(j\omega)$ can be obtained by replacing s by $j\omega$. Hence

$$G(j\omega) = \frac{1}{1 + j\omega\tau}$$

We can put this into a more convenient form by multiplying the top and bottom of the expression by $(1 - j\omega\tau)$ to give

$$G(j\omega) = \frac{1}{1 + j\omega\tau} \times \frac{1 - j\omega\tau}{1 - j\omega\tau} = \frac{1 - j\omega\tau}{1 + j^2\omega^2\tau^2}$$

But $j^2 = -1$, thus

$$G(j\omega) = \frac{1}{1 + \omega^2\tau^2} - j\frac{\omega\tau}{1 + \omega^2\tau^2}$$

This is of the form $x + jy$ and so, since $G(j\omega)$ is the output phasor divided by the input phasor, we have the size of the output phasor bigger than that of the input phasor by a factor which can be written as $|G(j\omega)|$, with

$$|G(j\omega)| = \sqrt{x^2 + y^2} = \sqrt{\left(\frac{1}{1 + \omega^2\tau^2}\right)^2 + \left(\frac{\omega\tau}{1 + \omega^2\tau^2}\right)^2} = \frac{1}{\sqrt{1 + \omega^2\tau^2}}$$

$|G(j\omega)|$ tells us how much bigger the amplitude of the output is than the amplitude of the input. It is generally referred to as the **magnitude** or **gain**. The phase difference ϕ between the output phasor and the input phasor is given by

$$\tan\phi = \frac{y}{x} = -\omega\tau$$

The negative sign indicates that the output phasor lags behind the input phasor by this angle.

The following examples illustrate the above.

1 Determine the frequency-response function, the magnitude and phase of a system (an electric circuit with a resistor in series with a capacitor across which the output is taken) that has a transfer function of

$$G(s) = \frac{1}{RCs + 1}$$

The frequency-response function can be obtained by substituting $j\omega$ for s and so

$$G(j\omega) = \frac{1}{j\omega RC + 1}$$

We can multiply the top and bottom of the above equation by $1 - j\omega RC$ and then rearrange the result to give

$$G(j\omega) = \frac{1}{1 + \omega^2(RC)^2} - j\frac{\omega(RC)}{1 + \omega^2(RC)^2}$$

Hence

$$|G(j\omega)| = \frac{1}{\sqrt{1 + \omega^2(RC)^2}}$$

and $\tan\phi = -\omega RC$.

2 Determine the magnitude and phase of the output from a system when subject to a sinusoidal input of $2\sin(3t + 60°)$ if it has a transfer function of

$$G(s) = \frac{4}{s + 1}$$

The frequency-response function is obtained by replacing s by $j\omega$. Thus

$$G(j\omega) = \frac{4}{j\omega + 1}$$

Multiplying the top and bottom of the equation by $(-j\omega + 1)$,

$$G(j\omega) = \frac{-j4\omega + 4}{\omega^2 + 1} = \frac{4}{\omega^2 + 1} - j\frac{4\omega}{\omega^2 + 1}$$

The magnitude is thus

$$|G(j\omega)| = \sqrt{x^2 + y^2} = \sqrt{\frac{4^2}{(\omega^2 + 1)^2} + \frac{4^2\omega^2}{(\omega^2 + 1)^2}} = \frac{4}{\sqrt{\omega^2 + 1}}$$

and the phase angle is given by $\tan\phi = y/x$ and so

$$\tan\phi = -\omega$$

For the specified input we have $\omega = 3$ rad/s. The magnitude is thus

$$|G(j\omega)| = \frac{4}{\sqrt{3^2 + 1}} = 1.3$$

and the phase is given by $\tan\phi = -3$. Thus $\phi = -72°$. This is the phase angle between the input and the output. Thus the output is $2.6\sin(3t - 12°)$.

21.3.2 Frequency response for a second-order system

Consider a second-order system with the transfer function (see Section 20.3)

$$G(s) = \frac{\omega_n^2}{s^2 + 2\zeta\omega_n s + \omega_n^2}$$

where ω_n is the natural angular frequency and ζ the damping ratio. The frequency-response function is obtained by replacing s by $j\omega$. Thus

$$G(j\omega) = \frac{\omega_n^2}{-\omega^2 + j2\zeta\omega\omega_n + \omega_n^2} = \frac{\omega_n^2}{(\omega_n^2 - \omega^2) + j2\zeta\omega_n}$$

$$= \frac{1}{\left[1 - \left(\dfrac{\omega}{\omega_n}\right)^2\right] + j2\zeta\left(\dfrac{\omega}{\omega_n}\right)}$$

Multiplying the top and bottom of the expression by

$$\left[1 - \left(\frac{\omega}{\omega_n}\right)^2\right] - j2\zeta\left(\frac{\omega}{\omega_n}\right)$$

gives

$$G(j\omega) = \frac{\left[1 - \left(\dfrac{\omega}{\omega_n}\right)^2\right] - j2\zeta\left(\dfrac{\omega}{\omega_n}\right)}{\left[1 - \left(\dfrac{\omega}{\omega_n}\right)^2\right]^2 + \left[2\zeta\left(\dfrac{\omega}{\omega_n}\right)\right]^2}$$

This is of the form $x + jy$ and so, since $G(j\omega)$ is the output phasor divided by the input phasor, we have the size or magnitude of the output phasor bigger than that of the input phasor by a factor which is given by $\sqrt{(x^2 + y^2)}$ as

$$|G(j\omega)| = \frac{1}{\sqrt{\left[1 - \left(\dfrac{\omega}{\omega_n}\right)^2\right]^2 + \left[2\zeta\left(\dfrac{\omega}{\omega_n}\right)\right]^2}}$$

The phase ϕ difference between the input and output is given by $\tan\phi = x/y$ and so

$$\tan\phi = -\frac{2\zeta\left(\dfrac{\omega}{\omega_n}\right)}{1 - \left(\dfrac{\omega}{\omega_n}\right)^2}$$

The minus sign is because the output phase lags behind the input.

21.4   Bode plots

The frequency response of a system is the set of values of the magnitude $|G(j\omega)|$ and phase angle ϕ that occur when a sinusoidal input signal is varied over a range of frequencies. This can be expressed as two graphs, one of the magnitude $|G(j\omega)|$ plotted against the angular frequency ω and the other of the phase ϕ plotted against ω. The magnitude and angular frequency are plotted using logarithmic scales. Such a pair of graphs is referred to as a **Bode plot**.

The magnitude is expressed in decibel units (dB):

$$|G(j\omega)| \text{ in dB} = 20\,\lg_{10}|G(j\omega)|$$

Thus, for example, a magnitude of 20 dB means that

$$20 = 20\,\lg_{10}|G(j\omega)|$$

so $1 = \lg_{10}|G(j\omega)|$ and $10^1 = |G(j\omega)|$. Thus a magnitude of 20 dB means that the magnitude is 10, and therefore the amplitude of the output is 10 times that of the input. A magnitude of 40 dB would mean a magnitude of 100 and so the amplitude of the output is 100 times that of the input.

21.4.1   Bode plot for $G(s) = K$

Consider the Bode plot for a system having the transfer function $G(s) = K$, where K is a constant. The frequency-response function is thus $G(j\omega) = K$. The magnitude $|G(j\omega)| = K$ and so, in decibels, $|G(j\omega)| = 20\,\lg K$. The

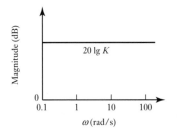

magnitude plot is thus a line of constant magnitude; changing K merely shifts the magnitude line up or down by a certain number of decibels. The phase is zero. Figure 21.3 shows the Bode plot.

21.4.2 Bode plot for $G(s) = 1/s$

Consider the Bode plot for a system having a transfer function $G(s) = 1/s$. The frequency-response function $G(j\omega)$ is thus $1/j\omega$. Multiplying this by j/j gives $G(j\omega) = -j/\omega$. The magnitude $|G(j\omega)|$ is thus $1/\omega$. In decibels this is $20 \lg(1/\omega) = -20 \lg \omega$. When $\omega = 1$ rad/s the magnitude is 0. When $\omega = 10$ rad/s it is -20 dB. When $\omega = 100$ rad/s it is -40 dB. For each 10-fold increase in angular frequency the magnitude drops by -20 dB. The magnitude plot is thus a straight line of slope -20 dB per decade of frequency which passes through 0 dB at $\omega = 1$ rad/s. The phase of such a system is given by

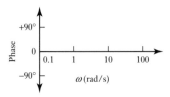

$$\tan \phi = \frac{-\dfrac{1}{\omega}}{0} = -\infty$$

Hence $\phi = -90°$ for all frequencies. Figure 21.4 shows the Bode plot.

Figure 21.3 Bode plot for $G(s) = K$.

21.4.3 Bode plot for a first-order system

Consider the Bode plot for a first-order system for which the transfer function is given by

$$G(s) = \frac{1}{\tau s + 1}$$

The frequency-response function is then

$$G(j\omega) = \frac{1}{j\omega t + 1}$$

The magnitude (see Section 21.2.1) is then

$$|G(j\omega)| = \frac{1}{\sqrt{1 + \omega^2 \tau^2}}$$

In decibels this is

$$20 \lg\left(\frac{1}{\sqrt{1 + \omega^2 \tau^2}}\right)$$

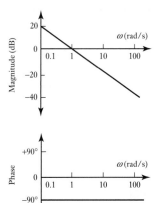

Figure 21.4 Bode plot for $G(s) = 1/s$.

When $\omega \ll 1/\tau$, then $\omega^2 \tau^2$ is negligible compared with 1 and so the magnitude is $20 \lg 1 = 0$ dB. Hence at low frequencies there is a straight line magnitude plot at a constant value of 0 dB. For higher frequencies, when $\omega \gg 1/\tau$, $\omega^2 \tau^2$ is much greater than 1 and so the 1 can be neglected. The magnitude is then $20 \lg(1/\omega\tau)$, i.e. $-20 \lg \omega\tau$. This is a straight line of slope -20 dB per decade of frequency which intersects the 0 dB line when $\omega\tau = 1$, i.e. when $\omega = 1/\tau$. Figure 21.5 shows these lines for low and high frequencies

Figure 21.5 Bode plot for first-order system.

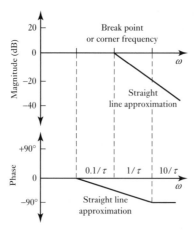

with their intersection, or so-called **break point** or **corner frequency**, at $\omega = 1/\tau$. The two straight lines are called the asymptotic approximation to the true plot. The true plot rounds off the intersection of the two lines. The difference between the true plot and the approximation is a maximum of 3 dB at the break point.

The phase for the first-order system (see Section 21.2.1) is given by $\tan \phi = -\omega\tau$. At low frequencies, when ω is less than about $0.1/\tau$, the phase is virtually $0°$. At high frequencies, when ω is more than about $10/\tau$, the phase is virtually $-90°$. Between these two extremes the phase angle can be considered to give a reasonable straight line on the Bode plot (Figure 21.5). The maximum error in assuming a straight line is $5.5°$.

An example of such a system is an RC filter (see Section 20.2.2), i.e. a resistance R in series with a capacitance C with the output being the voltage across the capacitor. It has a transfer function of $1/(RCs + 1)$ and so a frequency-response function of $1/(j\omega\tau + 1)$ where $\tau = RC$. The Bode plot is thus as shown in Figure 21.5.

21.4.4   Bode plot for a second-order system

Consider a second-order system with a transfer function of

$$G(s) = \frac{\omega_n^2}{s^2 + 2\zeta\omega_n s + \omega_n^2}$$

The frequency-response function is obtained by replacing s by $j\omega$:

$$G(j\omega) = \frac{\omega_n^2}{-\omega^2 + j2\zeta\omega_n\omega + \omega_n^2}$$

The magnitude is then (see Section 21.3.2)

$$|G(j\omega)| = \frac{1}{\sqrt{\left[1 - \left(\dfrac{\omega}{\omega_n}\right)^2\right]^2 + \left[2\zeta\left(\dfrac{\omega}{\omega_n}\right)\right]^2}}$$

Thus, in decibels, the magnitude is

$$20 \lg \frac{1}{\sqrt{\left[1 - \left(\frac{\omega}{\omega_n}\right)^2\right]^2 + \left[2\zeta\left(\frac{\omega}{\omega_n}\right)\right]^2}}$$

$$= -20 \lg \sqrt{\left[1 - \left(\frac{\omega}{\omega_n}\right)^2\right]^2 + \left[2\zeta\left(\frac{\omega}{\omega_n}\right)\right]^2}$$

For $(\omega/\omega_n) \ll 1$ the magnitude approximates to $-20 \lg 1$ or 0 dB and for $(\omega/\omega_n) \gg 1$ the magnitude approximates to $-20 \lg(\omega/\omega_n)^2$. Thus when ω increases by a factor of 10 the magnitude increases by a factor of $-20 \lg 100$ or -40 dB. Thus at low frequencies the magnitude plot is a straight line at 0 dB, while at high frequencies it is a straight line of -40 dB per decade of frequency. The intersection of these two lines, i.e. the break point, is at $\omega = \omega_n$. The magnitude plot is thus approximately given by these two asymptotic lines. The true value, however, depends on the damping ratio ζ. Figure 21.6 shows the two asymptotic lines and the true plots for a number of damping ratios.

Figure 21.6 Bode plot for a second-order system.

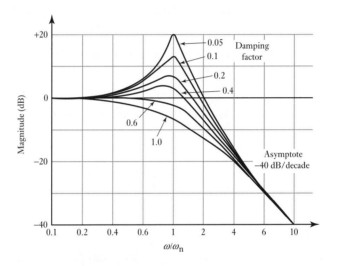

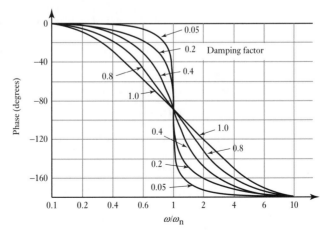

The phase is given by (see Section 21.3.2)

$$\tan \phi = -\frac{2\zeta\left(\dfrac{\omega}{\omega_n}\right)}{1 - \left(\dfrac{\omega}{\omega_n}\right)^2}$$

For $(\omega/\omega_n) \ll 1$, e.g. $(\omega/\omega_n) = 0.2$, then $\tan \phi$ is approximately 0 and so $\phi = 0°$. For $(\omega/\omega_n) \gg 1$, e.g. $(\omega/\omega_n) = 5$, $\tan \phi$ is approximately $-(-\infty)$ and so $\phi = -180°$. When $\omega = \omega_n$ then we have $\tan \phi = -\infty$ and so $\phi = -90°$. A reasonable approximation is given by a straight line drawn through $-90°$ at $\omega = \omega_n$ and the points $0°$ at $(\omega/\omega_n) = 0.2$ and $-180°$ at $(\omega/\omega_n) = 5$. Figure 21.6 shows the graph.

21.4.5 Building up Bode plots

Consider a system involving a number of elements in series. The transfer function of the system as a whole is given by (see Section 20.4)

$$G(s) = G_1(s)G_2(s)G_3(s) \ldots$$

Hence the frequency-response function for a two–element system, when s is replaced by $j\omega$, is

$$G(j\omega) = G_1(j\omega)G_2(j\omega)$$

We can write the transfer function $G_1(j\omega)$ as a complex number (see Section 21.2)

$$x + jy = |G_1(j\omega)| \ (\cos \phi_1 + j \sin \phi_1)$$

where $|G(j\omega)|$ is the magnitude and ϕ_1 the phase of the frequency-response function. Similarly we can write $G_2(j\omega)$ as

$$|G_2(j\omega)| \ (\cos \phi_2 + j \sin \phi_2)$$

Thus

$$G(j\omega) = |G_1(j\omega)| \ (\cos \phi_1 + j \sin \phi_1) \times |G_2(j\omega)| \ (\cos \phi_2 + j \sin \phi_2)$$

$$= |G_1(j\omega)| \, |G_2(j\omega)| \, [\cos \phi_1 \cos \phi_2$$

$$+ \, j(\sin \phi_1 \cos \phi_2 + \cos \phi_1 \sin \phi_2) + j^2 \sin \phi_1 \sin \phi_2]$$

But $j^2 = -1$ and, since $\cos \phi_1 \cos \phi_2 - \sin \phi_1 \sin \phi_2 = \cos(\phi_1 + \phi_2)$ and $\sin \phi_1 \cos \phi_2 + \cos \phi_1 \sin \phi_2 = \sin(\phi_1 + \phi_2)$, then

$$G(j\omega) = |G_1(j\omega)| \, |G_2(j\omega)| [\cos(\phi_1 + \phi_2) + j \sin(\phi_1 + \phi_2)]$$

The frequency-response function of the system has a magnitude, which is the product of the magnitudes of the separate elements, and a phase, which is the sum of the phases of the separate elements, i.e.

$$|G(j\omega)| = |G_1(j\omega)| \, |G_2(j\omega)| \, |G_3(j\omega)| \ \ldots$$

$$\phi = \phi_1 + \phi_2 + \phi_3 + \ \ldots$$

Now, considering the Bode plot where the logarithms of the magnitudes are plotted,

$$\lg |G(j\omega)| = \lg |G_1(j\omega)| + \lg |G_2(j\omega)| + \lg |G_3(j\omega)| + \ldots$$

Thus we can obtain the Bode plot of a system by adding together the Bode plots of the magnitudes of the constituent elements. Likewise the phase plot is obtained by adding together the phases of the constituent elements.

By using a number of basic elements, the Bode plots for a wide range of systems can be readily obtained. The following basic elements are used.

1 $G(s) = K$ gives the Bode plot shown in Figure 21.3.
2 $G(s) = 1/s$ gives the Bode plot shown in Figure 21.4.
3 $G(s) = s$ gives a Bode plot which is a mirror image of that in Figure 21.4. $|G(j\omega)| = 20$ dB per decade of frequency, passing through 0 dB at $\omega = 1$ rad/s. ϕ is constant at $90°$.
4 $G(s) = 1/(\tau s + 1)$ gives the Bode plot shown in Figure 21.5.
5 $G(s) = \tau s + 1$ gives a Bode plot which is a mirror image of that in Figure 21.5. For the magnitude plot, the break point is at $1/\tau$ with the line prior to it being at 0 dB and after it at a slope of 20 dB per decade of frequency. The phase is zero at $0.1/\tau$ and rises to $+90°$ at $10/\tau$.
6 $G(s) = \omega_n^2/(s^2 + 2\zeta\omega_n s + \omega_n^2)$ gives the Bode plot shown in Figure 21.6.
7 $G(s) = (s^2 + 2\zeta\omega_n s + \omega_n^2)/\omega_n^2$ gives a Bode plot which is a mirror image of that in Figure 21.6.

To illustrate the above, consider the drawing of the asymptotes of the Bode plot for a system having a transfer function of

$$G(s) = \frac{10}{2s + 1}$$

The transfer function is made up of two elements, one with a transfer function of 10 and one with transfer function $1/(2s + 1)$. The Bode plots can be drawn for each of these and then added together to give the required plot. The Bode plot for transfer function 10 will be of the form given in Figure 21.3 with $K = 10$ and that for $1/(2s + 1)$ like that given in Figure 21.5 with $\tau = 2$. The result is shown in Figure 21.7.

As another example, consider the drawing of the asymptotes of the Bode plot for a system having a transfer function of

$$G(s) = \frac{2.5}{s(s^2 + 3s + 25)}$$

The transfer function is made up of three components: one with a transfer function of 0.1, one with transfer function $1/s$ and one with transfer function $25/(s^2 + 3s + 25)$. The transfer function of 0.1 will give a Bode plot like that of Figure 21.3 with $K = 0.1$. The transfer function of $1/s$ will give a Bode plot like that of Figure 21.4. The transfer function of $25/(s^2 + 3s + 25)$ can be represented as $\omega_n^2/(s^2 + 2\zeta\omega_n s + \omega_n^2)$ with $\omega_n = 5$ rad/s and $\zeta = 0.3$. The break point will be when $\omega = \omega_n = 5$ rad/s. The asymptote for the phase passes through $-90°$ at the break point, and is $0°$ when we have $(\omega/\omega_n) = 0.2$ and $-180°$ when $(\omega/\omega_n) = 5$. Figure 21.8 shows the resulting plot.

Figure 21.7 Building up a
Bode diagram.

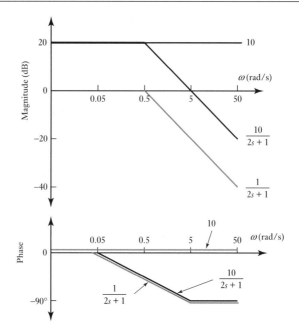

The above method of obtaining a Bode plot by building it up from its constituent elements, using the straight line approximations, was widely used but is now less necessary in the age of computers.

21.4.6 System identification

If we experimentally determine the Bode diagram for a system by considering its response to a sinusoidal input, then we can obtain the transfer function for the system. Basically we draw the asymptotes on the magnitude Bode plot and consider their gradients. The phase angle curve is used to check the results obtained from the magnitude analysis.

1   If the gradient at low frequencies prior to the first corner frequency is zero then there is no s or $1/s$ element in the transfer function. The K element in the numerator of the transfer function can be obtained from the value of the low-frequency magnitude; the magnitude in dB $= 20 \lg K$.

2   If the initial gradient at low frequencies is -20 dB/decade then the transfer function has a $1/s$ element.

3   If the gradient becomes more negative at a corner frequency by 20 dB/decade, there is a $(1 + s/\omega_c)$ term in the denominator of the transfer function, with ω_c being the corner frequency at which the change occurs. Such terms can occur for more than one corner frequency.

4   If the gradient becomes more positive at a corner frequency by 20 dB/decade, there is a $(1 + s/\omega_c)$ term in the numerator of the transfer function, with ω_c being the frequency at which the change occurs. Such terms can occur for more than one corner frequency.

5   If the gradient at a corner frequency becomes more negative by 40 dB/decade, there is a $(s^2/\omega_c^2 + 2\zeta s/\omega_c + 1)$ term in the denominator of the

Figure 21.8 Building up a Bode plot.

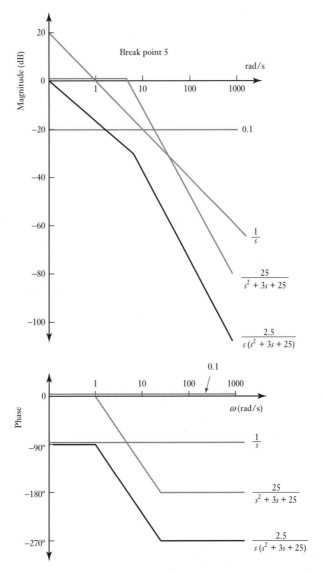

transfer function. The damping ratio ζ can be found by considering the detail of the Bode plot at a corner frequency, as in Figure 21.6.

6 If the gradient at a corner frequency becomes more positive by 40 dB/decade, there is a $(s^2/\omega_c^2 + 2\zeta s/\omega_c + 1)$ term in the numerator of the transfer function. The damping ratio ζ can be found by considering the detail of the Bode plot at a corner frequency, as in Figure 21.6.

7 If the low-frequency gradient is not zero, the K term in the numerator of the transfer function can be determined by considering the value of the low-frequency asymptote. At low frequencies, many terms in transfer functions can be neglected and the gain in dB approximates to $20 \lg(K/\omega^2)$. Thus, at $\omega = 1$ the gain in dB approximates to $20 \lg K$.

As an illustration of the above, consider the Bode magnitude plot shown in Figure 21.9. The initial gradient is 0 and so there is no $1/s$ or s term in the

Figure 21.9 Bode plot.

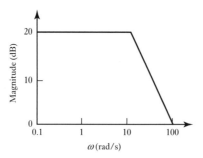

transfer function. The initial gain is 20 and so $20 = 20 \lg K$ and $K = 10$. The gradient changes by -20 dB/decade at a frequency of 10 rad/s. Hence there is a $(1 + s/10)$ term in the denominator. The transfer function is thus $10/(1 + 0.1s)$.

As a further illustration, consider Figure 21.10. There is an initial slope of -20 dB/decade and so a $1/s$ term. At the corner frequency 1.0 rad/s there is a -20 dB/decade change in gradient and so a $1/(1 + s/1)$ term. At the corner frequency 10 rad/s there is a further -20 dB/decade change in gradient and so a $1/(1 + s/10)$ term. At $\omega = 1$ the magnitude is 6 dB and so $6 = 20 \lg K$ and $K = 10^{6/20} = 2.0$. The transfer function is thus $2.0/s(1 + s)(1 + 0.1s)$.

Figure 21.10 Bode plot.

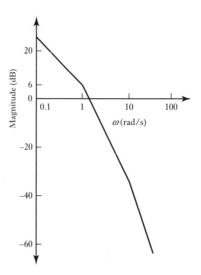

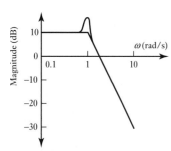

As a further illustration, Figure 21.11 shows a Bode plot which has an initial zero gradient which changes by -40 dB/decade at 10 rad/s. The initial magnitude is 10 dB and so $10 = 20 \lg K$ and $K = 10^{0.5} = 3.2$. The change of -40 dB/decade at 10 rad/s means that there is a $(s^2/10^2 + 2\zeta s/10 + 1)$ term in the denominator. The transfer function is thus $3.2/(0.01s^2 + 0.2\zeta s + 1)$. The damping factor can be obtained by comparison of the Bode plot at the corner frequencies with Figure 21.6. It rises by about 6 dB above the corner and this corresponds to a damping factor of about 0.2. The transfer function is thus $3.2/(0.01s^2 + 0.04s + 1)$.

Figure 21.11 Bode plot.

21.5 Performance specifications

The terms used to describe the performance of a system when subject to a sinusoidal input are peak resonance and bandwidth. The **peak resonance** M_p is defined as being the maximum value of the magnitude (Figure 21.12). A large value of the peak resonance corresponds to a large value of the maximum overshoot of a system. For a second-order system it can be directly related to the damping ratio by comparison of the response with the Bode plot of Figure 21.6, a low damping ratio corresponding to a high peak resonance.

Figure 21.12 Performance specifications.

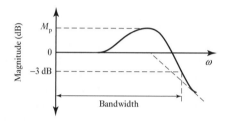

The **bandwidth** is defined as the frequency band between which the magnitude does not fall below -3 dB, the frequencies at which this occurs being termed the cut-off frequencies. With the magnitude expressed in decibel units (dB),

$$|G(j\omega)| \text{ in dB} = 20 \lg_{10} |G(j\omega)|$$

and so

$$-3 = 20 \log_{10} |G(j\omega)|$$

and $|G(j\omega)| = 0.707$ so the amplitude has dropped to 0.707 of its initial value. Since the power of a sinusoidal waveform is the square of its amplitude, then the power has dropped to $0.707^2 = 0.5$ of its initial value. Thus, the -3 dB cut-off is the decibel value at which the power of the input signal is attenuated to half the input value. For the system giving the Bode plot in Figure 21.12, the bandwidth is the spread between zero frequency and the frequency at which the magnitude drops below -3 dB. This is typical of measurement systems; they often exhibit no attenuation at low frequencies and the magnitude only degrades at high frequencies.

As an illustration, for the example described in Section 20.2.2, item 1, the magnitude of a system (an electric circuit with a resistor in series with a capacitor across which the output is taken) with a transfer function of

$$G(s) = \frac{1}{RCs + 1}$$

was determined as

$$|G(j\omega)| = \frac{1}{\sqrt{1 + \omega^2 (RC)^2}}$$

For this magnitude ratio to be 0.707, the cut-off frequency ω_c must be given by

$$0.707 = \frac{1}{\sqrt{1 + \omega_c^2 (RC)^2}}$$

$$1 + \omega_c^2 (RC)^2 = (1/0.707)^2 = 2$$

Hence $\omega_c = 1/RC$. Such a circuit is called a low-pass filter since lower frequencies are passed to the output with little attenuation and higher frequencies are attenuated.

21.6 Stability

When there is a sinusoidal input to a system, the output from that system is sinusoidal with the same angular frequency but can have an output with an amplitude and phase which differ from that of the input. Consider a closed-loop system with negative feedback (Figure 21.13) and no input to the system. Suppose, somehow, we have a half-rectified sinusoidal pulse as the error signal in the system and that it passes through to the output and is fed back to arrive at the comparator element with amplitude unchanged but delayed by just half a cycle, i.e. a phase change of 180° as shown in the figure. When this signal is subtracted from the input signal, we have a resulting error signal which just continues the initial half-rectified pulse. This pulse then goes back round the feedback loop and once again arrives just in time to continue the signal. Thus we have a self-sustaining oscillation.

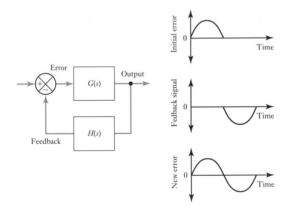

Figure 21.13 Self-sustaining oscillations.

For self-sustained oscillations to occur we must have a system which has a frequency-response function with a magnitude of 1 and a phase of −180°. The system through which the signal passes is $G(s)$ in series with $H(s)$. If the magnitude is less than 1 then each succeeding half-wave pulse is smaller in size and so the oscillation dies away. If the magnitude is greater than 1 then each succeeding pulse is larger than the previous one and so the wave builds up and the system is unstable.

1 A control system will oscillate with a constant amplitude if the magnitude resulting from the system $G(s)$ in series with $H(s)$ is 1 and the phase is $-180°$.

2 A control system will oscillate with a diminishing amplitude if the magnitude resulting from the system $G(s)$ in series with $H(s)$ is less than 1 and the phase is $-180°$.

3 A control system will oscillate with an increasing amplitude, and so is unstable, if the magnitude resulting from the system $G(s)$ in series with $H(s)$ is greater than 1 and the phase is $-180°$.

A good, stable control system usually requires that the magnitude of $G(s)H(s)$ should be significantly less than 1. Typically a value between 0.4 and 0.5 is used. In addition, the phase angle should be between about $-115°$ and $-125°$. Such values produce a slightly under-damped control system which gives, with a step input, about a 20 to 30% overshoot with a subsidence ratio of about 3 to 1 (see Section 19.5 for an explanation of these terms).

A concern with a control system is how stable it is and thus not likely to oscillate as a result of some small disturbance. The term **gain margin** is used for the factor by which the magnitude ratio must be multiplied when the phase is $-180°$ to make it have the value 1 and so be on the verge of instability. The term **phase margin** is used for the number of degrees by which the phase angle is numerically smaller than $-180°$ when the magnitude is 1. These rules mean a gain margin of between 2 and 2.5 and a phase margin between 45° and 65° for a good, stable control system.

Summary

We can convert from the s-domain to the **frequency domain** by replacing s by $j\omega$. The **frequency-response function** is the transfer function when transformed into the frequency domain.

The frequency response of a system is the set of values of the magnitude $|G(j\omega)|$ and phase angle ϕ that occur when a sinusoidal input signal is varied over a range of frequencies. This can be expressed as two graphs, one of the magnitude $|G(j\omega)|$ plotted against the angular frequency ω and the other of the phase ϕ plotted against ω. The magnitude and angular frequency are plotted using logarithmic scales. Such a pair of graphs is referred to as a **Bode plot**.

We can obtain the Bode plot of a system by adding together the Bode plots of the magnitudes of the constituent elements. Likewise the phase plot is obtained by adding together the phases of the constituent elements.

The **peak resonance** M_p is the maximum value of the magnitude. The **bandwidth** is the frequency band between which the magnitude does not fall below -3 dB, the frequencies at which this occurs being termed the cut-off frequencies.

For self-sustained oscillations to occur with a feedback system, i.e. it is on the verge of **instability**, we must have a system which has a frequency-response function with a magnitude of 1 and a phase of $-180°$ The **gain margin** is the factor by which the magnitude ratio must be multiplied when the phase is $-180°$ to make it have the value 1 and so be on the verge of instability. The **phase margin** is the number of degrees by which the phase angle is numerically smaller than $-180°$ when the magnitude is 1.

Problems

21.1 What are the magnitudes and phases of the systems having the following transfer functions?

(a) $\dfrac{5}{s + 2}$, (b) $\dfrac{2}{s(s + 1)}$, (c) $\dfrac{1}{(2s + 1)(s^2 + s + 1)}$

21.2 What will be the steady-state response of a system with a transfer function $1/(s + 2)$ when subject to the sinusoidal input $3 \sin(5t + 30°)$?

21.3 What will be the steady-state response of a system with a transfer function $5/(s^2 + 3s + 10)$ when subject to the input $2 \sin(2t + 70°)$?

21.4 Determine the values of the magnitudes and phase at angular frequencies of (i) 0 rad/s, (ii) 1 rad/s, (iii) 2 rad/s, (iv) ∞ rad/s for systems with the transfer functions (a) $1/[s(2s + 1)]$, (b) $1/(3s + 1)$.

21.5 Draw Bode plot asymptotes for systems having the transfer functions (a) $10/[s(0.1s + 1)]$, (b) $1/[(2s + 1)(0.5s + 1)]$.

21.6 Obtain the transfer functions of the systems giving the Bode plots in Figure 21.14.

Figure 21.14   Problem 21.6.

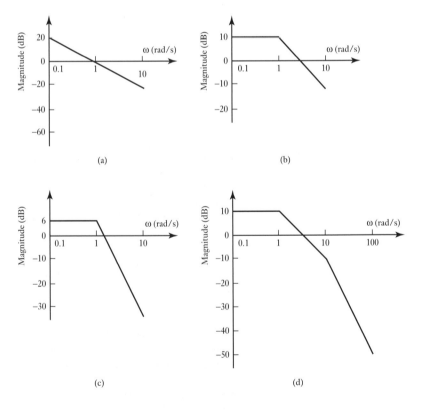

Chapter twenty-two

Closed-loop controllers

Objectives

The objectives of this chapter are that, after studying it, the reader should be able to:
- Explain the term steady-state error.
- Explain the operation of the two-step mode of control.
- Predict the behaviour of systems with proportional, integral, derivative, proportional plus integral, proportional plus derivative and PID control.
- Describe how digital controllers operate.
- Explain how controllers can be tuned.

22.1 Continuous and discrete control processes

Open-loop control is often just a switch on–switch off form of control, e.g. an electric fire is either switched on or off in order to heat a room. With **closed-loop control systems**, a controller is used to compare continuously the output of a system with the required condition and convert the error into a control action designed to reduce the error. The error might arise as a result of some change in the conditions being controlled or because the set value is changed, e.g. there is a step input to the system to change the set value to a new value. In this chapter we are concerned with the ways in which controllers can react to error signals, i.e. the **control modes** as they are termed, which occur with continuous processes. Such controllers might, for example, be pneumatic systems or operational amplifier systems. However, computer systems are rapidly replacing many of these. The term **direct digital control** is used when the computer is in the feedback loop and exercising control in this way. This chapter is about such closed-loop control.

Many processes not only involve controlling some variable, e.g. temperature, to a required value, but also involve the sequencing of operations. A domestic washing machine (see Section 1.5.5) where a number of actions have to be carried out in a predetermined sequence is an example. Another example is the manufacture of a product which involves the assembly of a number of discrete parts in a specific sequence by some controlled system. The sequence of operations might be **clock-based** or **event-based** or a combination of the two. With a clock-based system the actions are carried out at specific times; with an event-based system the actions are carried out when there is feedback to indicate that a particular event has occurred.

In many processes there can be a mixture of continuous and discrete control. For example, in the domestic washing machine there will be sequence control for the various parts of the washing cycle with feedback loop control of the temperature of the hot water and the level of the water.

22.1.1   Open- and closed-loop systems

Closed-loop systems differ from open-loop systems in having feedback. An open-loop system is one where the input signal does not automatically depend on the actual process output. With a closed-loop system, there is a feedback from the output to modify the input so that the system maintains the required output.

One consequence of having feedback is that there is a reduction of the effects of disturbance signals on the system. A disturbance signal is an unwanted signal which affects the output signal of a system. All physical systems are subject to some forms of extraneous signals during their operation. With an electric motor, this might be brush or commutator noise.

Consider the effect of external disturbances on the overall gain of an open-loop system. Figure 22.1 shows a two-element open-loop system with a disturbance which gives an input between the two elements. For a reference input $R(s)$ to the system, the first element gives an output of $G_1(s)R(s)$. To this is added the disturbance $D(s)$ to give an input of $G_1(s)R(s) + D(s)$. The overall system output $X(s)$ will then be

$$X(s) = G_2(s)[G_1(s)R(s) + D(s)] = G_1(s)G_2(s)R(s) + G_2(s)D(s)$$

Figure 22.1 Disturbance with an open-loop system.

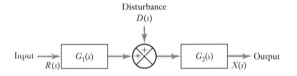

For the comparable system with negative feedback (Figure 22.2), the input to the first forward element $G_1(s)$ is $R(s) - H(s)X(s)$ and so its output is $G_1(s)[R(s) - H(s)X(s)]$. The input to $G_2(s)$ is $G_1(s)[R(s) - H(s)X(s)] + D(s)$ and so its output is

$$X(s) = G_2(s)\{G_1(s)[R(s) - H(s)X(s)] + D(s)\}$$

Thus

$$X(s) = \frac{G_1(s)G_2(s)}{1 + G_1(s)G_2(s)H(s)}R(s) + \frac{G_2(s)}{1 + G_1(s)G_2(s)H(s)}D(s)$$

Figure 22.2 Disturbance with closed-loop system.

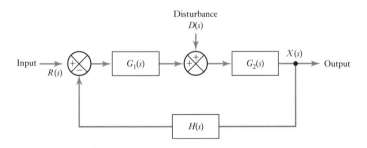

Comparing this with the equation for the open-loop system indicates that with the closed-loop system the effect of the disturbance on the output of the system has been reduced by a factor of $[1 + G_1(s)G_2(s)H(s)]$. The effect of a disturbance is reduced when there is feedback.

22.2 Terminology

The following are terms commonly used in discussing closed-loop controllers.

22.2.1 Lag

In any control system there are **lags**. Thus, for example, a change in the condition being controlled does not immediately produce a correcting response from the control system. This is because time is required for the system to make the necessary responses. For example, in the control of the temperature in a room by means of a central heating system, a lag will occur between the room temperature falling below the required temperature and the control system responding and switching on the heater. This is not the only lag. Even when the control system has responded there is a lag in the room temperature responding as time is taken for the heat to transfer from the heater to the air in the room.

22.2.2 Steady-state error

A closed-loop control system uses a measurement of the system output and a comparison of its value with the desired output to generate an error signal. We might get an error signal to the controller occurring as a result of the controlled variable changing or a change in the set value input. For example, we might have a ramp input to the system with the aim that the controlled variable increases steadily with time. When a change occurs, there are likely to be some transient effects; these, however, die away with time. The term **steady-state error** is used for the difference between the desired set value input and the output after all transients have died away. It is thus a measure of the accuracy of the control system in tracking the set value input. Whenever there is an error, the output is *not* at the desired output.

Consider a control system which has unity feedback (Figure 22.3). When there is a reference input of $R(s)$, there is an output of $X(s)$. The feedback signal is $X(s)$ and so the error signal is $E(s) = R(s) - X(s)$. If $G(s)$ is the forward-path transfer function, then for the unity feedback system as a whole:

$$\frac{X(s)}{R(s)} = \frac{G(s)}{1 + G(s)H(s)} = \frac{G(s)}{1 + G(s)}$$

Figure 22.3   Unity feedback.

Unity feedback

Hence

$$E(s) = R(s) - X(s) = R(s) - \frac{G(s)R(s)}{1 + G(s)} = \frac{1}{1 + G(s)}R(s)$$

The error thus depends on $G(s)$.

In order to determine the steady-state error we can determine the error e as a function of time and then determine the value of the error when all transients have died down and so the error as the time t tends to an infinite value. While we could determine the inverse of $E(s)$ and then determine its value when $t \to \infty$, there is a simpler method using the **final-value theorem** (see Appendix A); this involves finding the value of $sE(s)$ as s tends to a zero value:

$$e_{SS} = \lim_{t \to \infty} e(t) = \lim_{s \to 0} sE(s)$$

To illustrate the above, consider a unity feedback system with a forward-path transfer function of $k/(\tau s + 1)$ and subject to a unit step input of $1/s$:

$$e_{SS} = \lim_{s \to 0} sE(s) = \lim_{s \to 0} \left[s \frac{1}{1 + k/(\tau s + 1)} \frac{1}{s} \right] = \frac{1}{1 + k}$$

There is thus a steady-state error; the output from the system will never attain the set value. By increasing the gain k of the system then the steady-state error can be reduced.

The forward path might be a controller with a gain of k and a system with a transfer function $1/(\tau s + 1)$. Such a controller gain is termed a proportional controller. The steady-state error in this case is commonly termed **offset**. It can be minimised by increasing the gain.

However, if the unity feedback system had a forward-path transfer function of $k/s(\tau s + 1)$ and was subject to a step input, then the steady-state error would be

$$e_{SS} = \lim_{s \to 0} sE(s) = \lim_{s \to 0} \left[s \frac{1}{1 + k/s(\tau s + 1)} \frac{1}{s} \right] = 0$$

There is no steady-state error with this system. In this case, the forward path might be a controller with a gain of k/s and a system with a transfer function $1/(\tau s + 1)$. Such a controller gain is termed an integral controller and it gives no offset. Thus by combining an integral and a proportional controller it is possible to eliminate offset. Adding a derivative controller enables the controller to respond more rapidly to changes.

22.2.3 Control modes

There are a number of ways by which a control unit can react to an error signal and supply an output for correcting elements.

1 In the *two-step mode* the controller is essentially just a switch which is activated by the error signal and supplies just an on/off correcting signal.

2 The *proportional mode* (P) produces a control action that is proportional to the error. The correcting signal thus becomes bigger, the bigger the error. Thus as the error is reduced the amount of correction is reduced and the correcting process slows down.

3 The *derivative mode* (D) produces a control action that is proportional to the rate at which the error is changing. When there is a sudden change in the error signal the controller gives a large correcting signal; when there is a gradual change only a small correcting signal is produced. Derivative control can be considered to be a form of anticipatory control in that the existing rate of change of error is measured, a coming larger error is anticipated and correction is applied before the larger error has arrived. Derivative control is not used alone but always in conjunction with proportional control and, often, integral control.

4 The *integral mode* (I) which produces a control action that is proportional to the integral of the error with time. Thus a constant error signal will produce an increasing correcting signal. The correction continues to increase as long as the error persists. The integral controller can be considered to be 'looking back', summing all the errors and thus responding to changes that have occurred.

5 *Combinations of modes*: proportional plus derivative (PD) modes, proportional plus integral (PI) modes, proportional plus integral plus derivative (PID) modes. The term **three-term controller** is used for PID control.

These five modes of control are discussed in the following sections of the chapter. A controller can achieve these modes by means of pneumatic circuits, analogue electronic circuits involving operational amplifiers or by the programming of a microprocessor or computer.

22.3 Two-step mode

An example of the **two-step mode** of control is the bimetallic thermostat (see Figure 2.46) that might be used with a simple temperature control system. This is just a switch which is switched on or off according to the temperature. If the room temperature is above the required temperature then the bimetallic strip is in an off position and the heater is off. If the room temperature falls below the required temperature then the bimetallic strip moves into an on position and the heater is switched fully on. The controller in this case can be in only two positions, on or off, as indicated by Figure 22.4(a).

With the two-step mode the control action is discontinuous. A consequence of this is that oscillations of the controlled variable occur about the required condition. This is because of lags in the time that the control system and the process take to respond. For example, in the case of the temperature control for a domestic central heating system, when the room temperature drops below the required level, the time that elapses before the control system responds and switches the heater on might be very small in comparison with the time that

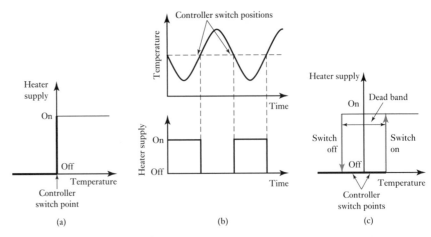

Figure 22.4 Two-step control.

elapses before the heater begins to have an effect on the room temperature. In the meantime the temperature has fallen even more. The reverse situation occurs when the temperature has risen to the required temperature. Since time elapses before the control system reacts and switches the heater off, and yet more time while the heater cools and stops heating the room, the room temperature goes beyond the required value. The result is that the room temperature oscillates above and below the required temperature (Figure 22.4(b)).

With the simple two-step system described above, there is the problem that when the room temperature is hovering about the set value the thermostat might be almost continually switching on or off, reacting to very slight changes in temperature. This can be avoided if, instead of just a single temperature value at which the controller switches the heater on or off, two values are used and the heater is switched on at a lower temperature than the one at which it is switched off (Figure 22.4(c)). The term **dead band** is used for the values between the on and off values. A large dead band results in large fluctuations of the temperature about the set temperature; a small dead band will result in an increased frequency of switching. The bimetallic element shown in Figure 2.46 has a permanent magnet for a switch contact; this has the effect of producing a dead band.

Two-step control action tends to be used where changes are taking place very slowly, i.e. with a process with a large capacitance. Thus, in the case of heating a room, the effect on the room temperature of switching the heater on or off is only a slow change. The result of this is an oscillation with a long periodic time. Two-step control is thus not very precise, but it does involve simple devices and is thus fairly cheap. On/off control is not restricted to mechanical switches such as bimetallic strips or relays; rapid switching can be achieved with the use of thyristor circuits (see Section 9.3.2); such a circuit might be used for controlling the speed of a motor, and operational amplifiers.

22.4 Proportional mode

With the two-step method of control, the controller output is either an on or an off signal, regardless of the magnitude of the error. With the **proportional mode,** the size of the controller output is proportional to the size of the error: the bigger the error, the bigger the output from the controller. This means that the correction element of the control system, e.g. a valve, will receive a signal which is proportional to the size of the correction required. Thus

$$\text{controller output} = K_\text{p}e$$

where e is the error and K_p a constant. Thus taking Laplace transforms,

$$\text{controller output} (s) = K_\text{p}E(s)$$

and so K_p is the transfer function of the controller.

22.4.1 Electronic proportional controller

A summing operational amplifier with an inverter can be used as a proportional controller (Figure 22.5). For a summing amplifier we have (see Section 3.2.3)

$$V_\text{out} = -R_\text{f}\left(\frac{V_0}{R_2} + \frac{V_\text{e}}{R_1}\right)$$

Figure 22.5 Proportional controller.

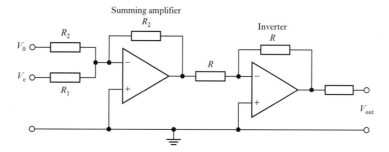

The input to the summing amplifier through R_2 is the zero error voltage value V_0, i.e. the set value, and the input through R_1 is the error signal V_e. But when the feedback resistor $R_\text{f} = R_2$, then the equation becomes

$$V_\text{out} = -\frac{R_2}{R_1}V_\text{e} - V_0$$

If the output from the summing amplifier is then passed through an inverter, i.e. an operational amplifier with a feedback resistance equal to the input resistance, then

$$V_\text{out} = \frac{R_2}{R_1}V_\text{e} + V_0$$

$$V_\text{out} = K_\text{p}V_\text{e} + V_0$$

where K_p is the proportionality constant. The result is a proportional controller.

As an illustration, Figure 22.6 shows an example of a proportional control system for the control of the temperature of a liquid in a container as liquid is pumped through it.

Figure 22.6 Proportional
controller for temperature
control.

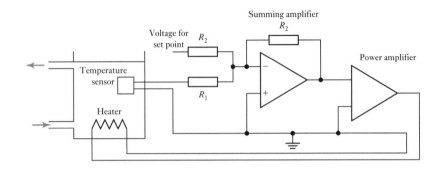

22.4.2 System response

With proportional control we have a gain element with transfer function K_P
in series with the forward-path element $G(s)$ (Figure 22.7). The error is thus

$$E(s) = \frac{K_P G(s)}{1 + K_P G(s)} R(s)$$

Figure 22.7 System with
proportional control.

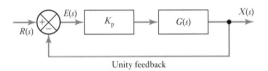

and so, for a step input, the steady-state error is

$$e_{SS} = \lim_{s \to 0} sE(s) = \lim_{s \to 0} \left[s \frac{1}{1 + 1/K_P G(s)} \frac{1}{s} \right]$$

This will have a finite value and so there is always a steady-state error. Low
values of K_P give large steady-state errors but stable responses. High values
of K_P give smaller steady-state errors but a greater tendency to instability.

22.5   Derivative
control

With the **derivative mode** of control the controller output is proportional
to the rate of change with time of the error signal. This can be represented
by the equation

$$\text{controller output} = K_D \frac{de}{dt}$$

K_D is the constant of proportionality. The transfer function is obtained by
taking Laplace transforms, thus

$$\text{controller output } (s) = K_D s E(s)$$

Hence the transfer function is $K_D s$.

With derivative control, as soon as the error signal begins to change,
there can be quite a large controller output since it is proportional to the

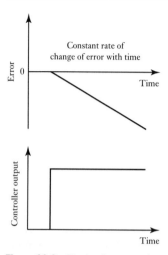

Figure 22.8 Derivative control.

Figure 22.9 Derivative controller.

rate of change of the error signal and not its value. Rapid initial responses to error signals thus occur. Figure 22.8 shows the controller output that results when there is a constant rate of change of error signal with time. The controller output is constant because the rate of change is constant and occurs immediately the deviation occurs. Derivative controllers do not, however, respond to steady-state error signals, since with a steady error the rate of change of error with time is zero. Because of this, derivative control is always combined with proportional control; the proportional part gives a response to all error signals, including steady signals, while the derivative part responds to the rate of change. Derivative action can also be a problem if the measurement of the process variable gives a noisy signal, the rapid fluctuations of the noise resulting in outputs which will be seen by the controller as rapid changes in error and so give rise to significant outputs from the controller.

Figure 22.9 shows the form of an electronic derivative controller circuit, the circuit involving an operational amplifier connected as a differentiator circuit followed by another operational amplifier connected as an inverter. The derivative time K_D is R_2C.

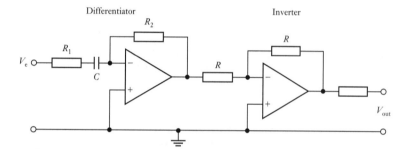

22.5.1 Proportional plus derivative (PD) control

Derivative control is never used alone because it is not capable of giving an output when there is a steady error signal and so no correction is possible. It is thus invariably used in conjunction with proportional control so that this problem can be resolved.

With proportional plus derivative control the controller output is given by

$$\text{controller output} = K_Pe + K_D\frac{de}{dt}$$

K_P is the proportionality constant and K_D the derivative constant, de/dt is the rate of change of error. The system has a transfer function given by

$$\text{controller output }(s) = K_PE(s) + K_DsE(s)$$

Hence the transfer function is $K_P + K_Ds$. This is often written as

$$\text{transfer function} = K_D\left(s + \frac{1}{T_D}\right)$$

where $T_D = K_D/K_P$ and is called the **derivative time constant**.

Figure 22.10   PD control.

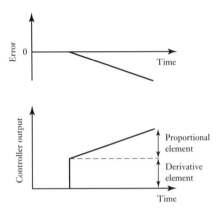

Figure 22.10 shows how the controller output can vary when there is a constantly changing error. There is an initial quick change in controller output because of the derivative action followed by the gradual change due to proportional action. This form of control can thus deal with fast process changes.

22.6   Integral control

The **integral mode** of control is one where the rate of change of the control output I is proportional to the input error signal e:

$$\frac{dI}{dt} = K_I e$$

K_I is the constant of proportionality and has units of $1/s$. Integrating the above equation gives

$$\int_{I_0}^{I_{out}} dI = \int_0^t K_I e\, dt$$

$$I_{out} - I_0 = \int_0^t K_I e\, dt$$

I_0 is the controller output at zero time, I_{out} is the output at time t.

The transfer function is obtained by taking the Laplace transform. Thus

$$(I_{out} - I_0)(s) = \frac{1}{s} K_I E(s)$$

and so

$$\text{transfer function} = \frac{1}{s} K_I$$

Figure 22.11 illustrates the action of an integral controller when there is a constant error input to the controller. We can consider the graphs in two ways. When the controller output is constant, the error is zero; when the controller output varies at a constant rate, the error has a constant value. The alternative way of considering the graphs is in terms of the area under the error graph:

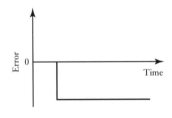

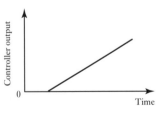

Figure 22.11   Integral control.

$$\text{area under the error graph between } t = 0 \text{ and } t = \int_0^t e\, dt$$

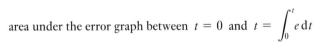

Thus up to the time when the error occurs the value of the integral is zero. Hence $I_{out} = I_0$. When the error occurs it maintains a constant value. Thus the area under the graph is increasing as the time increases. Since the area increases at a constant rate the controller output increases at a constant rate.

Figure 22.12 shows the form of the circuit used for an electronic integral controller. It consists of an operational amplifier connected as an integrator and followed by another operational amplifier connected as a summer to add the integrator output to that of the controller output at zero time. K_I is $1/R_IC$.

Figure 22.12 Integral controller.

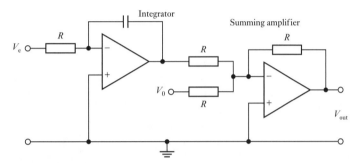

22.6.1 Proportional plus integral (PI) control

The integral mode of control is not usually used alone but is frequently used in conjunction with the proportional mode. When integral action is added to a proportional control system the controller output is given by

$$\text{controller output} = K_{P}e + K_I \int e\,dt$$

where K_P is the proportional control constant, K_I the integral control constant and e the error e. The transfer function is thus

$$\text{transfer function} = K_P + \frac{K_I}{s} = \frac{K_P}{s}\left(s + \frac{1}{T_I}\right)$$

where $T_I = K_P/K_I$ and is the **integral time constant**.

Figure 22.13(a) shows how the system reacts when there is an abrupt change to a constant error. The error gives rise to a proportional controller output which remains constant since the error does not change. There is then superimposed on this a steadily increasing controller output due to the integral action. Figure 22.13(b) shows the effects of the proportional action and the integral action if we create an error signal which is increased from the zero value and then decreased back to it again. With proportional action alone the controller mirrors the change and ends up back at its original set point value. With the integral action the controller output increases in proportion to the way the area under the error–time graph increases and since, even when the error has reverted back to zero, there is still a value for the area, there is a change in controller output which continues after the error has ceased.

Figure 22.13   PI control.

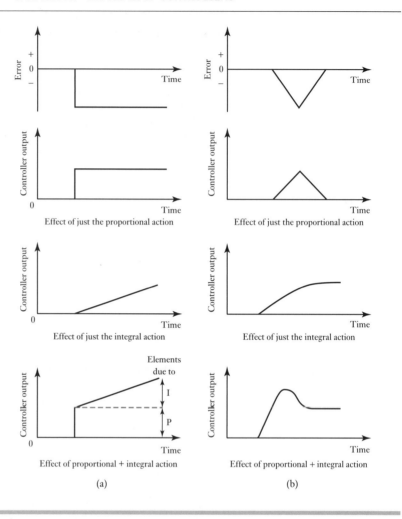

Effect of just the proportional action

Effect of just the integral action

Effect of proportional + integral action

(a)

(b)

22.7   PID controller

Combining all three modes of control (proportional, integral and derivative) gives a controller known as a **three-mode controller** or **PID controller**. The equation describing its action can be written as

$$\text{controller output} = K_{\text{p}}e + K_{\text{I}} \int e\,\text{d}t + K_{\text{D}}\frac{\text{d}e}{\text{d}t}$$

where K_{P} is the proportionality constant, K_{I} the integral constant and K_{D} the derivative constant. Taking the Laplace transform gives

$$\text{controller output}\,(s) = K_{\text{P}}E(s) + \frac{1}{s}K_{\text{I}}E(s) + sK_{\text{D}}(s)$$

and so

$$\text{transfer function} = K_{\text{p}}e + \frac{1}{s}K_{\text{I}} + sK_{\text{D}} = K_{\text{P}}\left(1 + \frac{1}{T_{\text{I}}s} + T_{\text{D}}s\right)$$

22.7.1 Operational amplifier PID circuits

A three-mode controller can be produced by combining the various circuits described earlier in this chapter for the separate proportional, derivative and integral modes. A more practical controller can, however, be produced with a single operational amplifier. Figure 22.14 shows one such circuit. The proportional constant K_P is $R_I/(R + R_D)$, the derivative constant K_D is $R_D C_D$ and the integral constant K_I is $1/R_I C_I$.

Figure 22.14 PID circuit.

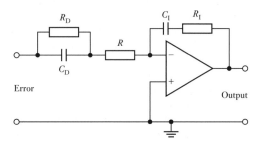

22.8 Digital controllers

Figure 22.22 shows the basis of a direct digital control system that can be used with a continuous process; the term **direct digital control** is used when the digital controller, basically a microprocessor, is in control of the closed-loop control system. The controller receives inputs from sensors, executes control programs and provides the output to the correction elements. Such controllers require inputs which are digital, process the information in digital form and give an output in digital form. Since many control systems have analogue measurements an analogue-to-digital converter (ADC) is used for the inputs. A clock supplies a pulse at regular time intervals and dictates when samples of the controlled variable are taken by the ADC. These samples are then converted to digital signals which are compared by the microprocessor with the set point value to give the error signal. The microprocessor can then initiate a control mode to process the error signal and give a digital output. The control mode used by the microprocessor is determined by the program of instructions used by the microprocessor for processing the digital signals, i.e. the *software*. The digital output, generally after processing by a digital-to-analogue converter (DAC) since correcting elements generally require analogue signals, can be used to initiate the correcting action.

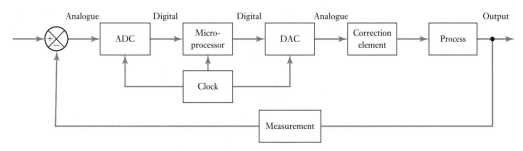

Figure 22.15 Digital closed-loop control system.

A digital controller basically operates the following cycle of events:

1 samples the measured value;
2 compares it with the set value and establishes the error;
3 carries out calculations based on the error value and stored values of previous inputs and outputs to obtain the output signal;
4 sends the output signal to the DAC;
5 waits until the next sample time before repeating the cycle.

Microprocessors as controllers have the advantage over analogue controllers that the form of the controlling action, e.g. proportional or three-mode, can be altered by purely a change in the computer software. No change in hardware or electrical wiring is required. Indeed the control strategy can be altered by the computer program during the control action in response to the developing situation.

They also have other advantages. With analogue control, separate controllers are required for each process being controlled. With a microprocessor many separate processes can be controlled by sampling processes with a multiplexer (see Section 4.4). Digital control gives better accuracy than analogue control because the amplifiers and other components used with analogue systems change their characteristics with time and temperature and so show drift, while digital control, because it operates on signals in only the on/off mode, does not suffer from drift in the same way.

22.8.1 Implementing control modes

In order to produce a digital controller which will give a particular mode of control it is necessary to produce a suitable program for the controller. The program has to indicate how the digital error signal at a particular instant is to be processed in order to arrive at the required output for the following correction element. The processing can involve the present input together with previous inputs and previous outputs. The program is thus asking the controller to carry out a difference equation (see Section 4.6).

The transfer function for a PID analogue controller is

$$\text{transfer function} = K_\text{P} + \frac{1}{s}K_\text{I} + sK_\text{D}$$

Multiplication by s is equivalent to differentiation. We can, however, consider the gradient of the time response for the error signal at the present instant of time as being (latest sample of the error e_n minus the last sample of the error e_{n-1})/(sampling interval T_s) (Figure 22.16).

Division by s is equivalent to integration. We can, however, consider the integral of the error at the end of a sampling period as being the area under the error–time graph during the last sampling interval plus the sum of the areas under the graph for all previous samples (Int_prev). If the sampling period is short relative to the times involved then the area during the last sampling interval is approximately $\frac{1}{2}(e_n + e_{n-1})/T_\text{s}$ (see Section 4.6 for another approximation known as Tustin's approximation). Thus we can

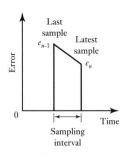

Figure 22.16   Error signals.

write for the controller output x_n at a particular instant the equivalent of the transfer function as

$$x_n = K_\mathrm{p}e_n + K_\mathrm{I}\left(\frac{(e_n + e_{n-1})T_\mathrm{s}}{2} + \mathrm{Int}_\mathrm{prev}\right) + K_\mathrm{D}\frac{e_n - e_{n-1}}{T_\mathrm{s}}$$

We can rearrange this equation to give

$$x_n = Ae_n + Be_{n-1} + C(\mathrm{Int}_\mathrm{prev})$$

where $A = K_\mathrm{P} + 0.5K_\mathrm{I}T_\mathrm{s} + K_\mathrm{D}/T_\mathrm{s}$, $B = 0.5K_\mathrm{I}T_\mathrm{s} - K_\mathrm{D}/T_\mathrm{s}$ and $C = K_\mathrm{I}$.
The program for PID control thus becomes:

1 Set the values of K_P, K_I and K_D.
2 Set the initial values of e_{n-1}, $\mathrm{Int}_\mathrm{prev}$ and the sample time T_s.
3 Reset the sample interval timer.
4 Input the error e_n.
5 Calculate y_n using the above equation.
6 Update, ready for the next calculation, the value of the previous area to $\mathrm{Int}_\mathrm{prev} + 0.5(e_n + e_{n-1})T_\mathrm{s}$.
7 Update, ready for the next calculation, the value of the error by setting e_{n-1} equal to e_n.
8 Wait for the sampling interval to elapse.
9 Go to step 3 and repeat the loop.

22.8.2 Sampling rate

When a continuous signal is sampled, for the sample values to reflect accurately the continuous signal, they have to be sufficiently close together in time for the signal not to fluctuate significantly between samples. During a sampling interval, no information is fed back to the controller about changes in the output. In practice this is taken to mean that the samples have to be taken at a rate greater than twice the highest frequency component in the continuous signal. This is termed Shannon's sampling theorem (see Section 4.2.1). In digital control systems, the sampling rate is generally much higher than this.

22.8.3 A computer control system

Typically a computer control system consists of the elements shown in Figure 22.15 with set points and control parameters being entered from a keyboard. The software for use with the system will provide the program of instructions needed, for example, for the computer to implement the PID control mode, provide the operator display, recognise and process the instructions inputted by the operator, provide information about the system, provide start-up and shut-down instructions, and supply clock/calendar information. An operator display is likely to show such information as the set point value, the actual measured value, the sampling interval, the error, the controller settings and the state of the correction element. The display is likely to be updated every few seconds.

22.9 **Control system performance**

The transfer function of a control system is affected by the mode chosen for the controller. Hence the response of the system to, say, a step input is affected. Consider the simple system shown in Figure 22.17.

Figure 22.17 Control system.

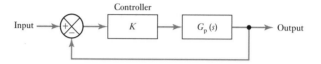

With proportional control the transfer function of the forward path is $K_pG(s)$ and so the transfer function of the feedback system $G(s)$ is

$$G(s) = \frac{K_pG_p(s)}{1 + K_pG_p(s)}$$

Suppose we have a process which is first order with a transfer function of $1/(\tau s + 1)$ where τ is the time constant (it might represent a d.c. motor, often modelled as a first-order system – see Section 20.5.1). With proportional control, and unity feedback, the transfer function of the control system becomes

$$G(s) = \frac{K_p/(\tau s + 1)}{1 + K_p/(\tau s + 1)} = \frac{K_p}{\tau s + 1 + K_p}$$

The control system remains a first-order system. The proportional control has had the effect of just changing the form of the first-order response of the process. Without the controller, the response to a unit step input was (see Section 20.2.1)

$$y = 1 - e^{-t/\tau}$$

Now it is

$$y = K_p\left(1 - e^{-t/(\tau/1 + K_p)}\right)$$

The effect of the proportional control has been to reduce the time constant from τ to $\tau/(1 + K_p)$, making it faster responding to the higher value of K_p. It also decreases the steady-state error.

With integral control we have a forward-path transfer function of $K_IG_p(s)/s$ and so the system transfer function is

$$G(s) = \frac{K_IG_p(s)}{s + K_IG_p(s)}$$

Thus, if we now have a process which is first order with a transfer function of $1/(\tau s + 1)$, with proportional control and unity feedback the transfer function of the control system becomes

$$G(s) = \frac{K_I/(\tau s + 1))}{s + K_I/(\tau s + 1)} = \frac{K_I}{s(\tau s + 1) + K_I} = \frac{K_I}{\tau s^2 + s + K_I}$$

The control system is now a second-order system. With a step input, the system will give a second-order response instead of a first-order response.

With a system having derivative control the forward-path transfer function is $sK_D G(s)$ and so, with unity feedback, the system transfer function is

$$G(s) = \frac{sK_D G_p(s)}{1 + K_D G_p(s)}$$

With a process which is first order with a transfer function of $1/(\tau s + 1)$, derivative control gives an overall transfer function of

$$G(s) = \frac{sK_D/(\tau s + 1)}{1 + sK_D/(\tau s + 1)} = \frac{sK_D}{\tau s + 1 + sK_D}$$

22.10 Controller tuning

The term **tuning** is used to describe the process of selecting the best controller settings. With a proportional controller this means selecting the value of K_P; with a PID controller the three constants K_P, K_I and K_D have to be selected. There are a number of methods of doing this; here just two methods will be discussed, both by Ziegler and Nichols. They assumed that when the controlled system is open loop a reasonable approximation to its behaviour is a first-order system with a built-in time delay. Based on this, they then derived parameters for optimum performance. This was taken to be settings which gave an under-damped transient response with a decay (subsidence) ratio of ¼., i.e. the second overshoot is ¼ of the first overshoot (see Section 19.5). This overshoot criterion gives a good compromise of small rise time, small settling time and a reasonable margin of stability.

22.10.1 Process reaction method

The process control loop is opened, generally between the controller and the correction unit, so that no control action occurs. A test input signal is then applied to the correction unit and the response of the controlled variable determined. The test signal should be as small as possible. Figure 22.18 shows the form of test signal and a typical response. The test signal is a step signal with a step size expressed as the percentage change P in the correction unit. The graph of the measured variable plotted against time is called the **process reaction curve**. The measured variable is expressed as the percentage of the full-scale range.

A tangent is drawn to give the maximum gradient of the graph. For Figure 22.18 the maximum gradient R is M/T. The time between the start of the test signal and the point at which this tangent intersects the graph time axis is termed the lag L. Table 22.1 gives the criteria recommended by Ziegler and Nichols for control settings based on the values of P, R and L.

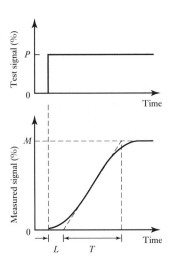

Figure 22.18 Process reaction curve.

Table 22.1 Process reaction curve criteria.

Control mode	K_P	T_I	T_D
P	P/RL		
PI	$0.9P/RL$	$3.33L$	
PID	$1.2P/RL$	$2L$	$0.5L$

Consider the following example. Determine the settings required for a three-mode controller which gave the process reaction curve shown in Figure 22.19 when the test signal was a 6% change in the control valve position. Drawing a tangent to the maximum gradient part of the graph gives a lag L of 150 s and a gradient R of $5/300 = 0.017/\text{s}$. Hence

$$K_P = \frac{1.2P}{RL} = \frac{1.2 \times 6}{0.017 \times 150} = 2.82$$

$$T_I = 2L = 300\,\text{s}$$

$$T_D = 0.5L = 0.5 \times 150 = 75\,\text{s}$$

Figure 22.19 Process curve example.

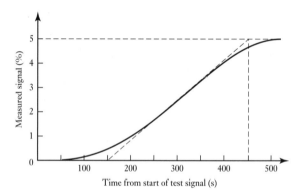

22.10.2 Ultimate cycle method

With this method, integral and derivative actions are first reduced to their minimum values. The proportional constant K_P is set low and then gradually increased. While doing this small disturbances are applied to the system. This is continued until continuous oscillations occur. The critical value of the proportional constant K_{Pc} at which this occurs is noted and the periodic time of the oscillations T_c measured. Table 22.2 shows how the Ziegler and

Table 22.2 Ultimate cycle criteria.

Control mode	K_P	T_I	T_D
P	$0.5K_{Pc}$		
PI	$0.45K_{Pc}$	$T_c/1.2$	
PID	$0.6K_{Pc}$	$T_c/2.0$	$T_c/8$

Nichols recommended criteria for controller settings are related to this value of K_{Pc}. The critical proportional band is $100/K_{Pc}$.

Consider the following example. When tuning a three-mode control system by the ultimate cycle method it was found that oscillations begin when K_{Pc} is 3.33. The oscillations have a periodic time of 500 s. What are the suitable settings for the controller? Using the criteria given in Table 22.2, then $K_P = 0.6K_{Pc} = 0.6 \times 3.33 = 2.0$, $T_I = T_c/2.0 = 500/2 = 2.5$ s, $T_D = T_c/8 = 500/8 = 62.5$ s.

22.11 Velocity control

Consider the problem of controlling the movement of a load by means of a motor. Because the motor system is likely to be of second order, proportional control will lead to the system output taking time to reach the required displacement when there is, say, a step input to the system and may oscillate for a while about the required value. Time will thus be taken for the system to respond to an input signal. A higher speed of response, with fewer oscillations, can be obtained by using PD rather than just P control. There is, however, an alternative of achieving the same effect and this is by the use of a second feedback loop which gives a measurement related to the rate at which the displacement is changing. This is termed **velocity feedback**. Figure 22.20 shows such a system; the velocity feedback might involve the use of a tachogenerator giving a signal proportional to the rotational speed of the motor shaft, and hence the rate at which the displacement is changing, and the displacement might be monitored using a rotary potentiometer.

22.12 Adaptive control

There are many control situations where the parameters of the plant change with time or, perhaps, load, e.g. a robot manipulator being used to move loads when the load is changed. If the transfer function of the plant changes then retuning of the system is desirable for the optimum values to be determined for proportional, derivative and integral constants. For the control systems so far considered, it has been assumed that the system once tuned retains its values of proportional, derivative and integral constants until the operator decides to retune. The alternative to this is an **adaptive control system** which 'adapts' to changes and changes its parameters to fit the circumstances prevailing.

The adaptive control system is based on the use of a microprocessor as the controller. Such a device enables the control mode and the control parameters used to be adapted to fit the circumstances, modifying them as the circumstances change.

An adaptive control system can be considered to have three stages of operation.

1 Starts to operate with controller conditions set on the basis of an assumed condition.
2 The desired performance is continuously compared with the actual system performance.
3 The control system mode and parameters are automatically and continuously adjusted in order to minimise the difference between the desired and actual system performance.

For example, with a control system operating in the proportional mode, the proportional constant K_P may be automatically adjusted to fit the

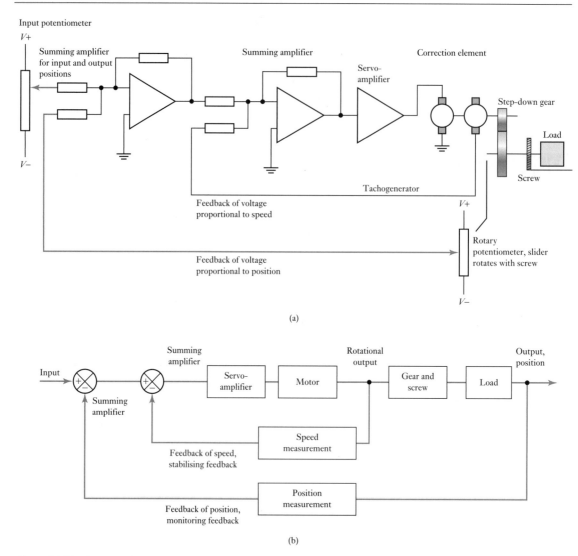

Figure 22.20 System with velocity feedback: (a) descriptive diagram of the system, (b) block diagram of the system.

circumstances, changing as they do. Adaptive control systems can take a number of forms. Three commonly used forms are:

1 gain–scheduled control;
2 self-tuning;
3 model-reference adaptive systems.

22.12.1 Gain-scheduled control

With **gain-scheduled control** or, as it is sometimes referred to, **pre-programmed adaptive control**, preset changes in the parameters of the controller are made on the basis of some auxiliary measurement of some

Figure 22.21 Gain-scheduled control.

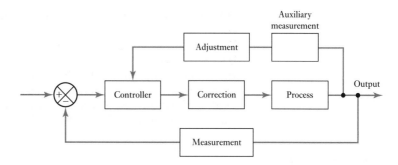

process variable. Figure 22.21 illustrates this method. The term gain-scheduled control was used because the only parameter originally adjusted was the gain, i.e. the proportionality constant K_P.

For example, for a control system used to control the positioning of some load, the system parameters could be worked out for a number of different load values and a table of values loaded into the memory of the controller. A load cell might then be used to measure the actual load and give a signal to the controller indicating a mass value which is then used by the controller to select the appropriate parameters.

A disadvantage of this system is that the control parameters have to be determined for many operating conditions so that the controller can select the one to fit the prevailing conditions. An advantage, however, is that the changes in the parameters can be made quickly when the conditions change.

22.12.2 Self-tuning

With **self-tuning control** the system continuously tunes its own parameters based on monitoring the variable that the system is controlling and the output from the controller. Figure 22.22 illustrates the features of this system.

Figure 22.22 Self-tuning.

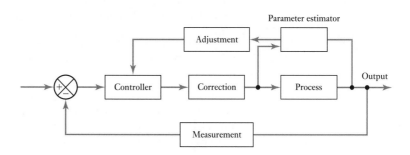

Self-tuning is often found in commercial PID controllers, generally then being referred to as **auto-tuning**. When the operator presses a button, the controller injects a small disturbance into the system and measures the response. This response is compared to the desired response and the control parameters adjusted, by a modified Ziegler–Nichols rule, to bring the actual response closer to the desired response.

22.12.3   Model-reference adaptive systems

With the **model-reference adaptive system** an accurate model of the system is developed. The set value is then used as an input to both the actual and the model systems and the difference between the actual output and the output from the model compared. The difference in these signals is then used to adjust the parameters of the controller to minimise the difference. Figure 22.23 illustrates the features of the system.

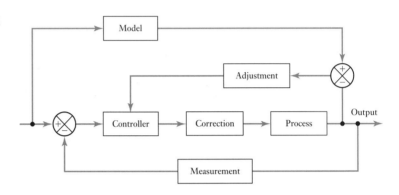

Figure 22.23   Model-referenced control.

Summary

The **steady-state error** is the difference between the desired set value input and the output after all transients have died away.

Control modes are **two-step** in which the controller supplies an on/off correcting signal, **proportional** (P) in which the correcting signal is proportional to the error, **derivative** (D) in which the correcting signal is proportional to the rate at which the error is changing, and **integral** (I) in which the correcting signal is proportional to the integral of the error with time. The transfer function for a PID system is

$$\text{transfer function} = K_P e + \frac{1}{s}K_I + sK_D = K_P\left(1 + \frac{1}{T_I s} + T_D s\right)$$

A **digital controller** basically operates by sampling the measured value, comparing it with the set value and establishing the error, carrying out calculations based on the error value and stored values of previous inputs and outputs to obtain the output signal, outputting and then waiting for the next sample.

The term **tuning** is used to describe the process of selecting the best controller settings, i.e. the values of K_P, K_I and K_D.

The term **adaptive control** is used for systems which 'adapt' to changes and change their parameters to fit the circumstances prevailing. Three commonly used forms are gain-scheduled control, self-tuning and model-reference adaptive systems.

Problems

22.1 What are the limitations of two-step (on/off) control and in what situation is such a control system commonly used?

22.2 A two-position mode controller switches on a room heater when the temperature falls to 20°C and off when it reaches 24°C. When the heater is on, the air in the room increases in temperature at the rate of 0.5°C per minute; when the heater is off, it cools at 0.2°C per minute. If the time lags in the control system are negligible, what will be the times taken for (a) the heater switching on to off, (b) the heater switching off to on?

22.3 A two-position mode controller is used to control the water level in a tank by opening or closing a valve which in the open position allows water at the rate of $0.4 \, \text{m}^3/\text{s}$ to enter the tank. The tank has a cross-sectional area of $12 \, \text{m}^2$ and water leaves it at the constant rate of $0.2 \, \text{m}^3/\text{s}$. The valve opens when the water level reaches 4.0 m and closes at 4.4 m. What will be the times taken for (a) the valve opening to closing, (b) the valve closing to opening?

22.4 A proportional controller is used to control the height of water in a tank where the water level can vary from 0 to 4.0 m. The required height of water is 3.5 m and the controller is to close a valve fully when the water rises to 3.9 m and open it fully when the water falls to 3.1 m. What transfer function will be required?

22.5 Describe and compare the characteristics of (a) proportional control, (b) proportional plus integral control, (c) proportional plus integral plus derivative control.

22.6 Determine the settings of K_P, T_I and T_D required for a three-mode controller which gave a process reaction curve with a lag L of 200 s and a gradient R of 0.010%/s when the test signal was a 5% change in the control valve position.

22.7 When tuning a three-mode control system by the ultimate cycle method it was found that oscillations began when the proportional critical value was 5. The oscillations had a periodic time of 200 s. What are the suitable values of K_P, T_I and T_D?

22.8 Explain the basis on which the following forms of adaptive control systems function: (a) gain-scheduled, (b) self-tuning, (c) model-reference.

22.9 A d.c. motor behaves like a first-order system with a transfer function of relating output position to which it has rotated a load with input signal of $1/s(1 + s\tau)$. If the time constant τ is 1 s and the motor is to be used in a closed-loop control system with unity feedback and a proportional controller, determine the value of the proportionality constant which will give a closed-loop response with a 25% overshoot.

22.10 The small ultrasonic motor used to move the lens for automatic focusing with a camera (see Section 24.2.3) drives the ring with so little inertia that the transfer function relating angular position with input signal is represented by $1/cs$, where c is the constant of proportionality relating the frictional torque and angular velocity. If the motor is to be controlled by a closed-loop system with unity feedback, what type of behaviour can be expected if proportional control is used?

Chapter twenty-three

Artificial intelligence

Objectives

The objectives of this chapter are that, after studying it, the reader should be able to:
- Explain what is meant by an intelligent machine and the capabilities of such machines.
- Explain the meaning of neural networks and their relevance to pattern recognition.
- Explain the term fuzzy logic.

23.1 What is meant by artificial intelligence?

What constitutes an intelligent machine? A dictionary definition of intelligent might be 'endowed with the ability to reason'. The more intelligent we think a person is, the more we consider he or she is able to learn, generalise from this acquired knowledge, be capable of reasoning and able to make predictions by considering what is possible, learning from any mistakes. We can apply the same criteria to a machine: an **intelligent machine** is one endowed with the ability to reason.

A central heating system makes decisions about its actions. For example, should the boiler switch on or off as a result of information from a thermostat? It is not, however, considered to be intelligent because it is not capable of reasoning and making decisions under a wide range of conditions. For example, it cannot recognise a pattern in inputs from a thermostat and so make predictions about whether to switch the boiler on or off. It just does what it is told to do. It does not 'think for itself'.

In this chapter we take a brief look at basic concepts associated with intelligent machines.

23.1.1 Self-regulation

We can consider the closed-loop feedback systems discussed in earlier chapters as being self-regulation systems in that they regulate the output of a system to a required value. Thus a thermostatically controlled central heating system is used to maintain the room temperature at the value set for the thermostat. Such systems cannot, however, be considered intelligent; they merely do what they were told to do.

23.2 Perception and cognition

Perception with an intelligent system is the collecting of information using sensors and the organising of the gathered information so that decisions can be made. For example, a control system used with a production line might have a video camera to observe components on a conveyor belt. The signals received from the camera enable a computed representation of the components to be made so that features can be identified. This will contain information about critical elements of the components. These can then be compared with representations of the components so that decisions can be made by the control system as to whether the component is correctly assembled or perhaps which component it is. Then action can be taken by the control system perhaps to reject faulty components or divert particular components to particular boxes.

Thus, with a mechatronics system, perception involves sensors gathering appropriate information about a system and its environment, decoding it and processing it to give useful information which can be used elsewhere in the system to make decisions.

23.2.1 Cognition

Once a machine has collected and organised information, it has to make decisions about what to do as a consequence of the information gathered. This is termed **cognition**. Vital to this perception and cognition is **pattern recognition**. What are the patterns in the data gathered?

Humans are very good at pattern recognition. Think of a security guard observing television monitors. He or she is able to look at the monitors and recognise unusual patterns, e.g. a person where there should be no person, an object having been moved, etc. This is the facility required of intelligent machines. An autopilot system on an aircraft monitors a lot of information and, on the basis of the patterns perceived in that data, makes decisions as to how to adjust the aircraft controls.

Machine pattern recognition can be achieved by the machine having a set of patterns in its memory and gathered patterns are then compared with these and a match sought. The patterns in its memory may arise from models or a process of training in which data is gathered for a range of objects or situations and these given identification codes. For example, for recognising coins, information may be gathered about diameter and colour. Thus a particular one-pound coin might be classified as having a diameter of 2.25 cm and a colour which is a particular degree of redness (it is a bronze coin). However, an intelligent machine will need to take account of worn and dirty coins and still be able to recognise the one-pound coin.

23.2.2 Neural networks

In the above example of the coins, only two dimensions were considered, namely diameter and colour. In more complex situations there may be many more dimensions. The human brain is faced with sorting and classifying multidimensional information and does this using **neural networks**. Artificial neural networks are now being used with intelligent machines. Such networks do not need to be programmed but can learn and generalise from examples and training. A neural network (Figure 23.1) is composed of a large number of interconnected processing units, the outputs from some units being inputs to others. Each processor in the network receives information at its inputs, and multiplies each by a weighting factor. If operating as AND, it

Figure 23.1   Neural network.

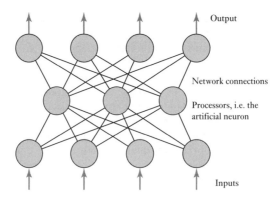

then sums these weighted inputs and gives an output of a 1 if the sum exceeds a certain value or is positive. For example, we might have an input of 1 with a weighting factor of -1.5 to give -1.5, another input of 1 with a weighting factor of 1.0 to give 1.0 and a third input of 1 with a weighting factor of 1.0 to give 1.0. The sum of these weighted inputs is thus $-1.5 + 1.0 + 1.0 = 0.5$ and so an output of 1 if the values are to be positive for an output. With these inputs as 1×-1.5, 0×1.0 and 0×1.0, the weighted sum is -1.5 and so an output of 0. The network can be programmed by learning from examples and so be capable of learning.

23.3   Reasoning

Reasoning is the process of going from what is known to what is not known. There are a number of mechanisms for carrying out reasoning.

23.3.1   Reasoning mechanisms

An example of **deterministic reasoning** is to use the 'if–then' rule. Thus, we might deduce that *if* a coin has a diameter of 1.25 cm *then* it is a pound coin. If the first part of the statement is true then the second part of the statement is true; if it is false then the second part is not true. In this form of reasoning we have a *true–false* situation and it is assumed that there is no default knowledge so that when the deduction is made there are no exceptions. Thus, in the above example, we cannot take account of there being a coin from another country with the same diameter.

Non-deterministic reasoning allows us to make predictions based on probability. If you toss a coin, there are two ways it can land: face upwards or face downwards. Out of the two ways there is just one way which will give face upwards. Hence, the probability of its landing face upwards is 1 in 2 or $1/2$. An alternative way of arriving at this value is to toss a coin repeatedly and, in the long run, in $1/2$ of the times it will end up with face upwards. Figure 23.2(a) shows how we can represent this as a probability tree. If we throw a six-sided die then the probability of its landing with a six uppermost is $1/6$. Figure 23.2(b) shows how we can represent this as a probability tree. On each limb of the tree the probability is written. The chance of a coin landing with either heads or tails is 1. Thus, for a tree, the total probability will be 1.

Figure 23.2 Probability trees: (a) a coin, (b) a die.

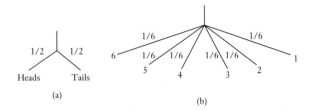

Thus in the example of the pound coin we might want to consider that there is a probability of 0.9 of a coin with diameter 1.25 cm being a pound coin. In the case of a mechatronics system we might monitor it for, say, 1000 hours and during that time the number of hours that the temperature has been high was found to be 3 hours. We can then say that the probability of the temperature being high is $3/1000 = 0.003$.

Sometimes we might know the probability of an event occurring and want to establish the probability that it will result in some other event. Thus, in a mechatronics system we might want to know what are the chances of, say, when a sensor detects a low pressure that the system will overheat, bearing in mind that there could be other reasons for a high temperature. We might represent this as the tree shown in Figure 23.3.

Figure 23.3 A conditional probability tree.

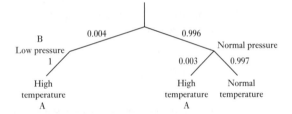

Bayes' rule can be used to solve this problem. This can be stated as

$$p(A \mid B) = \frac{p(B \mid A) \times p(A)}{p(B)}$$

$p(A \mid B)$ is the probability of A happening given that B has happened, $p(B \mid A)$ is the probability of B happening given that A has happened, $p \mid A \mid$ is the probability of A happening, $p \mid B \mid$ is the probability of B happening. Thus, if the probability for the system of a high temperature occurring $p \mid A \mid$ is 0.003, i.e. in 3 times in 1000 a high temperature occurs, and the probability of there being a low pressure $p \mid B \mid$ is 0.004, i.e. in 4 times in 1000 a low pressure occurs, then as we might be certain that the system will overheat if the pressure is low, i.e. $p(B \mid A)$ is 1, we must have a conditional probability of $(1 \times 0.003)/0.004 = 0.75$ that the system will overheat when a low pressure is detected.

23.3.2 Rule-based reasoning

At the heart of a **rule-based system** are a set of rules. These, when combined with facts, i.e. in mechatronics this could be inputs from sensors and users, enable inferences to be made which are then used to actuate actuators and control outputs. Figure 23.4 illustrates this sequence. The combination of fact

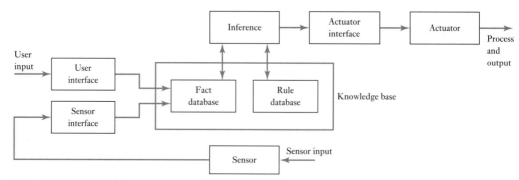

Figure 23.4 Rule-based system.

and rule databases is known as the knowledge base for a machine. Inference is where the reasoning takes place as a result of the input facts being combined with the rules and decisions made which are then fed to actuators.

The rules used are often in the form of 'if–then' statements. Thus we might have a group of rules for a central heating system of the form:

If boiler on
Then pump is on

If pump on AND room temperature less than 20°C
Then open valve

If boiler not on
Then pump not on

etc.

The fact database with such a system would contain the facts:

Room temperature < 20°C
Timer On
Valve Open
Boiler On
Pump On

The rules can also be in the form of propositions involving probability statements or fuzzy logic.

Lotfi Zadeh proposed in 1965 a form of reasoning which has become known as **fuzzy logic**. One of its main ideas is that propositions need not be classified as true or false, but their truth or falsehood can be weighted so that it can be classified between the two on a scale. A **membership function** is defined for a value as being whether it is a member of a particular set. Thus we might define one set of temperature values as being 0 to 20°C and another as 20 to 40°C. If the temperature is, say, 18°C then membership of the 0 to 20°C set is 1 and that of the 20 to 40°C set 0. However, with fuzzy logic we define overlapping sets, e.g. cold 0 to 20°C, warm 10 to 30°C and hot 20 to 40°C. A temperature of 18°C is thus a member of two sets. If the fuzzy set membership functions are defined as shown in Figure 23.5, then 18°C has a cold membership of 0.2, a warm membership of 0.8 and a hot membership of 0. On the basis of data such as this, rules can be devised to trigger appropriate action. For example, a cold membership of 0.2 might have heating switched on low, but a cold membership of 0.6 might have it switched on high.

Figure 23.5 Fuzzy set membership functions.

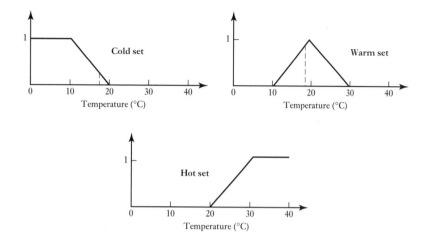

Fuzzy logic is now being used in many commonly encountered products. For example, domestic washing machines can sense the fabric type, dirtiness and load size and adjust the wash cycle accordingly.

23.4 Learning

Machines that can learn and extend their knowledge base have great advantages compared with machines that cannot learn. **Learning** can be thought of as adapting to the environment based on experience. With machines, learning can be accomplished in a number of ways.

A simple method of learning is by new data being inputted and accumulated in memory. Machines can also learn by the data they receive being used to modify parameters in the machine.

Another method of learning that can be used is when reasoning is defined in terms of probabilities and that is to update the probabilities used in the light of what happens. We can think of this in terms of a simple example. Say we have a bag containing 10 coloured balls, all being red apart from one black one. When we first draw a ball from the bag, the probability of pulling the black ball out is 1/10. If we find it is a red ball, then the next time we draw a ball out the probability that it will be a black ball is 1/9. Our 'machine' can learn from the first ball being red by adjusting its probability value for a black ball being drawn. Bayes' rule given in Section 23.3.1 can be used to update a machine, being now written as

$$p(H|E) = \frac{p(E|H) \times p(H)}{p(E)}$$

where H is the hypothesis that we start with and E the example now encountered. Then $p(H|E)$ is the probability of the hypothesis H being true given that the example E has happened, $p(E|H)$ is the probability of the example E happening given that the hypothesis H is true, $p|E|$ is the probability of an example E happening, $p|H|$ is the probability of the hypothesis H being true. This allows the machine to update the probability of H every time new information comes in.

Yet another method a machine can learn is from examples. This is when a machine generalises from a set of examples. These may be the result of training with examples being supplied to the machine so that is can build up its rules or as a consequence of events it has encountered. Pattern recognition generally involves this form of learning. Thus, given an example of the number 2 in an array of pixels, the machine can learn to recognise the number 2. Neural networks (Section 23.2.2) also involve learning by example.

A machine may also learn by drawing analogies between a problem it has solved before and a new problem.

Summary

An **intelligent machine** is one endowed with the ability to reason. **Perception** with an intelligent system is the collecting of information using sensors and the organising of the gathered information so that decisions can be made. **Reasoning** is the process of going from what is known to what is not known. An example of **deterministic reasoning** is to use the 'if–then' rule. **Non-deterministic reasoning** allows us to make predictions based on probability. With **fuzzy logic** propositions need not be classified as true or false, but their truth or falsehood can be weighted so that it can be classified between the two on a scale. **Learning** can be thought of as adapting to the environment based on experience.

Problems

23.1   Examine a range of coins of your country and produce a pattern recognition table.

23.2   What is the probability of (a) throwing a six with a single six-sided die, (b) throwing two dice and one of them giving a six, (c) taking a black ball out of a bag containing nine red balls and one black one?

23.3   If the probability of a mechatronics system showing a high temperature is 0.01, what is the probability it will not show a high temperature?

23.4   A machine has been monitored for 2000 hours and during that time the cooling system has only shown leaks for 4 hours. What is the probability of leaks occurring?

23.5   The probability of a cooling system of a machine leaking has been found to be 0.005 and the probability of the system showing a high temperature 0.008. If a leakage will certainly cause a high temperature, what is the probability that a high temperature will be caused by a cooling system leak?

23.6   The probability of there being a malfunction with a machine consisting of three elements A, B and C is 0.46. If the probability of element A being active is 0.50 and the probability a malfunction occurs with A is 0.70, what is the probability that A was responsible for a malfunction?

23.7   Propose 'if–then' rules for a temperature controller that is used to operate a boiler and has a valve allowing water to circulate round central heating radiators when it only operates at a certain time period.

Part VI
Conclusion

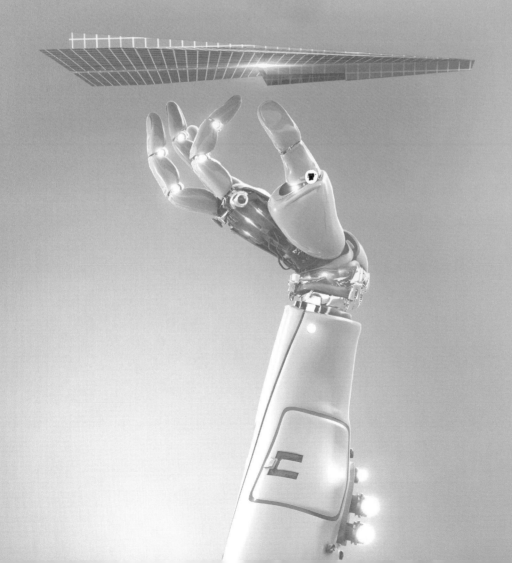

Chapter twenty-four

Mechatronic systems

Objectives

The objectives of this chapter are that, after studying it, the reader should be able to:
- Develop possible solutions to design problems when considered from the mechatronics point of view.
- Analyse case studies of mechatronics solutions.

24.1 Mechatronic designs

This chapter brings together many of the topics discussed in this book in the consideration of mechatronics solutions to design problems and gives case studies.

24.1.1 Timed switch

Consider a simple requirement for a device which switches on some actuator, e.g. a motor, for some prescribed time. Possible solutions might involve:

1 a rotating cam;
2 a programmable logic controller (PLC);
3 a microprocessor;
4 a microcontroller;
5 a timer, e.g. 555.

A mechanical solution could involve a rotating cam (Figure 24.1) (see Section 8.4). The cam would be rotated at a constant rate and the cam follower used to actuate a switch, the length of time for which the switch is closed depending on the shape of the cam. This is a solution that has widely been used in the past.

A PLC solution could involve the arrangement shown in Figure 24.2 with the given ladder program. This would have the advantage over the rotating cam of having off and on times which can be adjusted by purely changing the timer preset values in the program, whereas a different cam is needed if the times have to be changed with the mechanical solution. The software solution is much easier to implement than the hardware one.

A microprocessor-based solution could involve a microprocessor combined with a memory chip and input/output interfaces. The program is then used to switch an output on and then off after some time delay, with the time delay being produced by a block of program in which there is a timing loop. This generates a time delay by branching round a loop the number of

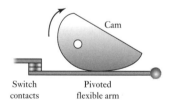

Switch contacts   Pivoted flexible arm

Figure 24.1 Cam-operated switch.

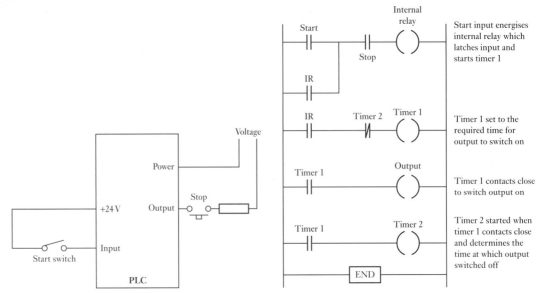

Figure 24.2 PLC timer system.

cycles required to generate the requisite time. Thus, in assembly language we might have:

```
DELAY     LDX     #F424      ; F424 is number of loops
LOOP      DEX
          BNE     LOOP
          RTS
```

DEX decrements the index register, and this and BNE, branch if not equal, each take 4 clock cycles. The loop thus takes 8 cycles and there will be n such loops until $8n + 3 + 5$ gives the number F424 (LDX takes 3 cycles and RTS takes 5 cycles). In C we could write the program lines using the while function.

Another possibility is to use the timer system in a microcontroller such as MC68HC11. This is based on a 16-bit counter TCNT operating from the system E-clock signal (Figure 24.3(a)). The system E-clock can be prescaled by setting bits in the timer interrupt mask register 2 (TMSK2),

Figure 24.3 (a) Generating 2 MHz internal clock, (b) prescale factor.

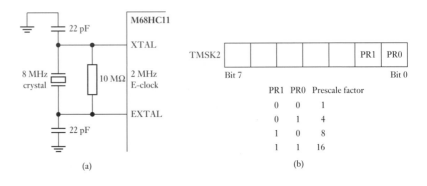

address $1024 (Figure 24.3(b)). The TCNT register starts at $0000 when the processor is reset and counts continuously until it reaches the maximum count of $FFFF. On the next pulse it overflows and reads $0000 again. When it overflows, it sets the timer overflow flag TOF (bit 7 in miscellaneous timer interrupt flag register 2, TFLG2 at address $1025). Thus with a prescale factor of 1 and an E-clock of 2 MHz, overflow occurs after 32.768 ms.

One way of using this for timing is for the TOF flag to be watched by polling. When the flag is set, the program increments its counter. The program then resets the flag, by writing a 1 to bit 7 in the TFLG2 register. Thus the timing operation just consists of the program waiting for the required number of overflag settings.

A better way of timing involves the use of the output-compare function. Port A of the microcontroller can be used for general inputs or outputs or for timing functions. The timer has output pins, OC1, OC2, OC3, OC4 and OC5, with internal registers TOC1, TOC2, TOC3, TOC4 and TOC5. We can use the output-compare function to compare the values in the TOC1 to TOC5 registers with the value in the free-running counter TCNT. This counter starts at 0000 when the CPU is reset and then runs continuously. When a match occurs between a register and the counter, then the corresponding OCx flag bit is set and output occurs through the relevant output pin. Figure 24.4 illustrates this. Thus by programming the TOCx registers, so the times at which outputs occur can be set. The output-compare function can generate timing delays with much higher accuracy than the timer overflow flag.

Figure 24.4 Output compare.

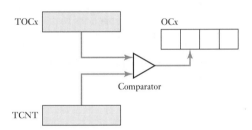

The following program illustrates how output compare can be used to produce a time delay. The longest delay that can be generated in one output-compare operation is 32.7 ms when the E-clock is 2 MHz. In order to generate longer delays, multiple output-compare operations are required. Thus we might have each output-compare operation producing a delay of 25 ms and repeating this 40 times to give a total delay of 1 s.

REGBAS	EQU	$1000	; Base address of registers
TOC2	EQU	$18	; Offset of TOC2 from REGBAS
TCNT	EQU	$0E	; Offset of TCNT from REGBAS
TFLG1	EQU	$23	; Offset of TFLGI from REGBAS
OC1	EQU	$40	; Mask to clear OC1 pin and OC1F flag
CLEAR	EQU	$40	; Clear OC2F flag
D25MS	EQU	50000	; Number of E-clock cycles to generate a 25 ms delay
NTIMES	EQU	40	; Number of output-compare operations needed to give 1 s delay
	ORG	$1000	
COUNT	RMB	1	; Memory location to keep track of the number of ; output-compare operations still to be carried out

```
          ORG      $C000           ; Starting address of the program
          LDX      #REGBAS
          LDAA     #OC1            ; Clear OC1 flag
          STAA     TFLG1,X
          LDAA     #NTIMES         ; Initialise the output-compare count
          STAA     COUNT
          LDD      TCNT,X
WAIT      ADDD     #D25MS          ; Add 25 ms delay
          STD      TOC2,X          ; Start the output-compare operation
          BRCLR    TFLG1,X OC1     ; Wait until the OC1F flag is set
          LDAA     #OC1            ; Clear the OC1F flag
          STAA     TFLG1,X
          DEC      COUNT           ; Decrement the output-compare counter
          BEQ      OTHER           ; Branch to OTHER if 1 s elapsed
          LDD      TOC2,X          ; Prepare to start the next compare operation
          BRA      WAIT
OTHER                             ; The other operations of the program which occur after the 1 s
                                   delay
```

Another possible method of producing a timed output signal is to use a timer module, e.g. 555. With the 555 timer, the timing intervals are set by external resistors and capacitors. Figure 24.5 shows the timer and the external circuitry needed to give an on output when triggered, the duration of the on output being $1.1RC$. Large times need large values of R and C. R is limited to about 1 MΩ, otherwise leakage becomes a problem, and C is limited to about 10 µF if electrolytic capacitors with the problems of leakage and low accuracy are to be avoided. Thus the circuit shown is limited to times less than about 10 s. The lower limit is about $R = 1$ kΩ and $C = 100$ pF, i.e. times of a fraction of a millisecond. For longer times, from 16 ms to days, an alternative timer such as the ZN1034E can be used.

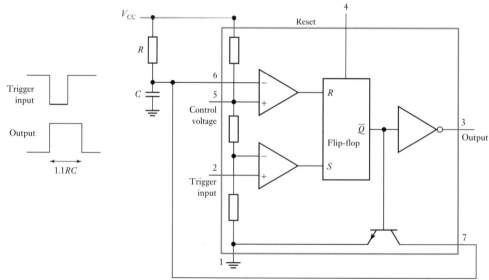

Figure 24.5 The 555 timer.

24.1.2 Windscreen-wiper motion

Consider a requirement for a device which will oscillate an arm back and forth in an arc like a windscreen wiper. Possible solutions might be:

1 mechanical linkage and a d.c. motor;
2 a stepper motor.

A mechanical solution is shown in Figure 24.6. Rotation of arm 1 by a motor causes arm 2 to impart an oscillatory motion to arm 3. Automobile windscreen wipers generally use such a mechanism with a d.c. permanent magnet motor.

Figure 24.6 Wiper mechanism.

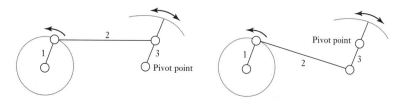

An alternative solution is to use a stepper motor. Figure 24.7 shows how a microprocessor with a peripheral interface adapter (PIA), or a microcontroller, might be used with a stepper. The input to the stepper is required to cause it to rotate a number of steps in one direction and then reverse to rotate the same number of steps in the other direction.

Figure 24.7 Interfacing a stepper.

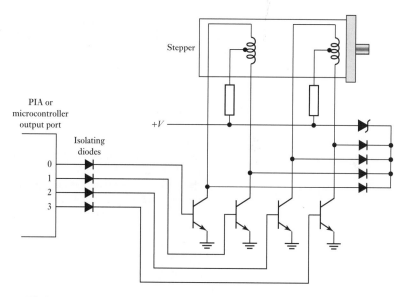

If the stepper is to be in the 'full-step' configuration then the outputs need to be as shown in Table 24.1(a). Thus to start and rotate the motor in a forward direction involves the sequence A, 9, 5, 6 and then back to the beginning with 1 again. To reverse we would use the sequence 6, 5, 9, A and then back to begin with 6 again. If 'half–step' configuration is used then the outputs need to be as shown in Table 24.1(b). Forward motion then involves the sequence A, 8, 9, 1, 5, 4, 6, 2 and then back to A, with reverse requiring 2, 6, 4, 5, 1, 9, 8, A and back to 2.

Table 24.1   (a) Full-step, (b) half-step configuration.

(a)

Step	Bit 3	Bit 2	Bit 1	Bit 0	Code
1	1	0	1	0	A
2	1	0	0	1	9
3	0	1	0	1	5
4	0	1	1	0	6
1	1	0	1	0	A

(b)

Step	Bit 3	Bit 2	Bit 1	Bit 0
1	1	0	1	0
2	1	0	0	0
3	1	0	0	1
4	0	0	0	1
5	0	1	0	1
6	0	1	0	0
7	0	1	1	0
8	0	0	1	0
1	1	0	1	0

The basic elements of a program could be:

Advance a step
Jump to time-delay routine to give time for the step to be completed
Loop or repeat the above until the requisite number of steps completed in the
     forward direction
Reverse direction
Repeat the above for the same number of steps in reverse direction

Such a program in C might, for three half-steps forward and three back, and following the inclusion of an appropriate header file, have the following elements:

```
main ( )
{
    portB = 0xa; /*first step*/
    delay ( ); /*incorporate delay program for, say, 20 ms*/
    portB = 0x8; /*second step*/
    delay ( ); /*incorporate delay program for 20 ms*/
    port B = 0x9; /*third step*/
    delay ( ); /*incorporate delay program for 20 ms*/
    port B = 0x8; /*reverse a step*/
    delay ( ); /*incorporate delay program for 20 ms*/
    port B = 0xa; /*reverse a further step*/
    delay ( ); /*incorporate delay program for 20 ms*/
    port B = 0x2; /*reverse back to where motor started*/
    delay ( ); /*incorporate delay program for 20 ms*/
}
```

Where there are many steps involved, a simpler program is to increment a counter with each step and loop until the counter value reaches the required number. Such a program would have the basic form of:

Advance a step
Jump to time-delay routine to give time for the step to be completed
Increment the counter

Loop or repeat the above with successive steps until the counter indicates the requisite number of steps completed in the forward direction
Reverse direction
Repeat the above for the same number of steps in reverse direction

Integrated circuits are available for step motor control and their use can simplify the interfacing and the software. Figure 24.8 shows how such a circuit can be used. All that is then needed is the requisite number of input pulses to the trigger, the motor stepping on the low-to-high transition of a high–low–high pulse. A high on the rotation input causes the motor to step anti-clockwise, while a low gives clockwise rotation. Thus we just need one output from the microcontroller for output pulses to the trigger and one output to rotation. An output to set is used to reset the motor back to its original position.

Figure 24.8 Integrated circuit SAA 1027 for stepper motor.

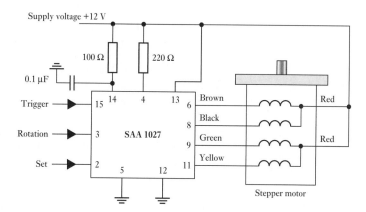

The above has indicated how we might use a stepper motor to give an angular rotation. But how will a stepper motor behave when given a voltage signal input? Can we expect it to rotate directly to the angle concerned with no overshoot and no oscillations before settling down to the required angle? As an illustration of how we can develop a model for the stepper motor system and so predict its behaviour, consider the following simplified analysis (for a more detailed analysis, *see Stepping Motors and their Microprocessor Controls* by T. Kenjo and A. Sugawara (Clarenden Press 1995)).

The system involving a stepper motor being driven by pulses from a microcontroller is an open-loop control system. The permanent magnet stepper motor (see Section 9.7) has a stator with a number of poles, the poles being energised by current being passed through coils wound on them. We can determine a model for how the rotor will rotate when there is a voltage pulse input to it by considering, for simplicity, a stepper with just a pair of poles and treat it in the same manner as the d.c. motor that was analysed in Section 18.3.2. If v is the voltage supplied to the motor pair of coils and v_b the back e.m.f. then

$$v - v_b = L\frac{di}{dt} + Ri$$

where L is the inductance of the circuit, R the resistance and i the circuit current. We will make the simplifying assumption that the inductance does not significantly change and so treat L as a constant.

The back e.m.f will be proportional to the rate at which the magnetic flux is changing for the pair of coils. This will depend on the angle θ of the rotor relative to the poles concerned. Thus we can write

$$v_b = -k_b \frac{d}{dt} \cos \theta = k_b \sin \theta \frac{d\theta}{dt}$$

where k_b is a constant. Thus

$$v - k_b \sin \theta \frac{d\theta}{dt} = L \frac{di}{dt} + Ri$$

Taking the Laplace transform of this equation gives

$$V(s) - k_b s \sin \theta \, \theta(s) = sL \, I(s) + R \, I(s) = (sL + R) \, I(s)$$

As with the d.c. motor, the current through a pair of coils will generate a torque (the torque on the magnet, i.e. the rotor, being the reaction resulting from the torque exerted on the coils – Newton's third law). The torque is proportional to the product of the flux density at the coil turns and the current through them. The flux density will depend on the angular position of the rotor and thus we can write

$$T = k_t i \sin \theta$$

where k_t is a constant. This torque will cause an angular acceleration α and since $T = J\alpha$, where J is the moment of inertia of the rotor,

$$T - J \frac{d^2\theta}{dt^2} = k_t i \sin \theta$$

Taking the Laplace transform of this equation gives

$$s^2 J\theta(s) = k_t \sin \theta \, I(s)$$

and so we can write

$$V(s) - k_b s \sin \theta \, \theta(s) = (sL + R)(s^2 J\theta(s)/k_t \sin \theta)$$

The transfer function between the input voltage and the resulting angular displacement is

$$G(s) = \frac{\theta(s)}{V(s)} = \frac{k_t \sin \theta}{J(sL + R)s^2 + k_b k_t s \sin^2 \theta}$$

$$= \frac{1}{s} \times \frac{k_t \sin \theta}{JLs^2 + JRs + k_b k_t \sin^2 \theta}$$

When there is a voltage impulse supplied to the motor coils, since for a unit impulse $V(s) = 1$,

$$\theta(s) = \frac{1}{s} \times \frac{k_t \sin \theta}{JLs^2 + JRs + k_b k_t \sin^2 \theta}$$

$$= \frac{1}{s} \times \frac{(k_t \sin \theta)/JL}{s^2 + (R/L)s + (k_b k_t \sin^2 \theta)/JL}$$

The quadratic equation in s is of the form $s^2 + 2\zeta\omega_n s + \omega_n^2$ (see Section 20.3.1) and thus has a natural frequency ω_n of $\sqrt{(k_b k_t \sin^2\theta / JL)}$ and a damping factor ζ of $(R/L)/2\omega_n$. The rotor will rotate to some angle and gives oscillations about that angle with the oscillations dying away with time.

The method generally used for automobile windscreen wipers is the mechanical linkage operated by a d.c. motor. A demonstration of its operation and a discussion of wiper systems are given in the Wikipedia website for Windscreen wipers.

24.1.3 Bathroom scales

Consider the design of a simple weighing machine, i.e. bathroom scales. The main requirements are that a person can stand on a platform and the weight of that person will be displayed on some form of readout. The weight should be given with reasonable speed and accuracy and be independent of where on the platform the person stands. Possible solutions can involve:

1 a purely mechanical system based on a spring and gearing;
2 a load cell and a microprocessor/microcontroller system.

One possible solution is to use the weight of the person on the platform to deflect an arrangement of two parallel leaf springs (Figure 24.9(a)). With such an arrangement the deflection is virtually independent of where on the platform the person stands. The deflection can be transformed into movement of a pointer across a scale by using the arrangement shown in Figure 24.9(b). A rack-and-pinion is used to transform the linear motion into a circular motion about a horizontal axis. This is then transformed into a rotation about a vertical axis, and hence movement of a pointer across a scale, by means of a bevel gear.

Figure 24.9 Bathroom scales.

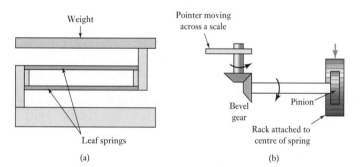

(a)

(b)

Another possible solution involves the use of a microprocessor. The platform can be mounted on load cells employing electrical resistance strain gauges. When the person stands on the platform the gauges suffer strain and change resistance. If the gauges are mounted in a four-active-arm Wheatstone bridge then the out-of-balance voltage output from the bridge is a measure of the weight of the person. This can be amplified by a differential operational amplifier. The resulting analogue signal can then be fed through a latched analogue-to-digital converter for inputting to the microprocessor, e.g. the Motorola 6820. Figure 24.10 shows the input interface. There will also be a need to provide a non-erasable memory and this can be provided by an

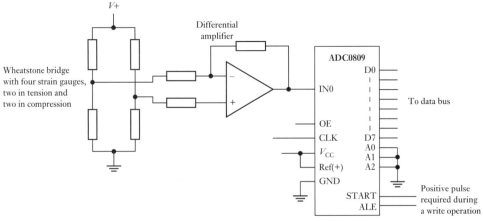

Figure 24.10   Input interface.

erasable and programmable ROM (EPROM) chip, e.g. Motorola 2716. The output to the display can then be taken through a PIA, e.g. Motorola 6821.

However, if a microcontroller is used then memory is present within the single microprocessor chip, and by a suitable choice of microcontroller, e.g. M68HC11, we can obtain analogue-to-digital conversion for the inputs. The system then becomes: strain gauges feeding through an operational amplifier a voltage to the port E (the ADC input) of the microcontroller, with the output passing through suitable drives to output through ports B and C to a decoder and hence an light-emitting diode (LED) display (Figure 24.11).

The program structure might be:

Initialisation by clearing LED displays and memory

Start
    Is someone on the scales? If not display 000
    If yes
        input data
        convert weight data into suitable output form
        output to decoder and LED display
        time delay to retain display
Repeat from start again to get new weight

In considering the design of the mechanical parts of the bathroom scales we need to consider what will happen when someone stands on the scales. We have a spring–damper–mass system comparable with that described in Figure 14.3(a) (see Section 14.2.1) and so can describe its behaviour by

$$m\frac{d^2x}{dt^2} + c\frac{dx}{dt} + kx = F$$

where x is the vertical deflection of the platform when a force F is applied. Taking the Laplace transform gives

$$ms^2\,X(s) + cs\,X(s) + kX(s) = F(s)$$

and so the system can be described by a transfer function of the form

$$G(s) = \frac{X(s)}{F(s)} = \frac{1}{ms^2 + cs + k}$$

Figure 24.11 Bathroom scales.

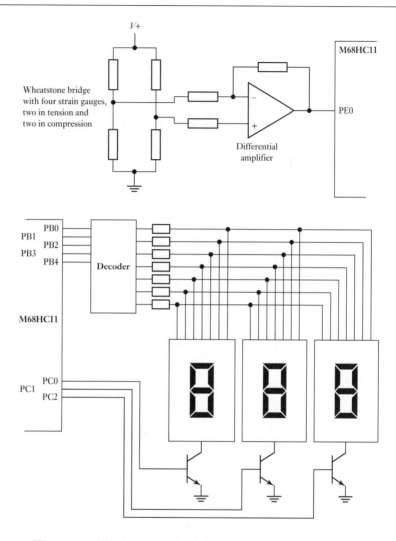

We can consider a person of weight W standing on the platform as a step input and so

$$X(s) = \frac{1}{ms^2 + cs + k} \times \frac{W}{s}$$

The quadratic term is of the form $s^2 + 2\zeta\omega_n s + \omega_n^2$ (see Section 20.3.1) and thus has a natural frequency ω_n of $\sqrt{(k/m)}$ and a damping factor ζ of $c/(2\sqrt{(mk)})$.

When a person stands on the scales he or she wants the scales to indicate quickly the weight value and not oscillate for a long time about the value. If the damping was adjusted to be critical it would take too long to reach the value and so the damping needs to be adjusted to allow some oscillations which are rapidly damped away. We might decide that a 2% settling time t_s (see Section 19.5) of, say, 4 s was desirable. Since $t_s = 4/\zeta\omega_n$ then we require $\zeta\omega_n = 1$ and so $\zeta = \sqrt{(m/k)}$. A simple way of altering the damping is thus by changing the mass.

The above indicates how we can use a mathematical model to predict the behaviour of a system and what factors we can then change to improve its performance.

24.2   Case studies

The following are outlines of examples of mechatronic systems.

24.2.1   A pick-and-place robot

Figure 24.12(a) shows the basic form of a pick-and-place robot unit. The robot has three axes about which motion can occur: rotation in a clockwise or anti-clockwise direction of the unit on its base, arm extension or contraction and arm up or down; also the gripper can open or close. These movements can be actuated by the use of pneumatic cylinders operated by solenoid-controlled valves with limit switches to indicate when a motion is completed. Thus the clockwise rotation of the unit might result from the piston in a cylinder being extended and the anti-clockwise direction by its retraction. Likewise the upward movement of the arm might result from the piston in a linear cylinder being extended and the downward motion from it retracting; the extension of the arm by the piston in another cylinder extending and its return movement by the piston retracting. The gripper can be opened or closed by the piston in a linear cylinder extending or retracting, Figure 24.12(b) showing a basic mechanism that could be used.

Figure 24.12   (a) Pick-and-place, (b) a gripper.

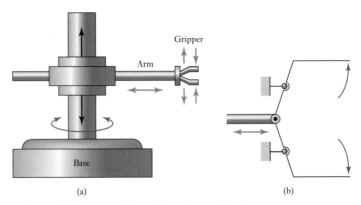

A typical program for such a robot might be:

1   Close an upright gripper on a component hanging from an overhead feeder.
2   Contract the arm so that the component is withdrawn from the feeder.
3   Rotate the arm in a horizontal plane so that it points in the direction of the workpiece.
4   Extend the arm so that the gripper is over the workpiece.
5   Rotate the wrist so that the component hangs downwards from the gripper.
6   Release the gripper so that the component falls into the required position.
7   Rotate the gripper into an upright position.
8   Contract the arm.
9   Rotate the arm to point towards the feeder.

Repeat the sequence for the next component.

Figure 24.13 shows how a microcontroller could be used to control solenoid valves and hence the movements of the robot unit.

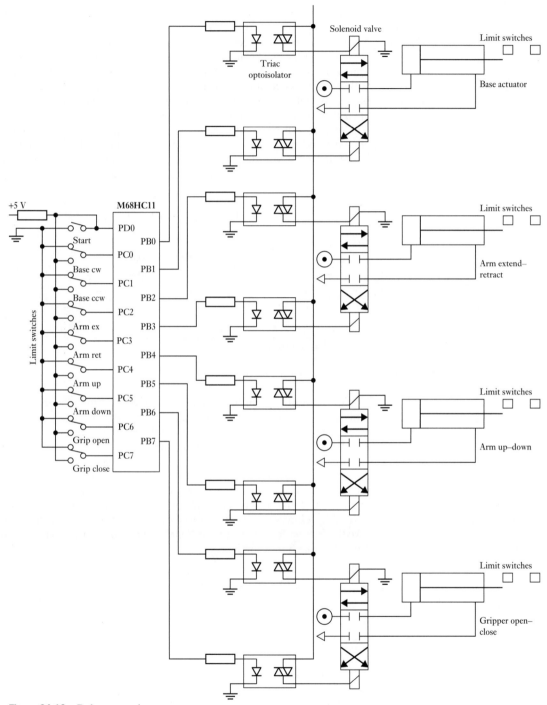

Figure 24.13 Robot control.

Hydraulic and pneumatic rams are widely used to drive robot arms since they can easily be controlled to move limbs at a relatively slow speed, while electric motors would need to operate through a gearbox.

The positions of the arm and gripper in Figure 24.13 are determined by limit switches. This means that only two positions can be accurately attained with each actuator and the positions cannot be readily changed without physically moving the positions of the switches. The arrangement is an open-loop control system. In some applications this may not be a problem.

However, it is more common to use closed-loop control with the positions of an arm and gripper being monitored by sensors and fed back to be compared in the controller with the positions required. When there is a difference from the required positions, the controller operates actuators to reduce the error. The angular position of a joint is often monitored by using an encoder (see Section 2.3.7), this being capable of high precision. Figure 24.14 shows a closed-loop arrangement that might be used for linear motion of a robot arm.

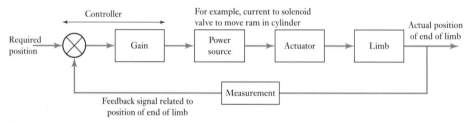

Figure 24.14 Closed-loop control for limb.

The output from the actuator is a force F applied to move the end of the limb. For a set position of y_s and an actual position y, the error signal will be $y_s - y$, assuming for simplicity that the measurement system has a gain of 1. If we consider the controller to have a gain of G_c and G_a to be that of the actuator assembly, then $F = G_c G_a(y_s - y)$. The masses to be accelerated by this force are the mass of the load that the arm is carrying, the mass of the arm and the mass of the moving parts of the actuator. If this is a total mass of m, then Newton's law gives $F = ma$, where the acceleration a can be written as $\mathrm{d}^2 y/\mathrm{d}t^2$. However, this does not take account of friction and since we can take the friction force to be proportional to the velocity, the frictional force is $k\,\mathrm{d}y/\mathrm{d}t$. Thus we can write

$$F = G_c G_a(y_s - y) = m\frac{\mathrm{d}^2 y}{\mathrm{d}t^2} + k\frac{\mathrm{d}y}{\mathrm{d}t}$$

and so

$$y_s = \frac{m}{G_c G_a}\frac{\mathrm{d}^2 y}{\mathrm{d}t^2} + \frac{k}{G_c G_a}\frac{\mathrm{d}y}{\mathrm{d}t} + y$$

This is a second-order differential equation and so the deflection y will be as described in Section 20.3.1 and the form it will take will depend on the damping factor. An under-damped system will have a natural angular frequency ω_n given by

$$\omega_n = \sqrt{\frac{G_c G_a}{m}}$$

This angular frequency will determine how fast the system responds to a change (see Section 19.5): the larger the angular frequency, the faster the system responds (the rise time is inversely proportional to the angular frequency). This means that increasing the controller gain or decreasing the mass can increase the speed of response. The damping factor ζ is given from the differential equation as

$$\zeta = \frac{k}{2\sqrt{G_c G_a m}}$$

The time taken for the oscillations to die away, i.e. the settling time (see Section 19.5), is inversely proportional to the damping factor and so, for example, increasing any part of the mass will result in a decrease in the damping factor and so the oscillations take longer to die away.

24.2.2 Car park barriers

As an illustration of the use of a PLC, consider the coin-operated barriers for a car park. The in-barrier is to open when the correct money is inserted in the collection box and the out-barrier is to open when a car is detected at the car park side of the barrier. Figure 24.15 shows the types of valve systems that can be used to lift and lower the pivoted barriers.

Figure 24.15 System for raising and lowering a barrier.

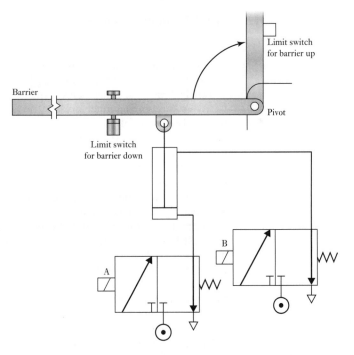

When a current flows through the solenoid of valve A, the piston in a cylinder moves upwards and causes the barrier to rotate about its pivot and raise to let a car through. When the current through the solenoid of valve A ceases, the return spring of the valve results in the valve position changing back to its original position. When the current is switched through the solenoid of valve B, the pressure is applied to lower the barrier. Limit switches are used to detect when the barrier is down and also when fully up.

Figure 24.16 PLC
connections.

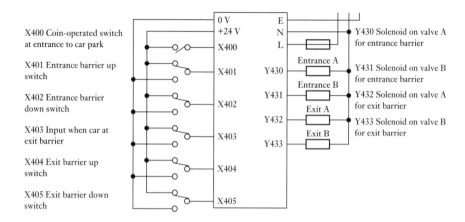

With two of the systems shown in Figure 24.15, one for the entrance barrier and one for the exit barrier, and the connections to PLC inputs and outputs shown in Figure 24.16, the ladder program can be of the form shown in Figure 24.17.

24.2.3 Digital camera

A digital camera is one that captures images and stores them in a digital format in a memory card, unlike the earlier film cameras where the image was stored in an analogue form as a chemical change on film. Figure 24.18 shows the basic elements of a less expensive digital camera.

When the photographer presses the shutter button to its first position, that of being partially depressed, a microcontroller calculates the shutter speed and aperture settings from the input from the metering sensor and displays them on the liquid crystal display (LCD) screen. At the same time, the microcontroller processes the input from the range sensor and sends signals to drive a motor to adjust the focusing of the lens. When the photographer presses the shutter button to its second position, that of being fully depressed, the microcontroller issues signals to change the aperture to that required, open the shutter for the required exposure time, and then, when the shutter has closed, process the image received at the image sensor and store it on the memory card. Also when the shutter button is partially depressed, the automatic focus control system is used to move the lens so that the image will be in focus (see Section 1.7.1 for details of autofocus mechanisms and later in this section for a discussion of the motor used to move the lens).

The light from the object being photographed passes through a lens system and is focused onto an image sensor. This is typically a charge-coupled device (CCD) (see Section 2.10) consisting of an array of many small light-sensitive cells, termed pixels, which are exposed to the light passing through the lens when the electromechanical shutter is opened for a brief interval of time. The light falling on a cell is converted into a small amount of electric charge which, when the exposure has been completed, is read and stored in a register before being processed and stored on the memory card.

The sensors are colour-blind and so, in order that colour photographs can be produced, a colour filter matrix is situated prior to the array of cells.

Figure 24.17 Ladder program.

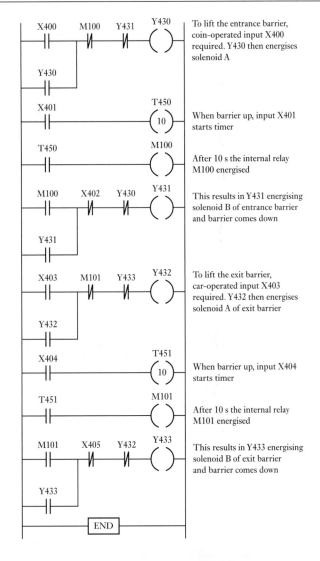

X400 M100 Y431 — Y430
To lift the entrance barrier, coin-operated input X400 required. Y430 then energises solenoid A

Y430

X401 — T450 (10)
When barrier up, input X401 starts timer

T450 — M100
After 10 s the internal relay M100 energised

M100 X402 Y430 — Y431
This results in Y431 energising solenoid B of entrance barrier and barrier comes down

Y431

X403 M101 Y433 — Y432
To lift the exit barrier, car-operated input X403 required. Y432 then energises solenoid A of exit barrier

Y432

X404 — T451 (10)
When barrier up, input X404 starts timer

T451 — M101
After 10 s the internal relay M101 energised

M101 X405 Y432 — Y433
This results in Y433 energising solenoid B of exit barrier and barrier comes down

Y433

END

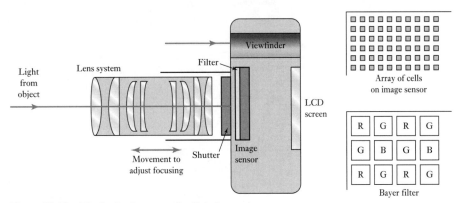

Figure 24.18 The basic elements of a digital camera.

There are separate filters, blue, green or red, for each cell. The most common design for the matrix is the Bayer array. This has the three colours arranged in a pattern so that no two filters of the same colour are next to each other and there are twice as many green filters as either red or blue, this being because green is roughly in the centre of the visible spectrum and gives more detail. The result at this stage is a mosaic of red, green and blue pixels. The files of the results for the pixels at this stage are termed RAW files in that no processing has been done to them. In order to give the full range of colours for a particular pixel, an algorithm is used in which the colour to be allocated to a particular pixel is determined by taking into account the intensities of the colours of neighbouring pixels.

The next stage in processing the signal is to compress the files so that they take up as little memory as possible. This way, more can be stored on a memory card than would be the case with RAW files. Generally, the compressed file format is JPEG, short for Joint Photographic Experts Group. JPEG compression uses the principle that in many photographs, many of the pixels in the same area are identical and so instead of storing the same information for each it can effectively store one and tell the others to just repeat it.

The exposure required is determined by a camera microcontroller in response to the output from a sensor such as a photodiode detecting the intensity of the light. It gives outputs which are used to control the aperture and the shutter speed. The aperture drive system with a digital camera can be a stepper motor which opens or closes a set of diaphragm blades according to the signal received from the microcontroller. The shutter mechanism used with a digital camera is generally of the form shown in Figure 24.19. The shutter involves two sets of curtains, each being controlled by a spring-loaded latch. In the absence of a current to the electromagnet, the spring forces the latch over to a position which has the upper set of curtains down to overlap with the lower set. When a current is passed through an electromagnet it causes the latch to rotate and in doing so lifts the upper set of curtains. The lower set of curtains is initially held down at the bottom by a current through its electromagnet holding the latch. When the current through the lower curtain latch is switched off, the curtains rise. Thus the opening of

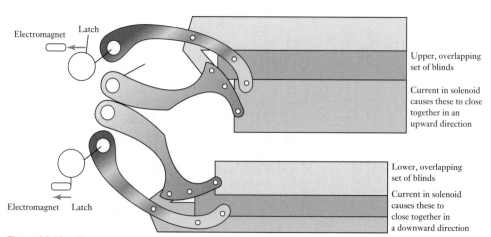

Figure 24.19 Shutter mechanism.

the aperture through to the image sensor is determined by the time between switching a current to the upper latch and switching it off at the lower latch.

The focusing requires a mechanism to move the lens. This is often an ultrasonic motor which consists of a series of piezoelectric elements, such as lead zirconium titanate (PZT). When a current is supplied to such a piezoelectric element it expands or contracts according to the polarity of the current (Figure 24.20(a)). PZT elements are bonded to both sides of a thin strip of spring steel and then, when a potential difference is applied across the strip, the only way the PZT elements can expand or contract is by bending the metal strip (Figure 24.20(b)). When opposite polarity is applied to alternating elements, they are made to bend in opposite directions (Figure 24.20(c)). Thus by using an alternating current with a sequence of such elements round a ring, a displacement wave can be made to travel around the piezoelectric ring of elements in either a clockwise or anti-clockwise direction. The amplitude of this displacement wave is only about 0.001 mm. There is a strip of material with minute cogs attached to the outside of the PZT elements and when the displacement wave moves round the PZT elements they are able to push against the lens mount (Figure 24.20(d)) and thus drive the focusing element.

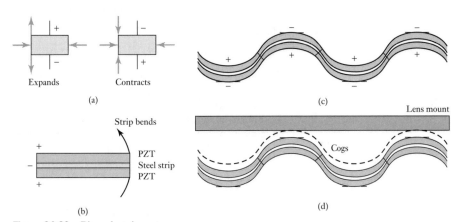

Figure 24.20 Piezoelectric motor.

As an illustration of the use of the modelling techniques discussed in earlier chapters of this book, consider this ultrasonic motor. The torque T generated by the motor is required to rotate the motor ring to some angular position θ. The ring is very light and so we neglect its inertia in comparison with the friction between the rings. Assuming the frictional force is proportional to the angular velocity ω, then $T = c\omega = c\,d\theta/dt$, where c is the friction constant. Integration then gives

$$\theta = \frac{1}{c} \int dt$$

and so a transfer function $G(s)$ of $1/cs$.

The control system for the ultrasonic motor is of the form shown in Figure 24.21. y_n is the nth input pulse and x_n the nth output pulse. With the microprocessor exercising proportional control gain K, the input to it is

Figure 24.21 Control system.

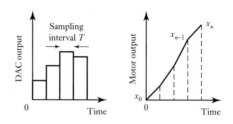

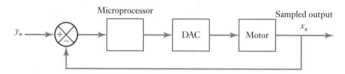

$y_n - x_n$ and the output is $K(y_n - x_n)$. This then passes through the DAC to give an analogue output consisting of a number of steps (Figure 24.21). The motor acts as an integrator and so its output will be $1/c$ times the progressive sum of the areas under the steps (Figure 24.21). Each step has an area of (DAC change in output for the step) × T. Thus

$$x_n - x_{n-1} = (\text{DAC output for } x_{n-1})T/c = K(y_{n-1} - x_{n-1})T/c$$

Hence

$$x_n = [1 - (KT/c)]x_{n-1} + (KT/c)y_{n-1}$$

Suppose we have $K/c = 5$ and a sampling interval of 0.1 s. Thus

$$x_n = 0.5y_{n-1} + 0.5x_{n-1}$$

If there is an input to the control system for the focusing of a sequence of pulses of constant size 1, prior to that there being no input, i.e. $y_0 = 1, y_1 = 1, y_2 = 1, \ldots$, then

$$x_0 = 0$$
$$x_1 = 0.5 \times 0 + 0.5 \times 1 = 0.5$$
$$x_2 = 0.5 \times 0.5 + 0.5 \times 1 = 0.75$$
$$x_3 = 0.5 \times 0.75 + 0.5 \times 1 = 0.875$$
$$x_4 = 0.5 \times 0.875 + 0.5 \times 1 = 0.9375$$
$$x_5 = 0.5 \times 0.9375 + 0.5 \times 1 = 0.968\,75$$
$$x_6 = 0.5 \times 0.96875 + 0.5 \times 1 = 0.984\,375$$
$$x_7 = 0.5 \times 0.984365 + 0.5 \times 1 = 0.992\,187\,5$$

and so on

The output thus takes about seven sampling periods, i.e. 0.7 s, for the focusing to be achieved. This is too long. Suppose, however, we choose values so that $KT/c = 1$. The difference equation then becomes $x_n = y_{n-1}$. Then we have

$$x_0 = 0$$
$$x_1 = 1$$
$$x_2 = 1$$
$$x_3 = 1$$

$\ldots$

This means that the output will reach the required position after just one sample. This is a much faster response. By using a high sampling rate a very fast response can be achieved. This form of response is termed a **deadbeat response**.

24.2.4 Automotive control systems

The modern automobile includes many electronic control systems. These can be grouped into the following categories.

1 **Power train control**. This term is used for the engine and transmission control systems. The engine control unit (ECU) aims to ensure that the engine operates at optimal conditions at all times and includes such items as fuel-injection control, carburettor control, spark-timing control, idle-speed control and anti-knock control. It does this by reading values from many sensors within the engine bay, interpreting the results and then adjusting engine actuators accordingly. Transmission control is primarily involved in automatic transmissions. Often a single engine control unit is used for both engine and transmission control. The engine control unit includes a microcontroller, with the operating software stored in EPROMs or flash memory. Figure 24.22 illustrates some of the basic inputs and outputs for an engine control system.

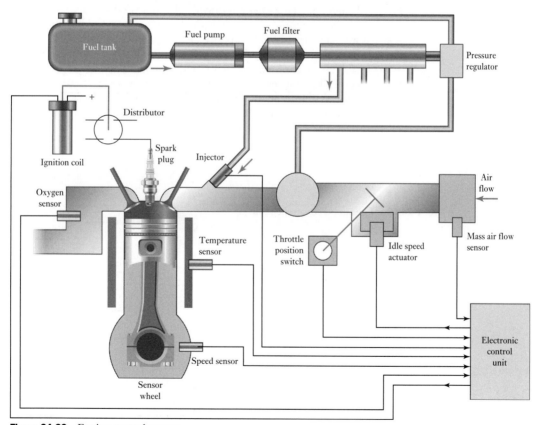

Figure 24.22 Engine control system.

2 **Vehicle control**. This includes suspension control, steering control, cruise control, braking control and traction control.

3 **Person control**. This includes systems such as air conditioning, instrument displays, security systems, communication systems, air-bag systems and rear-obstacle detection.

There are a number of control systems in an automobile, and a network is used to communicate information between them. The standard network used to allow microcontrollers and devices to communicate with each other is the **controller area network (CAN)** (see Section 15.6.4).

The following are discussions of some of the control systems involved in automotive control systems.

The **air–fuel ratio (AFR)** is the mass ratio of air to fuel present in an internal combustion engine. If exactly enough air is provided to completely burn all of the fuel, the ratio is known as the stoichiometric mixture. For gasoline fuel, the stoichiometric air–fuel mixture is about 14.7:1 and so, for every 1 g of fuel, 14.7 g of air are required. The air–fuel equivalence ratio, λ (lambda), is the ratio of actual AFR to that at stoichiometry for a given mixture. Thus if we have $\lambda = 1.0$, the mixture is at stoichiometry, for a rich mixture $\lambda < 1.0$, and for a lean mixture $\lambda > 1.0$. The exhaust gas oxygen sensor (EGO) is thus the key sensor in the engine fuel control feedback loop, producing a voltage output that is related to the amount of oxygen in the exhaust. The zirconium dioxide exhaust gas oxygen sensor is widely used. When hot, the zirconium dioxide element produces a voltage that depends on the amount of oxygen in the exhaust compared to that in the outside air. A stoichiometric field mixture gives an output of about 0.45 V, the voltage ranging from about 0.2 V when lean to 0.8 V when rich (see Figure 24.23).

The control system used with the lambda sensor is basically a PI controller of essentially a first-order engine system. Figure 24.24 shows the basic system model. The engine can be modelled basically as a first-order system (see Section 20.2) and so will have a transfer function G_e of the form:

$$G_e(s) = \frac{K}{\tau s + 1}$$

Figure 24.23   The lambda control circuit.

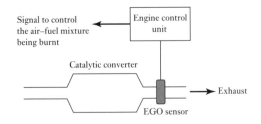

Figure 24.24   The lambda control system.

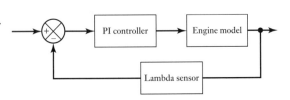

The PI controller will have a transfer function G_c of $K_P + K_I/s$, where K_P is the proportional control constant and K_I the integral control constant (see Section 15.6.1). Thus, taking the lambda sensor to have a transfer function of 1, the closed loop control system will have an overall transfer function $T(s)$ of:

$$T(s) = \frac{G_e(s)G_c(s)}{1 + G_e(s)G_c(s)}$$

This gives a second-order system with a natural frequency and damping ratio. We can then determine the performance of the system with a given input, and taking into account such factors as the rise time and settling time (see Section 19.5).

However, there is a time delay T_L in the lambda sensor responding to a change in oxygen level, typically 50–500 ms, and so to take account of this the transfer function for the engine can be modified by introducing a delay term:

$$G_e(s) = \frac{Ke^{-sT_L}}{\tau s + 1}$$

Another example of a control system used in automobiles is the **anti-lock braking system (ABS)**. The main component of an ABS system consists of two counters which alternately measure the speed of a wheel (Figure 24.25). If the earlier wheel speed exceeds the later wheel speed by a preset value then a skid condition is considered to have occurred. The ABS system then generates an electrical signal which reduces the hydraulic pressure sufficiently to eliminate the locked brake. The wheel speed sensor generally used is a variable reluctance tachogenerator (see Section 2.4.2) consisting of a ferrous toothed wheel which, as it rotates, produces pulses in a pick-up coil. The pulses when counted give a measure of the wheel speed. A magnetic sensor is chosen because, unlike optical sensor systems, it is not affected by the inevitable contamination of the wheel and sensor by mud and water.

Figure 24.25 Anti-lock braking system.

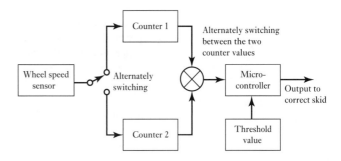

24.2.5 Bar code reader

The familiar scene at the check-out of a supermarket is of the purchases being passed in front of a light beam or a hand-held wand being passed over the goods so that the bar code can be read and the nature of the purchase and hence its price automatically determined. The code consists of a series of black and white bars of varying widths. For example, there is such a bar code on the back of this book.

Figure 24.26 Bar code.

ISBN   0-582-25634-8

Figure 24.26 shows the basic form of the bar code used in the retail trade. The bar code represents a series of numbers. There is a prefix which identifies the coding scheme being used; this is a single digit for the regular Universal Product Coding (UPC) scheme used in the United States and two digits for the European Article Number (EAN) scheme used in Europe. The UPC code uses a 0 prefix for grocery and a 3 for pharmaceuticals. The EAN prefix is from 00 to 09 and is such that the UPC code can be read within the EAN code. This is followed by five digits to represent the manufacturer, each manufacturer having been assigned a unique number. This brings up the centre of the code pattern which is identified by two taller bar patterns. The five-digit number that then follows represents the product. The final number is a check digit which is used to check that the code has been correctly read. A guard pattern of two taller bars at the start and end of the bar pattern is used to frame the bars.

Each number is coded as seven 0 or 1 digits. The codes used on either side of the centre line are different so that the direction of the scan can be determined. To the right the characters have an even number of ones and so even parity and, for UPC, to the left an odd number of ones and so odd parity, the EAN coding for the left being a mixture. Table 24.2 shows the UPC and EAN codings, UPC being the left A coding and the EAN using both left A and left B character codes.

Each 1 is entered as a dark bar and thus the right-hand character 2 would be represented 1101100 and, with the adjacent dark bars run together, it appears as a double-width dark bar followed by a narrow space and then another double-width dark wide bar followed by a double-width space. This

Table 24.2 UPC and EAN codings.

Decimal number	Left A characters	Left B characters	Right characters
0	0001101	0100111	1110010
1	0011001	0110011	1100110
2	0010011	0011011	1101100
3	0111101	0100001	1000010
4	0100011	0011101	0011100
5	0110001	0111001	0001110
6	0101111	0000101	1010000
7	0111011	0010001	1000100
8	0110111	0001001	1001000
9	0001011	0010111	1110100

11 0 11 00

Figure 24.27 Bar code for right-hand 2.

is illustrated in Figure 24.27. The guard pattern at the ends of the code represents 101 and the central band of bars is 01010.

The bar code shown in Figure 24.26 was that for the first edition of this book. It uses the EAN code and has the prefix 97 to identify it as a publication, 80582 to identify the publisher, 25634 to identify the particular book and a check digit of 7. Note that the bar code contains the relevant parts of the ISBN number, this also being a number to identify the publisher and the book concerned.

The following procedure is used for checking the check code digit.

1 Starting at the left, sum all the characters, excluding the check digit, in the odd positions, i.e. first, third, fifth, etc., and then multiply the sum by 3.
2 Starting at the left, sum all the characters in the even positions.
3 Add the results of steps 1 and 2. The check character is the smallest number which when added to this sum produces a multiple of 10.

As an illustration of the use of the check digit, consider the bar code for the book where we have 9780582256347. For the odd characters we have $9 + 8 + 5 + 2 + 5 + 3 = 32$ and when multiplied by 3 we have 96. For the even characters we have $7 + 0 + 8 + 2 + 6 + 4 = 27$. The sum is 123 and thus the check digit should be 7.

Reading the bar code involves determining the widths of the dark and light bands. This can involve a solid-state laser being used to direct an intense, narrow, beam of light at the code and detecting the reflected light by means of a photocell. Usually with the supermarket version the scanner is fixed and a spinning mirror is used to direct the light across the bar code and so scan all the bars. Signal conditioning involves amplification of the output of the photocell using operational amplifiers and then using an operational amplifier circuit as a comparator in order to give a high, i.e. 1, output when a black bar is scanned and a low, i.e. 0, output when a white space is scanned. This sequence of zeros and ones can then be an input to, say, a PIA connected to a Motorola 6800 microprocessor. The overall form of the microprocessor program is outlined below.

1 Initialisation to clear the various memory locations.
2 Recovering the data from the input. This involves continually testing the input to determine whether it is 0 or 1.
3 Processing the data to obtain the characters in binary format. The input is a serial signal consisting of different-duration zeros and ones depending on the width of the spaces of black bars. The microprocessor system is programmed to find the module time width by dividing the time of scan between the end marker bars by the number of modules, a module being a light or dark band to represent a single 0 or 1. The program can then determine whether a dark or light band is a single digit or more than one and hence interpret the scanner signal.
4 Process the binary outcome into a statement of the item purchased and cost.

24.2.6 Hard disk drive

Figure 24.28(a) shows the basic form of a hard disk drive. It consists of a disk coated with a metal layer which can be magnetised. The gap between

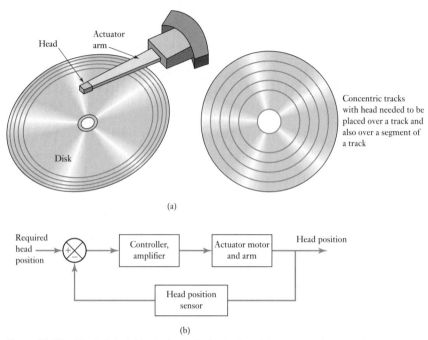

(a)

(b)

Figure 24.28 Hard disk: (a) basic form, (b) basic closed-loop control system for positioning of the read/write head.

the write/read head and the disk surface is very small, about 0.1 μm. The data is stored in the metal layer as a sequence of bit cells (see Section 6.3.1). The disk is rotated by a motor at typically 3600, 5400 or 7200 rev/min and an actuator arm has to be positioned so that the relevant concentric track and relevant part of the track come under the read/write head at the end of that arm. The head is controlled by a closed-loop system (Figure 24.28(b)) in order to position it. Control information is written onto a disk during the formatting process, this enabling each track and sector of track to be identified. The control process then involves the head using this information to go to the required part of the disk.

The actuator movement generally involves a voice coil actuator (Figure 24.29) to rotate the arm. This voice coil actuator is essentially a coil mounted in an iron core so that when a current is passed through the coil it moves, the arrangement being like that of a moving-coil loudspeaker, and so is able to move the actuator arm to position the head over the required track. The head element reads the magnetic field on the disk and provides a feedback signal to the control amplifier.

Figure 24.29 Voice coil actuator.

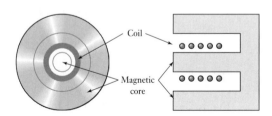

The voice coil actuator is a form of field–controlled permanent magnet d.c. motor and has a transfer function of the same form (see Section 20.5). Thus, since we are concerned with the transfer function relating displacement with time, i.e. the integral of the velocity time function given in Section 20.5, the voice coil actuator has a transfer function of the form

$$G(s) = \frac{k}{s(Ls + R)(Is + c)} = \frac{k/Rc}{s(\tau_L s + 1)(\tau s + 1)}$$

The $(\tau s + 1)$ term is generally close to 1 and so the transfer function approximates to

$$G(s) = \frac{k/Rc}{s(\tau_L s + 1)}$$

Thus the closed-loop control system in Figure 24.28(a), with a control amplifier having a proportional gain of K_a and the head position transfer a gain of 1, might have an overall transfer function giving the relationship between the output signal $X(s)$ and the input required signal $R(s)$ of

$$\frac{X(s)}{R(s)} = \frac{K_a G(s)}{1 + K_a G(S)}$$

Thus if we have, say, $G(s) = 0.25/s(0.05s + 1) = 5/s(s + 20)$ and $K_a = 40$, then

$$X(s) = \frac{200}{s^2 + 20s + 200} R(s)$$

Thus for a unit step input, i.e. $R(s) = 1/s$, the output will be described by

$$X(s) = \frac{200}{s(s^2 + 20s + 200)}$$

The quadratic term is of the form $s^2 + 2\zeta\omega_n s + \omega_n^2$ (see Section 20.3.1) and thus has a natural frequency ω_n of $\sqrt{(200)}$ and a damping factor ζ of $10/\sqrt{(200)}$. Thus we can work out what the response of this second-order system will be to step input signals and how long the system will need to settle down; for example, the 2% settling time (see Section 19.5) is $4/\zeta\omega_n$ and so $4/10 = 0.4$ s. This is rather a long time and so we would need to consider how it can be reduced to perhaps milliseconds. We might consider replacing the amplifier with its proportional gain by one exercising PD control.

24.3 Robotics

The term **robotics** is used for the technology involved in the design, construction, operation and application of robots. A **robot** is a device that can be construed as an intelligent machine. A non-intelligent machine can be termed a manual handling device and would be actuated by an operator. The intelligent machine would not be actuated by a human but by a computer and so operate automatically, not requiring a human to make decisions about what to do next but itself responding to signals from its environment.

The following are some of the key events in the development of robotics.

1922	The Czech author Karel Capek introduced the term *robot* in his play *Universal Robots* involving a factory which made artificial people he called robots. The term robot comes from the Slavic word *robota* which means labour.
1942	The science fiction writer Isaac Asimov in a story called Liar introduced the term robotics and then went on to develop his three laws of robotics.
1956	George Devol and Joe Engelberger established a company called Unimation and developed the first industrial robot arm which they called the Unimate.
1961	A US patent was issued to George Devol for 'Programmed Article Transfer', a basis for the Unimate robot, and the first industrial robot was installed by them. General Motors installed its first robot from Unimation on its production line, using it to sequence and stack hot pieces of diecast metal.
1967	General Motors installed the first spot-welding robots and so made it possible to automatic some 90% of the body welding operations.
1973	The first industrial robot with six electromechanically driven axes was developed by the KUKA Robot graph and named Famulus.
1974	The first microprocessor-controlled industrial robot was developed for ASEA, a mechanical engineering company in Sweden. Its arm movement mimicked that of a human arm. The microprocessor used was an Intel 8-bit.
1978	Unimation developed the Programmable Universal Machine for Assembly (PUMA) for a small parts handling assembly line at General Motors.
1979	Nachi in Japan developed, for spot welding, the first electromotor-driven robots, replacing the earlier hydraulic drive robots.
1986	Honda introduced its first humanoid robot.
1987	The International Federation of Robotics (IFR) was established.

24.3.1 The Three Laws of Robotics

The following **Three Laws** are a set of rules devised by the science fiction author Isaac Asimo in 1942.

1 A robot may not injure a human being or, through inaction, allow a human being to come to harm.
2 A robot must obey the orders given to it by human beings, except where such orders would conflict with the First Law.
3 A robot must protect its own existence as long as such protection does not conflict with the First or Second Law.

Robots do not inherently obey the Three Laws; if the rules are to be obeyed their human creators must choose to program them in. At present, robots are not intelligent enough to have the capacity to understand when they are causing pain or injury and know to stop. However, they can be constructed with physical safeguards such as bumpers, warning beepers, safety cages, etc., to prevent accidents.

Asimov later added a fourth, or zeroth law, to precede the others:

0 A robot may not harm humanity, or, by inaction, allow humanity to come to harm.

24.3.2 Robot components

The following are the basic elements involved in robots.

1 **Manipulator**. This is the main body of the robot, the structure doing the manipulation, and consists of links, joints and other structural elements.
2 **End effector**. This the part connected to the last joint of a manipulator and is concerned with handling objects or making connections to machines. It is the 'hands' of the robot. A common end effector is the gripper, consisting of just two fingers which can open and close to pick up and then let go of objects. The gripper may hold the object by means of friction or encompassing jaws. Pick and place robots used with large objects such as car windscreens often use vacuum grippers, which require a smooth surface to grip in order to ensure suction. The term robotic end effector includes not only robotic grippers but also such devices as robotic tool changers, robotic paint guns, robotic deburring tools, robotic arc welding guns, etc.
3 **Actuators**. These are the means by which the robot moves its joints and links and are rather like the 'muscles' of a robot. The most popular actuators for robots are electric motors that rotate a wheel or gear and linear actuators, powered by compressed air or oil, that cause pistons to move in cylinders.
4 **Sensors**. These are used by the robot to collect information about the environment and about the state of its joints and links. Sensors have been developed that mimic the mechanical properties and touch receptors of human fingers. For example, rotary potentiometers can be used to monitor the angles between joints in a hand and correspond with the equivalent knuckle joint. Force/touch-sensitive sensors can be used on the fingers and palm of the hand to give feedback on when contact has been made and the pressure between an object and the hand. Figure 24.30 shows an example of such a sensor array, one which involves plungers interrupting the light beams between LEDs and sensors. Another method that has been used involves the measurement of the resistance of a conductive elastomer, such as a carbon doped rubber, between two points. The resistance changes with the application of force (Figure 24.31). Such

Figure 24.30 An example of a touch sensor. Touching causes the plunger to interrupt the light beam and so the signal detected by the LED.

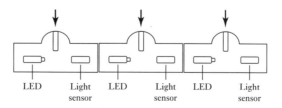

LED   Light   LED   Light   LED   Light
       sensor   sensor   sensor

Figure 24.31 The resistance of the doped elastomer between the contact points changes with the application of force.

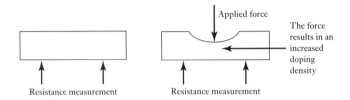

sensors have been developed using elastomer cords laid in a grid pattern with the resistances being measured at the points of intersection. The individual touch sensors detect different information about the size and shape of the object, depending on which ones are actuated. A touch-and-slip sensor can be used on the thumb to indicate when a gripped object is slipping and a tighter grip should be applied. This can be achieved by interpreting the outputs from an array of touch sensors or it may involve a specially designed slip sensor.

5 **Controller**. The mechanical structure of a robot needs to be controlled in order to perform tasks. This involves using sensor information, and processing that information into actuator commands and then controlled action. The 'brain' behind the controller is a computer.

24.3.3 Applications of robots

The following are some of the applications of robots.

1 **Machine loading**. Robots are used to supply other machines with parts and remove processed parts from machines.
2 **Pick and place operations**. The robot is used to pick up parts and place them on perhaps a pallet, or possibly to pick up two parts and assemble them, e.g. putting items in a box, placing parts in or removing them from an oven.
3 **Welding**. The end effector of the robot is a welding gun and is used to weld two parts together.
4 **Painting**. The end effector is a paint gun and is used to paint an assembly or part.
5 **Assembly**. This can involve locating and identifying parts and then assembling them.
6 **Inspection**. This could be inspection of parts using perhaps x-rays, ultrasonics or 'visually'.
7 **Assisting disabled individuals**. This could be via intelligent artificial limbs enabling individuals to carry out daily tasks.
8 **Operating in hazardous or inaccessible locations**. Robots can be designed to operate in environments which would not be practically feasible for a human.

24.3.4 Arduino robot

The Arduino website describes a basic wheeled robot which has been designed to operate with the Arduino control board and motor board (see

Section 10.3.4). Both motor and control boards have a microcontroller. Every element of the robot, i.e. hardware, software and documentation, is freely available and open-source. The robot can be programmed with the Arduino. The processors on the Arduino robot come pre-burned with a **bootloader** that allows new code to be uploaded to it without the use of an external hardware programmer. A simple program to move the robot repeatedly back and forth is as follows. Once you've uploaded it, unplug the USB cable for the robot as, for safety reasons, while the USB is connected the motors are disengaged. When the power is turned on, the robot springs into action.

```
#include <ArduinoRobot.h> // import the robot library
void setup()
{
Robot.begin(); // initialize the library
}
void loop()
{
    Robot.motorsWrite(255,255);// set the speeds of the motors to full speed
    delay(1000);// move for one second
    Robot.motorsWrite(0,0); // stop moving
    delay(1000);// stop for one second
    Robot.motorsWrite(-255,-255);// reverse both motors
    delay(1000); move backwards for one second
    Robot.motorsWrite(0,0); // stop moving
    delay(1000); stop for one second
}
```

A more complex program could involve the use of IR sensors for left, right and ahead movement, so that the robot can be programmed to move through a maze without bumping into walls.

Summary

Mechatronics is a co-ordinated, and concurrently developed, integration of mechanical engineering with electronics and intelligent computer control in the design and manufacture of products. It involves developing an integrated solution rather than a separate discipline approach. In developing solutions, models need to be considered in order to make predictions as to how solutions are likely to function.

Problems

24.1   Present outline solutions of possible designs for the following:

(a) a temperature controller for an oven;

(b) a mechanism for sorting small-, medium- and large-size objects moving along a conveyor belt so that they each are diverted down different chutes for packaging;

(c) an x–y plotter (such a machine plots graphs showing how an input to x varies as the input to y changes).

Research assignments

24.2   Research the anti-lock braking system used in cars and describe the principles of its operation.

24.3   Research the mechanism used in the dot matrix printer and describe the principles of its operation.

24.4   Research the control area network (CAN) protocol used with cars.

Design assignments

24.5   Design a digital thermometer system which will display temperatures between 0 and 99°C. You might like to consider a solution based on the use of a microprocessor with RAM and ROM chips or a microcontroller solution.

24.6   Design a digital ohmmeter which will give a display of the resistance of a resistor connected between its terminals. You might like to consider basing your solution on the use of a monostable multivibrator, e.g. 74121, which will provide an impulse with a width related to the time constant RC of the circuit connected to it.

24.7   Design a digital barometer which will display the atmospheric pressure. You might like to base your solution on the use of the MPX2100AP pressure sensor.

24.8   Design a system which can be used to control the speed of a d.c. motor. You might like to consider using the M68HC11 evaluation board.

24.9   Design a system involving a PLC for the placing on a conveyor belt of boxes in batches of four.

Appendices

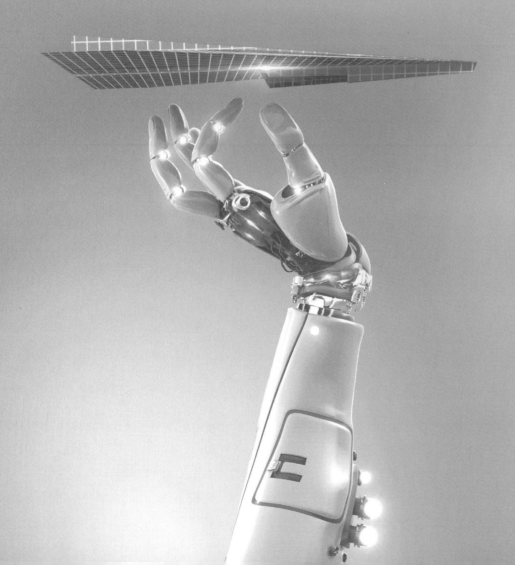

Appendix A: The Laplace transform

Consider a quantity which is a function of time. We can talk of this quantity being in the **time domain** and represent such a function as $f(t)$. In many problems we are only concerned with values of time greater than or equal to 0, i.e. $t \geq 0$. To obtain the Laplace transform of this function we multiply it by e^{-st} and then integrate with respect to time from zero to infinity. Here s is a constant with the unit of 1/time. The result is what we now call the **Laplace transform** and the equation is then said to be in the s-**domain**. Thus the Laplace transform of the function of time $f(t)$, which is written as $\mathcal{L}\{f(t)\}$, is given by

$$\mathcal{L}\{f(t)\} = \int_0^\infty e^{-st} f(t) \, dt$$

The transform is **one-sided** in that values are only considered between 0 and $+\infty$, and not over the full range of time from $-\infty$ to $+\infty$.

We can carry out algebraic manipulations on a quantity in the s-domain, i.e. adding, subtracting, dividing and multiplying, in the normal way we do on any algebraic quantities. We could not have done this on the original function, assuming it to be in the form of a differential equation, when in the time domain. By this means we can obtain a considerably simplified expression in the s-domain. If we want to see how the quantity varies with time in the time domain then we have to carry out the inverse transformation. This involves finding the time domain function that could have given the simplified s-domain expression.

When in the s-domain a function is usually written, since it is a function of s, as $F(s)$. It is usual to use a capital letter F for the Laplace transform and a lower case letter f for the time-varying function $f(t)$. Thus

$$\mathcal{L}\{f(t)\} = F(s)$$

For the inverse operation, when the function of time is obtained from the Laplace transform, we can write

$$f(t) = \mathcal{L}^{-1}\{F(s)\}$$

This equation thus reads as: $f(t)$ is the inverse transform of the Laplace transform $F(s)$.

A.1.1   The Laplace transform from first principles

To illustrate the transformation of a quantity from the time domain into the s-domain, consider a function that has the constant value of 1 for all values

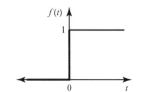

Figure A.1 Unit step function.

of time greater than 0, i.e. $f(t) = 1$ for $t \geq 0$. This describes a **unit step** function and is shown in Figure A.1.

The Laplace transform is then

$$\mathcal{L}\{f(t)\} = F(s) = \int_0^\infty 1e^{-st} \, dt = -\frac{1}{s}[e^{-st}]_0^\infty$$

Since with $t = \infty$ the value of e is 0 and with $t = 0$ the value of e^{-0} is -1, then

$$F(s) = \frac{1}{s}$$

As another example, the following shows the determination, from first principles, of the Laplace transform of the function e^{at}, where a is a constant. The Laplace transform of $f(t) = e^{at}$ is thus

$$F(s) = \int_0^\infty e^{at}e^{-st} \, dt = \int_0^\infty e^{-(s-a)t} \, dt = -\frac{1}{s-a}[e^{-(s-a)t}]_0^\infty$$

When $t = \infty$ the term in the square brackets becomes 0 and when $t = 0$ it becomes -1. Thus

$$F(s) = \frac{1}{s-a}$$

A.2   Unit steps and impulses

Common input functions to systems are the unit step and the impulse. The following indicates how their Laplace transforms are obtained.

A.2.1   The unit step function

Figure A.1 shows a graph of a unit step function. Such a function, when the step occurs at $t = 0$, has the equation

$f(t) = 1$ for all values of t greater than 0
$f(t) = 0$ for all values of t less than 0

The step function describes an abrupt change in some quantity from zero to a steady value, e.g. the change in the voltage applied to a circuit when it is suddenly switched on.

The unit step function thus cannot be described by $f(t) = 1$ since this would imply a function that has the constant value of 1 at all values of t, both positive and negative. The unit step function that switches from 0 to $+1$ at $t = 0$ is conventionally described by the symbol $u(t)$ or $H(t)$, the H being after the originator O. Heaviside. It is thus sometimes referred to as the **Heaviside function**.

The Laplace transform of this step function is, as derived in the previous section,

$$F(s) = \frac{1}{s}$$

The Laplace transform of a step function of height a is

$$F(s) = \frac{a}{s}$$

A.2.2 Impulse function

Consider a rectangular pulse of size $1/k$ that occurs at time $t = 0$ and which has a pulse width of k, i.e. the area of the pulse is 1. Figure A.2(a) shows such a pulse. The pulse can be described as

$$f(t) = \frac{1}{k} \text{ for } 0 \leq t < k$$

$$f(t) = 0 \text{ for } t > k$$

Figure A.2 (a) Rectangular pulse, (b) impulse.

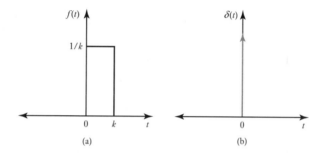

If we maintain this constant pulse area of 1 and then decrease the width of the pulse (i.e. reduce k), the height increases. Thus, in the limit as $k \to 0$ we end up with just a vertical line at $t = 0$, with the height of the graph going off to infinity. The result is a graph that is zero except at a single point where there is an infinite spike (Figure A.2(b)). Such a graph can be used to represent an impulse. The impulse is said to be a unit impulse because the area enclosed by it is 1. This function is represented by $\delta(t)$, the **unit impulse function** or the **Dirac delta function**.

The Laplace transform for the unit area rectangular pulse in Figure A.2(a) is given by

$$F(s) = \int_0^\infty f(t)\mathrm{e}^{-st}\,\mathrm{d}t = \int_0^k \frac{1}{k}\mathrm{e}^{-st}\,\mathrm{d}t + \int_k^\infty 0\,\mathrm{e}^{-st}\mathrm{d}t$$

$$= \left[-\frac{1}{sk}\mathrm{e}^{-st}\right]_0^k = -\frac{1}{sk}(\mathrm{e}^{-sk} - 1)$$

To obtain the Laplace transform for the unit impulse we need to find the value of the above in the limit as $k \to 0$. We can do this by expanding the exponential term as a series. Thus

$$\mathrm{e}^{-sk} = 1 - sk + \frac{(-sk)^2}{2!} + \frac{(-sk)^3}{3!} + \cdots$$

and so we can write

$$F(s) = 1 - \frac{sk}{2!} + \frac{(sk)^2}{3!} + \cdots$$

Thus in the limit as $k \to 0$ the Laplace transform tends to the value 1:

$$\mathcal{L}\{\delta(t)\} = 1$$

Since the area of the above impulse is 1 we can define the size of such an impulse as being 1. Thus the above equation gives the Laplace transform for a unit impulse. An impulse of size a is represented by $a\delta(t)$ and the Laplace transform is

$$\mathcal{L}\{a\delta(t)\} = a$$

A.3	**Standard Laplace transforms**

In determining the Laplace transforms of functions it is not usually necessary to evaluate integrals since tables are available that give the Laplace transforms of commonly occurring functions. These, when combined with a knowledge of the properties of such transforms (see the next section), enable most commonly encountered problems to be tackled. Table A.1 lists some of the commoner time functions and their Laplace transforms. Note that in the table $f(t) = 0$ for all negative values of t and the $u(t)$ terms have been omitted from most of the time functions and have to be assumed.

Table A.1   Laplace transforms.

Time function $f(t)$	Laplace transform $F(s)$
1 $\delta(t)$, unit impulse	1
2 $\delta(t - T)$, delayed unit impulse	e^{-sT}
3 $u(t)$, a unit step	$\dfrac{1}{s}$
4 $u(t - T)$, a delayed unit step	$\dfrac{e^{-sT}}{s}$
5 t, a unit ramp	$\dfrac{1}{s^2}$
6 t^n, nth-order ramp	$\dfrac{n!}{s^{n+1}}$
7 e^{-at}, exponential decay	$\dfrac{1}{s + a}$
8 $1 - e^{-at}$, exponential growth	$\dfrac{a}{s(s + a)}$
9 te^{-at}	$\dfrac{1}{(s + a)^2}$
10 $t^n e^{-at}$	$\dfrac{n!}{(s + a)^{n+1}}$

(Continued)

Table A.1   (*Continued*)

Time function $f(t)$	Laplace transform $F(s)$
11 $t - \dfrac{1 - \mathrm{e}^{-at}}{a}$	$\dfrac{a}{s^2(s + a)}$
12 $\mathrm{e}^{-at} - \mathrm{e}^{-bt}$	$\dfrac{b - a}{(s + a)(s + b)}$
13 $(1 - at)\mathrm{e}^{-at}$	$\dfrac{s}{(s + a)^2}$
14 $1 - \dfrac{b}{b - a}\mathrm{e}^{-at} + \dfrac{a}{b - a}\mathrm{e}^{-bt}$	$\dfrac{ab}{s(s + a)(s + b)}$
15 $\dfrac{\mathrm{e}^{-at}}{(b - a)(c - a)} + \dfrac{\mathrm{e}^{-bt}}{(c - a)(a - b)} + \dfrac{\mathrm{e}^{-ct}}{(a - c)(b - c)}$	$\dfrac{1}{(s + a)(s + b)(s + c)}$
16 $\sin \omega t$, a sine wave	$\dfrac{\omega}{s^2 + \omega^2}$
17 $\cos \omega t$, a cosine wave	$\dfrac{s}{s^2 + \omega^2}$
18 $\mathrm{e}^{-at} \sin \omega t$, a damped sine wave	$\dfrac{\omega}{(s + a)^2 + \omega^2}$
19 $\mathrm{e}^{-at} \cos \omega t$, a damped cosine wave	$\dfrac{s + a}{(s + a)^2 + \omega^2}$
20 $1 - \cos \omega t$	$\dfrac{\omega^2}{s(s^2 + \omega^2)}$
21 $t \cos \omega t$	$\dfrac{s^2 - \omega^2}{(s^2 + \omega^2)^2}$
22 $t \sin \omega t$	$\dfrac{2\omega s}{(s^2 + \omega^2)^2}$
23 $\sin(\omega t + \theta)$	$\dfrac{\omega \cos \theta + s \sin \theta}{s^2 + \omega^2}$
24 $\cos(\omega t + \theta)$	$\dfrac{s \cos \theta - \omega \sin \theta}{s^2 + \omega^2}$
25 $\dfrac{\omega}{\sqrt{1 - \zeta^2}}\mathrm{e}^{-\zeta \omega t} \sin \omega \sqrt{1 - \zeta^2}\, t$	$\dfrac{\omega^2}{s^2 + 2\zeta \omega s + \omega^2}$
26 $1 - \dfrac{1}{\sqrt{1 - \zeta^2}}\mathrm{e}^{-\zeta \omega t} \sin(\omega \sqrt{1 - \zeta^2}\, t + \phi),\ \cos \phi = \zeta$	$\dfrac{\omega^2}{s(s^2 + 2\zeta \omega s + \omega^2)}$

A.3.1   Properties of Laplace transforms

In this section the basic properties of the Laplace transform are outlined. These properties enable the table of standard Laplace transforms to be used in a wide range of situations.

Linearity property

If two separate time functions, e.g. $f(t)$ and $g(t)$, have Laplace transforms then the transform of the sum of the time functions is the sum of the two separate Laplace transforms:

$$\mathcal{L}\{af(t) + bg(t)\} = a\mathcal{L}f(t) + b\mathcal{L}g(t)$$

a and b are constants. Thus, for example, the Laplace transform of $1 + 2t + 4t^2$ is given by the sum of the transforms of the individual terms in the expression. Thus, using items 1, 5 and 6 in Table A.1,

$$F(s) = \frac{1}{s} + \frac{2}{s^2} + \frac{8}{s^3}$$

The s-domain shifting property

This property is used to determine the Laplace transform of functions that have an exponential factor and is sometimes referred to as the **first shifting property**. If $F(s) = \mathcal{L}\{f(t)\}$ then

$$\mathcal{L}\{e^{at}f(t)\} = F(s - a)$$

For example, the Laplace transform of $e^{at}t^n$ is, since the Laplace transform of t^n is given by item 6 in Table A.1 as $n!/s^{n+1}$, given by

$$\mathcal{L}\{e^{at}t^n\} = \frac{n!}{(s - a)^{n+1}}$$

Time domain shifting property

If a signal is delayed by a time T then its Laplace transform is multiplied by e^{-sT}. If $F(s)$ is the Laplace transform of $f(t)$ then

$$\mathcal{L}\{f(t - T)u(t - T)\} = e^{-st}F(s)$$

This delaying of a signal by a time T is referred to as the **second shift theorem**.

The time domain shifting property can be applied to all Laplace transforms. Thus for an impulse $\delta(t)$ which is delayed by a time T to give the function $\delta(t - T)$, the Laplace transform of $\delta(t)$, namely 1, is multiplied by e^{-sT} to give $1e^{-sT}$ as the transform for the delayed function.

Periodic functions

For a function $f(t)$ which is a periodic function of period T, the Laplace transform of that function is

$$\mathcal{L}f(t) = \frac{1}{1 - e^{-sT}} F_1(s)$$

where $F_1(s)$ is the Laplace transform of the function for the first period. Thus, for example, consider the Laplace transform of a sequence of periodic rectangular pulses of period T, as shown in Figure A.3. The Laplace transform

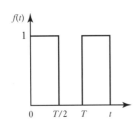

Figure A.3 Rectangular pulses.

of a single rectangular pulse is given by $(1/s)(1 - e^{-sT/2})$. Hence, using the above equation, the Laplace transform is

$$\frac{1}{1 - e^{-sT}} \times \frac{1}{s}(1 - e^{-sT/2}) = \frac{1}{s(1 + e^{-sT/2})}$$

Initial- and final-value theorems

The initial-value theorem can be stated as: if a function of time $f(t)$ has a Laplace transform $F(s)$ then in the limit as the time tends to zero the value of the function is given by

$$\lim_{t \to 0} f(t) = \lim_{s \to \infty} sF(s)$$

For example, the initial value of the function giving the Laplace transform $3/(s + 2)$ is the limiting value of $3s/(s + 2) = 3/(1 + 2/s)$ as s tends to infinity and so is 3.

The final-value theorem can be stated as: if a function of time $f(t)$ has a Laplace transform $F(s)$ then in the limit as the time tends to infinity the value of the function is given by

$$\lim_{t \to \infty} f(t) = \lim_{s \to 0} sF(s)$$

Derivatives

The Laplace transform of a derivative of a function $f(t)$ is given by

$$\mathcal{L}\left\{\frac{d}{dt}f(t)\right\} = sF(s) - f(0)$$

where $f(0)$ is the value of the function when $t = 0$. For example, the Laplace transform of $2(dx/dt) + x = 4$ is $2[sX(s) - x(0)] + X(s) = 4/s$ and if we have $x = 0$ at $t = 0$ then it is $2sX(s) + X(s) = 4/s$ or $X(s) = 4/[s(2s + 1)]$.

For a second derivative

$$\mathcal{L}\left\{\frac{d^2}{dt^2}f(t)\right\} = s^2F(s) - sf(0) - \frac{d}{dt}f(0)$$

where $df(0)/dt$ is the value of the first derivative at $t = 0$.

Integrals

The Laplace transform of the integral of a function $f(t)$ which has a Laplace transform $F(s)$ is given by

$$\mathcal{L}\left\{\int_0^t f(t)\, dt\right\} = \frac{1}{s}F(s)$$

For example, the Laplace transform of the integral of the function e^{-t} between the limits 0 and t is given by

$$\mathcal{L}\left\{\int_0^t e^{-t}\, dt\right\} = \frac{1}{s}\mathcal{L}\{e^{-t}\} = \frac{1}{s(s + 1)}$$

The inverse Laplace transformation is the conversion of a Laplace transform $F(s)$ into a function of time $f(t)$. This operation can be written as

$$\mathcal{L}^{-1}\{F(s)\} = f(t)$$

The inverse operation can generally be carried out by using Table A.1. The linearity property of Laplace transforms means that if we have a transform as the sum of two separate terms then we can take the inverse of each separately and the sum of the two inverse transforms is the required inverse transform:

$$\mathcal{L}^{-1}\{aF(s) + bG(s)\} = a\mathcal{L}^{-1}F(s) + b\mathcal{L}^{-1}G(s)$$

Thus, to illustrate how rearrangement of a function can often put it into the standard form shown in the table, the inverse transform of $3/(2s + 1)$ can be obtained by rearranging it as

$$\frac{3(1/2)}{s + (1/2)}$$

The table (item 7) contains the transform $1/(s + a)$ with the inverse of e^{-at}. Thus the inverse transformation is just this multiplied by the constant $(3/2)$ with $a = (1/2)$, i.e. $(3/2)\, e^{-t/2}$.

As another example, consider the inverse Laplace transform of $(2s + 2)/(s^2 + 1)$. This expression can be rearranged as

$$2\left[\frac{s}{s^2 + 1} + \frac{1}{s^2 + 1}\right]$$

The first term in the square brackets has the inverse transform of $\cos t$ (item 17 in Table A.1) and the second term $\sin t$ (item 16 in Table A.1). Thus the inverse transform of the expression is $2 \cos t + 2 \sin t$.

A.4.1   Partial fractions

Often $F(s)$ is a ratio of two polynomials and cannot be readily identified with a standard transform in Table A.1. It has to be converted into simple fraction terms before the standard transforms can be used. The process of converting an expression into simple fraction terms is called decomposing into **partial fractions**. This technique can be used provided the degree of the numerator is less than the degree of the denominator. The degree of a polynomial is the highest power of s in the expression. When the degree of the numerator is equal to or higher than that of the denominator, the denominator must be divided into the numerator until the result is the sum of terms with the remainder fractional term having a numerator of lower degree than the denominator.

We can consider there to be basically three types of partial fractions.

1   The denominator contains factors which are only of the form $(s + a)$, $(s + b)$, $(s + c)$, etc. The expression is of the form

$$\frac{f(s)}{(s + a)(s + b)(s + c)}$$

and has the partial fractions of

$$\frac{A}{(s + a)} + \frac{B}{(s + b)} + \frac{C}{(s + c)}$$

2   There are repeated $(s + a)$ factors in the denominator, i.e. the denominator contains powers of such a factor, and the expression is of the form

$$\frac{f(s)}{(s + a)^n}$$

This then has partial fractions of

$$\frac{A}{(s + a)^1} + \frac{B}{(s + a)^2} + \frac{C}{(s + a)^3} + \cdots + \frac{N}{(s + a)^n}$$

3   The denominator contains quadratic factors and the quadratic does not factorise without imaginary terms. For an expression of the form

$$\frac{f(s)}{(as^2 + bs + c)(s + d)}$$

the partial fractions are

$$\frac{As + B}{as^2 + bs + c} + \frac{C}{s + d}$$

The values of the constants A, B, C, etc., can be found by making use of the fact either that the equality between the expression and the partial fractions must be true for all values of s or that the coefficients of s^n in the expression must equal those of s^n in the partial fraction expansion. The use of the first method is illustrated by the following example where the partial fractions of

$$\frac{3s + 4}{(s + 1)(s + 2)}$$

are

$$\frac{A}{s + 1} + \frac{B}{s + 2}$$

Then, for the expressions to be equal, we must have

$$\frac{3s + 4}{(s + 1)(s + 2)} = \frac{A(s + 2) + B(s + 1)}{(s + 1)(s + 2)}$$

and consequently $3s + 4 = A(s + 2) + B(s + 1)$. This must be true for all values of s. The procedure is then to pick values of s that will enable some of the terms involving constants to become zero and so enable other constants to be determined. Thus if we let $s = -2$ then we have $3(-2) + 4 = A(-2 + 2) + B(-2 + 1)$ and so $B = 2$. If we now let $s = -1$ then $3(-1) + 4 = A(-1 + 2) + B(-1 + 1)$ and so $A = 1$. Thus

$$\frac{3s + 4}{(s + 1)(s + 2)} = \frac{1}{s + 1} + \frac{2}{s + 2}$$

Problems

A.1   Determine the Laplace transforms of (a) $2t$, (b) $\sin 2t$, (c) a unit impulse at time $t = 2$ s, (d) $4\, dx/dt$ when $x = 2$ at $t = 0$, (e) $3\, d^2x/dt^2$ when $x = 0$ and $dx/dt = 0$ at $t = 0$, (f) the integral between t and 0 of e^{-t}.

A.2   Determine the inverses of the Laplace transforms (a) $1/s^2$, (b) $5s/(s^2 + 9)$, (c) $(3s - 1)/[s(s - 1)]$, (d) $1/(s + 3)$.

A.3   Determine the initial value of the function with the Laplace transform $5/(s + 2)$.

Appendix B: Number systems

The **decimal system** is based on the use of 10 symbols or digits: 0, 1, 2, 3, 4, 5, 6, 7, 8, 9. When a number is represented by this system, the digit position in the number indicates that the weight attached to each digit increases by a factor of 10 as we proceed from right to left:

$$\ldots \quad 10^3 \qquad 10^2 \qquad 10^1 \qquad 10^0$$

thousands   hundreds   tens   units

The **binary system** is based on just two symbols or states: 0 and 1. These are termed *bi*nary dig*its* or **bits**. When a number is represented by this system, the digit position in the number indicates that the weight attached to each digit increases by a factor of 2 as we proceed from right to left:

$$\ldots \quad 2^3 \qquad 2^2 \qquad 2^1 \qquad 2^0$$

bit 3   bit 2   bit 1   bit 0

For example, the decimal number 15 in the binary system is 1111. In a binary number the bit 0 is termed the **least significant bit** (LSB) and the highest bit the **most significant bit** (MSB).

The **octal system** is based on eight digits: 0, 1, 2, 3, 4, 5, 6, 7. When a number is represented by this system, the digit position in the number indicates that the weight attached to each digit increases by a factor of 8 as we proceed from right to left:

$$\ldots \quad 8^3 \qquad 8^2 \qquad 8^1 \qquad 8^0$$

For example, the decimal number 15 in the octal system is 17.

The **hexadecimal system** is based on 16 digits/symbols: 0, 1, 2, 3, 4, 5, 6, 7, 8, 9, A, B, C, D, E, F. When a number is represented by this system, the digit position in the number indicates that the weight attached to each digit increases by a factor of 16 as we proceed from right to left:

$$\ldots \quad 16^3 \qquad 16^2 \qquad 16^1 \qquad 16^0$$

For example, the decimal number 15 is F in the hexadecimal system. This system is generally used in the writing of programs for microprocessor-based systems since it represents a very compact method of entering data.

The **Binary-Coded Decimal system** (BCD system) is a widely used system with computers. Each decimal digit is coded separately in binary. For example, the decimal number 15 in BCD is 0001 0101. This code is useful for outputs from microprocessor-based systems where the output has to drive decimal displays, each decimal digit in the display being supplied by the microprocessor with its own binary code.

Table B.1   Number systems.

Decimal	Binary	BCD	Octal	Hexadecimal
0	0000	0000 0000	0	0
1	0001	0000 0001	1	1
2	0010	0000 0010	2	2
3	0011	0000 0011	3	3
4	0100	0000 0100	4	4
5	0101	0000 0101	5	5
6	0110	0000 0110	6	6
7	0111	0000 0111	7	7
8	1000	0000 1000	10	8
9	1001	0000 1001	11	9
10	1010	0001 0000	12	A
11	1011	0001 0001	13	B
12	1100	0001 0010	14	C
13	1101	0001 0011	15	D
14	1110	0001 0100	16	E
15	1111	0001 0101	17	F

Table B.1 gives examples of numbers in the decimal, binary, BCD, octal and hexadecimal systems.

B.2   Binary mathematics

Addition of binary numbers follows the following rules:

$$0 + 0 = 0$$
$$0 + 1 = 1 + 0 = 1$$
$$1 + 1 = 10 \qquad \text{i.e. } 0 + \text{carry } 1$$
$$1 + 1 + 1 = 11 \qquad \text{i.e. } 1 + \text{carry } 1$$

In decimal numbers the addition of 14 and 19 gives 33. In binary numbers this addition becomes

Augend	01110
Addend	10111
Sum	100001

For bit 0, $0 + 1 = 1$. For bit 1, $1 + 1 = 10$ and so we have 0 with 1 carried to the next column. For bit 3, $1 + 0 + \text{carried } 1 = 10$. For bit 4, $1 + 0 + \text{carried } 1 = 10$. We continue this through the various bits and end up with the sum plus a carry 1. The final number is thus 100001. When adding binary numbers A and B to give C, i.e. $A + B = C$, then A is termed the **augend**, B the **addend** and C the **sum**.

Subtraction of binary numbers follows the following rules:

$$0 - 0 = 0$$
$$1 - 0 = 1$$
$$1 - 1 = 0$$
$$0 - 1 = 10 - 1 + \text{borrow} = 1 + \text{borrow}$$

When evaluating $0 - 1$, a 1 is borrowed from the next column on the left containing a 1. The following example illustrates this. In decimal numbers the subtraction of 14 from 27 gives 13.

Minuend	11011
Subtrahend	01110
Difference	01101

For bit 0 we have $1 - 0 = 1$. For bit 1 we have $1 - 1 = 0$. For bit 2 we have $0 - 1$. We thus borrow 1 from the next column and so have $10 - 1 = 1$. For bit 3 we have $0 - 1$; remember that we borrowed the 1. Again borrowing 1 from the next column, we then have $10 - 1 = 1$. For bit 4 we have $0 - 0 = 0$; remember that we borrowed the 1. When subtracting binary numbers A and B to give C, i.e. we have $A - B = C$, then A is termed the **minuend**, B the **subtrahend** and C the **difference**.

The subtraction of binary numbers is more easily carried out electronically when an alternative method of subtraction is used. The subtraction example above can be considered to be the addition of a positive number and a negative number. The following techniques indicate how we can specify negative numbers and so turn subtraction into addition. It also enables us to deal with negative numbers in any circumstances.

The numbers used so far are referred to as **unsigned**. This is because the number itself contains no indication whether it is negative or positive. A number is said to be **signed** when the most significant bit is used to indicate the sign of the number, a 0 being used if the number is positive and a 1 if it is negative. When we have a positive number then we write it in the normal way with a 0 preceding it. Thus a positive binary number of 10010 would be written as 010010. A negative number of 10010 would be written as 110010. However, this is not the most useful way of representing negative numbers for ease of manipulation by computers.

A more useful way of representing negative numbers is to use the two's complement method. A binary number has two complements, known as the **one's complement** and the **two's complement**. The one's complement of a binary number is obtained by changing all the ones in the unsigned number into zeros and the zeros into ones. The two's complement is then obtained by adding 1 to the one's complement. When we have a negative number then we obtain the two's complement and then sign it with a 1, the positive number being signed by a 0. Consider the representation of the decimal number -3 as a signed two's complement number. We first write the binary number for the unsigned 3 as 0011, then obtain the one's complement of 1100, add 1 to give the unsigned two's complement of 1101, and finally sign it with a 1 to indicate it is negative. The result is thus 11101. The following is another example, the signed two's complement being obtained as an 8-bit number for -6:

Unsigned binary number	000 0110
One's complement	111 1001
Add 1	1
Unsigned two's complement	111 1010
Signed two's complement	1111 1010

Table B.2 Signed numbers.

Denary number	Signed number		Denary number	Signed number	
+127	0111 1111	Just the binary	−1	1111 1111	The two's
...		number signed	−2	1111 1110	complement
+6	0000 0110	with a 0	−3	1111 1101	signed with a 1
+5	0000 0101		−4	1111 1100	
+4	0000 0101		−5	1111 1011	
+3	0000 0011		−6	1111 1010	
+2	0000 0010		...		
+1	0000 0001		−127	1000 0000	
+0	0000 0000				

When we have a positive number then we write it in the normal way with a 0 preceding it. Thus a positive binary number of 100 1001 would be written as 01001001. Table B.2 shows some examples of numbers on this system.

Subtraction of a positive number from a positive number involves obtaining the signed two's complement of the subtrahend and then adding it to the signed minuend. Hence, for the subtraction of the decimal number 6 from the decimal number 4 we have

Signed minuend	0000 0100
Subtrahend, signed two's complement	1111 1010
Sum	1111 1110

The most significant bit of the outcome is 1 and so the result is negative. This is the signed two's complement for −2.

Consider another example, the subtraction of 43 from 57. The signed positive number of 57 is 00111001. The signed two's complement for −43 is given by

Unsigned binary number for 43	010 1011
One's complement	101 0100
Add 1	1
Unsigned two's complement	101 0101
Signed two's complement	1101 0101

Thus we obtain by the addition of the signed positive number and the signed two's complement number

Signed minuend	0011 1001
Subtrahend, signed two's complement	1101 0101
Sum	0000 1110 + carry 1

The carry 1 is ignored. The result is thus 0000 1110 and since the most significant bit is 0, the result is positive. The result is the decimal number 14.

If we wanted to add two negative numbers then we would obtain the signed two's complement for each number and then add them. Whenever a number is negative, we use the signed two's complement; when positive, just the signed number.

| **B.3** | **Floating numbers** |

In the decimal number system, large numbers such as 120 000 are often written in **scientific notation** as 1.2×10^5 or perhaps 120×10^3 and small numbers such as 0.000 120 as 1.2×10^{-4} rather than as a number with a fixed location for the decimal point. Numbers in this form of notation are written in terms of 10 raised to some power. Likewise we can use such notation for binary numbers but with them written in terms of 2 raised to some power. For example, we might have 1010 written as 1.010×2^3 or perhaps 10.10×2^2. Because the binary point can be moved to different locations by a choice of the power to which the 2 is raised, this notation is termed **floating point**.

A floating-point number is in the form $a \times r^e$, where a is termed the **mantissa**, r the **radix** or **base** and e the **exponent** or **power**. With binary numbers the base is understood to be 2, i.e. we have $a \times 2^e$. The advantage of using floating-point numbers is that, compared with fixed-point representation, a much wider range of numbers can be represented by a given number of digits.

Because with floating-point numbers it is possible to store a number in several different ways, e.g. 0.1×10^2 and 0.01×10^3, with computing systems such numbers are **normalised**, i.e. they are all put in the form $0.1 \times r^e$. Hence, with binary numbers we have 0.1×2^e and so if we had 0.00001001 it would become 0.1001×2^{-4}. In order to take account of the sign of a binary number we then add a sign bit of 0 for a positive number and 1 for a negative number. Thus the number 0.1001×2^{-4} becomes 1.1001×2^{-4} if negative and 0.1001×2^{-4} if positive.

If we want to add 2.01×10^3 and 10.2×10^2 we have to make the power (the term exponent is generally used) the same for each. Thus we can write $2.01 \times 10^3 + 1.02 \times 10^3$. We can then add them digit by digit, taking account of any carry, to give 2.03×10^3. We adopt a similar procedure for binary floating-point numbers. Thus if we want to add 0.101100×2^4 and 0.111100×2^2 we first adjust them to have the same exponents, e.g. 0.101100×2^4 and 0.001111×2^4, and then add them digit by digit to give 0.111011×2^4.

Likewise for subtraction, digit-by-digit subtraction of floating-point numbers can only occur between two numbers when they have the same exponent. Thus 0.1101100×2^{-4} minus 0.1010100×2^{-5} can be written as $0.01010100 \times 2^{-4} - 0.101010 \times 2^{-4}$ and the result given as 0.1000010×2^{-4}.

| **B.4** | **Gray code** |

Consider two successive numbers in binary code 0001 and 0010 (denary 2 and 3); 2 bits have changed in the code group in going from one number to the next. Thus if we had, say, an absolute encoder (see Section 2.3.7) and assigned successive positions to successive binary numbers then two changes have to be made in this case. This can present problems in that both changes must be made at exactly the same instant; if one occurs fractionally before the other then there can momentarily be another number indicated. Thus in going from 0001 to 0010 we might momentarily have 0011 or 0000. Thus an alternative method of coding is likely to be used.

The **Gray code** is such a code: only 1 bit in the code group changes in going from one number to the next. The Gray code is unweighted in that

Table B.3   Gray code.

Decimal number	Binary code	Gray code	Decimal number	Binary code	Gray code
0	0000	0000	8	1000	1100
1	0001	0001	9	1001	1101
2	0010	0011	10	1010	1111
3	0011	0010	11	1011	1110
4	0100	0110	12	1100	1010
5	0101	0111	13	1101	1011
6	0110	0101	14	1110	1001
7	0111	0100	15	1111	1000

the bit positions in the code group do not have any specific weight assigned to them. It is thus not suited to arithmetic operations but is widely used for input/output devices such as absolute encoders. Table B.3 lists decimal numbers and their values in the binary code and in Gray code.

Problems

B.1   What is the largest decimal number that can be represented by the use of an 8-bit binary number?

B.2   Convert the following binary numbers to decimal numbers: (a) 1011, (b) 10 0001 0001.

B.3   Convert the decimal numbers (a) 423, (b) 529 to hex.

B.4   Convert the BCD numbers (a) 0111 1000 0001, (b) 0001 0101 011 1 to decimal.

B.5   What are the two's complement representations of the decimal numbers (a) −90, (b) −35?

B.6   What even-parity bits should be attached to (a) 100 1000, (b) 100 1111?

B.7   Subtract the following decimal numbers using two's complements: (a) 21 − 13, (b) 15 − 3.

Appendix C: Boolean algebra

C.1

Laws of Boolean algebra

Boolean algebra involves the binary digits 1 and 0 and the operations $\cdot$, $+$ and the inverse. The laws of this algebra are outlined below.

1. Anything ORed with itself is equal to itself: $A + A = A$.
2. Anything ANDed with itself is equal to itself: $A \cdot A = A$.
3. It does not matter in which order we consider inputs for OR and AND gates, e.g.

$$A + B = B + A \text{ and } A \cdot B = B \cdot A$$

4. As the following truth table indicates:

$$A + (B \cdot C) = (A + B) \cdot (A + C)$$

A	B	C	$B \cdot C$	$A + B \cdot C$	$A + B$	$A + C$	$(A + B) \cdot (A + C)$
0	0	0	0	0	0	0	0
0	0	1	0	0	0	1	0
0	1	0	0	0	1	0	0
0	1	1	1	1	1	1	1
1	0	0	0	1	1	1	1
1	0	1	0	1	1	1	1
1	1	0	0	1	1	1	1
1	1	1	1	1	1	1	1

5. Likewise we can use a truth table to show that we can treat bracketed terms in the same way as in ordinary algebra, e.g.

$$A \cdot (B + C) = A \cdot B + A \cdot C$$

6. Anything ORed with its own inverse equals 1:

$$A + \overline{A} = 1$$

7. Anything ANDed with its own inverse equals 0:

$$A \cdot \overline{A} = 0$$

8. Anything ORed with a 0 is equal to itself; anything ORed with a 1 is equal to 1. Thus $A + 0 = A$ and $A + 1 = 1$.

9. Anything ANDed with a 0 is equal to 0; anything ANDed with a 1 is equal to itself. Thus $A \cdot 0 = 0$ and $A \cdot 1 = A$.

As an illustration of the use of the above to simplify Boolean expressions, consider simplifying

$$(A + B) \cdot \overline{C} + A \cdot C$$

Using item 5 for the first term gives

$$A \cdot \overline{C} + B \cdot \overline{C} + A \cdot C$$

We can regroup this and use item 6 to give

$$A \cdot (\overline{C} + C) + B \cdot \overline{C} = A \cdot 1 + B \cdot \overline{C}$$

Hence, using item 9 the simplified expression becomes

$$A + B \cdot \overline{C}$$

C.2   De Morgan's laws

As illustrated above, the laws of Boolean algebra can be used to simplify Boolean expressions. In addition we have what are known as **De Morgan's laws**:

1 The inverse of the outcome of ORing A and B is the same as when the inverses of A and B are separately ANDed. The following truth table shows the validity of this.

$$\overline{A + B} = \overline{A} \cdot \overline{B}$$

A	B	$A + B$	$\overline{A + B}$	$\overline{A}$	$\overline{B}$	$\overline{A} \cdot \overline{B}$
0	0	0	1	1	1	1
0	1	1	0	1	0	0
1	0	1	0	0	1	0
1	1	1	0	0	0	0

2 The inverse of the outcome of ANDing A and B is the same as when the inverses of A and B are separately ORed. The following truth table shows the validity of this:

$$\overline{A \cdot B} = \overline{A} + \overline{B}$$

A	B	$A \cdot B$	$\overline{A \cdot B}$	$\overline{A}$	$\overline{B}$	$\overline{A} + \overline{B}$
0	0	0	1	1	1	1
0	1	0	1	1	0	1
1	0	0	1	0	1	1
1	1	1	0	0	0	0

As an illustration of the use of De Morgan's laws, consider the simplification of the logic circuit shown in Figure C.1.

Figure C.1 Circuit simplification.

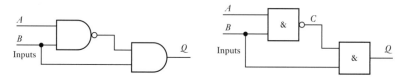

The Boolean equation for the output in terms of the input is

$$Q = \overline{A \cdot B} \cdot B$$

Applying the second law from above gives

$$Q = (\overline{A} + \overline{B}) \cdot B$$

We can write this as

$$Q = \overline{A} \cdot B + \overline{B} \cdot B = \overline{A} \cdot B + 0 = \overline{A} \cdot B$$

Hence the simplified circuit is as shown in Figure C.2.

Figure C.2 Circuit simplification.

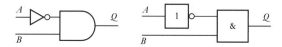

C.3   Boolean function generation from truth tables

Given a situation where the requirements of a system can be specified in terms of a truth table, how can a logic gate system using the minimum number of gates be devised to give that truth table?

Boolean algebra can be used to manipulate switching functions into many equivalent forms, some of which take many more logic gates than others; the form, however, to which most are minimised is AND gates driving a single OR gate or vice versa. Two AND gates driving a single OR gate (Figure C.3(a)) give

$$A \cdot B + A \cdot C$$

This is termed the **sum of products** form.

Figure C.3 (a) Sum of products, (b) product of sums.

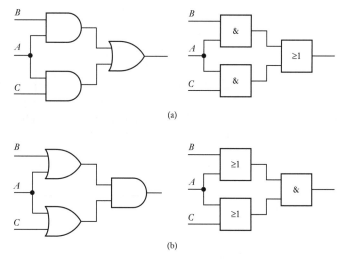

(a)

(b)

For two OR gates driving a single AND gate (Figure C.3(b)), we have

$$(A + B) \cdot (A + C)$$

This is known as the **product of sums** form. Thus in considering what minimum form might fit a given truth table, the usual procedure is to find the sum of products or the product of sums form that fits the data. Generally the sum of products form is used. The procedure used is to consider each row of the truth table in turn and find the product that would fit a row. The overall result is then the sum of all these products.

Suppose we have a row in a truth table of

$$A = 1, B = 0 \text{ and output } Q = 1$$

When A is 1 and B is not 1 then the output is 1, thus the product which fits this is

$$Q = A \cdot \overline{B}$$

We can repeat this operation for each row of a truth table, as the following table indicates:

A	B	Output	Products
0	0	0	$\overline{A} \cdot \overline{B}$
0	1	0	$\overline{A} \cdot B$
1	0	1	$A \cdot \overline{B}$
1	1	0	$A \cdot B$

However, only the row of the truth table that has an output of 1 need be considered, since the rows with 0 output do not contribute to the final expression; the result is thus

$$Q = A \cdot \overline{B}$$

The logic gate system that will give this truth table is thus that shown in Figure C.4.

Figure C.4 Logic gates for truth table.

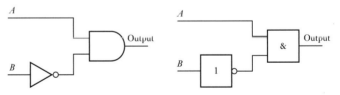

As a further example, consider the following truth table, only the products terms giving a 1 output being included:

A	B	C	Output	Products
0	0	0	1	$\overline{A} \cdot \overline{B} \cdot \overline{C}$
0	0	1	0	
0	1	0	1	$\overline{A} \cdot B \cdot \overline{C}$
0	1	1	0	
1	0	0	0	
1	0	1	0	
1	1	0	0	
1	1	1	0	

Thus the sum of products which fits this table is

$$Q = \overline{A} \cdot \overline{B} \cdot \overline{C} + \overline{A} \cdot B \cdot \overline{C}$$

This can be simplified to give

$$Q = \overline{A} \cdot \overline{C} \cdot (\overline{B} + B) = \overline{A} \cdot \overline{C}$$

The truth table can thus be generated by just a NAND gate.

C.4   Karnaugh maps

The **Karnaugh map** is a graphical method that can be used to produce simplified Boolean expressions from sums of products obtained from truth tables. The truth table has a row for the value of the output for each combination of input values. With two input variables there are four lines in the truth table, with three input variables six lines and with four input variables sixteen lines. Thus with two input variables there are four product terms, with three input variables there are six and with four input variables sixteen. The Karnaugh map is drawn as a rectangular array of cells, with each cell corresponding to a particular product value. Thus with two input variables there are four cells, with three input variables six cells and with four input variables sixteen cells. The output values for the rows are placed in their cells in the Karnaugh map, though it is usual to indicate only the 1 output values and leave the cells having 0 output as empty.

Figure C.5(a) shows the map for two input variables. The cells are given the output values for the following products:

the upper left cell $\overline{A} \cdot \overline{B}$,
the lower left cell $A \cdot \overline{B}$,
the upper right cell $\overline{A} \cdot B$,
the lower right cell $A \cdot B$.

Figure C.5 Two input variable map.

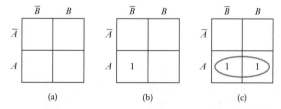

(a)   (b)   (c)

The arrangement of the map squares is such that horizontally adjacent squares differ only in one variable and, likewise, vertically adjacent squares differ in only one variable. Thus horizontally with our two–variable map the variables differ only in A and vertically they differ only in B.

For the following truth table, if we put the values given for the products in the Karnaugh map, only indicating where a cell has a 1 value and leaving blank those with a 0 value, then the map shown in Figure C.5(b) is obtained:

A	B	Output	Products
0	0	0	$\overline{A} \cdot \overline{B}$
0	1	0	$\overline{A} \cdot B$
1	0	1	$A \cdot \overline{B}$
1	1	0	$A \cdot B$

Because the only 1 entry is in the lower right square, the truth table can be represented by the Boolean expression

$$\text{output} = A \cdot \overline{B}$$

As a further example, consider the following truth table:

A	B	Output	Products
0	0	0	$\overline{A} \cdot \overline{B}$
0	1	0	$\overline{A} \cdot B$
1	0	1	$A \cdot \overline{B}$
1	1	1	$A \cdot B$

It gives the Karnaugh map shown in Figure C.5(c). This has an output given by

$$\text{output} = A \cdot \overline{B} + A \cdot B$$

We can simplify this to

$$A \cdot \overline{B} + A \cdot B = A \cdot (\overline{B} + B) = A$$

When two cells containing a 1 have a common vertical edge we can simplify the Boolean expression to just the common variable. We can do this by inspection of a map, indicating which cell entries can be simplified by drawing loops round them, as in Figure C.5(c).

Figure C.6(a) shows the Karnaugh map for the following truth table having three input variables:

A	B	C	Output	Products
0	0	0	1	$\overline{A} \cdot \overline{B} \cdot \overline{C}$
0	0	1	0	$\overline{A} \cdot \overline{B} \cdot C$
0	1	0	1	$\overline{A} \cdot B \cdot \overline{C}$
0	1	1	0	$\overline{A} \cdot B \cdot C$
1	0	0	0	$A \cdot \overline{B} \cdot \overline{C}$
1	0	1	0	$A \cdot \overline{B} \cdot C$
1	1	0	0	$A \cdot B \cdot \overline{C}$
1	1	1	0	$A \cdot B \cdot C$

Figure C.6   (a) Three–input–variable map, (b) four–input–variable map.

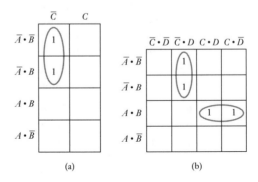

(a)   (b)

As before we can use looping to simplify the resulting Boolean expression to just the common variable. The result is

$$\text{output} = \overline{A} \cdot \overline{C}$$

Figure C.6(b) shows the Karnaugh map for the following truth table having four input variables. Looping simplifies the resulting Boolean expression to give

$$\text{output} = \overline{A} \cdot \overline{C} \cdot D + A \cdot B \cdot C$$

A	B	C	D	Output	Products
0	0	0	0	0	
0	0	0	1	1	$\overline{A} \cdot \overline{B} \cdot \overline{C} \cdot D$
0	0	1	0	0	
0	0	1	1	0	
0	1	0	0	0	
0	1	0	1	1	$\overline{A} \cdot B \cdot \overline{C} \cdot D$
0	1	1	0	0	
0	1	1	1	0	
1	0	0	0	0	
1	0	0	1	0	
1	0	1	0	0	
1	0	1	1	0	
1	1	0	0	0	
1	1	0	1	0	
1	1	1	0	1	$A \cdot B \cdot C \cdot \overline{D}$
1	1	1	1	1	$A \cdot B \cdot C \cdot D$

The above represents just some simple examples of Karnaugh maps and the use of looping. Note that, in looping, adjacent cells can be considered to be those in the top and bottom rows of the left- and right-hand columns. Think of opposite edges of the map being joined together. Looping a pair of adjacent ones in a map eliminates the variable that appears in complemented and uncomplemented form. Looping a quad of adjacent ones eliminates the two variables that appear in both complemented and uncomplemented form. Looping an octet of adjacent ones eliminates the three variables that appear in both complemented and uncomplemented form.

As a further illustration, consider an automated machine that will only start when two of three sensors A, B and C give signals. The following truth table fits this requirement and Figure C.7(a) shows the resulting three-variable Karnaugh diagram. The Boolean expression which fits the map and thus describes the outcome from the machine is

$$\text{outcome} = A \cdot B + B \cdot C + A \cdot C$$

Figure C.7(b) shows the logic gates that could be used to generate this Boolean expression. $A \cdot B$ describes an AND gate for the inputs A and B. Likewise $B \cdot C$ and $A \cdot C$ are two more AND gates. The + signs indicate that the outputs from the three AND gates are then the inputs to an OR gate.

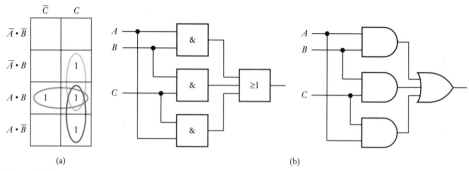

Figure C.7   Automated machine.

A	B	C	Output	Products
0	0	0	0	
0	0	1	0	
0	1	0	0	
0	1	1	1	$\overline{A} \cdot B \cdot C$
1	0	0	0	
1	0	1	1	$A \cdot \overline{B} \cdot C$
1	1	0	1	$A \cdot B \cdot \overline{C}$
1	1	1	1	$A \cdot B \cdot C$

In some logic systems there are some input variable combinations for which outputs are not specified. They are termed 'don't care states'. When entering these on a Karnaugh map, the cells can be set to either 1 or 0 in such a way that the output equations can be simplified.

Problems

C.1   State the Boolean functions that can be used to describe the following situations.

(a) There is an output when switch A is closed and either switch B or switch C is closed.

(b) There is an output when either switch A or switch B is closed and either switch C or switch D is closed.

(c) There is an output when either switch A is opened or switch B is closed.

(d) There is an output when switch A is opened and switch B is closed.

C.2   State the Boolean functions for each of the logic circuits shown in Figure C.8.

C.3   Construct a truth table for the Boolean equation $Q = (A \cdot C + B \cdot C) \cdot (A + C)$.

C.4   Simplify the following Boolean equations:

(a) $Q = A \cdot C + A \cdot C \cdot D + C \cdot D$

(b) $Q = A \cdot \overline{B} \cdot D + A \cdot \overline{B} \cdot \overline{D}$

(c) $Q = A \cdot B \cdot C + C \cdot D + C \cdot D \cdot E$

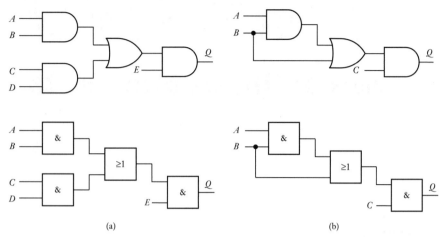

(a)   (b)

Figure C.8   Problem C.2.

C.5   Use De Morgan's laws to show that a NOR gate with inverted inputs is equivalent to an AND gate.

C.6   Draw the Karnaugh maps for the following truth tables and hence determine the simplified Boolean equation for the outputs:

(a)

A	B	Q
0	0	1
0	1	1
1	0	1
1	1	1

(b)

A	B	C	Q
0	0	0	0
0	0	1	1
0	1	0	1
0	1	1	1
1	0	0	0
1	0	1	1
1	1	0	0
1	1	1	1

C.7   Simplify the following Boolean equations by the use of Karnaugh maps:

(a) $Q = \overline{A} \cdot \overline{B} \cdot C + \overline{A} \cdot \overline{B} \cdot \overline{C} + \overline{A} \cdot B \cdot \overline{C}$

(b) $Q = \overline{A} \cdot B \cdot \overline{C} \cdot D + A \cdot \overline{B} \cdot \overline{C} \cdot D + \overline{A} \cdot \overline{B} \cdot C \cdot D$

$\qquad + A \cdot B \cdot \overline{C} \cdot D + A \cdot B \cdot \overline{C} \cdot \overline{D} + A \cdot B \cdot C \cdot D$

C.8   Devise a system which will allow a door to be opened only when the correct combination of four push-buttons is pressed, any incorrect combination sounding an alarm.

Appendix D: Instruction sets

The following are the instructions used with the Motorola M68HC11, Intel 8051 and PIC16Cxx microcontrollers.

M68HC11

Instruction	Mnemonic	Instruction	Mnemonic
Loading		*Rotate / shift*	
Load accumulator A	LDAA	Rotate bits in memory left	ROL
Load accumulator B	LDAB	Rotate bits in accumulator A left	ROLA
Load double accumulator	LDD	Rotate bits in accumulator B left	ROLB
Load stack pointer	LDS	Rotate bits in memory right	ROR
Load index register X	LDX	Rotate bits in accumulator A right	RORA
Load index register Y	LDY	Rotate bits in accumulator B right	RORB
Pull data from stack and load acc. A	PULA	Arithmetic shift bits in memory left	ASL
Pull data from stack and load acc. B	PULB	Arithmetic shift bits in acc. A left	ASLA
Pull index register X from stack	PULX	Arithmetic shift bits in acc. B left	ASLB
Pull index register Y from stack	PULY	Arithmetic shift bits in memory right	ASR
Transfer registers		Arithmetic shift bits in acc. A right	ASRA
Transfer from acc. A to acc. B	TAB	Arithmetic shift bits in acc. B right	ASRB
Transfer from acc. B to acc. A	TBA	Logical shift bits in memory left	LSL
From stack pointer to index reg. X	TSX	Logical shift bits in acc. A left	LSLA
From stack pointer to index reg. Y	TSY	Logical shift bits in acc. B left	LSLB
From index reg. X to stack pointer	TXS	Logical shift bits in acc. D left	LSLD
From index reg. Y to stack pointer	TYS	Logical shift bits in memory right	LSR
Exchange double acc. and index reg. X	XGDX	Logical shift bits in acc. A right	LSRA
Exchange double acc. and index reg. Y	XGDY	Logical shift bits in acc. B right	LSRB
Decrement / increment		Logical shift bits in acc. C right	LSRD
Subtract 1 from contents of memory	DEC	*Data test with setting of condition codes*	
Subtract 1 from contents of acc. A	DECA	Logical test AND between acc. A & memory	BITA
Subtract 1 from contents of acc. B	DECB		
Subtract 1 from stack pointer	DES	Logical test AND between acc. B & memory	BITB
Subtract 1 from index register X	DEX		
Subtract 1 from index register Y	DEY	Compare accumulator A to accumulator B	CBA
Add 1 to contents of memory	INC	Compare accumulator A and memory	CMPA
Add 1 to contents of accumulator A	INCA	Compare accumulator B and memory	CMPB
Add 1 to contents of accumulator B	INCB	Compare double accumulator with memory	CPD
Add 1 to stack pointer	INS	Compare index register X with memory	CPX
Add 1 to index register X	INX	Compare index register Y with memory	CPY
Add 1 to index register Y	INY	Subtract $00 from memory	TST

(Continued)

Instruction	Mnemonic	Instruction	Mnemonic
Subtract $00 from accumulator A	TSTA	Subtract mem. from acc. B with carry	SBCB
Subtract $00 from accumulator B	TSTB	Subtract mem. from accumulator A	SUBA
Interrupt		Subtract mem. from accumulator B	SUBB
Clear interrupt mask	CLI	Subtract mem. from double acc.	SUBD
Set interrupt mask	SEI	Replace acc. A with two' complement	NEGA
Software interrupt	SWI	Replace acc. B with two' complement	NEGB
Return from interrupt	RTI	Multiply unsigned acc. A by acc. B	MUL
Wait for interrupt	WAI	Unsigned integer divide D by index reg. X	IDIV
Complement and clear		Unsigned fractional divide D by index reg. X	FDIV
Clear memory	CLR	*Conditional branch*	
Clear A	CLRA	Branch if minus	BMI
Clear B	CLRB	Branch if plus	BPL
Clear bits in memory	BCLR	Branch if overflow set	BVS
Set bits in memory	BSET	Branch if overflow clear	BVC
Store registers		Branch if less than zero	BLT
Store contents of accumulator A	STAA	Branch if greater than or equal to zero	BGE
Store contents of accumulator B	STAB	Branch if less than or equal to zero	BLE
Store contents of double acc.	STD	Branch if greater than zero	BGT
Store stack pointer	STS	Branch if equal	BEQ
Store index register X	STX	Branch if not equal	BNE
Store index register Y	STY	Branch if higher	BHI
Push data from acc. A onto stack	PSHA	Branch if lower or same	BLS
Push data from acc. B onto stack	PSHB	Branch if higher or same	BHS
Push index reg. X contents onto stack	PSHX	Branch if lower	BLO
Push index reg. Y contents onto stack	PSHY	Branch if carry clear	BCC
Logic		Branch if carry set	BCS
AND with contents of accumulator A	ANDA	*Jump and branch*	
AND with contents of accumulator B	ANDB	Jump to address	JMP
EXCLUSIVE-OR with contents of acc. A	EORA	Jump to subroutine	JSR
EXCLUSIVE-OR with contents of acc. B	EORB	Return from subroutine	RTS
OR with contents of accumulator A	ORAA	Branch to subroutine	BSR
OR with contents of accumulator B	ORAB	Branch always	BRA
Replace memory with one's complement	COM	Branch never	BRN
Replace acc. A with one's complement	COMA	Branch bits set	BRSET
Replace acc. B with one's complement	COMB	Branch bits clear	BRCLR
Arithmetic		*Condition code*	
Add contents of acc. A to acc. B	ABA	Clear carry	CLC
Add contents of acc. B to index reg. X	ABX	Clear overflow	CLV
Add contents of acc. B to index reg. Y	ABY	Set carry	SEC
Add memory to acc. A without carry	ADDA	Set overflow	SEV
Add memory to acc. B without carry	ADDB	Transfer from acc. A to condition code reg.	TAP
Add mem. to double acc. without carry	ADDD	Transfer from condition code reg. to acc. A	TPA
Add memory to acc. A with carry	ADCA	*Miscellaneous*	
Add memory to acc. B with carry	ADCB	No operation	NOP
Decimal adjust	DAA	Stop processing	STOP
Subtract contents of acc. B from acc. A	SBA	Special test mode	TEST
Subtract mem. from acc. A with carry	SBCA		

Note: The number of bits in a register depends on the processor. An 8-bit microprocessor generally has 8-bit registers. Sometimes two of the data registers may be used together to double the number of bits. Such a combined register is referred to as a doubled register.

Intel 8051

Instruction	Mnemonic	Instruction	Mnemonic
Data transfer		Jump if carry is set	JC rel
Move data to accumulator	MOV A, #data	Jump if carry not set	JNC rel
Move register to accumulator	MOV A, Rn	Jump if direct bit is set	JB bit, rel
Move direct byte to accumulator	MOV A, direct	Jump if direct bit is not set	JNB bit, rel
Move indirect RAM to accumulator	MOV A, @Ri	Jump if direct bit is set and	JBC bit, rel
Move accumulator to direct byte	MOV direct, A	clear bit	
Move accumulator to external RAM	MOVX @Ri, A	*Subroutine call*	
Move accumulator to register	MOV Rn, A	Absolute subroutine call	ACALL addr 11
Move direct byte to indirect RAM	MOV @Ri, direct	Long subroutine call	LCALL addr 16
Move immediate data to register	MOV Rn, #data	Return from subroutine	RET
Move direct byte to direct byte	MOV direct, direct	Return from interrupt	RETI
Move indirect RAM to direct byte	MOV direct, @Ri	*Bit manipulation*	
Move register to direct byte	MOV direct, Rn	Clear carry	CLR C
Move immediate data to direct byte	MOV direct, #data	Clear bit	CLR bit
Move immediate data to indirect RAM	MOV @Ri, #data	Set carry but	SETB C
Load data pointer with a 16-bit constant	MOV DPTR, #data16	Set bit	SETB bit
		Complement carry	CPL C
Move code byte relative to DPTR to acc.	MOV A, @A+DPTR	AND bit to carry bit	ANL C,bit
		AND complement of bit to carry bit	ANL C,/bit
Move external RAM, 16-bit addr., to acc.	MOVX A, @DPTR	OR bit to carry bit	ORL C,bit
Move acc. to external RAM, 16-bit addr.	MOVX @DPTR, A	OR complement of bit to carry bit	ORL C,/bit
Exchange direct byte with accumulator	XCH A, direct	Move bit to carry	MOV C,bit
Exchange indirect RAM with acc.	XCH A, @Ri	Move carry bit to bit	MOV bit,C
Exchange register with accumulator	XCH A, Rn	*Logical operations*	
Push direct byte onto stack	PUSH direct	AND accumulator to direct byte	ANL direct, A
Pop direct byte from stack	POP direct	AND immediate data to direct byte	ANL direct, #data
Branching		AND immediate data to acc.	ANL A, #data
Absolute jump	AJMP addr 11	AND direct byte to accumulator	ANL A, direct
Long jump	LJMP addr 16	AND indirect RAM to accumulator	ANL A, @Ri
Short jump, relative address	SJMP rel	AND register to accumulator	ANL A, Rn
Jump indirect relative to the DPTR	JMP @A+DPTR	OR accumulator to direct byte	ORL direct, A
Jump if accumulator is zero	JZ rel	OR immediate data to direct byte	ORL direct, #data
Jump if accumulator is not zero	JNZ rel	OR immediate data to accumulator	ORL A, #data
Compare direct byte to acc. and	CJNE A, direct, rel	OR direct byte to accumulator	ORL A, direct
jump if not equal		OR indirect RAM to accumulator	ORL A, @Ri
Compare immediate to acc. and	CJNE A, #data, rel	OR register to accumulator	ORL A, Rn
jump if not equal		XOR accumulator to direct byte	XRL direct, A
Compare immediate to register	CJNE Rn, #data, rel	XOR immediate data to acc.	XRL direct, #data
and jump if not equal		XOR immediate data to acc.	XRL A, #data
Compare immediate to indirect	CJNE @Ri, #data, rel	XOR direct byte to accumulator	XRL A, direct
and jump if not equal		XOR indirect RAM to accumulator	XRL A, @Ri
Decrement register and jump if	DJNZ Rn, rel	XOR register to accumulator	XRL A, Rn
not zero		*Addition*	
Decrement direct byte, jump if	DJNZ A, direct, rel	Add immediate data to acc.	ADD A, #data
not zero			

(Continued)

Instruction	Mnemonic	Instruction	Mnemonic
Add direct byte to accumulator	ADD A, direct	Swap nibbles within the acc.	SWAP A
Add indirect RAM to accumulator	ADD A, @Ri	Decimal adjust accumulator	DA A
Add register to accumulator	ADD A, Rn	*Increment and decrement*	
Add immediate data to acc. with carry	ADDC A, #data	Increment accumulator	INC A
		Increment direct byte	INC direct
Add direct byte to acc. with carry	ADDC A, direct	Increment indirect RAM	INC @Ri
Add indirect RAM to acc. with carry	ADDC A, @Ri	Increment register	INC Rn
		Decrement accumulator	DEC A
Add register to acc. with carry	ADDC A, Rn	Decrement direct byte	DEC direct
Subtraction		Decrement indirect RAM	DEC @Ri
Subtract immediate data from acc. with borrow	SUBB A, #data	Decrement register	DEC Rn
		Increment data pointer	INC DPTR
Subtract direct byte from acc. with Borrow	SUBB A, 29	*Clear and complement operations*	
		Complement accumulator	CPL A
Subtract indirect RAM from acc. with borrow	SUBB A, @Ri	Clear accumulator	CLR A
		Rotate operations	
Multiplication and division		Rotate accumulator right	RR A
Multiply A and B	MUL AB	Rotate accumulator right thro. C	RRC A
Divide A by B	DIV AB	Rotate accumulator left	RL A
Decimal maths operations		Rotate accumulator left through C	RLC A
Exchange low-order digit indirect		*No operation*	
RAM with accumulator	XCHD A, @Ri	No operation	NOP

Note: A value preceded by # is a number, #data16 is a 16-bit constant; Rn refers to the contents of a register; @Ri refers to the value in memory where the register points; DPTR is the data pointer; direct is the memory location where data used by an instruction can be found.

PIC16Cxx

Instruction	Mnemonic	Instruction	Mnemonic
Add a number with number in working reg.	addlw number	Move (copy) the number in a file reg. into the working reg.	movf FileReg,w
Add number in working reg. to number in file register and put number in file register	addwf FileReg,f	Move (copy) number into working reg.	movlw number
Add number in working reg. to number in file register and put number in working reg.	addwf FileReg,w	Move (copy) the number in file reg. into the working reg.	movwf FileReg
AND a number with the number in the working reg. and put result in working reg.	andlw number	No operation	nop
AND a number in the working reg. with the number in file reg., and put result in file reg.	andwf FileReg,f	Return from a subroutine and enable global interrupt enable bit	refie
Clear a bit in a file reg., i.e. make it 0	bcf FileReg,bit	Return from a subroutine with a number in the working register	retlw number
Set a bit in a file reg., i.e. make it 1	bsf FileReg,bit	Return from a subroutine	return
Test a bit in a file reg. and skip the next instruction if the bit is 0	btfsc FileReg,bit	Rotate bits in file reg. to the left through the carry bit	rlf FileReg,f
Test a bit in a file reg. and skip the next instruction if the bit is 1	btfss FileReg,bit	Rotate bits in file reg. to the right through the carry bit	rrf FileReg,f
Call a subroutine, after which return to where it left off	call AnySub	Send the PIC to sleep, a low-power-consumption mode	sleep
Clear, i.e. make 0, the number in file reg.	clrf FileReg	Subtract the number in working reg. from a number	sublw number
Clear, i.e. make 0, the no. in working reg.	clrw	Subtract the no. in working reg. from number in file reg., put result in file reg.	subwf FileReg,f
Clear the number in the watchdog timer	clrwdt		
Complement the number in file reg. and leave result in file register	comf FileReg,f	Swap the two halves of the 8 bit no. in a file reg, leaving result in file reg.	swapf FileReg, f
Decrement a file reg., result in file reg.	decf FileReg,f	Use the number in working reg. to specify which bits are input or output	tris PORTX
Decrement a file reg. and if result zero skip the next instruction	decfsz FileReg,f		
Go to point in program labelled	gotot label	XOR a number with number in working register	xorlf number
Increment file reg. and put result in file reg.	Incf FileReg,f		
OR a number with number in working reg.	iorlw number	XOR the number in working reg. with number in file reg. and put result in the file reg.	xorwf FileReg,f
OR the number in working reg. with the number in file reg., put result in file reg.	iorwf FileReg,f		

Note: f is used for the file register, w for the working register and b for bit. The mnemonics indicate the types of operand involved, e.g. movlw indicates the move operation with the lw indicating that a literal value, i.e. a number, is involved in the working register w; movwf indicates the move operation when the working register and a file register are involved.

Appendix E: C library functions

The following are some common C library functions. This is not a complete list of all the functions within any one library or a complete list of all the libraries that are likely to be available with any one compiler.

<ctype.h>

isalnum	int isalnum(int ch)	Tests for alphanumeric characters, returning non-zero if argument is either a letter or a digit or a 0 if it is not alphanumeric.
isalpha	int isalpha(int ch)	Tests for alphabetic characters, returning non-zero if a letter of the alphabet, otherwise 0.
iscntrl	int iscntrl(int ch)	Tests for control character, returning non-zero if between 0 and 0x1F or is equal to 0x7F (DEL), otherwise 0.
isdigit	int isdigit(int ch)	Tests for decimal digit character, returning non-zero if a digit (0 to 9), otherwise 0.
isgraph	int isgraph(int ch)	Tests for a printable character (except space), returning non-zero if printable, otherwise 0.
islower	int islower(int ch)	Tests for lower case character, returning non-zero if lower case, otherwise 0.
isprint	int isprint(int ch)	Tests for printable character (including space), returning non-zero if printable, otherwise 0.
ispunct	int ispunct(int ch)	Tests for punctuation character, returning non-zero if a punctuation character, otherwise 0.
isspace	int isspace(int ch)	Tests for space character, returning non-zero if a space, tab, form feed, carriage return or new-line character, otherwise 0.
isupper	int isupper(int ch)	Tests for upper case character, returning non-zero if upper case, otherwise 0.
isxdigit	int isxdigit(int ch)	Tests for hexadecimal character, returning non-zero if a hexadecimal digit, otherwise 0.

<math.h>

acos	double acos(double arg)	Returns the arc cosine of the argument.
asin	double asin(double arg)	Returns the arc sine of the argument.
atan	double atan(double arg)	Returns the arc tangent of the argument. Requires one argument.

atan2	double atan2(double y, double x)	Returns the arc tangent of y/x.
ceil	double ceil(double num)	Returns the smallest integer that is not less than num.
cos	double cos(double arg)	Returns the cosine of arg. The value of arg must be in radians.
cosh	double cosh(double arg)	Returns the hyperbolic cosine of arg.
exp	double exp(double arg)	Returns e^x where x is arg.
fabs	double fabs(double num)	Returns the absolute value of num.
floor	double floor(double num)	Returns the largest integer not greater than num.
fmod	double fmod(double x, double y)	Returns the floating-point remainder of x/y.
ldexp	double ldexp(double x, int y)	Returns x times 2^y.
log	double log(double num)	Returns the natural logarithm of num.
log10	double log10(double num)	Returns the base 10 logarithm of num.
pow	double pow(double base, double exp)	Returns base raised to the exp power.
sin	double sin(double arg)	Returns the sine of arg.
sinh	double sinh(double arg)	Returns the hyperbolic sine of arg.
sqrt	double sqrt(double num)	Returns the square root of num.
tan	double tan(double arg)	Returns the tangent of arg.
tanh	double tanh(double arg)	Returns the hyperbolic tangent of arg.

<stdio.h>

getchar	int getchar(void)	Returns the next character typed on the keyboard.
gets	char gets(char *str)	Reads characters entered at the keyboard until a carriage return is read and stores them in the array pointed to by str.
printf	int printf(char *str, ...)	Outputs the string pointed to by str.
puts	int puts(char *str)	Outputs the string pointed to by str.
scanf	int scanf(char *str, ...)	Reads information into the variables pointed to by the arguments following the control string.

<stdlib.h>

abort	void abort(void)	Causes immediate termination of a program.
abs	int abs(int num)	Returns the absolute value of the integer num.
bsearch	void bsearch(const void *key, const void *base, size_t num, size_t size, int(*compare)(const void *, const void *))	Performs a binary search on the sorted array pointed to by base and returns a pointer to the first member that matches the key pointed to by key. The number of the elements in the array is specified by num and the size in bytes of each element by size.

calloc	void *calloc(size_t num, size_t size)	Allocates sufficient memory for an array of num objects of size given by size, returning a pointer to the first byte of the allocated memory.
exit	void exit(int status)	Causes immediate normal termination of a program. The value of the status is passed to the calling process.
free	void free(void *ptr)	Frees the allocated memory pointed to by ptr.
labs	long labs(long num)	Returns the absolute value of the long int num.
malloc	void *malloc(size_t size)	Returns a pointer to the first byte of memory of size given by size that has been allocated.
qsort	void qsort(void *base, size_t num, size_t size, int(*compare)(const void*, const void*))	Sorts the array pointed to by base. The number of elements in the array is given by num and the size in bytes of each element by size.
realloc	void *realloc(void *ptr, size_t size)	Changes the size of the allocated memory pointed to by ptr to that specified by size.

Note: size_t is the type for 'size of' variables and usually represents the size of another parameter or object.

<time.h>

asctime	char *asctime(const struct tm *ptr)	Converts time from a structure form to a character string appropriate for display, returning a pointer to the string.
clock	clock_t clock(void)	Returns the number of clock cycles that have occurred since the program began execution.
ctime	char *ctime(const time_t *time)	Returns a pointer to a string of the form day month date hours:minutes:seconds year\n\0 given a pointer to the numbers of seconds elapsed since 00:00:00 Greenwich Mean Time.
difftime	double difftime(time_t time 2, time_t time 1)	Returns the difference in seconds between time 1 and time 2.
gmtime	struct tm *gmtime (const time_t *time)	Returns a pointer to time converted from long inter form to a structure form.
localtime	struct tm *localtime (const time_t *time)	Returns a pointer to time converted from long inter form to structure form in local time.
time	time_t time(time_t *system)	Returns the current calendar time of the system.

Note: time_t and clock_t are used as the type for 'time of' and 'number of cycles of' variables.

Appendix F: MATLAB and SIMULINK

F.1 MATLAB

Computer software can be used to aid computation and modelling of systems; a program that is often used is MATLAB. The following is a brief introduction to MATLAB (registered trademark of the Mathworks Inc.). For additional information you are referred to the user guide or textbooks such as Hahn, B., *Essential MATLAB for Engineers and Scientists*, 5th edn, Elsevier 2012 or Moore, H., *MATLAB for Engineers*, Pearson 2013.

Commands are entered by typing them in after the prompt and then pressing the enter or return key in order that the command can be executed. In the discussion of the commands that follow, this pressing of the enter or return key will not be repeated but should be assumed in all cases. To start MATLAB, in Windows or the Macintosh systems, click on the MATLAB icon, otherwise type matlab. The screen will then produce the MATLAB prompt ≫. To quit MATLAB type quit or exit after the prompt. Because MATLAB is case sensitive, lower case letters should be used throughout for commands.

Typing help after the prompt, or selecting help from the menu bar at the top of the MATLAB window, displays a list of MATLAB broad help topics. To get help on a particular topic in the list, e.g. exponentials, type help exp. Typing lookfor plus some topic will instruct MATLAB to search for information on that topic, e.g. lookfor integ will display a number of commands which could be considered for integration.

In general, mathematical operations are entered into MATLAB in the same way as they would be written on paper. For example,

$$\gg a = 4/2$$

yields the response

$$a =$$
$$2$$

and

$$\gg a = 3*2$$

yields the response

$$a =$$
$$6$$

Operations are carried out in the following order: $\wedge$ power operation, $*$ multiplication, $/$ division, $+$ addition, $-$ subtraction. Precedence of operators is from left to right but brackets ( ) can be used to affect the order. For example,

> $\gg$ a $= 1 + 2\wedge3/4*5$

yields the response

> a $=$
>
> 11

because we have $2^3/4$ which is multiplied by 5 and then added to 1, whereas

> $\gg$ a $= 1 + 2\wedge3/(4*5)$

yields the response

> a $=$
>
> 1.4

because we have 2^3 divided by the product of 4 and 5, and then added to 1.

The following are some of the mathematical functions available with MATLAB:

abs(x)	Gives the absolute value of x, i.e. $\lvert x \rvert$
exp(x)	Gives the exponential of x, i.e. e^x
log(x)	Gives the natural logarithm of x, i.e. $\ln x$
log10(x)	Gives the base 10 logarithm of x, i.e. $\log_{10} x$
sqrt(x)	Gives the square root of x, i.e. $\sqrt{x}$
sin(x)	Gives $\sin x$ where x is in radians
cos(x)	Gives $\cos x$ where x is in radians
tan(x)	Gives $\tan x$ where x is in radians
asin(x)	Gives arcsin x, i.e. $\sin^{-1} x$
acos(x)	Gives arccos x, i.e. $\cos^{-1} x$
atan(x)	Gives arctan x, i.e. $\tan^{-1} x$
csc(x)	Gives $1/\sin x$
sec(x)	Gives $1/\cos x$
cot(x)	Gives $1/\tan x$

π is entered by typing pi.

Instead of writing a series of commands at the prompt, a text file can be written and then the commands executed by referring MATLAB to that file. The term M-file is used since these text files, containing a number of consecutive MATLAB commands, have the suffix .m. In writing such a file, the first line must begin with the word function followed by a statement identifying the name of the function and the input and output in the form

> function [output] $=$ function name [input]

e.g. function y $=$ cotan(x) which is the file used to determine the value of y given by cotan x. Such a file can be called up in some MATLAB sequence of commands by writing the name followed by the input, e.g. cotan(x). It is in fact already included in MATLAB and is used when the cotangent of x is required. However, the file could have been user written. A function that has multiple inputs should list all of them in the function statement.

Likewise, a function that is to return more than one value should list all the outputs.

Lines that start with % are comment lines; they are not interpreted by MATLAB as commands. For example, supposing we write a program to determine the root-mean-square values of a single column of data points, the program might look like

```
function y=rms(x)
% rms Root mean square
% rms(x) gives the root mean square value of the
% elements of column vector x.
xs=x^2;
 s=size(x);
 y=sqrt(sum(xs)/s);
```

We have let xs be the square values of each x value. The command s=size(x) obtains the size, i.e. number of entries, in the column of data. The command y=sqrt(sum(xs)/s (1)) obtains the square root of the sum of all the xs values divided by s. The ; command is used at the end of each program line.

MATLAB supplies a number of toolboxes containing collections of M-files. Of particular relevance to this book is the Control System toolbox. It can be used to carry out time responses of systems to impulses, steps, ramps, etc., along with Bode and Nyquist analysis, root locus, etc. For example, to carry out a Bode plot of a system described by a transfer function $4/(s^2 + 2s + 3)$, the program is

```
%Generate Bode plot for G(s)=4/(s^2 + 2s + 3)
num=4
den=[1 2 3];
bode(num,den)
```

The command bode (num,den) produces the Bode plot of gain in dB against frequency in rad/s on a log scale and phase in degrees against frequency in rad/s on a log scale.

F.1.1   Plotting

Two-dimensional linear plots can be produced by using the plot(x,y) command; this plots the values of x and y. For example, we might have

```
x=[0 1 2 3 4 5];
y=[0 1 4 9 16 25];
plot(x,y)
```

To plot a function, whether standard or user defined, we use the command fplot(function name,lim), where lim determines the plotting interval, i.e. the minimum and maximum values of x.

The command semilogx(x,y) generates a plot of the values of x and y using a logarithmic scale for x and a linear scale for y. The command semilogy(x,y) generates a plot of the values of x and y using a linear scale for x and a logarithmic scale for y. The command loglog(x,y) generates a plot of the values of x and y using logarithmic scales for both x and y. The command polar(theta,r) plots in polar co-ordinates with theta being the argument in radians and r the magnitude.

The subplot command enables the graph window to be split into subwindows and plots to be placed in each. For example, we might have

```
x=(0 1 2 3 4 5 6 7);
y=expx;
subplot(2,1,1);plot(x,y);
subplot(2,1,2);semilogy(x,y);
```

Three integers m, n, p are given with the subplot command; the digits m and n indicate the graph window is to be split into an $m \times n$ grid of smaller windows, where m is the number of rows and n is the number of columns, and the digit p specifies the window to be used for the plot. The subwindows are numbered by row from left to right and top to bottom. Thus the above sequence of commands divides the window into two, with one plot above the other; the top plot is a linear plot and the lower plot is a semilogarithmic plot.

The number and style of grid lines, the plot colour and the adding of text to a plot can all be selected. The command print is used to print a hard copy of a plot, either to a file or a printer. This can be done by selecting the file menu-bar item in the figure window and then selecting the print option.

F.1.2 Transfer functions

The following lines in a MATLAB program illustrate how a transfer function can be entered and displayed on screen:

```
% G(s)=4(s+10)/(s+5)(s+15)
num=4*[1 10];
den=conv([1 5],[1 15]);
printsys(num,den,'s')
```

The command num is used to indicate the numerator of the transfer function, in descending powers of s. The command den is used to indicate the denominator in descending powers of s for each of the two polynomials in the denominator. The command conv multiplies two polynomials, in this case they are $(s + 5)$ and $(s + 15)$. The printsys command displays the transfer function with the numerator and denominator specified and written in the s-domain.

Sometimes we may be presented with a transfer function as the ratio of two polynomials and need to find the poles and zeros. For this we can use

```
% Finding poles and zeros for the transfer function
% G(s)=(5s^2 + 3s + 4)/(s^3 + 2s^2 + 4s + 7)
num=[5 3 4];
den=[1 2 4 7];
[z,p,k]=tf2zp(num,den)
```

[z,p,k]=tf2zp(num,den) is the command to determine and display the zeros (z), poles (p) and gain (k) of the zero–pole–gain transfer function entered.

MATLAB can be used to give graphs showing the response of a system to different inputs. For example, the following program will give the response of the system to a unit step input $u(t)$ with a specified transfer function:

```
% Display of response to a step input for a system with
% transfer function G(s )=5/(s^2 + 3s + 12)
```

```
num=5;
den=[1 3 12];
step(num,den)
```

F.1.3   Block diagrams

Control systems are often represented as a series of interconnected blocks, each block having specific characteristics. MATLAB allows systems to be built up from interconnected blocks. The commands used are cloop when a block with a given open-loop transfer function has unity feedback. If the feedback is not unity the command feedback is used, e.g. with Figure F.1 we have the program

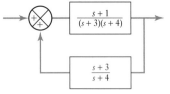

Figure F.1   Block diagram.

```
% System with feedback loop
ngo=[1 1];
dgo=conv([1 3],[1 4]);
nh=[1 3];
dh=[1 4];
[ngc2,dgc2]=feedback(ngo,dgo,nh,dh)
printsys(ngc2,dgc2,'s')
```

ngo and dgo indicate the numerator and denominator of open-loop transfer function $G_0(s)$, nh and dh the numerator and denominator of the feedback loop transfer function $H(s)$. The program results in the display of the transfer function for the system as a whole.

The command scrics is uscd to indicate that two blocks are in series in a particular path; the command parallel indicates that they are in parallel.

F.2   SIMULINK

SIMULINK is used in conjunction with MATLAB to specify systems by 'connecting' boxes on the screen rather than, as above, writing a series of commands to generate the description of the block diagram. Once MATLAB has been started, SIMULINK is selected using the command ≫ simulink. This opens the SIMULINK control window with its icons and pull-down menus in its header bar. Click on file, then click on new from the drop-down menu. This opens a window in which a system can be assembled.

To start assembling the blocks required, go back to the control window and double click on the linear icon. Click and then drag the transfer Fcn icon into the untitled window. If you require a gain block, click and drag the gain icon into the untitled window. Do the same for a sum icon and perhaps an integrator icon. In this way, drag all the required icons into the untitled window. Then double click on the Sources icon and select the appropriate source from its drop-down menu, e.g. step input, and drag it into the untitled window. Now double click on the sinks icon and drag the graph icon into the untitled window. To connect the icons, depress the mouse button while the mouse arrow is on the output symbol of an icon and drag to it the input symbol of the icon to which it is to be connected. Repeat this for all the icons until the complete block diagram is assembled.

To give the transfer Fcn box a transfer function, double click in the box. This will give a dialogue box in which you can use MATLAB commands

Figure F.2   Example of use of
SIMULINK.

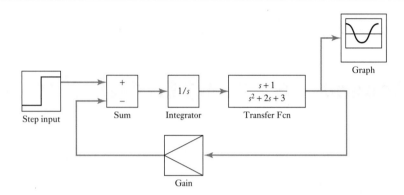

for numerator and denominator. Click on the numerator and type in
[1 1] if $(s + 1)$ is required. Click on the denominator and type in [1 2 3] if
$(s^2 + 2s + 3)$ is required. Then click the done icon. Double click on the
gain icon and type in the gain value. Double click on the sum icon and set
the signs to + or − according to whether positive or negative feedback is
required. Double click on the graph icon and set the parameters for the
graph. You then have the complete simulation diagram on screen. Figure F.2
shows the form it might take. To delete any block or connection, select them
by clicking and then press the <DEL> key.

To simulate the behaviour of the system, click on Simulation to pull
down its menu. Select Parameters and set the start and stop times for the
simulation. From the Simulation menu, select Start. SIMULINK will then
create a graph window and display the output of the system. The file can be
saved by selecting File and clicking on SAVE AS in the drop-down menu.
Insert a file name in the dialogue box then click on Done.

Appendix G: Electrical circuit analysis

<table>
<tr><td>G.1</td><td>Direct current circuits</td></tr>
</table>

The fundamental laws used in circuit analysis are Kirchhoff's laws.

1 **Kirchhoff's current law** states that at any junction in an electrical circuit, the current entering it equals the current leaving it.
2 **Kirchhoff's voltage law** states that around any closed path in a circuit, the sum of the voltage drops across all the components is equal to the sum of the applied voltage rises.

While circuits containing combinations of series- and parallel-connected resistors can often be reduced to a simple circuit by systematically determining the equivalent resistance of series or parallel connected resistors and reducing the analysis problem to a very simple circuit, the following techniques are likely to be needed for more complex circuits.

G.1.1 Node analysis

A **node** is a point in a circuit where two or more devices are connected together, i.e. it is a junction at which we have current entering and current leaving. A **principal node** is a point where three or more elements are connected together. Thus in Figure G.1, just b and d are principal nodes. One of the principal nodes is chosen to be a reference node so that the potential differences at the other nodes are then considered with reference to it. For the following analysis with Figure G.1, d has been taken as the reference node. Kirchhoff's current law is then applied to each non-reference node. The procedure is thus as outlined below.

1 Draw a labelled circuit diagram and mark the principal nodes.
2 Select one of the principal nodes as a reference node.
3 Apply Kirchhoff's current law to each of the non-reference nodes, using Ohm's law to express the currents through resistors in terms of node voltages.
4 Solve the resulting simultaneous equations. If there are n principal nodes there will be $(n - 1)$ equations.
5 Use the derived values of the node voltages to determine the currents in each branch of the circuit.

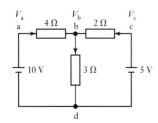

Figure G.1 Circuit for node analysis.

As an illustration, consider Figure G.1. The nodes are a, b, c and d with b and d being principal nodes. Take node d as the reference node.

If V_a, V_b and V_c are the node voltages relative to node d then the potential difference across the 4 Ω resistor is $(V_a - V_b)$, that across the 3 Ω resistor is V_b and that across the 2 Ω resistor is $(V_c - V_b)$. Thus the current through the 4 Ω resistor is $(V_a - V_b)/4$, that through the 3 Ω resistor is $V_b/3$ and that through the 2 Ω resistor is $(V_c - V_b)/2$. Thus, applying Kirchhoff's current law to node b gives:

$$\frac{V_a - V_b}{4} + \frac{V_c - V_b}{2} = \frac{V_b}{3}$$

However, $V_a = 10$ V and $V_c = 5$ V and so:

$$\frac{10 - V_b}{4} + \frac{5 - V_b}{2} = \frac{V_b}{3}$$

Thus $V_b = 4.62$ V. The potential difference across the 4 Ω resistor is thus $10 - 4.62 = 5.38$ V and so the current through it is $5.38/4 = 1.35$ A. The potential difference across the 3 Ω resistor is 4.62 V and so the current through it is $4.62/3 = 1.54$ A. The potential difference across the 2 Ω resistor is $5 - 4.62 = 0.38$ V and so the current through it is $0.38/2 = 0.19$ A.

G.1.2 Mesh analysis

The term **loop** is used for a sequence of circuit elements that form a closed path. A **mesh** is a circuit loop which does not contain any other loops within it. Mesh analysis involves defining a current as circulating round each mesh. The same direction must be chosen for each mesh current and the usual convention is to make all the mesh currents circulate in a clockwise direction. Having specified mesh currents, Kirchhoff's voltage law is then applied to each mesh. The procedure is thus as outlined below.

1 Label each of the meshes with clockwise mesh currents.
2 Apply Kirchhoff's voltage law to each of the meshes, the potential differences across each resistor being given by Ohm's law in terms of the current through it and in the opposite direction to the current. The current through a resistor which borders just one mesh is the mesh current; the current through a resistor bordering two meshes is the algebraic sum of the mesh currents through the two meshes.
3 Solve the resulting simultaneous equations to obtain the mesh currents. If there are n meshes there will be n equations.
4 Use the results for the mesh currents to determine the currents in each branch of the circuit.

As an illustration, for the circuit shown in Figure G.2 there are three loops – ABCF, CDEF and ABCDEF – but only the first two are meshes. We can define currents I_1 and I_2 as circulating in a clockwise direction in these meshes.

Figure G.2   Circuit illustrating mesh analysis.

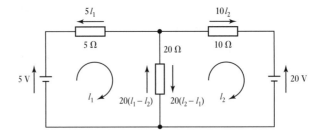

For mesh 1, applying Kirchhoff's voltage law gives $5 - 5I_1 - 20 (I_1 - I_2) = 0$. This can be rewritten as:

$$5 = 25I_1 - 20I_2$$

For mesh 2, applying Kirchhoff's voltage law gives $-10I_2 - 20 - 20 (I_2 - I_1) = 0$. This can be rewritten as:

$$20 = 20I_1 - 30I_2$$

We now have a pair of simultaneous equations and so $I_2 = -1.14$ A and $I_1 = -0.71$ A. The minus signs indicate that the currents are in the opposite directions to those indicated in the figure. The current through the 20 Ω resistor is thus in the direction of I_1 and $-0.71 + 1.14 = 0.43$ A.

G.1.3   Thévenin's theorem

The equivalent circuit for any two-terminal network containing a voltage or current source is given by **Thévenin's theorem**:

> Any two-terminal network (Figure G.3(a)) containing voltage or current sources can be replaced by an equivalent circuit consisting of a voltage equal to the open-circuit voltage of the original circuit in series with the resistance measured between the terminals when no load is connected between them and all independent sources in the network are set equal to zero (Figure G.3(b)).

Figure G.3   (a) The network, (b) its equivalent.

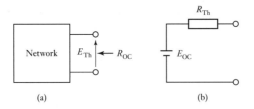

(a)   (b)

If we have a linear circuit, to use Thévenin's theorem, we have to divide it into two circuits, A and B, connected at a pair of terminals. We can then use Thévenin's theorem to replace, say, circuit A by its equivalent circuit. The open-circuit Thévenin voltage for circuit A is that given when circuit B is disconnected and the Thévenin resistance for A is the resistance looking into the terminals of A with all its independent sources set equal to zero. Figure G.4 illustrates this sequence of steps.

Figure G.4   Step-by-step approach for circuit analysis.

1. Identify the two parts A and B of the circuit and separate them by terminals.

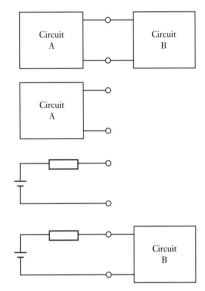

2. Separate part A from part B.

3. Replace A by its Thévenin equivalent, i.e. a voltage source with a series resistance.

4. Reconnect circuit B and carry out the analysis.

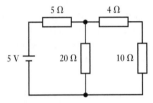

Figure G.5   Example circuit illustrating the use of Thévenin's theorem.

As an illustration, consider the use of Thévenin's theorem to determine the current through the 10 Ω resistor in the circuit given in Figure G.5.

Since we are interested in the current through the 10 Ω resistor, we identify it as network B and the rest of the circuit as network A, connecting them by terminals (Figure G.6(a)). We then separate A from B (Figure G.6(b)) and determine the Thévenin equivalent for it.

The open-circuit voltage is that across the 20 Ω resistor, i.e. the fraction of the total voltage drop across the 20 Ω:

$$E_{Th} = 5 \frac{20}{20 + 5} = 4\,\text{V}$$

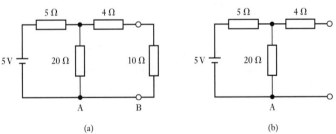

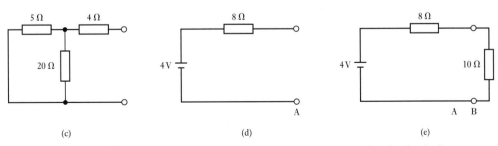

Figure G.6   The Thévenin analysis: (a) picking the terminal points; (b) separating the circuit elements; (c) resistance looking in the terminals; (d) equivalents circuit; (e) the complete circuit.

The resistance looking into the terminals when voltage source is equal to zero is that of 4 Ω in series with a parallel arrangement of 5 Ω and 20 Ω (Figure G.6(c)):

$$R_{\text{Th}} = 4 + \frac{20 \times 5}{20 + 5} = 8\Omega$$

Thus the equivalent Thévenin circuit is as shown in Figure G.6(d) and when network B is connected to it we have the circuit shown in Figure G.6(e). Hence the current through the 10 Ω resistor is $I_{10} = 4/(8 + 10) = 0.22$ A.

G.1.4   Norton's theorem

In a manner similar to Thévenin's theorem, we can have an equivalent circuit for any two-terminal network containing voltage or current sources in terms of an equivalent network of a current source shunted by a resistance. This is known as **Norton's theorem**:

> Any two-terminal network containing voltage or current sources can be replaced by an equivalent network consisting of a current source, equal to the current between the terminals when they are short-circuited, in parallel with the resistance measured between the terminals when there is no load between them and all independent sources in the network are set equal to zero.

> If we have a linear circuit we have to divide it into two circuits, A and B, connected at a pair of terminals (Figure G.7). We can then use Norton's theorem to replace, say, circuit A by its equivalent circuit. The short-circuit Norton current for circuit A is that given when circuit B is disconnected and the Norton resistance for A is the resistance looking into the terminals of A with all its independent sources set equal to zero.

Figure G.7 Step-by-step approach for circuit analysis using Norton's theorem.

1. Identify the two parts A and B of the circuit and separate by terminals.

2. Separate A from B.

3. Replace A by its Norton equivalent.

4. Reconnect circuit B and carry out the analysis.

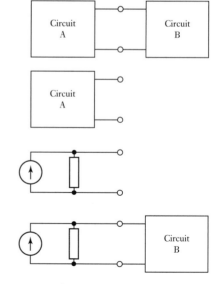

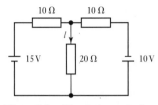

Figure G.8 Circuit for analysis using Norton's theorem.

As an illustration of the use of Norton's theorem, consider the determination of the current I through the 20 Ω resistor in Figure G.8.

We can redraw the circuit in the form shown in Figure G.9(a) as two connected networks A and B with network B selected to be the 20 Ω resistor through which we require the current. We then determine the Norton equivalent circuit for network A (Figure G.9(b)). Short-circuiting the terminals of network A gives the circuit shown in Figure G.9(c). The short circuit current will be the sum, taking into account directions, of the currents from the two branches of the circuits containing the voltage sources, i.e. $I_{sc} = I_1 - I_2$. The current $I_1 = 15/10 = 1.5$ A, since the other branch of the network is short-circuited, and $I_2 = 10/10 = 1.0$ A. Thus $I_{sc} = 0.5$ A. The Norton resistance is given by that across the terminals when all the sources are set to zero (Figure G.9(d)). It is thus:

$$R_N = \frac{10 \times 10}{10 + 10} = 5 \; \Omega$$

Thus the Norton equivalent circuit is that shown in Figure G.9(e). Hence when we put this with network B (Figure G.9(f)), we can obtain the current I. The p.d. across the resistors is $0.5 \times R_{total}$ and so the current I is this p.d. divided by 20. Hence:

$$I = 0.5 \times \frac{5}{5 + 20} = 0.1 \; \text{A}$$

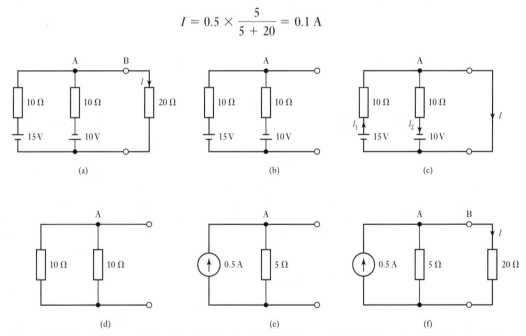

Figure G.9 The Norton analysis: (a) redrawing the circuit; (b) network A; (c) short-circuiting the terminals; (d) sources set to zero; (e) Norton equivalent; (f) the combined parts of the circuits.

G.2 Alternating current circuits

We can generate a sinusoidal waveform by rotating a radius line OA at a constant angular velocity ω (Figure G.10(a)), the vertical projection of the line AB varying with time in a sinusoidal manner. The angle θ of the line AB at

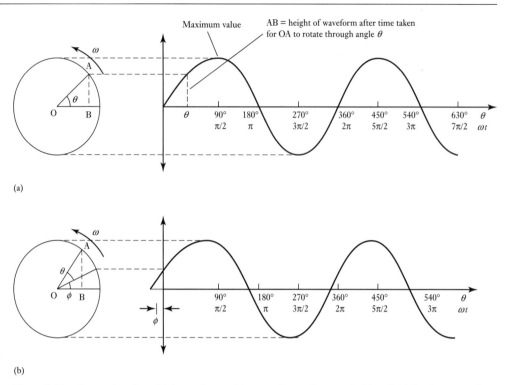

(a)

(b)

Figure G.10   Generating sinusoidal waveforms: (a) zero value at time $t = 0$, (b) an initial value at $t = 0$.

a time t is ωt. The frequency f of rotation is $1/T$, where T is the time taken for one complete rotation, and so $\omega = 2\pi f$. In Figure G.10(a) the rotating line OA was shown as starting from the horizontal position at time $t = 0$. Figure G.10(b) shows that the line OA at $t = 0$ is already at some angle ϕ. As the line OA rotates with an angular velocity ω, then in a time t the angle swept out is ωt and thus at time t the angle with respect to the horizontal is $\omega t + \phi$. Sinusoidal alternating currents and voltages can be described by such rotating lines and hence by the equations $i = I_m \sin \omega t$ and $v = V_m \sin \omega t$ for currents and voltages with zero values at time $t = 0$ and for those starting at some initial angle ϕ by $i = I_m \sin (\omega t + \phi)$ and $v = V_m \sin (\omega t + \phi)$. Lowercase symbols are used for the current and voltage terms that change with time, uppercase being reserved for the non-varying terms.

With alternating current circuits there is a need to consider the relationship between an alternating current through a component and the alternating voltage across it. If we take the alternating current as the reference for a series circuit and consider it to be represented by $i = I_m \sin \omega t$, then the voltage may be represented by $v = V_m \sin (\omega t + \phi)$. There is said to be a **phase difference** of ϕ between the current and the voltage. If ϕ has a positive value then the voltage is said to be **leading** the current (as in Figure G.10 if (a) represents the current and (b) the voltage); if it has a negative value then it is said to be **lagging** the current.

We can describe a sinusoidal alternating current by just specifying the rotating line in terms of its length and its initial angle relative to a horizontal reference line. The term **phasor**, being an abbreviation of the term phase vector, is used for such lines. The length of the phasor can represent the maximum value of the sinusoidal waveform or the root-mean-square (r.m.s.) value, since the maximum value is proportional to the r.m.s value. Because currents and voltages in the same circuit will have the same frequency and thus the phasors used to represent them will rotate with the same angular velocity and maintain the same phase angles between them at all times, we do not need to bother about drawing the effects of their rotation but can just draw phasor diagrams giving the relative angular positions of the phasors and ignore their rotations.

The following summarises the main points about phasors.

1 A phasor has a length that is directly proportional to the maximum value of the sinusoidally alternating quantity or, because the maximum value is proportional to the r.m.s. value, a length proportional to the r.m.s. value.

2 Phasors are taken to rotate anti-clockwise and have an arrow-head at the end which rotates.

3 The angle between two phasors shows the phase angle between their waveforms. The phasor which is at a larger anti-clockwise angle is said to be leading, the one at the lesser anti-clockwise angle lagging.

4 The horizontal line is taken as the reference axis and one of the phasors is given that direction; the others have their phase angles given relative to this reference axis.

G.2.1 Resistance, inductance and capacitance in a.c. circuits

Consider a sinusoidal current $i = I_m \sin \omega t$ passing through a **pure resistance**. A pure resistance is one that has only resistance and no inductance or capacitance. Since we can assume Ohm's law to apply, then the voltage v across the resistance must be $v = Ri$ and so $v = RI_m \sin \omega t$. The current and the voltage are thus in phase. The maximum voltage will be when $\sin \omega t = 1$ and so $V_m = RI_m$.

Consider a sinusoidal current $i = I_m \sin \omega t$ passing through a **pure inductance**. A pure inductance is one which has only inductance and no resistance or capacitance. With an inductance, a changing current produces a back e.m.f. $L \, di/dt$, where L is the inductance. The applied e.m.f. must overcome this back e.m.f. for a current to flow. Thus the voltage v across the inductance is $L \, di/dt$ and so

$$v = L\frac{di}{dt} = L\frac{d}{dt}(I_m \sin \omega t) = \omega L I_m \cos \omega t$$

Since $\cos \omega t = \sin(\omega t + 90°)$, the current and the voltage are out of phase with the voltage leading the current by 90°. The maximum voltage is when $\cos \omega t = 1$ and so we have $V_m = \omega L I_m$. V_m/I_m is called the **inductive reactance** X_L. Thus $X_L = V_m/I_m = \omega L$. Since $\omega = 2\pi f$ then $X_L = 2\pi f L$ and so the reactance is proportional to the frequency f. The higher the frequency the greater the opposition to the current.

Consider a circuit having just **pure capacitance** with a sinusoidal voltage $v = V_m \sin \omega t$ being applied across it. A pure capacitance is one which has only capacitance and no resistance or inductance. The charge q on the plates of a capacitor is related to the voltage v by $q = Cv$. Thus, since current is the rate of movement of charge dq/dt, we have $i = $ rate of change of q = rate of change of (Cv) = $C \times$ (rate of change of v), i.e. $i = C$ dv/dt. Thus

$$i = \frac{dq}{dt} = \frac{d}{dt}(Cv) = C\frac{d}{dt}(V_m \sin \omega t) = \omega C V_m \cos \omega t$$

Since $\cos \omega t = \sin(\omega t + 90°)$, the current and the voltage are out of phase, the current leading the voltage by 90°. The maximum current occurs when $\cos \omega t = 1$ and so $I_m = \omega C V_m$. V_m/I_m is called the **capacitive reactance** X_C. Thus $X_C = V_m/I_m = 1/\omega C$. The reactance has the unit of ohms and is a measure of the opposition to the current. The bigger the reactance the greater the voltage has to be to drive the current through it. Since $\omega = 2\pi f$, the reactance is inversely proportional to the frequency f and so the higher the frequency the smaller the opposition to the current. With d.c., i.e. zero frequency, the reactance is infinite and so no current flows.

In summary, Figure G.11 shows the voltage and current phasors for (a) pure resistance, (b) pure inductance, (c) pure capacitance.

Figure G.11   Phasors with (a) pure resistance, (b) pure inductance, (c) pure capacitance.

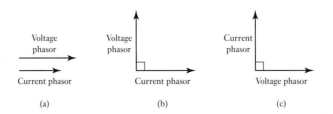

G.2.2   Series a.c. circuits

For a series circuit, the total voltage is the sum of the p.d.s across the series components, though the p.d.s may differ in phase. This means that if we consider the phasors, they will rotate with the same angular velocity but may have different lengths and start with a phase angle between them. We can obtain the sum of two series voltages by using the **parallelogram law** of vectors to add the two phasors:

If two phasors are represented in size and direction by adjacent sides of a parallelogram, then the diagonal of that parallelogram is the sum of the two (Figure G.12).

Figure G.12   Adding phasors 1 and 2 which have a phase angle ϕ between them.

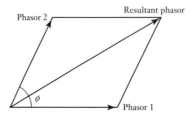

If the phase angle between the two phasors of sizes V_1 and V_2 is 90°, then the resultant can be calculated by the use of Pythagoras theorem as having a size V given by $V^2 = V_1^2 + V_2^2$ and a phase angle ϕ relative to the phasor for V_1 given by $\tan \phi = V_1/V_2$.

Figure G.13   RL series circuit.

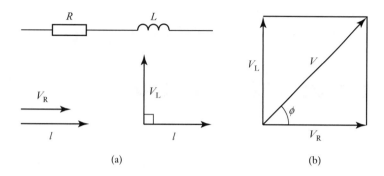

(a)   (b)

As an illustration of the use of the above, consider an alternating current circuit having resistance in series with inductance (Figure G.13(a)). For such a circuit, the voltage for the resistance is in phase with the current and the voltage for the inductor leads the current by 90°. Thus the phasor for the sum of the voltage drops across the two series components is given by Figure G.13(b) as a voltage phasor with a phase angle ϕ. We can use Pythagoras theorem to give the magnitude V of the voltage, i.e. $V^2 = V_R^2 + V_L^2$, and trigonometry to give the phase angle ϕ, i.e. the angle by which the voltage leads the current as $\tan \phi = V_L/V_R$ or $\cos \phi = V_R/V$.

Since $V_R = IR$ and $V_L = IX_L$ then $V^2 = (IR)^2 + (IX_L)^2 = I^2(R^2 + X_L^2)$. The term **impedance** Z is used for the opposition of a circuit to the flow of current, being defined as $Z = V/I$ with the unit of ohms. Thus, for the resistance and inductance in series, the circuit impedance is given by

$$Z = \sqrt{R^2 + X_L^2} = \sqrt{R^2 + (\omega L)^2}$$

Further information

The following is a short list of texts which can usefully provide further information of relevance to a study of mechatronics.

Sensors and signal conditioning

Boyes, W., *Instrumentation Reference Book*, Newnes 2002

Clayton, G. B. and Winder, S., *Operational Amplifiers*, Newnes 2003

Figliola, R.S. and Beasley, D. E., *Theory and Design for Mechanical Measurements*, John Wiley 2000, 2005, 2011

Fraden, J., *Handbook of Modern Sensors*, Springer 2001, 2004, 2010

Gray, P. R., Hurst, P. J., Lewis, S.H. and Meyer, R. G., *Analysis and Design of Analog Integrated Circuits*, Wiley 2009

Holdsworth, B., *Digital Logic Design*, Newnes 2000

Johnson, G. W. and Jennings, R., *LabVIEW Graphical Programming*, McGraw-Hill 2006

Morris, A.S., *Measurement and Instrumentation Principles*, 3rd edition, Newnes 2001

Park, J. and Mackay, S., *Practical Data Acquisition for Instrumentation and Control Systems*, Elsevier 2003

Travis, J. and Kring, J., *LabVIEW for Everyone*, Prentice-Hall 2006

Actuation

Bolton, W., *Mechanical Science*, Blackwell Scientific Publications 1993, 1998, 2006

Gottlieb, I. M., *Electric Motors and Control Techniques*, TAB Books, McGraw-Hill 1994

Kenjo, T. and Sugawara, A., *Stepping Motors and their Microprocessor Controls*, Clarenden Press 1995

Manring, N., *Hydraulic Control Systems*, Wiley 2005

Norton, R. L., *Design of Machinery*, McGraw-Hill 2003

Pinches, M. J. and Callear, B. J., *Power Pneumatics*, Prentice-Hall 1996

Wildi, T., *Electrical Machines, Drives and Power Systems*, Pearson 2005

System models

Åstrom, K. J. and Wittenmark, B., *Adaptive Control*, Dover 1994, 2008

Attaway, S., *Matlab: A Practical Introduction to Programming and Problem Solving*, Butterworth-Heinemann 2009

Bennett, A., *Real-time Computer Control*, Prentice-Hall 1993

Bolton, W., *Laplace and z-Transforms*, Longman 1994

Bolton, W., *Control Engineering*, Longman 1992, 1998

Bolton, W., *Control Systems*, Newnes 2002

D'Azzo, J. J., Houpis, C. H. and Sheldon, N., *Linear Control System Analysis and Design with Matlab*, CRC Press 2003

Dorf, R. C. and Bishop, H., *Modern Control Systems*, Pearson 2007

Fox, H. and Bolton, W., *Mathematics for Engineers and Technologists*, Butterworth-Heinemann 2002

Hahn, B., *Essential MATLAB for Engineers and Scientists*, 5th ed. Elsevier 2012

Moore, H., *MATLAB for Engineers*, Pearson 2013

Microprocessor systems

Arduino web site, www.arduino.cc

Barnett, R. H., *The 8051 Family of Microcontrollers*, Prentice-Hall 1994

Barrett, S. F., *Arduino Microcontroller Processing for Everyone!*, Morgan & Claypool Publishers 2013

Bates, M., *PIC Microcontrollers*, Newnes 2000, 2004

Blum, J., *Exploring Arduino: Tools and Techniques for Engineering Wizardy*, Wiley 2013

Bolton, W., *Microprocessor Systems*, Longman 2000

Bolton, W., *Programmable Logic Controllers*, Newnes 1996, 2003, 2006, 2009

Cady, F. M., *Software and Hardware Engineering: Motorola M68HC11*, OUP 2000

Calcutt, D., Cowan, F. and Parchizadeh, H., *8051 Microcontrollers: An Application Based Introduction*, Newnes 2004

Ibrahim, D., *PIC Basic: Programming and Projects*, Newnes 2001

Johnsonbaugh, R. and Kalinn, M., *C for Scientists and Engineers*, Prentice Hall 1996

Lewis, R. W., *Programming Industrial Control Systems Using IEC 1131-3*, The Institution of Electrical Engineers 1998

Monk, S., *Programming Arduino*, McGraw Hill 2012

Morton, J., *PIC: Your Personal Introductory Course*, Newnes 2001, 2005

Parr, E. A., *Programmable Controllers*, Newnes 1993, 1999, 2003

Pont, M. J., *Embedded C*, Addison-Wesley 2002

Predko, M., *Programming and Customizing the PIC Microcontroller*, Tab Electronics 2007

Rohner, P., *Automation with Programmable Logic Controllers*, Macmillan 1996

Spasov, P., *Microcontroller Technology: The 68HC11*, Prentice-Hall 1992, 1996, 2001

Vahid, F. and Givargis, T., *Embedded System Design*, Wiley 2002

Van Sickle, T., *Programming Microcontrollers in C*, Newnes 2001

Yeralan, S. and Ahluwalia A., *Programming and Interfacing the 8051 Microcontroller*, Addison-Wesley 1995

Zurrell, K., *C Programming for Embedded Systems*, Kindle Edition 2000

Electronic systems

Storey, N., *Electronics A Systems Approach* 5th Edition Pearson 2013

This text consists of two parts: Electrical circuits and components, and Electronic systems.

Student resources specifically written to complement the text can be viewed at www.pearsoned.co.uk/storey-elec. Video tutorials that can be accessed by clicking on their titles include some of particular relevance to mechatronics.

3A: Kirchhoff's laws
3B: Nodal analysis
3C: Mesh analysis
3D: Selecting circuit analysis techniques
4A: Capacitors in series and in parallel
5A: Transformers
5B: Applications of inductive sensors
6A: Alternating voltages and currents
6B: Using complex impedance
7A: Power in circuits with resistance and reactance
7B: Power factor correction for an electric motor
8A: Bode diagrams
9A: Initial and final value theorems
9B: Determination of a circuit's time constant
10A: Selection of a motor for a given application
11A: Top-down system design
11B: Identification of system inputs and outputs
12A: Optical position sensors
12B: Selecting a sensor for a given application
13A: A comparison of display techniques
13B: Selecting an actuator for a given application
14A: Modelling the characteristics of an amplifier
14B: Power gain
14C: Differential amplifiers
14D: Specifying an audio amplifier
15A: Feedback systems
15B: Negative feedback
15C: Open-loop and closed-loop systems
16A: Basic op-amp circuits
16B: Some additional useful op-amp circuits
16C: Frequency response of op-amp circuits
16D: Input and output resistance of op-amp circuits
16E: Analysis of op-amp circuits
18A: Simple FET amplifiers
18B: Small signal equivalent circuit of an FET amplifier
18C: A negative feedback amplifier based on a DE MOSFET
18D: A switchable gain amplifier
19A: A simple bipolar transistor amplifier
19B: Analysis of a simple bipolar transistor amplifier
19C: Analysis of a feedback amplifier based on a bipolar transistor
19D: A common collector amplifier
19E: Design of a phase splitter
20A: Push-pull amplifiers
20B: Amplifier classes
20C: Power amplifiers
20D: Design of an electrically operated switch
21A: Simplified circuit of a bipolar operational amplifier
21B: Simplified circuit of a CMOS operational amplifier

Answers

The following are answers to numerical problems and brief clues as to possible answers with descriptive problems.

Chapter 1

1.1 (a) Sensor, mercury; signal conditioner, fine bore stem; display, marks on the stem; (b) Sensor, curved tube; signal conditioner, gears; display, pointer moving across a scale

1.2 See text

1.3 Comparison/controller, thermostat; correction, perhaps a relay; process, heat; variable, temperature; measurement, a temperature-sensitive device, perhaps a bimetallic strip

1.4 See Figure P.1

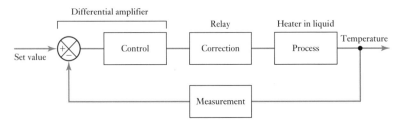

Figure P.1 Problem 1.4.

1.5 See text

1.6 See text

1.7 For example: water in, rinse, water out, water in, heat water, rinse, water out, water in, rinse, water out

1.8 Traditional: bulky, limited functions, requires rewinding. Mechatronics: compact, many functions, no rewinding, cheaper

1.9 Bimetallic element: slow, limited accuracy, simple functions, cheap. Mechatronics: fast, accurate, many functions, getting cheaper

Chapter 2

2.1 See the text for explanation of the terms

2.2 −3.9%

2.3   76.4 s
2.4   0.73%
2.5   0.105 Ω
2.6   Incremental, angle from some datum, not absolute; absolute, unique identification of an angle
2.7   162
2.8   (a)±1.2° (b) 3.3 mV
2.9   See text
2.10 2.8 kPa
2.11 19.6 kPa
2.12 −0.89%
2.13 +1.54°C
2.14 Yes
2.15 −9.81 N, −19.62 N, e.g. strain gauges
2.16 For example, orifice plate with differential pressure cell
2.17 For example, differential pressure cell
2.18 For example, LVDT displacement sensor

Chapter 3

3.1   As Figure 3.2 with $R_2/R_1 = 50$, e.g. $R_1 = 1$ kΩ, $R_2 = 50$ kΩ
3.2   200 kΩ
3.3   Figure 3.5 with two inputs, e.g. $V_A = 1$ V, $V_B = 0$ to 100 mV, $R_A = R_2 = 40$ kΩ, $R_B = 1$ kΩ
3.4   Figure 3.11 with $R_1 = 1$ kΩ and $R_2 = 2.32$ kΩ
3.5 $V = K\sqrt{I}$
3.6   100 kΩ
3.7   80 dB
3.8   Fuse to safeguard against high current, limiting resistor to reduce currents, diode to rectify a.c., Zener diode circuit for voltage and polarity protection, low-pass filter to remove noise and interference, optoisolator to isolate the high voltages from the microprocessor
3.9   0.234 V
3.10 2.1×10^{-4} V
3.11 As given in the problem

Chapter 4

4.1   24.4 mV
4.2   9
4.3   0.625 V
4.4   1, 2, 4, 8
4.5   12 µs
4.6   See text
4.7   Buffer, digital-to-analogue converter, protection
4.8   0.33 V, 0.67 V, 1.33 V, 2.67 V
4.9 $32\,768R$
4.10 15.35 ms
4.11 Factor of 315

Chapter 5

5.1 For example: (a) ticket selected AND correct money in, correct money decided by OR gates analysis among possibilities, (b) AND with safety guards, lubricant, coolant, workpiece, power, etc., all operating or in place, (c) Figure P.2, (d) AND

5.2 (a) Q, (b) P

5.3 AND

5.4 A as 1, B as 0

5.5 See Figure P.3

5.6 See Figure P.4

5.7 As in the text, Section 5.3.1, for cross-coupled NOR gates

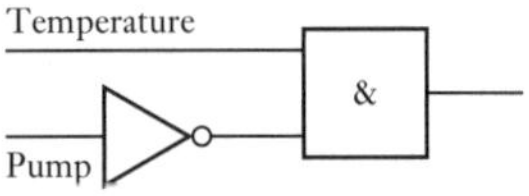

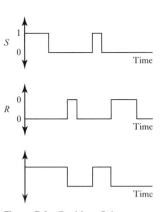

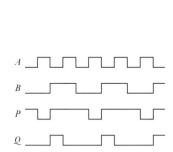

Figure P.2 Problem 5.1(c).

Figure P.3 Problem 5.5.

Figure P.4 Problem 5.6.

Chapter 6

6.1 See text

6.2 See Section 6.1

6.3 For example: (a) a recorder, (b) a moving-coil meter, (c) a hard disk or CD, (d) a storage oscilloscope or a hard disk or CD

6.4 Could be four-active-arm bridge, differential operational amplifier, display of a voltmeter. The values of components will depend on the thickness chosen for the steel and the diameter of a load cell. You might choose to mount the tank on three cells

6.5 Could be as in Figure 3.8 with cold junction compensation by a bridge (see Section 3.5.2). Linearity might be achieved by suitable choice of thermocouple materials

6.6 Could be thermistors with a sample and hold element followed by an analogue-to-digital converter for each sensor. This would give a digital signal for transmission, so reducing the effects of possible interference. Optoisolators could be used to isolate high voltages/currents, followed by a multiplexer feeding digital meters

6.7 This is based on Archimedes' principle: the upthrust on the float equals the weight of fluid displaced

6.8 Could use an LVDT or strain gauges with a Wheatstone bridge

6.9 For example: (a) Bourdon gauge, (b) thermistors, galvanometric chart recorder, (c) strain-gauged load cells, Wheatstone bridge, differential amplifier, digital voltmeter, (d) tachogenerator, signal conditioning to shape pulses, counter

Chapter 7

7.1	See Section 7.3	7.7	0.0057 m^2
7.2	See Section 7.3.2	7.8	124 mm
7.3	See Section 7.4	7.9	1.27 MPa, 3.9×10^{-5} m^3/s
7.4	See (a) Figure 7.14, (b) Figure 7.8(b),	7.10	(a) 0.05 m^3/s, (b) 0.10 m^3/s
	(c) Figure 7.10, (d) Figure 7.13	7.11	(a) 0.42 m^3/s, (b) 0.89 m^3/s
7.5	A+, B+, A−, B−	7.12	960 mm
7.6	See Figure P.5		

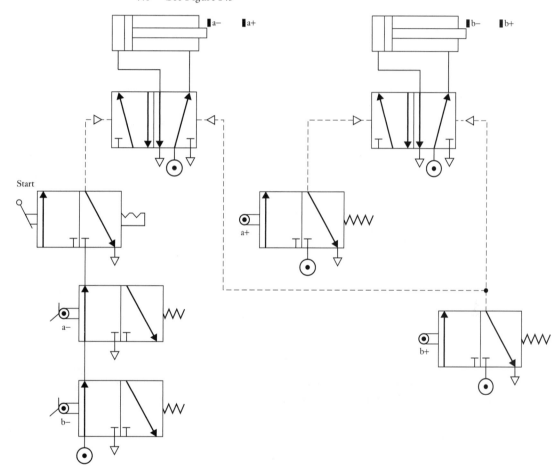

Figure P.5 Problem 7.6.

Chapter 8

8.1   (a) A system of elements arranged to transmit motion from one form to another form
(b) A sequence of joints and links to provide a controlled output in response to a
supplied input motion

8.2   See Section 8.3.1

8.3   (a) 1, (b) 2, (c) 1, (d) 1, (e) 3

8.4   (a) Pure translation, (b) pure translation, (c) pure rotation, (d) pure rotation, (e) translation plus rotation

8.5   Quick-return

8.6   Sudden drop in displacement followed by a gradual rise back up again

8.7   60 mm

8.8   Heart-shaped with distance from axis of rotation to top of heart 40 mm and to base 100 mm. See Figure 8.14(a)

8.9   For example: (a) cams on a shaft, (b) quick-return mechanism, (c) eccentric cam, (d) rack-and-pinion, (e) belt drive, (f) bevel gears

8.10   1/24

Chapter 9

9.1   It acts as a flip-flop

9.2   See text and Figure 9.7

9.3   (a) Series wound, (b) shunt wound

9.4   (a) d.c. shunt wound, (b) induction or synchronous motor with an inverter, (c) d.c., (d) a.c.

9.5   See Section 9.5.4

9.6   See Section 9.7

9.7   480 pulses/s

9.8   9°

9.9   (a) 4 kW, (b) 800 W, (c) 31.8 Nm

9.10   0.65 Nm

9.11   2

9.12   3.6 Nm

Chapter 10

10.1   See Section 10.2

10.2   256

10.3   64K × 8

10.4   See Section 10.3

10.5   See Figure 10.9 and associated text

10.6   (a) E, (b) C, (c) D, (d) B

10.7   256

10.8   (a) 0, (b) 1

10.9   See Section 10.3.1, item 6

10.10 See Section 10.3.2, item 5

10.11 High to reset pin

10.12 (a) IF A
               THEN
                  BEGIN B
                  END B
               ELSE
                  BEGIN C
                  END C
               ENDIF A

    (b) WHILE A
        BEGIN B
          END B
       ENDWHILE A

Chapter 11

11.1   (a) 89, (b) 99

11.2   No address has to be specified since the address is implied by the mnemonic

11.3   (a) CLRA, (b) STAA, (c) LDAA, (d) CBA, (e) LDX

11.4   (a) LDAA $20, (b) DECA, (c) CLR $0020, (d) ADDA $0020

11.5   (a) Store accumulator B value at address 0035, (b) load accumulator A with data F2, (c) clear the carry flag, (d) add 1 to value in accumulator A, (e) compare C5 to value in accumulator A, (f) clear address 2000, (g) jump to address given by index register plus 05

11.6   (a)

```
DATA1    EQU     $0050
DATA2    EQU     $0060
DIFF     EQU     $0070
         ORG     $0010
         LDAA    DATA1      ; Get minuend
         SUBA    DATA2      ; Subtract subtrahend
         STAA    DIFF       ; Store difference
         SWI                ; Program end
```

(b)

```
MULT1    EQU     $0020
MULT2    EQU     $0021
PROD     EQU     $0022
         ORG     $0010
         CLR     PROD       ; Clear product address
         LDAB    MULT1      ; Get first number
SUM      LDAA    MULT2      ; Get multiplicand
         ADDA    PROD       ; Add multiplicand
         STAA    PROD       ; Store result
         DECB               ; Decrement acc. B
         BNE     SUM        ; Branch if adding not complete
         WAI                ; Program end
```

(c)

```
FIRST    EQU     $0020
         ORG     $0000
         CLRA               ; Clear accumulator
         LDX     #0
MORE     STAA    $20,X
         INX                ; Increment index reg.
         INCA               ; Increment accumulator
         CMPA    #$10       ; Compare with number 10
         BNE     MORE       ; Branch if not zero
         WAI                ; Program end
```

(d)

```
            ORG     $0100
            LDX     #$2000      ; Set pointer
    LOOP    LDA A   $00,X       ; Load data
            STA A   $50,X       ; Store data
            INX                 ; Increment index register
            CPX     $3000       ; Compare
            BNE     LOOP        ; Branch
            SWI                 ; Program end
```

11.7

```
    YY      EQU     $??         ; Value chosen to give required time delay
    SAVEX   EQU     $0100
            ORG     $0010
            STA     SAVEX       ; Save accumulator A
            LDAA    YY          ; Load accumulator A
    LOOP    DECA                ; Decrement acc. A
            BNE     LOOP        ; Branch if not zero
            LDA     SAVEX       ; Restore accumulator
            RTS                 ; Return to calling program
```

11.8

```
    LDA     $2000       ; Read input data
    AND A   #$01        ; Mask off all bits but bit 0
    BEQ     $03         ; If switch low, branch over JMP which is 3
                        ; program lines
    JMP     $3000       ; If switch high no branch and so execute JMP
    Continue
```

Chapter 12

12.1 (a) The variable counter is an integer, (b) the variable num is assigned the value 10, (c) the word name will be displayed, (d) the display is Number 12, (e) include the file stdio.h

12.2 (a) Calls up the library necessary for the printf( ) function, (b) indicates the beginning and end of a group of statements, (c) starts a new line, (d) problem 3

12.3 The number is 12

12.4 # include <stdio.h>

```
int main(void);
{
    int len, width;
    printf("Enter length: ");
    scanf("%d", &len);
    printf("Enter width: ");
    scanf("%d", &width);
    printf("Area is %d", lens * width);
    return 0;
{
```

12.5 Similar to program given in Section 12.3, item 4

12.6 Divides first number by second number unless second is 0

Chapter 13

13.1 See Section 13.3
13.2 See Section 13.3. A parallel interface has the same number of input/output lines as the microprocessor. A serial interface has just a single input/output line
13.3 See Section 13.2
13.4 See Section 13.4
13.5 See Section 13.4 and Figure 13.10
13.6 See Section 13.4.1
13.7 See Section 13.3.3. Polling involves the interrogation of all peripherals at frequent intervals, even when some are not activated. It is thus wasteful of time. Interrupts are only initiated when a peripheral requests it and so is more efficient
13.8 CRA 00110100, CRB 00101111
13.9 As the program in 18.4.2 with LDAA #$05 replaced by LDAA #$34 and LDAA #$34 replaced by LDAA #$2F
13.10 As the program in Section 13.4.2 followed by
READ LDAA $2000 ; Read port A
Perhaps after some delay program there may then be
BRA READ

Chapter 14

14.1 (a) AND, (b) OR
14.2 (a) Figure 14.9(b), (b) Figure 14.10(b), (c) a latch circuit, Figure 14.16, with Input 1 the start and Input 2 the stop switches
14.3 0 LD X400, 1 LD Y430, 2 ORB, 3 ANI X401, 4 OUT Y430
14.4 0 LD X400, 1 OR Y430, 3 OUT Y430, 4 OUT T450, 5 K 50; delay-on timer
14.5 0 LD X400, 1 OR Y430, 2 ANI M100, 3 OUT Y430, 4 LD X401, 5 OUT M100; reset latch
14.6 As in Figure 14.28 with Timer 1 having $K = 1$ for 1 s and Timer 2 with $K = 20$ for 20s
14.7 Figure P.6
14.8 Figure P.7

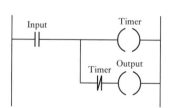

Figure P.6 Problem 14.7.

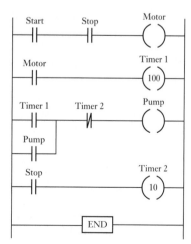

Figure P.7 Problem 14.8.

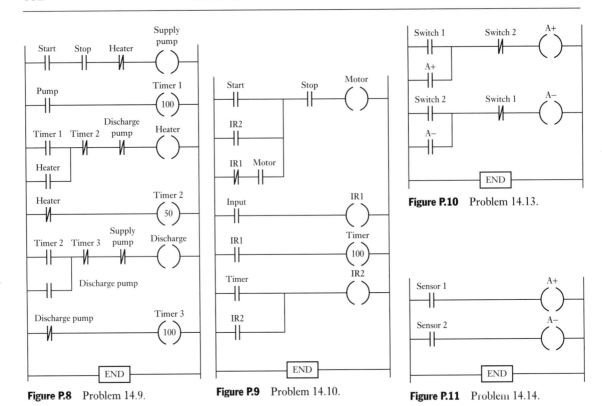

Figure P.8   Problem 14.9.

Figure P.9   Problem 14.10.

Figure P.10   Problem 14.13.

Figure P.11   Problem 14.14.

14.9   Figure P.8
14.10 Figure P.9
14.11 An output would come on, as before, but switch off when the next input occurs
14.12 See Section 14.10
14.13 Two latch circuits, as in Figure P.10
14.14 Figure P.11

Chapter 15

15.1   See Section 15.2
15.2   See Section 15.3
15.3   Bus
15.4   Broadband
15.5   See Section 15.5.1
15.6   See Section 15.4
15.7   See Section 15.3.1
15.8   NRFD to PD0, DAV to STRA and IRQ, NDAC to STRB, data to Port C
15.9   TTL to RS-232C signal-level conversion
15.10 See Section 15.7.1

Chapter 16

16.1 See Section 16.1
16.2 See Section 16.2
16.3 See Section 16.2
16.4 See Section 16.5.3 for programmable checks and checksum and Section 16.2 for watchdog timer
16.5 See Section 16.5.3

Chapter 17

17.1 (a) $m\dfrac{d^2x}{dt^2} + c\dfrac{dx}{dt} = F$, (b) $m\dfrac{d^2x}{dt^2} + c\dfrac{dx}{dt} + (k_1 + k_2)x = F$

17.2 As in Figure 17.3(a)

17.3 $c\dfrac{d\theta_i}{dt} = c\dfrac{d\theta_o}{dt} + k\theta_o$

17.4 Two torsional springs in series with a moment of inertia block,

$$T = I\dfrac{d^2\theta}{dt^2} + k_1(\theta_1 - \theta_2) = m\dfrac{d^2\theta}{dt^2} + \dfrac{k_1 k_2}{k_1 + k_2}\theta_1$$

17.5 $v = v_R + \dfrac{1}{RC}\displaystyle\int v_R\, dt$

17.6 $v = \dfrac{L}{R}\dfrac{dv_R}{dt} + \dfrac{1}{CR}\displaystyle\int v_R\, dt + v_R$

17.7 $v = R_1 C\dfrac{dv_C}{dt} + \left(\dfrac{R_1}{R_2} + 1\right)v_C$

17.8 $RA_2\dfrac{dh_2}{dt} + h_2\rho g = h_1$

17.9 $RC\dfrac{dT}{dt} + T = T_r$. Charged capacitor discharging through a resistor

17.10 $RC\dfrac{dT_1}{dt} = Rq - 2T_1 + T_2 + T_3$, $RC\dfrac{dT_2}{dt} = T_1 - 2T_2 + T_3$

17.11 $pA = m\dfrac{d^2x}{dt^2} + R\dfrac{dx}{dt} + \dfrac{1}{C}x$, $R =$ resistance to stem movement, $c =$ capacitance of spring

17.12 $T = \left(\dfrac{I_1}{n} + n\right)\dfrac{d^2\theta}{dt^2} + \left(\dfrac{c_1}{n} + nc_2\right)\dfrac{d\theta}{dt} + \left(\dfrac{k_1}{n} + nk_2\right)\theta$

Chapter 18

18.1 $\dfrac{IR}{k_1 k_2} \dfrac{d\omega}{dt} + \omega = \dfrac{1}{k_2} v$

18.2 $(L_a + L_L)\dfrac{di_a}{dt} + (R_a + R_L)i_a - k_1 \dfrac{d\theta}{dt} = 0,\ I\dfrac{d^2\theta}{dt^2} + B\dfrac{d\theta}{dt} + k_2 i_a = T$

18.3   Same as armature-controlled motor

Chapter 19

19.1 $4\dfrac{dx}{dt} + x = 6y$

19.2   (a) 59.9°C, (b) 71.9°C

19.3   (a) $i = \dfrac{V}{R}\left(1 - e^{-Rt/L}\right)$, (b) L/R, (c) V/R

19.4   (a) Continuous oscillations, (b) under-damped, (c) critically damped, (d) over-damped

19.5   (a) 4 Hz, (b) 1.25, (c) $i = I\left(\tfrac{1}{3}e^{-8t} - \tfrac{4}{3}e^{-2t} + 1\right)$

19.6   (a) 5 Hz, (b) 1.0, (c) $x = (-32 + 6t)e^{-5t} + 6$

19.7   (a) 9.5%, (b) 0.020 s

19.8   (a) 4 Hz, (b) 0.625, (c) 1.45 Hz, (d) 0.5 s, (e) 8.1%, (f) 1.4 s

19.9   (a) 0.59, (b) 0.87

19.10   2.4

19.11   0.09

19.12   3.93 rad/s, 0.63 Hz

Chapter 20

20.1   (a) $\dfrac{1}{As + \rho g/R}$, (b) $\dfrac{1}{ms^2 + cs + k}$, (c) $\dfrac{1}{LCs^2 + RCs + 1}$

20.2   (a) 3 s, (b) 0.67 s

20.3   (a) $1 + e^{-2t}$, (b) $2 + 2e^{-5t}$

20.4   (a) Over-damped, (b) under-damped, (c) critically damped, (d) under-damped

20.5 $t\,e^{-3t}$

20.6 $2e^{-4t} - 2e^{-3t}$

20.7   (a) $\dfrac{4s}{s^2(s + 1) + 4}$, (b) $\dfrac{2(s + 2)}{(s + 1)(s + 2) + 2}$,

(c) $\dfrac{4}{(s+2)(s+3)+20}$, (d) $\dfrac{2}{s(s+2)+20}$

20.8　$5/(s+53)$

20.9　$5s/(s^2+s+10)$

20.10　$2/(3s+1)$

20.11　$-1, -2$

20.12　(a) Stable, (b) unstable, (c) unstable, (d) stable, (e) unstable

Chapter 21

21.1　(a) $\dfrac{5}{\sqrt{\omega^2+4}}, \dfrac{\omega}{2}$, (b) $\dfrac{2}{\sqrt{\omega^4+\omega^2}}, \dfrac{1}{\omega}$,

　　　(c) $\dfrac{1}{\sqrt{4\omega^6-3\omega^4+3\omega^2+1}}, \dfrac{\omega(3-2\omega^2)}{1-3\omega^2}$

21.2　$0.56\sin(5t-38°)$

21.3　$1.18\sin(2t+25°)$

21.4　(a) (i) ∞, $90°$, (ii) 0.44, $450°$, (iii) 0.12, $26.6°$, (iv) 0, $0°$,

　　　(b) (i) 1, $0°$, (ii) 0.32, $-71.6°$, (iii) 0.16, $-80.5°$, (iv) 0, $-90°$.

21.5　See Figure P.12

21.6　(a) $1/s$, (b) $3.2/(1+s)$, (c) $2.0/(s^2+2\zeta s+1)$,

　　　(d) $3.2/[(1+s)(0.01s^2+0.2\zeta s+1)]$

Chapter 22

22.1　See Section 22.3
22.2　(a) 8 min, (b) 20 min
22.3　(a) 12 s, (b) 24 s
22.4　5
22.5　See the text. In particular P offset, PI and PID no offset
22.6　3, 666 s, 100 s
22.7　3, 100 s, 25 s
22.8　See Sections (a) 22.12.1, (b) 22.12.2, (c) 22.12.3
22.9　1.6
22.10　First-order response with time constant c/K_P

Chapter 23

23.1　For example, try diameter and degree of redness. You might also consider weight. Your results need to be able to distinguish clearly between denominations of coins, whatever their condition

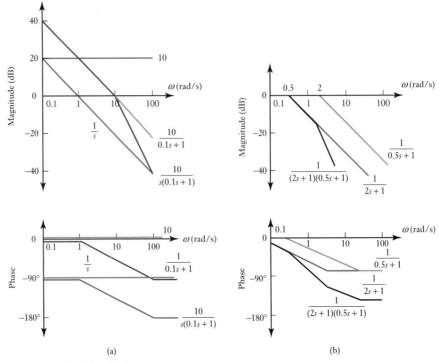

(a)   (b)

Figure P.12   Problem 21.5.

23.2   (a) 1/6, (b) 1/36, (c) 1/10
23.3   0.99
23.4   0.002
23.5   0.625
23.6   0.761
23.7   For example, if room temperature < 20°C and timer ON, then boiler ON; if boiler
       ON, then pump ON; if pump ON and room temperature < 20°C, then valve ON;
       if timer NOT ON, then boiler NOT ON; if room temperature NOT < 20°C, then
       valve NOT ON; if boiler NOT ON, then pump NOT ON. You might also refine
       this by considering there to be a restriction in that the boiler is restricted to operating
       below 60°C

Chapter 24

24.1   Possible solutions might be: (a) thermocouple, cold junction compensation, amplifier,
       ADC, PIA, microprocessor, DAC, thyristor unit to control the oven heating element,
       (b) light beam sensors, PLC, solenoid-operated delivery chute deflectors, (c) closed-
       loop control with, for movement in each direction, a d.c. motor as actuator for
       movement of pen, microprocessor as comparator and controller, and feedback from an
       optical encoder.

Research assignments

The following are brief indications of the type of information which might be contained in an answer.

24.2 A typical ABS system has sensors, inductance types, sensing the speeds of each of the car wheels, signal conditioning to convert the sensor signals into 5 V pulses, a microcontroller with a program to calculate the wheels' speed and rate of deceleration during braking so that when a set limit is exceeded the microcontroller gives an output to solenoid valves in the hydraulic modulator unit either to prevent an increase in braking force or if necessary to reduce it.

24.3 The carriage motor moves the printer head sideways while the print head prints the characters. After printing a line the paper feed motor advances the paper. The print head consists of solenoids driving pins, typically a row of nine, to impact on an ink ribbon. A microcontroller can be used to control the outputs. For more details, see *Microcontroller Technology: The 68HC11* by P. Spasov (Prentice Hall 1992, 1996).

24.4 The CAN bus operates with signals which have a start bit, followed by the name which indicates the message destination and its priority, followed by control bits, followed by the data being sent, followed by CRC bits, followed by confirmation of reception bits, and concluding with end bits.

Design assignments

The following are brief indications of possible solutions.

24.5 A digital thermometer using a microprocessor might have a temperature sensor such as LM35, an ADC, a ROM chip such as the Motorola MCM6830 or Intel 8355, a RAM chip such as the Motorola MCM6810 or Intel 8156, a microprocessor such as Motorola M6800 or Intel 8085A and a driver with LED display. With a microcontroller such as the Motorola MC68HC11 or Intel 8051 there might be just the temperature sensor, with perhaps signal conditioning, the microcontroller and the driver with LED display.

24.6 A digital ohmmeter might involve a monostable multivibrator which provides an impulse with a duration of $0.7RC$. A range of different fixed capacitors could be used to provide different resistance ranges. The time interval might then be determined using a microcontroller or a microprocessor plus memory, and then directed through a suitable driver to an LED display.

24.7 This might involve a pressure sensor, e.g. the semiconductor transducer Motorola MPX2100AP, signal conditioning to convert the small differential signal from the sensor to the appropriate level, e.g. an instrumentation amplifier using operational amplifiers, a microcontroller, e.g. MC68HC11, an LCD driver, e.g. MC145453, and a four-digit LCD display.

24.8 This could be tackled by using the M68HC11EVM board with a PWM output to the motor. Where feedback is wanted an optical encoder might be used.

24.9 The arrangement might be for each box to be loaded by current being supplied to a solenoid valve to operate a pneumatic cylinder to move a flap and allow a box down a chute. The box remains in the chute which is closed by a flap. Its presence is detected by a sensor which then indicates the next box can be allowed into the chute. This continues until four boxes are counted as being in the chute. The flap at the end of the chute might then be activated by another solenoid valve operating a cylinder and so allowing the boxes onto the belt. The arrival of the boxes on the belt might be indicated by a sensor mounted on the end of the chute. This can then allow the entire process to be repeated.

Appendix A

A.1 (a) $2/s^2$, (b) $2/(s^2 + 4)$, (c) e^{-2s}, (d) $sX(s) - 2$, (e) $3s^2X(s)$, (f) $1/[s(s + 1)]$

A.2 (a) t, (b) $5 \cos 3t$, (c) $1 + 2e^t$, (d) e^{-3t}

A.3 5

Appendix B

B.1 255

B.2 (a) 11, (b) 529

B.3 (a) 1A7, (b) 211

B.4 (a) 781, (b) 157

B.5 (a) 1010 0110, (b) 1101 1101

B.6 (a) 0, (b) 1

B.7 (a) 8, (b) 12

Appendix C

C.1 (a) $A \cdot (B + C)$, (b) $(A + B) \cdot (C + D)$, (c) $\overline{A} + B$, (d) $\overline{A} \cdot B$

C.2 (a) $Q = (A \cdot B + C \cdot D) \cdot E$, (b) $Q = (A \cdot B + B) \cdot C$

C.3

A	B	C	Q
0	0	0	0
0	0	1	0
0	1	0	0
0	1	1	1
1	0	0	0
1	0	1	1
1	1	0	0
1	1	1	1

C.4 (a) $Q = C \cdot (A + D)$, (b) $Q = A \cdot B$, (c) $Q = A \cdot \overline{B} \cdot C + C \cdot D$

C.5 As given in the problem

C.6 (a) $Q = A + B$, (b) $Q = C + \overline{A} \cdot C$

C.7 (a) $Q = \overline{A} \cdot \overline{B} + \overline{A} \cdot \overline{C}$, (b) $Q = A \cdot B \cdot D + A \cdot B \cdot \overline{C} + \overline{C} \cdot D$

C.8 Four input AND gates with two NOT gates if correct combination is 1, 1, 0, 0:
$Q = A \cdot B \cdot \overline{C} \cdot \overline{D}$

Index